Traffic Flow Dynamics

Martin Treiber · Arne Kesting

Traffic Flow Dynamics

Data, Models and Simulation

Translated by Martin Treiber and Christian Thiemann

Springer

Martin Treiber
Institut für Wirtschaft und Verkehr
TU Dresden
Dresden
Germany

Arne Kesting
TomTom Development Germany GmbH
Berlin
Germany

ISBN 978-3-642-32459-8 ISBN 978-3-642-32460-4 (eBook)
DOI 10.1007/978-3-642-32460-4
Springer Heidelberg New York Dordrecht London

Library of Congress Control Number: 2012944963

Printed on acid-free paper

Springer is part of Springer Science+Business Media (www.springer.com)

Preface

In order to keep people moving in times of rising traffic and limited resources, science is challenged to find intelligent solutions. Over the past few years, contributions from engineers, physicists, mathematicians, and behavioral psychologists have lead to a better understanding of driver behavior and vehicular traffic flow. This interdisciplinary field will surely produce further advances in the future. The focus is on new applications ranging from novel driver-assistance systems, to intelligent approaches to optimizing traffic flow, to the precise detection of traffic jams and the short-term forecasting of traffic for dynamic navigation aids.

This textbook offers a comprehensive and didactic account of the different aspects of vehicular traffic flow dynamics and how to describe and simulate them with mathematical models. We hope to make this fascinating field accessible to a broader readership; to date, it has only been documented in specialized scientific papers and monographs.

Part I describes how to obtain and interpret traffic flow data, the basis of any quantitative modeling. The second and main part is devoted to the different approaches and models used to mathematically describe traffic flow. The starting point of most models are the basic concepts of physics—many-particle systems, hydrodynamics, and classical Newtonian mechanics—augmented by behavioral aspects and traffic rules. At the website[1] accompanying this book, the reader can interactively run a selection of traffic models and reproduce some of the simulation results displayed in the figures. Part III gives an overview of major applications including traffic-state estimation, fuel consumption, and emission modeling, determining travel times (the basis of dynamic navigation), and how to optimize traffic flow.

The book is written for students, lecturers, and professionals of engineering and transportation sciences and for interested students in general. It also offers material for project work in programming, numerical methods, simulation, and mathematical modeling at college and university level. The reference implementations in the

[1] see: www.traffic-flow-dynamics.org

multi-model open-source vehicular traffic simulator *MovSim*[2] can be used as a starting point for the reader's own simulation experiments and model development.

This work originates from the lecture notes of courses in traffic flow dynamics and modeling at the Dresden University of Technology, Germany; these have been previously published, by the same publisher, in the German book "Verkehrsdynamik und Simulation". The English edition has been updated and significantly extended to include new topics, e.g., on model calibration. To underline its textbook character, it contains many problems with elaborated solutions.

We thank all colleagues at our Department for Traffic Econometrics and Modeling at the Dresden University of Technology, particularly Dirk Helbing, for various scientific discussions and stimulations. We would also like to thank Marietta Seifert, Christian Thiemann, and Stefan Lämmer for suggestions and corrections. Special thanks go to Martin Budden for reviewing the manuscript as a native English speaker. He is also one of the main contributors to *MovSim*. Finally, we would like to thank Martina Seifert, Christine and Hanskarl Treiber, Ingrid, Bernd, and Dörte Kesting, Claudia Perlitius, and Ralph Germ who contributed to the book with valuable suggestions.

Dresden, June 2012

Martin Treiber
Arne Kesting

[2] see: www.movsim.org

Contents

Chapter 1
Introduction

> *I was like a boy playing on the sea-shore, and diverting myself now and then finding a smoother pebble or a prettier shell than ordinary, whilst the great ocean of truth lay all undiscovered before me.*
>
> Isaac Newton

Abstract In this textbook, we describe the dynamics of vehicular traffic flow in terms of mathematical models. In the field of natural sciences, the mathematical approach has been eminently successful.

In this textbook, we describe the dynamics of vehicular traffic flow in terms of mathematical models. In the field of natural sciences, the mathematical approach has been eminently successful. Galileo Galilei is reported to have said "Mathematics is the language with which God has written the universe." In more recent times, human decisions and actions have been described in mathematical terms as well. At first sight, this appears to be paradoxical. After all, humans and their individual decisions certainly cannot be described by a formula.

There are several aspects why a mathematical description of traffic flow dynamics nevertheless makes sense. Firstly, a huge amount of traffic flow data is available ranging from the acceleration characteristics of single drivers and vehicles to macroscopic data obtained by stationary detectors, supplemented by a rapidly growing amount of data obtained by GPS, wireless LAN, and mobile phones inside the vehicles. The associated measurements—corresponding to experiments in the fields of the natural sciences—serve as the basis of any mathematical modeling (cf. Fig. 1.1). By comparing a model's predictions with the data and changing the values of the model parameters to obtain a maximum fit, a model can be *calibrated* which is a prerequisite for any meaningful application.

Secondly, traffic dynamics describes the interplay of many vehicles and drivers. Moreover, the interaction of the vehicles and drivers, technically termed *driver-vehicle units*, leads to new *collective effects* that do not depend on the details of

M. Treiber and A. Kesting, *Traffic Flow Dynamics*,
DOI: 10.1007/978-3-642-32460-4_1,

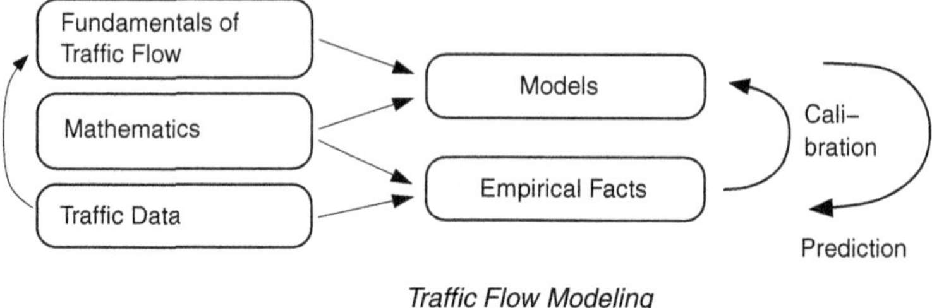

Fig. 1.1 Traffic flow models describe the dynamics of vehicles and drivers in terms of mathematical equations. Predictions are obtained by running the model simulation. The values of the model's parameters are chosen so that the simulation produces a best fit with the data (model calibration). Once calibrated, the model can be used for traffic flow prediction and other applications

individuals. Examples include the formation of stop-and-go waves but also more complicated spatiotemporal patterns of congested traffic. In all theses cases, individual details average out and are, therefore, not relevant. The classical analog in the field of physical sciences is the thermodynamic description of liquids and gases. For example, in order to describe sound waves or the pressure-temperature diagram of air, it is not necessary to know the motion and interactions of individual oxygen, nitrogen, or CO_2 molecules. In this sense, vehicles and drivers assume the role of molecules in gases or liquids. There are even models that are explicitly based on this analogy, see Chap. 9.

Finally, driving dynamics is subject to certain limitations. For example, drivers are typically restricted to interacting with their direct neighbors—again in analogy with gases. Furthermore, accelerations and decelerations are limited physically: After all, vehicles (but not drivers) are physical objects in the classical sense.

Delimitation of traffic flow dynamics. One can distinguish traffic flow dynamics from other fields of traffic science by the time scales given in Table 1.1. Traffic flow dynamics includes time scales ranging from about one second to a few hours. Human reaction times and the time gap between two vehicles following each other are of the order of 1 s while braking and acceleration maneuvers typically take several seconds. In city traffic, the period of one red-green cycle of traffic lights is of the order of 1 min while, on freeways, the period of traffic oscillations and stop-and-go waves is between 5 and 20 min. Finally, the traffic demand serving as exogenous variable (model input) for traffic flow models varies on time scales of one hour, as illustrated by the term "rush hour".

Longer time scales ranging from hours to years are the domain of *transportation planning*. This includes the very long time scales of variations in traffic demand caused by demographic change. Transportation planning and traffic flow dynamics complement each other: The endogenous variables (model output) of the classical four-step scheme of transportation planning[1] and its modern dynamical variants are

[1] The four steps are *trip generation*, *trip distribution*, *mode choice*, and *route assignment*.

Table 1.1 Delimitation of traffic flow dynamics from vehicular dynamics and transportation planning

Time scale	Field	Models	Aspect of traffic (examples)
<0.1 s	Vehicle dynamics	Sub-microscopic	Control of engine and brakes
1 s			Reaction time, time gap
10 s	Traffic flow dynamics	Car-following models	Acceleration and deceleration
1 min		Macroscopic models	Cycle period of traffic lights
10 min			Stop-and-go waves
1 h			Peak hour
1 day		Route assignment traffic demand	Daily demand pattern
1 year	Transportation planning		Building/changing infrastructure
5 years		Statistics age pyramid	Socioeconomic structure
50 years			Demographic change

the traffic demand (vehicles per hour) on each link of the considered network. For traffic flow simulations, in turn, these variables are exogenous (externally given), typically in form of boundary conditions.

Transport logistics operates on the same time scales as traffic flow dynamics but takes a different point of view: Freight transport operations are optimized while traffic flow itself takes the role of a (typically disturbing) external condition.

Dynamics on time scales smaller than one second is the realm of *vehicular dynamics*. This field is mainly relevant for car manufacturers. Typical applications include the control of vehicle components such as engine, brakes, and transmission, the dynamics of skidding, and the operation of various assistance systems such as electronic stability programs (ESP), airbags, and adaptive cruise control (ACC).

In the last few years, we have seen a growing overlap between these fields. For example, models for *agent-based dynamic traffic assignment* combine the route assignment step of classical transportation planning with traffic flow models. The new generation of connected navigation systems inside cars couple the dynamics of traffic flow (jam formation) with that of traffic demand (traffic-dependent routing). Or, when modeling the effects of driver-assistance systems on traffic flow, one needs to simultaneously model aspects of vehicular and traffic dynamics, see Sect. 21.5.

Applications. There are numerous applications for traffic flow dynamics and simulation including the following:

- generation of surrounding traffic in driving simulators,
- model-based online traffic-state recognition and short-term prediction as input for traffic information channels,
- determining the optimal routes in connected (traffic-dependent) navigation systems,

- planning and optimizing the logic of external traffic control such as variable message signs or ramp metering,
- optimizing the logic behind the operation of traffic lights, e.g., enabling robust "green traffic waves" by a progressive signal system,
- assessing traffic-related effects of advanced driver-assistance systems and telematic applications in the field of Intelligent Transportation Systems (ITS),
- simulating in great detail the environmental effects of traffic operations such as fuel consumption and CO_2 emission.

Outline. The book consists of three parts. The first part deals with traffic data. After introducing the main data categories, we present methods for reconstructing the spatiotemporal traffic state and for combining heterogeneous data sources (*data fusion*). Finally, we present a data-based overview of the phenomenology of traffic flow dynamics.

The second part can be considered the core of the book. Here, we describe the mathematics and simulation of traffic flow models. After an overview of the different classes of models, we treat in detail the main categories, macroscopic and microscopic models of longitudinal (acceleration) dynamics. While microscopic models describe traffic flow from the point of individual drivers and vehicles, macroscopic models describe the collective state in terms of spatiotemporal fields for the local density, speed, and flow. For the microscopic model classes, we subsequently present models for lane changes and other discrete-choice situations such as entering a priority road. Finally, we present and comprehensively analyze the different kinds of *traffic instabilities*.

In the third part, we present selected applications of the methods and models of traffic flow dynamics. We discuss the factors of traffic breakdown and principles of the spatiotemporal evolution of congested traffic, methods of traffic-state recognition and travel-time estimation, modal emission models, and ITS applications.

The most important definitions, equations, and formulas are in highlight boxes. Numerous figures illustrate the concepts. The small in-text questions and problems act as an initial test of the reader's understanding. Each chapter ends with suggestions for further reading and a series of problems that are solved in the Appendix at the end of the book.

Last but not least, the open-source simulation software *Movsim*[2] provides reference implementations for many models presented in the book. Detailed information is provided on the book's website.[3]

---

[2] Multi-model open-source vehicular traffic simulator written in Java, see: www.movsim.org
[3] see: www.traffic-flow-dynamics.org

Part I
Traffic Data

Chapter 2
Trajectory and Floating-Car Data

Measure what is measurable, and make measurable what is not so.

Galileo Galilei

Abstract Different aspects of traffic dynamics are captured by different measurement methods. In this chapter, we discuss trajectory data and floating-car data, both providing space-time profiles of vehicles. While trajectory data captures all vehicles within a selected measurement area, floating-car data only provides information on single, specially equipped vehicles. Furthermore, trajectory data is measured externally while, as the name implies, floating-car data is captured inside the vehicle.

2.1 Data Collection Methods

Traffic can be directly observed by cameras on top of a tall building or mounted on an airplane. Tracking software extracts *trajectories* $x_\alpha(t)$, i.e. the positions of each vehicle α over time, from the video footage (or a series of photographs). If *all* vehicles within a given road section (and time span) are captured in this way, the resulting dataset is called *trajectory data*.

Thus, trajectory data is the most comprehensive traffic data available. It is also the only type that allows direct and unbiased measurement of the traffic density (see Sect. 3.3) and lane changes. However, camera-based methods involve complex and error-prone procedures which require automated and robust algorithms for the vehicle tracking, and thus are often the most expensive option for data collection. Furthermore, a simple camera can cover a road section of at most a few hundred meters since smaller vehicles are occluded behind larger ones if the viewing angle is too low.

A different method uses probe vehicles which "float" in the traffic flow. Such cars collect geo-referenced coordinates via GPS receivers which are then "map-matched" to a road on a map—the speed is a derived quantity determined from the spacing (on a map) between two GPS points. This type of data is called *floating-car data*

M. Treiber and A. Kesting, *Traffic Flow Dynamics*,
DOI: 10.1007/978-3-642-32460-4_2, © Springer-Verlag Berlin Heidelberg 2013

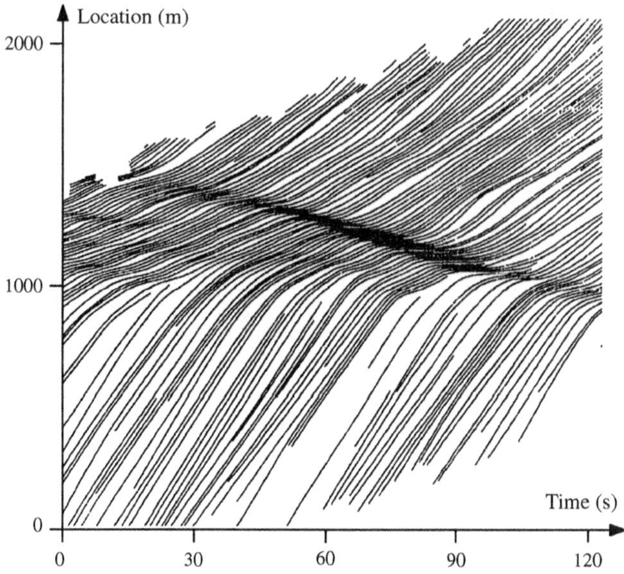

Fig. 2.1 Trajectories with moving stop-and-go waves on a British motorway segment [Adapted from: Treiterer et al. (1970)]

(FCD). Some more recent navigation systems also record (anonymized) trajectories and send them to the manufacturer. The probe vehicles can be equipped with other sensors (e.g. radar) to record distance to the leading vehicle and its speed (however, such equipment is expensive). FCD augmented in this way are also referred to as *extended floating-car data* (xFCD). One problem of FCD is that many equipped vehicles are taxis or trucks/vans of commercial transport companies which, due to their lower speeds, are not representative for the traffic as a whole. Fortunately, this bias vanishes just when the FCD information becomes relevant: In congested situations, free-flow speed differences do not matter.

Both trajectory and floating-car data record the vehicle location $x_\alpha(t)$ as a function of time, yet they differ substantially:

- Trajectory data records the spatiotemporal location of *all* vehicles within a given road segment and time interval while FCD only collects data on a few probe vehicles.
- Contrary to trajectory data, FCD does not record which lane a vehicle is using since present GPS accuracy is not sufficient for lane-fine map-matching.
- FCD may contain additional information such as the distance to the leading vehicle, position of the gas/brake pedals, activation of turning signals, or the rotation angle of the steering wheel (xFCD). In principle, every quantity available via the *CAN-bus*[1] can be recorded as a time-series. This kind of data is naturally missing in trajectory data due to the optical recording method.

[1] The CAN-bus is a micro-controller communication interface present in all modern vehicles.

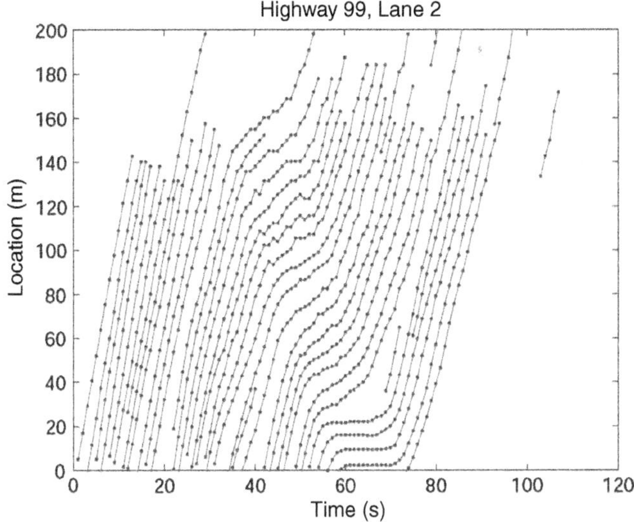

Fig. 2.2 Trajectories with moving stop-and-go waves on the California State Route 99 [From: www.ece.osu.edu/~coifman/shock]

2.2 Time-Space Diagrams

Figures 2.1 and 2.2 are examples of trajectory data of a single lane visualized in a *space-time diagram*. By convention, we will always plot time on the x-axis vs. space on the y-axis. The following information can be easily read off the diagrams:

- The local speed at (front-bumper) position x and time t is given by the gradient of the trajectory. A horizontal trajectory corresponds to a standing vehicle.
- The *time headway*, or simply *headway*, Δt_α between the front bumpers of two vehicles following each other (see Sect. 3.1) is the horizontal distance between two trajectories.[2]
- *Traffic flow*, defined as the number of vehicles passing a given location per time unit, is the number of trajectories crossing a horizontal line denoting this time interval. It is equal to the inverse of the time mean of the headways.
- The *distance headway* between two vehicles is the vertical distance of their trajectories. It is composed of the distance gap between the front and the rear bumpers plus the length of the leading vehicle.
- The *traffic density*, defined as the number of vehicles on a road segment at a given time, is the number of trajectories crossing a vertical line in the diagram and thus the inverse of the *space mean* of the distance headways (cf. Sect. 3.3).
- Lane changes to and from the observed lane are marked by beginning and ending trajectories, respectively.

[2] The time headway is composed of the (rear-bumper-to-front-bumper) time gap plus the occupancy time interval of the leading vehicle.

- The gradient of the boundary of a high-density area indicates the propagation velocity of a traffic jam. The congestions in the Figs. 2.1 and 2.2 are stop-and-go waves which are moving upstream and thus have a negative propagation speed.

If not only the longitudinal positions $x_\alpha(t)$ (along the road) but also the *lateral positions* $y_\alpha(t)$ (across the lanes) are recorded, one can generate a two-dimensional trajectory diagram from which one can deduce lateral accelerations and the duration of lane changes.

> Is it possible to estimate the time needed to pass through a given road segment using trajectory data? How would you calculate the travel time increase caused by a traffic jam? What additional assumption is needed to estimate the total time loss of all persons driving through the congestion?

Problems

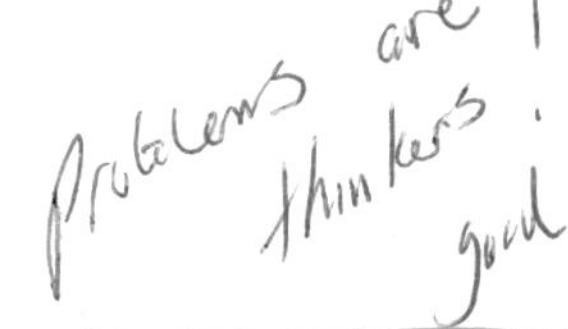

2.1 Floating-Car Data
Assume that some vehicles with GPS systems (accurate to approximately 20 m) send their (anonymized) locations to a traffic control center in fixed time intervals. Can this data be used to reconstruct (1) trajectories of single vehicles, (2) location and time of lane changes, (3) traffic density (vehicles per kilometer), (4) traffic flow (vehicles per hour), (5) vehicle speed, and (6) length and position of traffic jams? Justify your answers.

2.2 Analysis of Empirical Trajectory Data
Consider the trajectory data visualized in Fig. 2.2:

1. Determine the traffic density (vehicles per kilometer), traffic flow (vehicles per hour), and speed in different spatiotemporal sections, for example [10, 30 s] × [20, 80 m] (free traffic) and [50, 70 s] × [20, 100 m] (congested traffic).
2. Find the propagation velocity of the stop-and-go wave. Is it traveling with or against the direction of traffic flow?
3. Estimate the travel time increase incurred by the vehicle that is at $x = 0\,m$ at time $t \approx 50\,s$ due to the stop-and-go wave.
4. Estimate the average *lane-changing rate* (lane changes per kilometer and per hour) in the spatiotemporal area covered by the dataset. (Assume six trajectory beginnings or endings within [0, 80 s] × [0, 140 m].)

2.3 Trajectory Data of "Obstructed" Traffic Flow

Consider the trajectory data of city traffic shown in the diagram below:

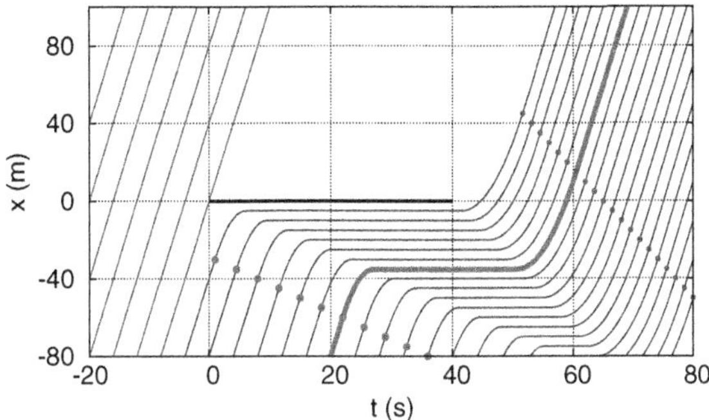

1. What situation is shown? What does the horizontal bar beginning at $x = t = 0$ mean?
2. Determine the traffic demand, i.e. the inflow for $t \leq 20$ s.
3. Determine the density and speed in the free traffic regime upstream of the "obstacle".
4. Determine the density within the traffic jam.
5. Determine the outflow after the "obstacle" disappears. Also find the density and speed in the outflow regime after the initial acceleration (the end of which is marked by smaller blue dots).
6. Determine the propagation speed of the transitions "free traffic → jam" and "jam → free traffic".
7. What travel time delay is imposed on a vehicle entering the scene at $t = 20$ s and $x = -80$ m?
8. Find the acceleration and deceleration values (assuming they are constant). The start of the deceleration phase and the end of the acceleration phase of each vehicle are marked by dots.

Further Reading

- May, A.D.: Traffic Flow Fundamentals. Prentice Hall, Eaglewood Cliffs, N.Y. (1990)
- Treiterer, J., et al.: Investigation of traffic dynamics by aerial photogrammetric techniques. Interim report EES 278-3, Ohio State University, Columbus, Ohio (1970)
- Thiemann, C., Treiber, M., Kesting, A.: Estimating acceleration and lane-changing dynamics from next generation simulation trajectory data. Transportation Research Record: Journal of the Transportation Research Board **2088** (2008) 90–101

- Schäfer, R.P., Lorkowski, S., Witte, N., Palmer, J., Rehborn, H., Kerner B.S.: A study of TomTom's probe vehicle data with three-phase traffic theory. Traffic Engineering and Control **52** (2011) 225–230

Chapter 3
Cross-Sectional Data

Nature loves to hide.
Heraklit

Abstract Cross-sectional data is captured by stationary induction loops, radar, or infrared sensors. The collected information is provided either directly as single-vehicle data or aggregated into macroscopic quantities. In this chapter we define the measurable and derived quantities characterizing both data formats, with special attention on the difference between temporal and spatial averages. Traffic density, a spatially defined quantity, cannot be directly measured using cross-sectional detectors, but several estimation methods are presented and discussed. Speed estimation methods are introduced to overcome the inability of single-loop detectors to directly measure vehicle speed.

3.1 Microscopic Measurement: Single-Vehicle Data

Cross-sectional data, measured at a fixed cross-section on the road, can be captured by laying pneumatic tubes across the road, by radar, or optically with infra-red sensors or light barriers. Most commonly, however, *induction loops* are installed beneath the road surface. They detect whether a metallic object (such as a car) is above them (Fig. 3.1). A single-loop detector can directly measure (only) the following quantities:

- The time $t_\alpha = t_\alpha^0$ at which the front of vehicle α passes the detector (voltage drop in Fig. 3.1).
- The time t_α^1 at which the rear end of the vehicle passes the detector (voltage rise in Fig. 3.1).

It is impossible for single-loop detectors to measure vehicle speed, but we can obtain an estimate in the case of relatively uniform speed values by assuming an average vehicle length l. However, this estimate is prone to large errors, as we will see in Sect. 3.4.

M. Treiber and A. Kesting, *Traffic Flow Dynamics*,
DOI: 10.1007/978-3-642-32460-4_3, © Springer-Verlag Berlin Heidelberg 2013

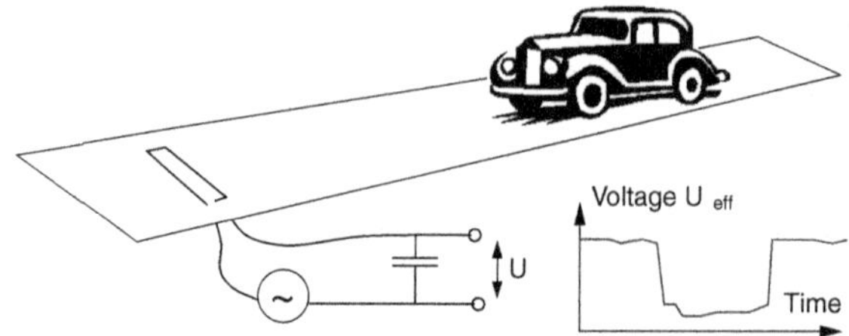

Fig. 3.1 The induction loop is part of an LC circuit (complemented by an external capacitor and an AC voltage source) tuned to be in resonance if the loop is "unoccupied", yielding a high voltage U_{eff}. The metallic parts of a vehicle will increase the inductance of the loop upon driving over it. This puts the circuit out of tune and decreases the voltage U_{eff}

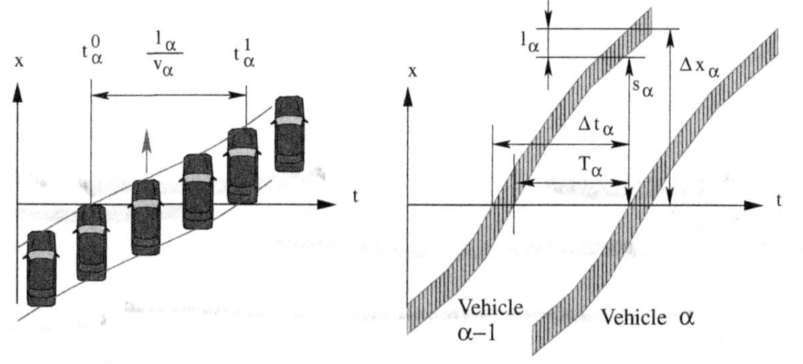

Fig. 3.2 Single-vehicle data as measured by an induction loop (or any other cross-sectional detector). The shaded area indicate the "detector occupancy" at different times

Double-loop detectors are composed of two (or more) induction loops separated by a fixed distance, e.g. 1 m. The time difference between passing the first and the second loop yields a direct measurement of the vehicle speed v_α.

From these directly measured quantities we can derive *secondary microscopic quantities* (cf. Fig. 3.2):

- *Length* of each vehicle α,

$$l_\alpha = v_\alpha(t_\alpha^1 - t_\alpha^0), \tag{3.1}$$

- vehicle type (motorcycle, car, truck, etc.) by classifying the vehicle length,
- *time headway* (sometimes also called simply *headway*) between the front bumpers of successive vehicles (the smaller index $\alpha - 1$ denotes the leading vehicle),

$$\Delta t_\alpha = t_\alpha^0 - t_{\alpha-1}^0, \tag{3.2}$$

- *time gap* between the rear and front bumpers

$$T_\alpha = t_\alpha^0 - t_{\alpha-1}^1 = \Delta t_\alpha - \frac{v_{\alpha-1}}{l_{\alpha-1}}, \tag{3.3}$$

- *distance headway*

$$d_\alpha = v_{\alpha-1} \Delta t_\alpha, \tag{3.4}$$

- and *distance gap* between the rear and front bumpers (sometimes denoted simply as *gap*)

$$s_\alpha = d_\alpha - l_{\alpha-1}. \tag{3.5}$$

All spatial quantities (vehicle length, distances) are only exact if the speed is constant during the measurement, which is a reasonable assumption.

3.2 Aggregated Data

Most detectors *aggregate* the microscopic single-vehicle data by averaging over fixed time intervals Δt and transmit only the *macroscopic data* (*aggregated data*) to the traffic control center. This saves both bandwidth in the transmission and disk space when archiving the data, but of course all the microscopic information is lost. Time intervals vary between 20 s and 5 min, the most common being $\Delta t = 60$ s. Averages over a fixed number of vehicles (e.g. $\Delta N = 50$ veh) are rarely used, even though they are statistically more meaningful. One or more of the following quantities are sent to the traffic control center:

Traffic flow. The traffic flow is defined as the number of vehicles ΔN passing the cross-section at location x within a time interval Δt:

$$\boxed{Q(x,t) = \frac{\Delta N}{\Delta t}.} \tag{3.6}$$

It is usually given in units of vehicles per hour (veh/h) or vehicles per minute. In terms of the microscopic quantities, the traffic flow Q can be considered as the inverse of the time mean of the headways, $Q = 1/\langle \Delta t_\alpha \rangle$.[1]

Sometimes, the inverse of the headway is called *microscopic flow*,

$$q_\alpha = \frac{1}{\Delta t_\alpha}, \tag{3.7}$$

and the scatter plot of q_α versus v_α the *microscopic flow-density diagram*.[2]

We emphasize that the traffic flow Q can be considered as the *harmonic* mean of the microscopic flow

$$Q = \frac{1}{\langle \Delta t_\alpha \rangle} = \frac{1}{\langle 1/q_\alpha \rangle}, \tag{3.8}$$

---

[1] The notation $\langle \cdot \rangle$ is used for the *arithmetic average* in the context of measurements and for the *expected value* in the context of statistical considerations (Sect. 3.3).

[2] Notice that the term *microscopic fundamental diagram* generally denotes the gap as a function of the speed for steady-state traffic flow as given by microscopic models.

Generally, the harmonic mean of a series of values x_α is defined as the inverse of the arithmetic mean of the inverse, $X_\mathrm{H} = 1/\langle 1/x_\alpha \rangle$.

Occupancy. The dimensionless occupancy is the fraction of the aggregation interval during which the cross-section is occupied by a vehicle:

$$O(x,t) = \frac{1}{\Delta t} \sum_{\alpha=\alpha_0}^{\alpha_0+\Delta N-1} (t_\alpha^1 - t_\alpha^0). \tag{3.9}$$

Arithmetic mean speed. The arithmetic mean speed is the average speed of the ΔN vehicles passing the cross-section during the aggregation interval:

$$V(x,t) = \langle v_\alpha \rangle = \frac{1}{\Delta N} \sum_{\alpha=\alpha_0}^{\alpha_0+\Delta N-1} v_\alpha. \tag{3.10}$$

We use V for the macroscopic speed to distinguish it from the (microscopic) speed v_α of single vehicles. To emphasize that the speed is measured at a fixed location for a time interval, V is sometimes called *time mean speed*.
Harmonic mean speed. The harmonic mean speed is defined as

$$V_\mathrm{H}(x,t) = \frac{1}{\left\langle \frac{1}{v_\alpha} \right\rangle} = \frac{\Delta N}{\sum_{\alpha=\alpha_0}^{\alpha_0+\Delta N-1} \frac{1}{v_\alpha}}. \tag{3.11}$$

When neglecting accelerations, V_H corresponds approximatively to the (spatial) average of the speed at a fixed time instant (cf. Sect. 3.3.2). Therefore, V_H is sometimes called (not completely correctly) *space mean speed*. One can show that always $V_\mathrm{H} \le V$ where the equal sign only holds if all speeds are identical. The harmonic mean speed and the following two quantities are rarely available although they would be useful for a less biased traffic density estimate (see Sects. 3.3 and 4.4).

Arithmetic time mean of microscopic flow. The arithmetic time mean of microscopic flow is defined by

$$Q^*(x,t) = \langle q_\alpha \rangle = \left\langle \frac{1}{\Delta t_\alpha} \right\rangle = \frac{1}{\Delta N} \sum_{\alpha=\alpha_0}^{\alpha_0+\Delta N-1} \frac{1}{\Delta t_\alpha}. \tag{3.12}$$

As will be shown in Sect. 4.4, this quantity is very useful in estimating the density when no microscopic data are available.

Speed variance. The speed variance

$$\mathrm{Var}(v) = \sigma_v^2(x,t) = \langle (v_\alpha - \langle v_\alpha \rangle)^2 \rangle = \langle v_\alpha^2 \rangle - \langle v_\alpha \rangle^2 \tag{3.13}$$

is a measure of the spread of the speed values within the aggregation interval. The spread is given by the standard deviation σ_V, the square root of the variance. The dimensionless *coefficient of variation* σ_V/V quantifies the relative spread of the speed values (cf. Fig. 9.7).

Considering the mean speed in a highly heterogeneous traffic flow, what is the advantage of averaging over a fixed number of vehicles instead of over fixed time intervals?

3.3 Estimating Spatial Quantities from Cross-Sectional Data

While the macroscopic quantities flow Q, occupancy O, and (in the case of double-loop detectors) the arithmetic mean speed V are measured directly, other important quantities can only be estimated by making some assumptions. The *traffic density* is defined as a *spatial* average at a fixed time (the number of vehicles on a given road segment) but cross-sectional detectors can only measure *temporal* averages at a fixed location (the cross-section). Contrary to flow and density, the macroscopic speed can be defined both as a temporal and a spatial average. However, these two definitions are not equivalent.

3.3.1 Traffic Density

The traffic density $\rho(x,t)$ can be estimated using the hydrodynamic relation

$$\rho(x,t) = \frac{Q(x,t)}{V(x,t)} = \frac{\text{flow}}{\text{speed}}. \tag{3.14}$$

However, this equation implicitly assumes that the speed V is a *spatial* average (because the density is defined as a spatial quantity). Using the *temporal* averages obtained from cross-sectional detectors induces systematic errors: Faster vehicles are "seen" more frequently by detectors than slower vehicles, yielding a bias towards larger speed values. Figure 3.3 shows a two-lane road where vehicles on the left lane drive twice as fast as vehicles on the right lane. The flow is equal on both lanes, thus the detector "sees" the same number of vehicles during the aggregation interval and reports the temporal mean speed $\langle v_\alpha \rangle$ =108 km/h. However, the space mean speed is

$$\frac{2}{3}\,72\,\text{km/h} + \frac{1}{3}\,144\,\text{km/h} = 96\,\text{km/h}.$$

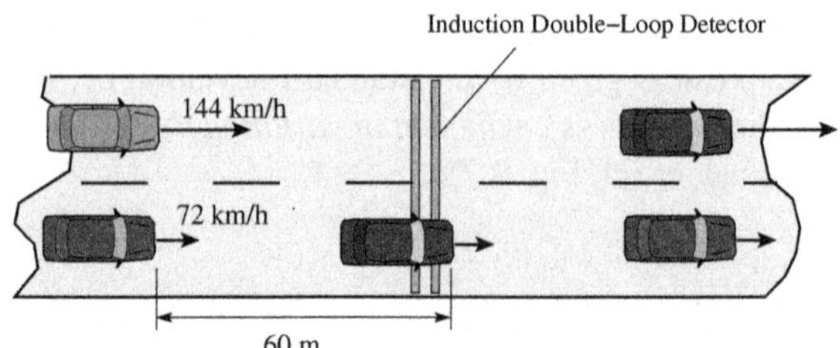

Fig. 3.3 Vehicles on the *left* lane drive twice as fast but with the same time headway

Thus, the density as obtained by the hydrodynamic relation (3.14) underestimates the real density by a factor of 8/9.

We can obtain a better estimate for the density from its definition "vehicles per distance", which can be expressed in terms of microscopic quantities as the inverse of the space mean of the distance headways,

$$\rho(x,t) = \frac{1}{\langle d_\alpha \rangle} = \frac{\Delta N}{\sum_\alpha d_\alpha}. \tag{3.15}$$

Similarly, the flow ("vehicles per time") can be written as the inverse of the time mean of the headways. For a given fixed time interval

$$\Delta t = \sum_{\alpha=\alpha_0}^{\alpha_0+\Delta N-1} \Delta t_\alpha = \Delta N \langle \Delta t_\alpha \rangle$$

the flow is given by

$$Q = \frac{\Delta N}{\Delta t} = \frac{1}{\langle \Delta t_\alpha \rangle}. \tag{3.16}$$

In the following section we discuss two different ways for expressing the density in terms of the measurable quantities Δt_α and v_α.

3.3.1.1 Derivation from the Expected Value of Traffic Density

Inserting Eq. (3.4) into the definition of the expected density (3.15) yields

$$\begin{aligned}
\frac{1}{\rho} &= \langle d_\alpha \rangle = \langle v_{\alpha-1} \Delta t_\alpha \rangle \\
&\approx \langle v_\alpha \Delta t_\alpha \rangle \\
&= \langle v_\alpha \rangle \langle \Delta t_\alpha \rangle + \mathrm{Cov}(v_\alpha, \Delta t_\alpha) \\
&= \frac{V}{Q} + \mathrm{Cov}(v_\alpha, \Delta t_\alpha),
\end{aligned}$$

and solving for ρ gives us

$$\boxed{\rho = \frac{Q}{V}\left(\frac{1}{1 + \frac{Q}{V}\mathrm{Cov}(v_\alpha, \Delta t_\alpha)}\right).} \tag{3.17}$$

Here $\mathrm{Cov}(\cdot, \cdot)$ denotes the *covariance*, defined for two random variables x and y as

$$\mathrm{Cov}(x, y) = \langle (x - \langle x \rangle)(y - \langle y \rangle) \rangle = \langle xy \rangle - \langle x \rangle \langle y \rangle. \tag{3.18}$$

The covariance is positive if both variables are *positively correlated*, i.e. larger values of x tend to be accompanied by proportionally larger values of y. The significance of such a linear relationship is quantified by the *correlation coefficient*

$$r_{x,y} = \frac{\mathrm{Cov}(x, y)}{\sigma_x \sigma_y}. \tag{3.19}$$

For uncorrelated x and y (that is, the variables have no *linear* relationship) the coefficient is 0. Its value is bounded between -1 (x and y are perfectly anti-correlated, $x \propto -y$) and $+1$ (x and y are perfectly correlated, $x \propto y$). The correlation coefficient allows us rewrite Eq. (3.17) as *Wardrop's equation*[3]:

$$\rho = \frac{Q}{V}\left(\frac{1}{1 + \frac{\sigma_V}{V}\frac{Q}{\sigma_Q} r_{v_\alpha, \Delta t_\alpha}}\right). \tag{3.20}$$

Thus, the real density equals the (widely used) estimate "flow divided by arithmetic mean speed" multiplied by a correction factor that captures the correlation between speed and headway, $r_{v,\Delta t}$, as well as the (relative) variance of vehicle speed and flow, σ_V/V and σ_Q/Q. In free traffic $r_{v,\Delta t}$ is near zero since every driver is able to choose his or her speed independently. In congested traffic, however, the headway Δt_α usually increases with *decreasing* speed and tends to infinity as the speed approaches zero. Therefore $r_{v,\Delta t}$ is negative in this case and the correction factor is greater than 1. Thus, the relation Q/V systematically *underestimates* the real density in congested traffic (cf. Fig. 4.10).

3.3.1.2 Derivation from the Expected Value of Traffic Flow

A different approach to derive the density from measurable quantities combines the expected value of the flow (3.16) with Eq. (3.4):

[3] Not to be confused with the Wardrop equilibrium, a concept in transportation planning where routes are chosen according to the *user equilibrium*, i.e., no user is better off when choosing a different route.

$$\frac{1}{Q} = \langle \Delta t_\alpha \rangle = \left\langle \frac{d_\alpha}{v_{\alpha-1}} \right\rangle \tag{3.21}$$

$$\approx \left\langle \frac{d_\alpha}{v_\alpha} \right\rangle = \langle d_\alpha \rangle \left\langle \frac{1}{v_\alpha} \right\rangle + \text{Cov}\left(d_\alpha, \frac{1}{v_\alpha} \right) \tag{3.22}$$

$$= \frac{1}{\rho V_\text{H}} + \text{Cov}\left(d_\alpha, \frac{1}{v_\alpha} \right). \tag{3.23}$$

Again solving for ρ we obtain

$$\boxed{\rho = \frac{Q}{V_\text{H}} \left(\frac{1}{1 - Q\,\text{Cov}\left(d_\alpha, 1/v_\alpha\right)} \right)} \tag{3.24}$$

where V_H is the harmonic mean speed (3.11) that gives stronger weight to small speed values. Since the distance headway d_α usually increases with v_α (and decreases with $1/v_\alpha$), $\text{Cov}(d_\alpha, 1/v_\alpha)$ is negative and the correction factor smaller than 1. Thus, Q/V_H generally *overestimates* the real density.

3.3.1.3 Discussion of the Two Approximations

In practice, the covariances in Eqs. (3.20) and (3.24) are usually assumed to be zero and

$$\rho^{(1)} = \frac{Q}{V} \quad \text{or} \quad \rho^{(2)} = \frac{Q}{V_\text{H}} \tag{3.25}$$

is used to calculate the density (both relations can be applied to multi-lane traffic as well). The following statements help in assessing the errors of the two estimates:

1. If all vehicle speeds v_α are the same, then $V = V_\text{H}$ and thus $\rho = \rho^{(1)} = \rho^{(2)}$.
2. If all headways Δt_α are the same, then $\text{Cov}(v_\alpha, \Delta t_\alpha) = 0$ and thus $\rho = \rho^{(1)} = Q/V$ holds exactly (cf. Fig. 3.3). Otherwise, $\rho^{(1)}$ most likely underestimates the real density as $\text{Cov}(v_\alpha, \Delta t_\alpha)$ is usually negative.
3. If all distance headways d_α are the same, then $\text{Cov}\left(d_\alpha, \frac{1}{v_\alpha}\right) = 0$ and thus $\rho = \rho^{(2)} = Q/V_\text{H}$ holds exactly (again, cf. Fig. 3.3). Otherwise, $\rho^{(2)}$ most likely overestimates the real density since $\text{Cov}\left(d_\alpha, \frac{1}{v_\alpha}\right)$ is usually negative as well.

Why is it not possible to measure the density of stopped traffic using stationary detectors of any kind?

3.3.2 *Space Mean Speed*

The space mean speed (instantaneous mean) $\langle V(t)\rangle$ is the arithmetic mean of the speed of all vehicles within a given road segment at time t (in Fig. 3.4 this is a segment of length L around the detector),

$$\langle V(t)\rangle = \frac{1}{n(t)} \sum_{\alpha=1}^{n(t)} v_\alpha(t). \tag{3.26}$$

In general (that is, with multiple lanes and arbitrary speeds and accelerations), aggregated detector data is unsuitable for determining the space mean speed because the number and identities of vehicles used in the average in Eq. (3.26) changes within the aggregation interval Δt. Also, it is possible that $n(t) = 0$.

We get a more suitable definition by averaging the instantaneous mean over the aggregation interval Δt. Furthermore, we choose the reference length L small enough so that no vehicles are on the reference road segment at time t or $t + \Delta t$ and the vehicle speed does not change significantly during the time needed for passing the segment, $\tau_\alpha \approx L/v_\alpha$. Averaging Eq. (3.26) over time gives us[4]

$$\begin{aligned}
\langle V\rangle &= \frac{\int_t^{t+\Delta t} n(t')\langle V(t')\rangle\,\mathrm{d}t'}{\int_t^{t+\Delta t} n(t')\,\mathrm{d}t'} = \frac{\sum_\alpha \int_{t_\alpha}^{t_\alpha+\tau_\alpha} v_\alpha(t')\,\mathrm{d}t'}{\sum_\alpha \tau_\alpha} \\
&\approx \frac{\sum_\alpha \tau_\alpha v_\alpha}{\sum_\alpha \tau_\alpha} \approx \frac{n\,L}{\sum_\alpha L/v_\alpha} \\
&= \frac{n}{\sum_{\alpha=1}^{n} \frac{1}{v_\alpha}},
\end{aligned}$$

Here n denotes the total number of vehicles that have passed the detector within the interval Δt (not to be confused with $n(t)$, the number of vehicles on the referenced road segment). The speed values v_α are those obtained from the detector (i.e., measured at the same location but different times, as opposed to measured simultaneously at different locations).

Thus, the time-averaged (over an aggregation interval) and space-averaged (over a road segment) speed is given by the *harmonic mean*,

$$\langle V\rangle = V_\mathrm{H}. \tag{3.27}$$

Although not exact, the harmonic mean V_H of temporal speed data obtained at a fixed location (stationary detectors) is often equated with the instantaneous mean, also called *space mean speed* in the literature:

---

[4] Note that the denominator equals the total travel time (vehicle-minutes) of all vehicles in the referenced spatiotemporal interval.

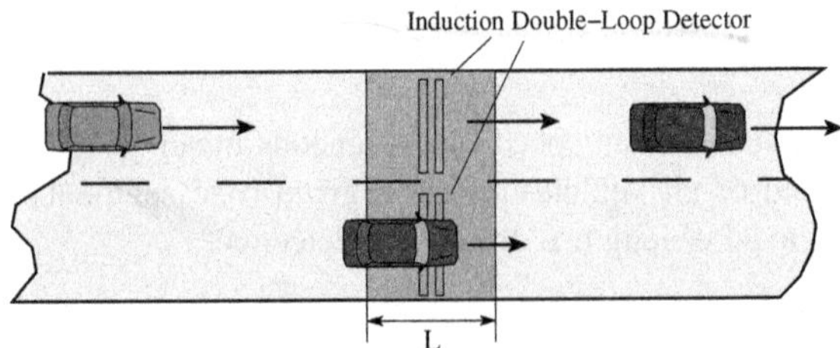

Fig. 3.4 Derivation of the space mean speed (3.27)

The harmonic time mean speed is approximately equal to the (arithmetic) space mean speed.

3.4 Determining Speed from Single-Loop Detectors

Single-loop detectors only measure the entry and exit times t^0_α and t^1_α of each vehicle α. If the vehicle length l_α was known, we could obtain the speed from $v_\alpha = l_\alpha/(t^1_\alpha - t^0_\alpha)$. However, single-loop detectors cannot measure vehicle length. Yet we can assume an *average* vehicle length $\langle l_\alpha \rangle$ and use the definition of the occupancy (3.9) to derive an estimate of the average speed:

$$\begin{aligned}
O &= \frac{1}{\Delta t} \sum_\alpha (t^1_\alpha - t^0_\alpha) \\
&= \frac{1}{\Delta t} \sum_\alpha \frac{l_\alpha}{v_\alpha} \\
&= \frac{n}{\Delta t} \left[\langle l_\alpha \rangle \left\langle \frac{1}{v_\alpha} \right\rangle + \text{Cov}\left(l_\alpha, \frac{1}{v_\alpha} \right) \right] \\
&= Q \left[\langle l_\alpha \rangle \left\langle \frac{1}{v_\alpha} \right\rangle + \text{Cov}\left(l_\alpha, \frac{1}{v_\alpha} \right) \right].
\end{aligned}$$

Solving for $V_H = 1/\langle 1/v_\alpha \rangle$ we get

$$\boxed{V_H = \frac{Q \langle l_\alpha \rangle}{O \left[1 - \frac{Q}{O} \text{Cov}(l_\alpha, 1/v_\alpha) \right]}.} \tag{3.28}$$

For large densities the covariance $\text{Cov}(l_\alpha, 1/v_\alpha)$ is nearly zero because all vehicles drive with approximately the same speed. Thus, the estimate for V_H simplifies to $Q\langle l_\alpha \rangle / O$. In free traffic, however, longer vehicles (trucks) usually drive more slowly than shorter vehicles (cars), thus $\text{Cov}(l_\alpha, 1/v_\alpha) > 0$. In this case $Q\langle l_\alpha \rangle / O$

systematically underestimates the harmonic mean of the speed values. However, since the harmonic mean is always less than the arithmetic mean for any data with finite variance, $Q\langle l_\alpha \rangle / O$ may be a good estimate for the arithmetic mean. If all vehicle lengths are equal, the simple relation between occupancy and harmonic mean speed is exact (for *arbitrary* speed values). For all these cases the traffic density can be easily estimated as well, yielding

$$V_H = \frac{Q}{\tilde{\rho}} \quad \text{with} \quad \tilde{\rho} = \frac{O}{\langle l_\alpha \rangle}. \tag{3.29}$$

To apply these equations only the average vehicle length $\langle l_\alpha \rangle$ must be known.

Problems

3.1 Data Aggregation at a Cross-Section

Consider the following 30 s excerpt from single-vehicle data of a cross-sectional detector:

Time (in s)	Speed (in m/s)	Lane (1 = right, 2 = left)	Vehicle length (in m)
2	26	1	5
7	24	1	12
7	32	2	4
10	32	2	5
12	29	1	4
18	28	1	4
20	34	2	5
21	22	1	15
25	26	1	3
29	38	2	5

1. Aggregate the data and calculate the macroscopic traffic flow and speed (arithmetic mean), separately for both lanes.
2. Calculate the traffic density in each lane assuming that speed and time headway of two succeeding vehicles are uncorrelated (which is realistic for free traffic).
3. Determine the flow, speed, and density of both lanes combined.
4. What percentage of the vehicles on the right lane (and in total) are trucks?

3.2 Determining Macroscopic Quantities from Single-Vehicle data

On a two-lane highway all vehicles drive with distance headway 60 m. The vehicles on the left lane all drive at speed 144 km/h, on the right lane at 72 km/h. A stationary detector captures single-vehicle data (cf. Fig. 3.3) and aggregates them using $\Delta t = 60\text{s}$.

1. What are the time headways Δt_α on both lanes? What are the time gaps, assuming all vehicles are 5 m long?
2. Find the traffic flow, occupancy, and average speed (both arithmetic and harmonic) separately for both lanes (i.e., each lane is captured by its own detector) and also for both lanes combined (i.e., one detector captures vehicles on both lanes). For which type of averaging does the following statement hold: The average speed of all vehicles in both lanes is equal to the *arithmetic* mean of the average speed of each lane?
3. Calculate the speed variance. Show that the speed variance of all vehicles (on both lanes) is

$$\sigma_V^2 = p_1 \left(\sigma_{V1}^2 + (V_1 - V)^2\right) + (1 - p_1) \left(\sigma_{V2}^2 + (V_2 - V)^2\right),$$

where $p_1 = \Delta n_1/(\Delta n_1 + \Delta n_2)$ is the fraction of vehicles that are detected on the right lane (within the time interval Δt). The speed variances of the single lanes are denoted by σ_{V1}^2 (right lane) and σ_{V2}^2 (left lane), and V_1 and V_2 are the corresponding arithmetic means. Finally, $V = p_1 V_1 + (1 - p_1) V_2$ is the average over both lanes. How does the equation simplify for $p_1 = 1/2$?

Further Reading

- Leutzbach, W.: Introduction to the Theory of Traffic Flow. Springer, Berlin (1988)
- Helbing, D.: Traffic and related self-driven many-particle systems. Reviews of Modern Physics **73** (2001) 1067–1141
- Cassidy, M.J.: Traffic Flow and Capacity. International Series in Operations Research & Management Science. In: Handbook of Transportation Science. Springer New York (2003) 155–191

Chapter 4
Representation of Cross-Sectional Data

The marvelous thing about traffic flow is the fact that you can jam it anywhere at any time with so little effort.
Siegfried Wache

Abstract In this chapter we discuss different visualizations of microscopic and macroscopic cross-sectional data and the possible conclusions that one can draw from them. Time series of aggregated quantities such as speed, flow, and density show temporal developments, while speed-density and flow-density diagrams allow us to make statements about the average driving behavior on the observed road segment. Particularly the flow-density diagram contains so much information about the traffic dynamics that its idealized form is also called *fundamental diagram of traffic flow*. If single-vehicle data is available, we can also obtain distributions of microscopic quantities (vehicle speeds, time gaps, etc.).

4.1 Time Series of Macroscopic Quantities

One way of representation are *time series* of some aggregated quantity, which has been measured at a cross-section. Flow, speed, and density time series of a few hours' data tell us about traffic breakdowns, types of traffic congestion (oscillatory or essentially stationary), and the capacity drop after a breakdown (Fig. 4.1).

From the specific daily patterns of traffic *demand* (Fig. 4.2), the reader can easily recognize whether it was recorded on a Monday, Tuesday/Wednesday/Thursday, Friday or on a weekend.[1] However, these daily traffic-demand plots are used primarily in transportation planning and are beyond the scope of this book.

---

[1] School and national holidays as well as holidays and associated "long weekends" are special cases with their own characteristic patterns.

DOI: 10.1007/978-3-642-32460-4_4,

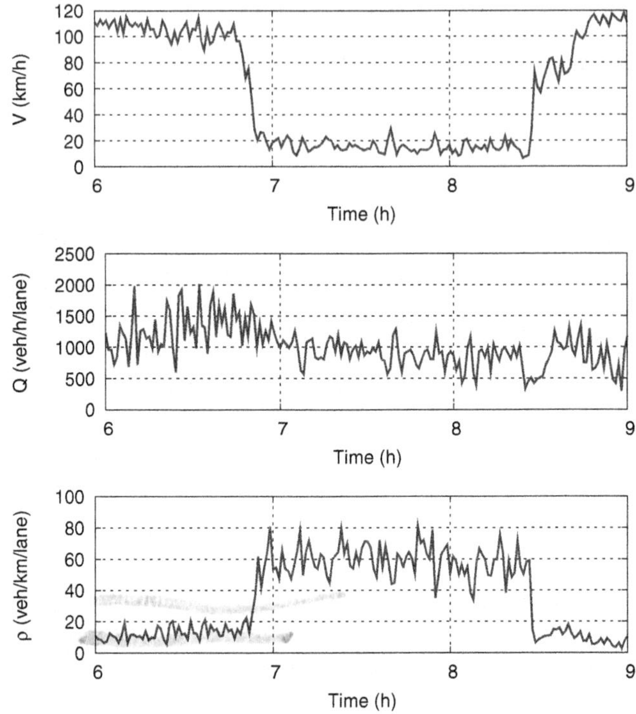

Fig. 4.1 Time series during the morning peak-hour from one-minute data. From top to bottom: arithmetic mean speed V, flow Q, and estimated density $\rho = Q/V$ (see Sect. 3.3.1)

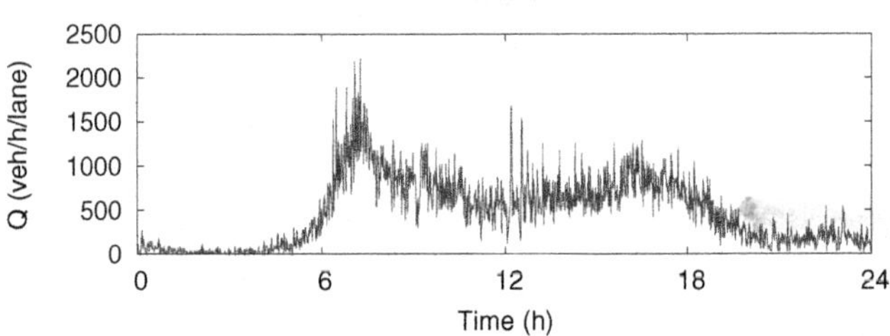

Fig. 4.2 Typical daily time series of the traffic flow (demand) on a weekday (Wednesday)

It is very easy to draw incorrect conclusions when interpreting traffic jam dynamics using single time series, as the following exercise illustrates:

Why is it wrong to conclude from the time series in Fig. 4.1 that the traffic breakdown occurred at around 7 a.m.? Can we at least conclude (from the figure) that vehicles near the cross-section at 7 a.m. decelerate, or that

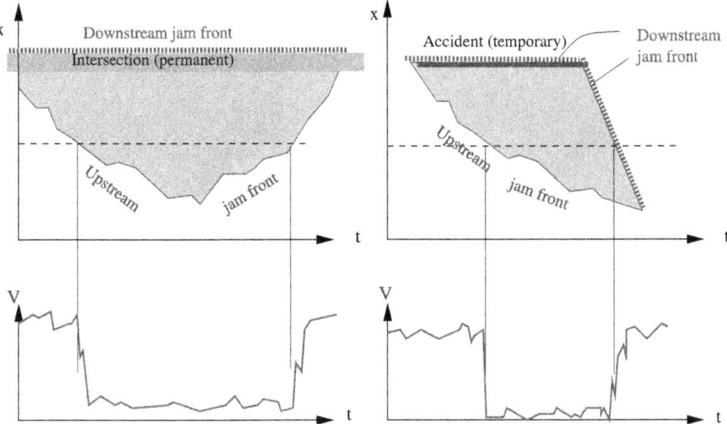

Fig. 4.3 Sketches of speed time series at a cross-section and possible spatiotemporal traffic patterns causing them

vehicles near the cross-section at 8.30 a.m. accelerate? If not, what are alternative explanations for the observed patterns?

Solution. According to Fig. 4.3 the speed drop shortly before 7 a.m. is an upstream jam front that is moving upstream. Alternatively, it could be a downstream jam front moving downstream (with the driving direction) that is caused by a *moving bottleneck*, e.g., by an oversize load. However, this case is rather unlikely, so we assume that it is an upstream jam front and vehicles are braking to avoid a rear-end collision.

The rise in speed at 8.30 a.m. can be explained by two different scenarios: (i) It is a downstream-moving upstream front, i.e. the traffic jam shrinks. This would imply that, after 8.30 a.m., vehicles are *braking* shortly after passing the detector, while the time series indicates an *acceleration*. (ii) Alternatively, it could be an upstream-moving downstream front, caused for example by a disappearing temporary bottleneck (road block, traffic light, etc.) as the waiting vehicles subsequently start to move again. In this case, the vehicles *accelerate* as indicated by the time series. For both scenarios, we can estimate the jam front velocity directly from the *fundamental diagram* (see Sect. 4.4 and Part II).

4.2 Speed-Density Relation

If we plot the aggregated vehicle speed over traffic density we obtain a *speed-density diagram* (cf. Fig. 4.4). We see that the average speed is lower in denser traffic. Furthermore, the diagram reflects the *average* behavior of a (typical) *driver-vehicle*

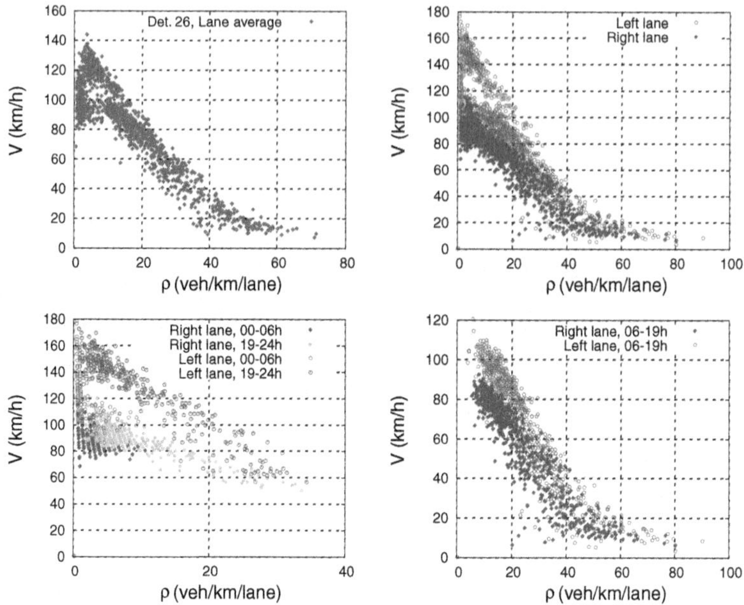

Fig. 4.4 Speed-density relation obtained from one-minute data collected on the Autobahn A9 near Munich, Germany, using the average over both lanes (*top left*), individual averages of both lanes (*top right*), and individual lane averages conditioned on night (*bottom left*) and day hours (*bottom right*)

unit in different densities and external influences such as speed limits, weather conditions, etc.

In very low-density traffic, the drivers are usually not influenced by other vehicles and we obtain the average *free speed* V_0 for $\rho \to 0$ (cf. Fig. 4.5). This speed is the minimum of (i) the actual desired speed of the drivers, (ii) the physically possible attainable speed (especially relevant for trucks on uphill slopes), and possibly (iii) an administrated speed limit (plus the drivers' average speeding). However, V_0 is often directly referred to as the *desired speed*.

To approximatively obtain the distribution of desired speeds from empirical data, we can use the speed distributions in single-vehicle data of low-density traffic (cf. Sect. 3.1 and Fig. 4.6). In this case, there are few interactions between the drivers and most of the drivers can be expected to drive at their desired speed. The distributions of speeds on the left and middle lane are symmetric and approximately Gaussian, while speeds on the right lane are distributed *bimodally*, showing the superposition of the different speed distributions of trucks and passenger cars. Figure. 4.7 shows average speed differences between lanes. In denser traffic, the speed difference tends towards zero, leading to a *speed synchronization* of the lanes.

Speed-density diagrams might show *heterogeneous traffic* and different external conditions, which has to be considered when interpreting them. Examples include a varying percentage of trucks at different times of the day, different weather conditions

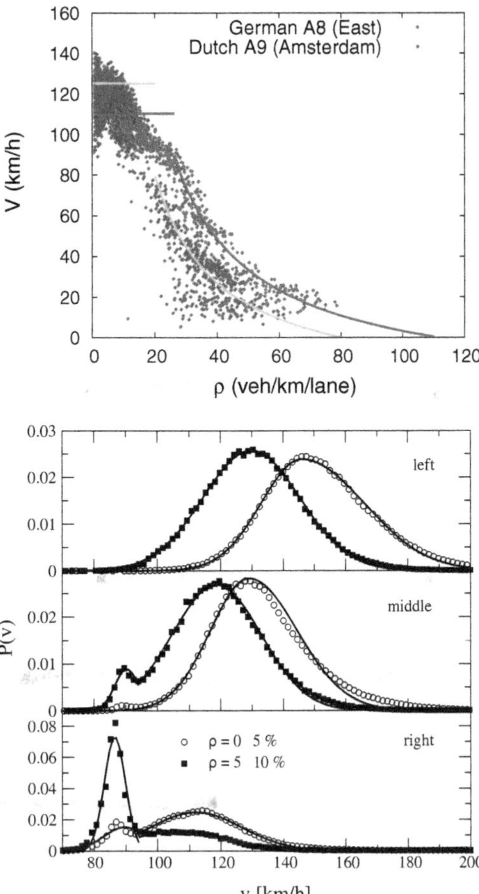

Fig. 4.5 Speed-density diagrams, averaged over all lanes, for segments of the Dutch A9 (Haarlem to Amsterdam) and the German A8 (Munich to Salzburg, Austria)

Fig. 4.6 Probability distributions of the vehicle speed, $P(v)$, in low-density traffic on the German Autobahn A3 (three lanes in each direction) [From: Knospe et al., Physical Review E 65, S. 56133 (2002)]

(lighting, precipitation), and time-dependent speed limits issued by traffic control systems. This also applies to the flow-density diagrams which will be discussed in Sect. 4.4.

(1) In the upper left (V, ρ)-diagram of Fig. 4.4, the average speed decreases again for very small densities. Does this imply that drivers are "afraid of the free road"? Explain this observation statistically.
(2) The upper right panel of Fig. 4.4 shows two point clusters, around 100 km/h and 125 km/h, in the left lane (red open circles). Give a possible explanation for this bimodality. Consider the diagrams in the bottom panels (a traffic control system issuing traffic-dependent speed limits by variable message signs is installed on this road segment).

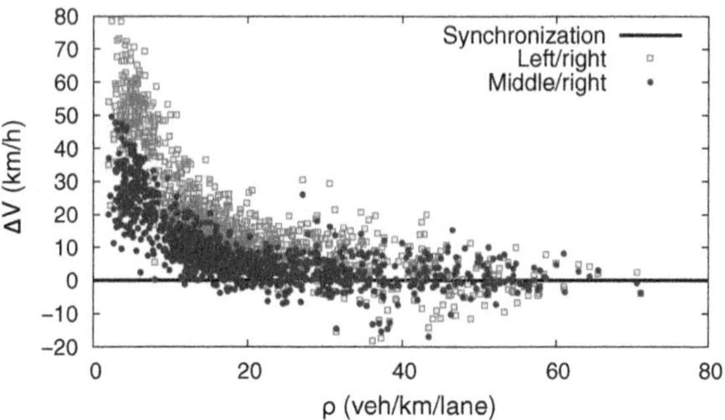

Fig. 4.7 Difference in average speed between neighboring lanes (A9-South near Munich, Germany)

4.3 Distribution of Time Gaps

Using single-vehicle data, we can also obtain the distributions of time gaps (cf. Eq. 3.3), as shown in Fig. 4.8 for two different speed ranges corresponding to free and congested traffic. The time-gap distributions exhibit the following properties:

1. Time gaps are broadly scattered—it is not unusual to see standard deviations larger than the arithmetic mean $\langle T \rangle$, i.e., a coefficient of variation greater than 1.
2. The distributions are strongly asymmetric. Both in free and congested traffic we observe time gaps longer than 10 s.
3. In free traffic (with speeds larger than some critical speed V_c) the most probable time gap $\hat{T}$ (the statistical *mode*) is significantly smaller than in congested traffic. In both speed regimes, $\hat{T}$ is significantly smaller than the recommended safe time gap in the USA ("leave one car length for every ten miles per hour of speed"), or in Europe ("safety distance (in meters) equals speed (in km/h) divided by two", corresponding to 1.8 s).
4. The arithmetic mean is also significantly smaller in dense free traffic than in congested traffic.

The mean flow is equal to the inverse of the arithmetic mean of the time headways. Thus, we can also determine the *flow decrease after a traffic breakdown* from the distributions in Fig. 4.8. Traffic jams usually do not dissolve quickly once they have emerged, due to this *capacity drop*.

Most of the observed time-gap distributions are not identical to the distribution of the drivers' *desired* time gaps, but provide an upper bound only. The real time gap is larger in free traffic because most vehicles are not actually following another vehicle. With a flow of, e.g., 360 veh/h per lane (corresponding to a mean headway of 10 s), the mode of the time-gap distribution is still below 1 s. There are also dynamic influences, since the followed vehicle might be "getting away" if the following vehicle cannot accelerate any further (or its driver does not want to). These effects explain, at least partially, the strong asymmetry of the distributions.

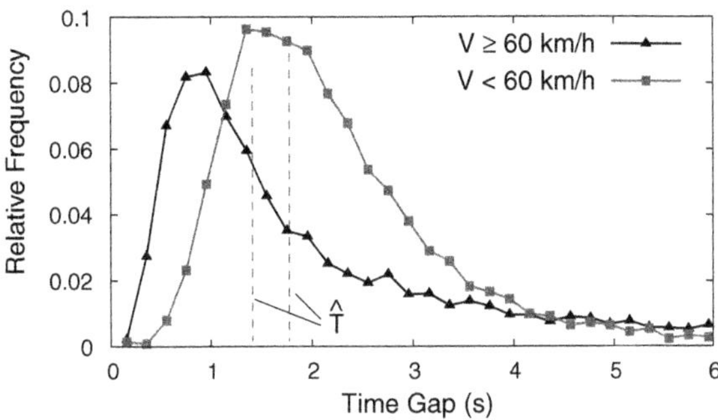

Fig. 4.8 Distribution of the time gaps in two speed regimes (free and congested traffic), measured on the Dutch A9

4.4 Flow-Density Diagram

The flow-density diagram, i.e., plotting traffic flow against density, allows us to make a number of statements on the *macroscopic* (i.e., average) behavior of a driver-vehicle unit. In its idealized form, i.e., steady state equilibrium of identical driver-vehicle units, it is also called *fundamental diagram*. The following quantities can be derived from the fundamental diagram:

1. The *desired speed* equals the asymptotic gradient $Q'(0)$ of the fit $Q(\rho)$ for $\rho = 0$. This quantity can be more accurately determined using speed-density diagrams (cf. Sect. 4.2).
2. The actual *mean speed* for a defined density is given by the slope $Q(\rho)/\rho$ of the secant through $(0, 0)$ and $(\rho, Q(\rho))$.
3. The maximum value of $Q(\rho)$ is the *road capacity* per lane.
4. The inverse of the smallest nonzero density ρ_{max} for which $Q(\rho_{max}) = 0$ equals the *average vehicle length* plus the average gap between stopped vehicles.
5. The mean *time gap* T can be determined from the (negative) slope of $Q(\rho)$ at large densities (see Chap. 8).
6. The slopes of flow-density diagrams also allow to read off the *propagation velocities* of jam fronts and variations of macroscopic quantities (this is also discussed in Chap. 8).

Bias with respect to the fundamental diagram. It is important to carefully distinguish between measured flow-density data and the fundamental diagram.

> The *fundamental diagram* describes the theoretical relation between density and flow in stationary homogeneous traffic, i.e., the steady state equilibrium of identical driver-vehicle units. The *flow-density diagram* represents aggregated empirical data that generally describes non-stationary heterogeneous traffic, i.e., different driver-vehicle units far from equilibrium.

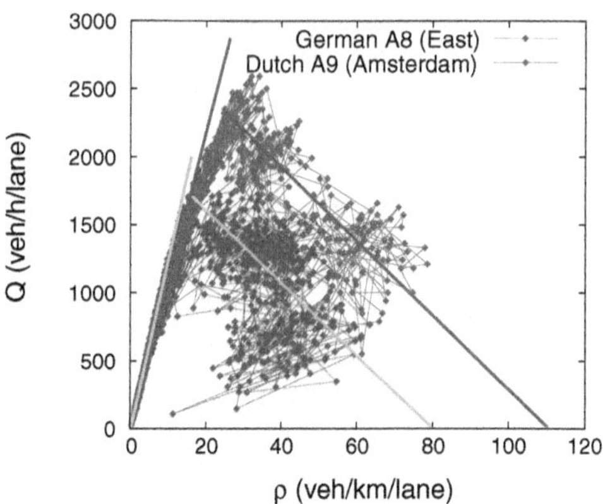

Fig. 4.9 Flow-density diagram (averaged over all lanes) for sections of the Dutch A9 (Haarlem to Amsterdam) and the German A8-East (Munich to the Austrian border) near Irschenberg

There are multiple reasons for flow-density data not to coincide with the fundamental diagram:

- The measurements process induces systematic errors (Sect. 3.3).
- The traffic flow is not at equilibrium.
- The traffic flow has spatial inhomogeneities or contains non-identical driver-vehicle units.

The statements on traffic jam dynamics and driving behavior derived in the above enumeration are exact for the fundamental diagram, only. Since *each* of the aforementioned factors can cause significant differences between the density obtained from Eq. (3.14) and the theoretical expectation in the fundamental diagram (it is not unusual to see discrepancies by a factor of two), deriving statements from flow-density data is quite error-prone. In the following examples of empirical flow-density relations shown in the Figs. 4.9, 4.11 and 4.12 (upper left panel), the maximum traffic density obtained by extrapolation is unrealistically small, while the front propagation velocities derived from the trend of flow-density point clouds of congested regions are too large in magnitude (and the point clouds do not always show a clear trend).

To estimate the effects of the errors mentioned above, we can use traffic simulations that also simulate the measurement process using *virtual cross-sectional detectors*. Fig. 4.10 shows that the flow-density diagram depends strongly on the method of averaging for obtaining the macroscopic speed and the flow (cf. Sect. 3.2), at least at large densities. Particularly, all methods yield estimated densities that strongly deviate from the actual density, which is, of course, available in the simulation. Remarkably, plotting the flow Q against the density estimate

$$\rho^* = \frac{Q^*}{V_H} \tag{4.1}$$

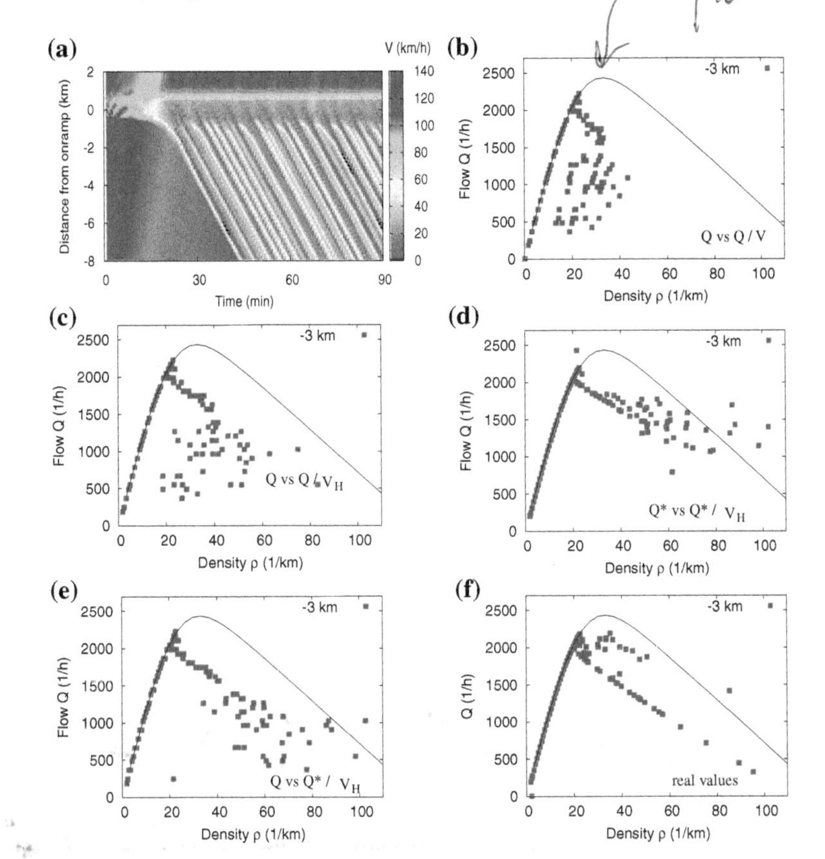

Fig. 4.10 **a** Microscopic simulation of a traffic breakdown and stop-and-go waves caused by an on-ramp. Shown is the local speed. **b–e** flow-density data where the measurement process was simulated using data of "virtual" detectors and different aggregation methods. **b** Flow $Q = 1/\langle\Delta t_\alpha\rangle$ versus density Q/V (the standard procedure), **c** flow Q versus density Q/V_H, **d** flow $Q^* = \langle 1/\Delta t_\alpha\rangle$ versus density Q^*/V_H, **e** flow Q versus density Q^*/V_H. For comparison, plot **f** displays the point cloud obtaining by using the actual local values of flow and density, and the fundamental diagram is plotted as *solid line* in **b–f**

(Fig. 4.10e) consistently yields the least biased result in the simulations although the unbiased flow is given by the harmonic mean Q (Eq. 3.6) of the microscopic flow, and not by the arithmetic average Q^* (Eq. 3.12). In any case, the difference between the true flow-density points (f) and the data shown in (b)–(e) is caused by the measurement process. The difference between the flow-density data (f) and the fundamental diagram, however, is due solely to non-equilibrium effects. This can be concluded since identical driver-vehicle units were simulated (for details, see Fig. 11.4 in Part II where this simulation is discussed in detail).

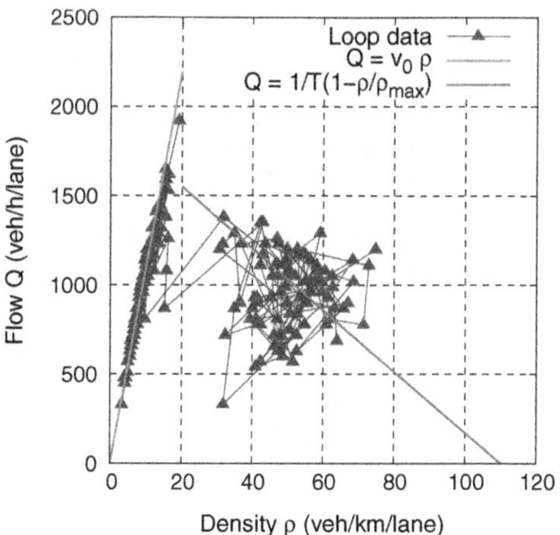

Fig. 4.11 Flow-density diagram describing hysteretic traffic dynamics. Time series of these data are shown in Fig. 4.1

We finally notice that quantities that are derived purely from measurements of the flow, such as the capacity and the hysteresis effects to be discussed in the next paragraph, are less subjected to errors.

Capacity drop and hysteresis. Sometimes, a sudden drop of the maximum possible traffic flow (*capacity drop*) is observed with a traffic breakdown (cf. Fig. 4.11 and 4.12). In this case the traffic shows hysteresis effects, i.e., the dynamics does not only depend on the traffic demand but also on the history of the system. When the traffic breaks down, the system state switches from the "free branch" onto the "congested branch", lowering the maximum possible flow. This implies that once a traffic jam has emerged, the traffic demand has to fall to a much lower value to dissolve the jam. The flow-density diagram describing this phenomenon is also said to have an *inverse-λ form* (due to its resemblance of a mirrored Greek letter lambda, λ).

Wide scattering. The strong variation of time gaps (cf. Sect. 4.3) partially explains the strong scattering of the flow-density data in congested traffic: While in free traffic the variations of density and time gaps both cause variations of the flow-density data *along* the one-dimensional curve $Q \approx \rho V_0$, variations of density in congested traffic lead to changes in the flow-density data which are *orthogonal* to those caused by variation in the time gaps. Both effects combined lead to a chaotic behavior of the flow-density data in congested traffic (cf. Figs. 4.11 and 4.12).

Finally, variations in the time gaps are not only caused by heterogeneous traffic (i.e., different *desired* time gaps of the individual drivers), but also by non-equilibrium traffic dynamics (i.e., the *actual* time gap is not equal to the *desired* time gap) and the systematic aggregation errors discussed above (Fig. 4.10).

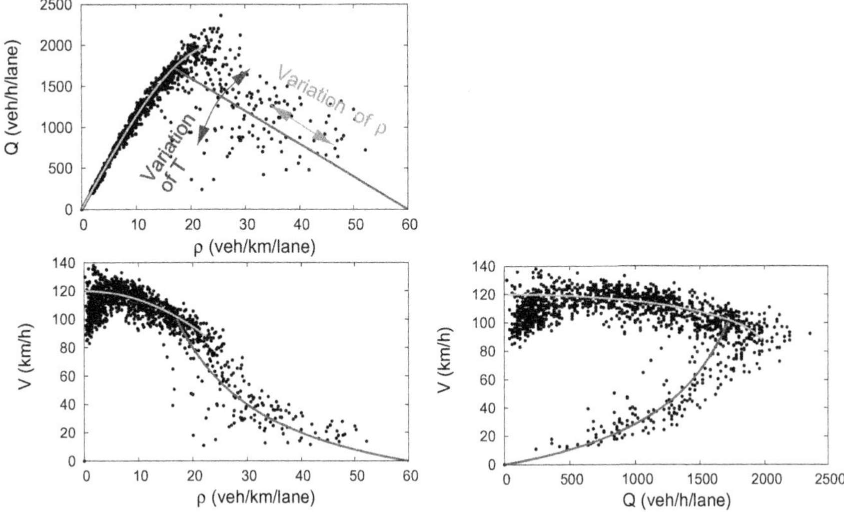

Fig. 4.12 Flow-density, speed-density, and speed-flow diagrams of the 1-minute data captured on the Autobahn A5 near Frankfurt, Germany using harmonic mean speed. The lines show the fit of a traffic-stream model (see Sect. 6.2.2)

4.5 Speed-Flow Diagram

Plotting vehicle speed against traffic flow is also possible, of course. However, this diagram is not as fundamental for modeling as the flow-density diagram and not as demonstrative as the speed-density diagram. It does have the advantage of showing only directly observed quantities, Nevertheless, it is also affected by the systematic errors in the speed aggregation. By the hydrodynamic relation $Q = \rho V$, all three diagram types are equivalent (cf. Fig. 4.12).

Problems

4.1 Analytical fundamental diagram
Derive and sketch both the speed-density diagram and the fundamental diagram, subject to the following idealized assumptions: (i) All vehicles are of length $l = 5$ m. (ii) In free traffic (speed does not depend on other vehicles), all vehicles drive at their desired speed $V_0 = 120$ km/h. (iii) In congested traffic (speed is the same as the speed of the leading vehicle), drivers keep a gap of $s(v) = s_0 + vT$ to the leading vehicle, with the minimum gap $s_0 = 2$ m and the time gap $T = 1.6$ s.

4.2 Flow-density diagram of empirical data
Considering the speed-density diagram (Fig. 4.5) and flow-density diagram (Fig. 4.9) of the German A8-East and the Dutch A9, determine the desired speed V_0, time gap T, maximum density $\rho_{\max}$, and the capacity drop on both highways from the fitted

curves. Which statements can you make about the driving behavior of German and Dutch drivers (at least on these specific highways at the time of measurement)?

Further Reading

- Hall, F.: Traffic stream characteristics. Traffic Flow Theory. US Federal Highway Administration (1996)
- Daganzo, C.: Fundamentals of Transportation and Traffic Operations. Pergamon-Elsevier, Oxford, U.K. (1997)
- McShane, W.R., Prassas, E.S., Roess, R.P.: Traffic Engineering. Prentice Hall (2010)
- Gazis, D.C.: Traffic Theory. Springer, USA (2002)
- Knospe, W., Santen, L., Schadschneider, A., Schreckenberg, M.: Single-vehicle data of highway traffic: Microscopic description of traffic phases. Physical Review E **65**, 056133 (2002)

Chapter 5
Spatiotemporal Reconstruction of the Traffic State

Amazement is the beginning of knowledge.
Plato

Abstract A detailed representation of the traffic state in space and time allows us to analyze various aspects of traffic dynamics. However, since traffic data are only available for a small subset of locations and times, the full traffic state can only be reconstructed by spatiotemporal interpolation, which can be formulated in terms of a convolution integral. Since naive "isotropic" interpolation is inadequate for traffic data, we introduce a more refined interpolation method. This *adaptive smoothing method* yields a detailed and plausible reconstruction of the traffic state. Finally, we discuss the combination and weighting of multiple, heterogeneous data sources for estimating the traffic state (data fusion).

5.1 Spatiotemporal Interpolation

The purpose of the two-dimensional spatiotemporal interpolation algorithm described below is to estimate the *speed field*, i.e., the continuous function of *local speed average* $V(x, t)$ given only *discrete* speed measurements v_i at discrete locations x_i and times t_i (Figs. 5.1 and 5.2). In most cases, data are available in the form of aggregated minute by minute data of speed and flow recorded by stationary detectors. Furthermore, floating cars transmitting "data telegrams" of their positions and time-mean speeds become increasingly relevant. The output is the traffic-state estimator in the form of a continuous speed field (and possibly other fields) as a function of space and time. Traffic-state reconstruction methods using interpolation techniques are useful for offline analysis of historical highway traffic flow data. Real-time estimation is described in Chap. 18

DOI: 10.1007/978-3-642-32460-4_5,

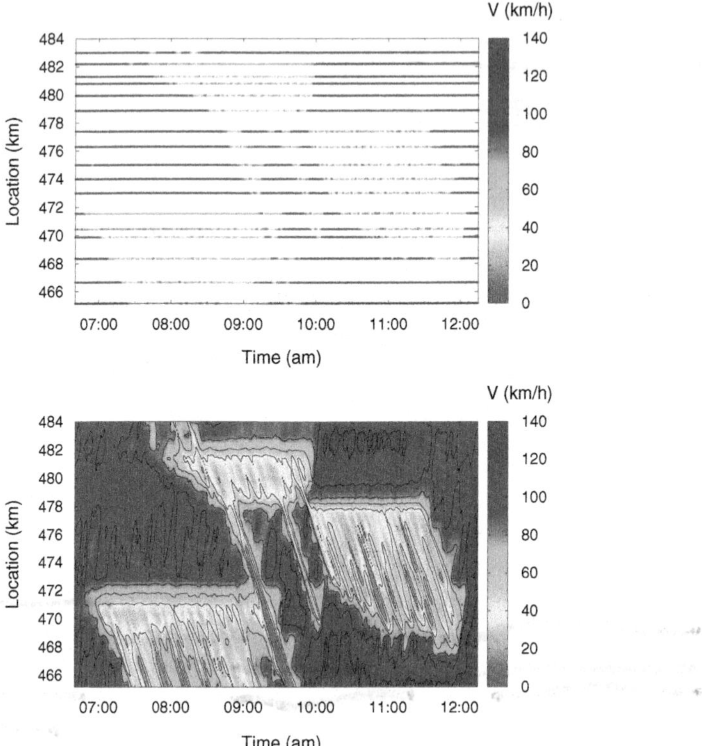

Fig. 5.1 Color-coded visualization of the speed measured by stationary detectors (*top*) and the speed field reconstructed by the adaptive smoothing method (Sect. 5.2). The data are from a section of the Autobahn A5 near Frankfurt/Main, Germany (south-bound, recorded May 28, 2001). The "active" bottlenecks causing congestions are the on-ramps of two highway junctions, and an accident at location 478 km restricting the local capacity at this location between 10:00 and 11:30 am

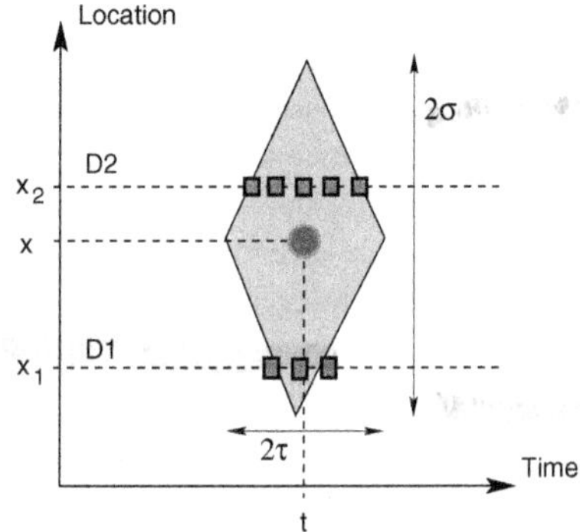

Fig. 5.2 Spatiotemporal interpolation as a way of reconstructing the traffic state at location x and time t using the isotropic weighting kernel (5.2)

The basis of spatiotemporal interpolation and smoothing is a *discrete convolution* with a kernel ϕ_0 that includes all data points i:

$$V(x,t) = \frac{1}{\mathcal{N}(x,t)} \sum_i \phi_0(x - x_i, t - t_i)\, v_i. \tag{5.1}$$

In principle, we can use any function as the weighting kernel $\phi_0(x,t)$. To avoid artifacts, the kernel should have following properties:

- It should be localized, i.e., $\phi_0(x,t)$ tends to zero for sufficiently large values of $|x|$ and $|t|$.
- The maximum of $\phi_0(x,t)$ should be at $x = 0$ and $t = 0$.
- $\phi_0(x,t)$ should be a continuous function of x and t.
- $\phi_0(x,t)$ should be monotonically decreasing with $|x|$ and $|t|$.

For our purposes, the symmetric exponential has proved itself useful[1]:

$$\phi_0(x - x_i, t - t_i) = \exp\left[-\left(\frac{|x - x_i|}{\sigma} + \frac{|t - t_i|}{\tau}\right)\right]. \tag{5.2}$$

Here, σ and τ are the smoothing widths in the spatial and temporal coordinates, respectively. The denominator $\mathcal{N}$ of Eq. 5.1 denotes the normalization of the weighting function, given by the sum of all discrete weights:

$$\mathcal{N}(x,t) = \sum_i \phi_0(x - x_i, t - t_i). \tag{5.3}$$

The exponential function operates as a *low-pass filter*, smoothing temporal variations on a scale smaller than τ and spatial fluctuations on a scale smaller than σ. For the interpolation of detector data with aggregation intervals of 1 min, good values for τ are between 30 and 60 s. The spatial smoothing with σ should be of the order of half of the average distance between detectors.

If the distance between two detectors is larger than half of the spatial distance between two stop-and-go waves, the *isotropic kernel* (5.2) produces artifacts such as the "egg-carton pattern" in Fig. 5.3. These introduce ambiguities into the interpretation of stop-and-go waves (Fig. 5.4).

Why does the isotropic reconstruction of stop-and-go traffic produce artifacts such as wrong propagation velocities, or even wrong propagation *directions* of congestion waves, if the detectors are further apart than half of the wavelength? Clarify the situation by drawing idealized, regular stop-and-go waves with wavelength λ and multiple cross-sectional detectors (all separated by a distance Δx) in a space-time diagram (cf. Fig. 5.4).

Sampling error

[1] One could also use a bivariate Gaussian.

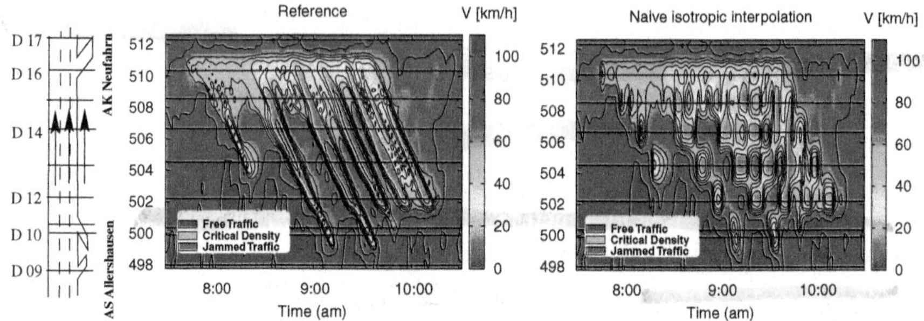

Fig. 5.3 Contour plots of spatiotemporally interpolated speed measurements of stationary detectors on a section of the Autobahn A9 north of Munich, Germany. The stop-and-go waves are distorted by the isotropic smoothing, while a traffic-adaptive smoothing yields a detailed reconstruction

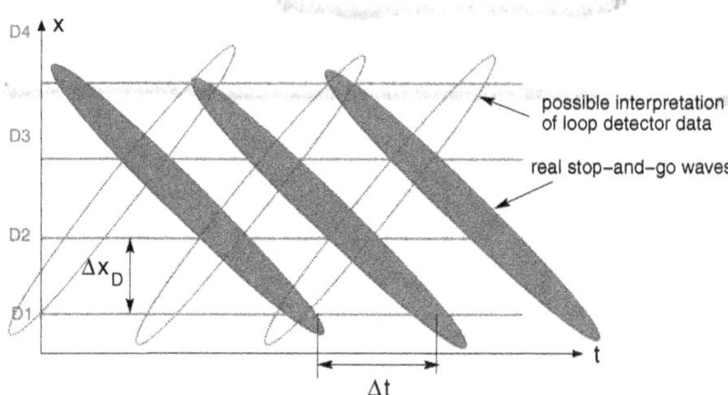

Fig. 5.4 Ambiguity in the interpretation of stop-and-go waves using data from cross-sectional detectors: Aside from the "true" stop-and-go waves (*solid patches*), a different interpretation is possible (*dotted outlines*). Since stop-and-go waves always propagate upstream, we can rule out the second interpretation

5.2 Adaptive Smoothing Method

The wavelengths of stop-and-go waves are usually of the order of 2 km while most stationary detectors or floating cars in the real world are separated by more than 1 km. Therefore isotropic smoothing using the kernel (5.2) is not suitable for typical situations. In the following, we introduce a traffic-adaptive smoothing method for the reconstruction of the spatiotemporal traffic dynamics, which is able to provide a more detailed and plausible reconstruction than the isotropic method (Fig. 5.3).

Like the isotropic smoothing procedure, the adaptive smoothing method is based on a two-dimensional interpolation in space and time. In contrast to the former, it takes into account the two typical velocities of information propagation in free and congested traffic.

First, the spatiotemporally averaged speeds (and other macroscopic quantities) are calculated using two different weighting kernels ("filters", Sect. 5.2.1) accounting for the different propagation velocities of macroscopic density and speed changes in free and congested traffic.

Then, the two filters are used to determine the traffic state at the spatiotemporal location (x, t) by calculating the "degree of congestion", w, which can take values between 0 and 1. The final speed field $V(x, t)$ is a superposition of the two filters, weighted by w (Sect. 5.2.2). To validate the smoothing method we will reconstruct the traffic state from a reduced dataset (in which data from some detectors has been omitted) and compare it to the reconstruction obtained from the full dataset (Sect. 5.2.4). Finally, we investigate the robustness of the method (Sect. 5.2.5).

5.2.1 Characteristic Propagation Velocities

The *adaptive smoothing method* takes into account that all perturbations to the traffic flow, i.e., "patterns" in the spatiotemporal speed diagram, are either stationary or moving with one of two distinct (remarkably universal) velocities: (i) In *free traffic* it has been observed that perturbations usually propagate *with* the traffic flow (downstream) at a characteristic velocity slightly below the local speed of the vehicles. This is due to the weak interactions between the vehicles. (ii) In *congested traffic*, perturbations propagate *against* the traffic flow (upstream) due to the reaction of the drivers to their respective leading vehicles. Empirical data shows that the propagation velocity of approximately $c_{\text{cong}} = -15\,\text{km/h}$ is universal in congested traffic situations, see Fig. 2.1. This includes the propagation of downstream fronts of individual stop-and-go waves (cf. Chap. 18), or the dissolution of a queue behind a traffic light once it turns green (cf. the figure in Problem 2.3). The only exception to this rule is the propagation velocity of upstream fronts of congestions which is above c_{cong}.

To account for these fundamental properties of the traffics dynamics, *two* smoothed speed fields with different propagation velocities in free and congested traffic, c_{free} and c_{cong}, are considered (cf. Fig. 5.5)[2]:

$$V_{\text{free}}(x,t) = \frac{1}{\mathcal{N}_{\text{free}}(x,t)} \sum_i \phi_0\left(x - x_i, t - t_i - \frac{x - x_i}{c_{\text{free}}}\right) v_i, \qquad (5.4)$$

$$V_{\text{cong}}(x,t) = \frac{1}{\mathcal{N}_{\text{cong}}(x,t)} \sum_i \phi_0\left(x - x_i, t - t_i - \frac{x - x_i}{c_{\text{cong}}}\right) v_i. \qquad (5.5)$$

The normalization constants $\mathcal{N}_{\text{free}}$ and $\mathcal{N}_{\text{cong}}$ are determined analogously to (5.3). According to the reasoning above, the propagation velocity in free traffic is set slightly less than the average free-flow vehicle speed, e.g., $c_{\text{free}} = 70\,\text{km/h}$ on highways. The propagation velocity in congested situations, $c_{\text{cong}} = -15\,\text{km/h}$, represents

[2] We use c to denote propagation velocities, and V and v for the macroscopic and microscopic vehicle speeds, respectively.

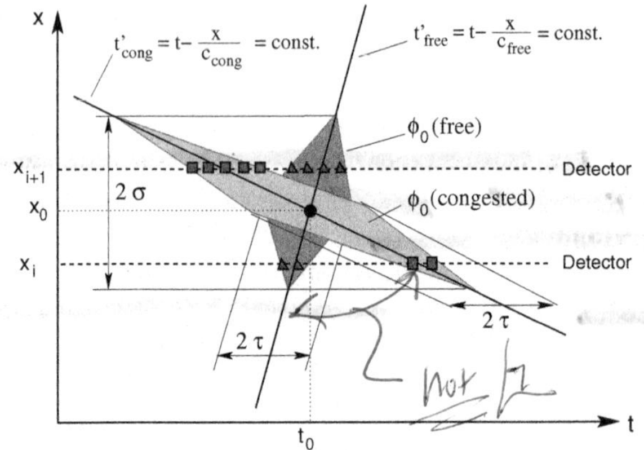

Fig. 5.5 Speed filters of the *adaptive smoothing method* for the reconstruction of free and congested traffic states at location x_0 and time t_0. The colored parallelograms indicate the spatiotemporal region for which the weighting function ϕ_0 is significantly different from zero. The shear of the parallelograms is determined by the propagation velocities of perturbations in free and congested traffic, respectively. The symbols along the *dotted lines*, which mark the positions of cross-sectional detectors, show the (aggregated) data which are most influential in the interpolation

the movement of traffic waves moving *against* the traffic flow (indicated by the negative sign). The space-dependent shifts of the time coordinate by $(x - x_i)/c_{\text{free}}$ and $(x - x_i)/c_{\text{cong}}$ given by Eqs. 5.4 and 5.5, respectively, represent the transitions from a coordinate system comoving with the propagation velocities to a stationary system. In effect, the transformations "shear" the smoothing kernel $\phi_0(x - x_i, t - t_i)$ with the gradients $1/c_{\text{free}}$ and $1/c_{\text{cong}}$. In the limit $c_{\text{free}} = c_{\text{cong}} \to \infty$ the adaptive interpolation is equivalent to the isotropic interpolation (5.2).

5.2.2 Nonlinear Adaptive Speed Filter

The result of the adaptive smoothing method—the average speed $V(x, t)$—is a superposition of the two speed fields V_{free} and V_{cong}:

$$V(x,t) = w(x,t)\, V_{\text{cong}}(x,t) + [1 - w(x,t)]\, V_{\text{free}}(x,t). \tag{5.6}$$

The weight $w(x, t)$ depends on both V_{free} and V_{cong}. Obviously, we want $w \approx 1$ for low speeds and $w \approx 0$ for high speeds. The continuous transition between the two extremes is characterized by an s-shaped (sigmoid) nonlinear function:

$$w(x,t) = \frac{1}{2}\left[1 + \tanh\left(\frac{V_c - V^*}{\Delta V}\right)\right]. \tag{5.7}$$

Table 5.1 Parameters of the adaptive smoothing method (ASM) and typical values for highway traffic

Parameter	Value (highway traffic)
Spatial smoothing width σ	$\Delta x/2$ (of the order of 1 km)
Temporal smoothing width τ	$\Delta t/2$ (of the order of 30 s)
Propagation velocity of perturbations in free traffic c_{free}	70 km/h
Propagation velocity of perturbations in congested traffic c_{cong}	−15 km/h
Threshold between free and congested traffic V_c	50 km/h
Width of the transition between free and congested traffic ΔV	10 km/h

The predictor $V^*(x,t) = \min\left[V_{\text{free}}, V_{\text{cong}}\right]$ is defined such that congested traffic states are represented more accurately than free traffic. The parameter ΔV determines the transition width around V_c, which is the threshold between free and congested traffic. Good parameters values are, for example, $V_c = 55$ km/h and $\Delta V = 10$ km/h.

5.2.3 *Parameters*

Table 5.1 summarizes the six parameters of the adaptive smoothing method (ASM) and suggests suitable values for highway traffic. Strictly speaking, the parameters must be estimated by appropriate calibration techniques as described in Chap. 16 below. However, the ASM is very robust in the sense that the resulting speed field is insensitive to the precise values as long as the order of magnitude is correct (cf. Sect. 5.2.5). In particular, the values of Table 5.1 are expected to yield good results for any highway traffic situation.

In the following, we show—using the ASM as an example—how one can test the validity and robustness of methods for traffic-state reconstruction.

5.2.4 *Testing the Predictive Power: Validation*

Traffic-state recognition methods can be validated by applying them to a subset of given detector data and comparing the results with the full dataset. Ideally, the full data set serving as reference is so dense that it can be regarded as representing the *ground truth* (Fig. 5.6).

The validation procedure consists in applying the ASM with standard parameters of Table 5.1 to input data chosen from just a small selection of the available detectors. The interpolated speed field $V(x,t)$ is then compared to speed data at detectors which are half way between those whose data has been used in the reconstruction. For example, at a spacing of 1 km corresponding to Fig. 5.7a, the prediction quality of the method is based on the differences between the predicted and measured data

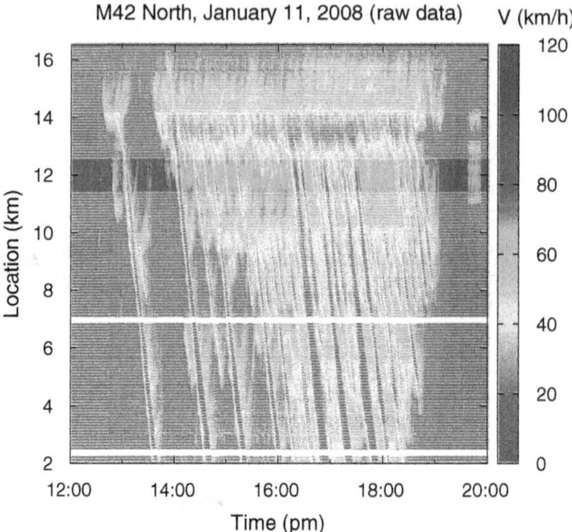

Fig. 5.6 Stationary detector data of traffic waves on the English motorway M42. The inter-detector spacing of the stationary detectors is 100 m (40 m in the vicinity of $x = 12$ km). The data are visualized as a spatiotemporal scatter plot. Each data point corresponds to the local speed aggregated over all lanes and over 1 min. No further data processing has been applied

at $x = 2.5$ km, $x = 3.5$ km and so forth. Figures 5.7b–d display the reconstructed traffic states with increasingly reduced sets of loop detectors. In summary, the most important features are identified even when the detector spacing is increased to 4 km. Generally, the method can be considered as valid for inter-detector distances at or below 2–3 km. At this distance, the ASM produces similar results as the naive isotropic interpolation would do for inter-detector distances of 1 km (Fig. 5.3).

5.2.5 *Testing the Robustness: Sensitivity Analysis*

By varying the parameters listed in Table 5.1 we can check the "robustness" of the traffic state reconstruction, i.e., the sensitivity of the result to changes in the parameter values. Figure 5.8 shows the reconstructed speed field of traffic waves on the German A9 when changing the transition parameters V_c and ΔV (top, right), or the propagation velocities c_{free} and c_{cong} (bottom row). The resulting average speed, especially the distinction between free and congested traffic, does not change strongly in either case. The most important factor is the propagation velocity in congested traffic: Values less than -20 km/h or greater than -12 km/h produce artificial, discrete steps. For $V_c \to \pm\infty$, the ASM reverts to the isotropic smoothing method (Fig. 5.3 right).

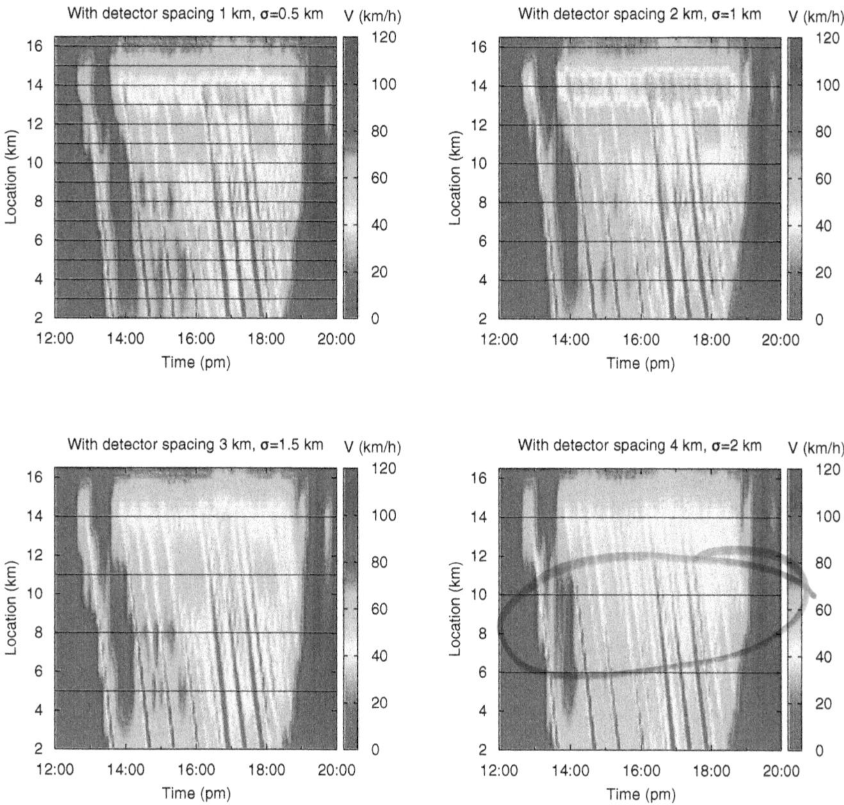

Fig. 5.7 Reconstruction of the reference situation of Fig. 5.6 by the adaptive smoothing method applied on reduced data sets with detector spacings between 1 and 4 km. The locations of detectors whose data has been used in the reconstruction are indicated by *horizontal lines*

5.3 Data Fusion

The term *data fusion* refers to the process of combining data from multiple, heterogeneous data sources such as cross-sectional data, floating-car data, "floating-phone data", police reports, etc. In general, each of these categories of data describes different aspects of the traffic situation and might even contradict each other. The goal of data fusion is to maximize the utility of the available information (cf. Figs. 5.9 and 5.10).

Real-time applications of the traffic state estimation, e.g., information on current traffic congestions, are a particular challenge, since data points in the future are, of course, not available.

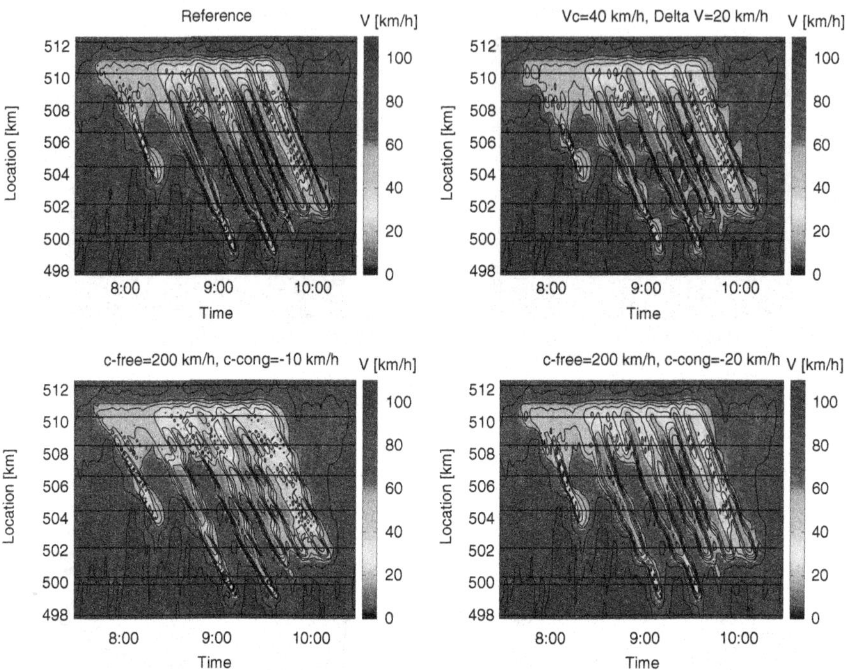

Fig. 5.8 Influence of parameter variations when applying the adaptive smoothing method on the detector data of Fig. 5.3

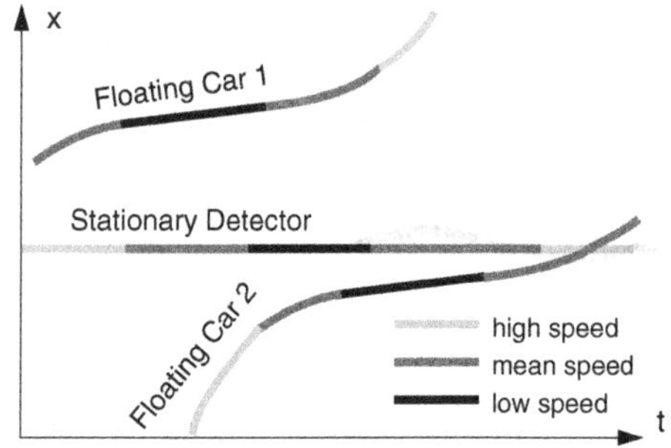

Fig. 5.9 Example of heterogeneous data sources: cross-sectional detectors and floating-car data

What are possible causes for inconsistencies between heterogeneous data sources? Where are visible inconsistencies in Fig. 5.9?

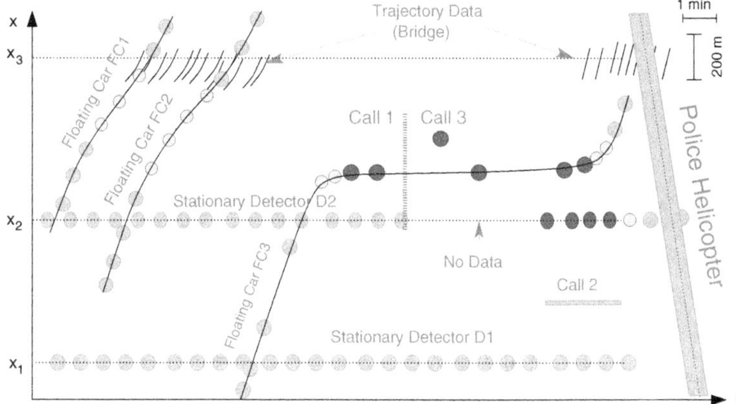

Fig. 5.10 Another example of diverse data sources used in the spatiotemporal reconstruction of the traffic state. The *horizontal dotted lines* represent two stationary detectors at locations x_1 and x_2, which send data every minute (green circles: free traffic, yellow: dense traffic, red: traffic jam). Three floating cars cross the road segment in question and also send data, though not in fixed intervals but event-based. A camera on a bridge at location x_3 reports trajectory data over a small road segment (*black curves*). An accident was reported via cell phone (*call 1*) but the caller was only able to give the approximate location (*vertical orange line*). Caller 2 was standing on a bridge and observed free traffic over some period of time. Caller 3 reported standing in a traffic jam at time 2:55 p.m. and location 435.5 km. Finally, a helicopter (flying against the driving direction) observed free traffic

5.3.1 Model-Based Validation of a Data Fusion Procedure

The adaptive smoothing method introduced in Sect. 5.2 can be used as an algorithm for data fusion if all data sources provide spatiotemporally resolved point measurements of the local speed, i.e., data sets $\{x_i, t_i, v_i\}$. This includes stationary detector data (SDD) and floating-car data (FCD). To test and validate this application of the adaptive smoothing method, one needs congested traffic situations where (i) SDD, (ii) FCD, (iii) a sufficient approximation to the ground truth are available. To date, such test cases are rarely available. We therefore demonstrate how to validate data-fusion procedures based on models and simulations. For this purpose, we simulate traffic waves with a model of human drivers that can reproduce the waves realistically (Fig. 5.12a, see Chap. 12 for a model description). As input for the adaptive smoothing method, we generate virtual SDD and FCD from the simulation (Fig. 5.11) and apply the method with the standard parameters.

The prediction quality of the method is assessed by comparing the reconstructed speed fields shown in Fig. 5.12b–d with the reference of Fig. 5.12a. It becomes evident that both data sources contribute to the reconstruction.

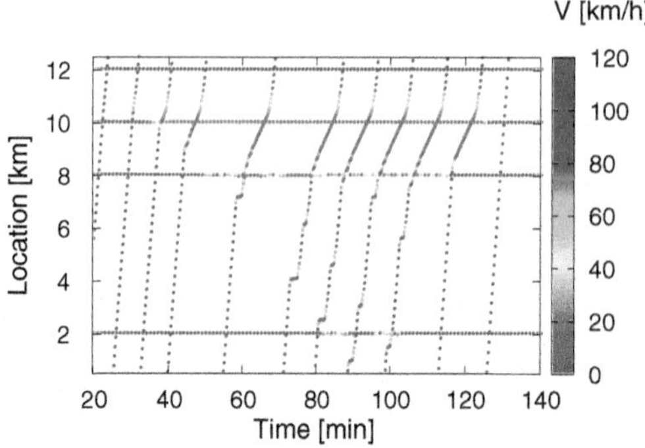

Fig. 5.11 Scatter plot of virtual floating-car and stationary detector speed data generated by simulating a bottleneck situation with a model for human drivers (described in Chap. 12)

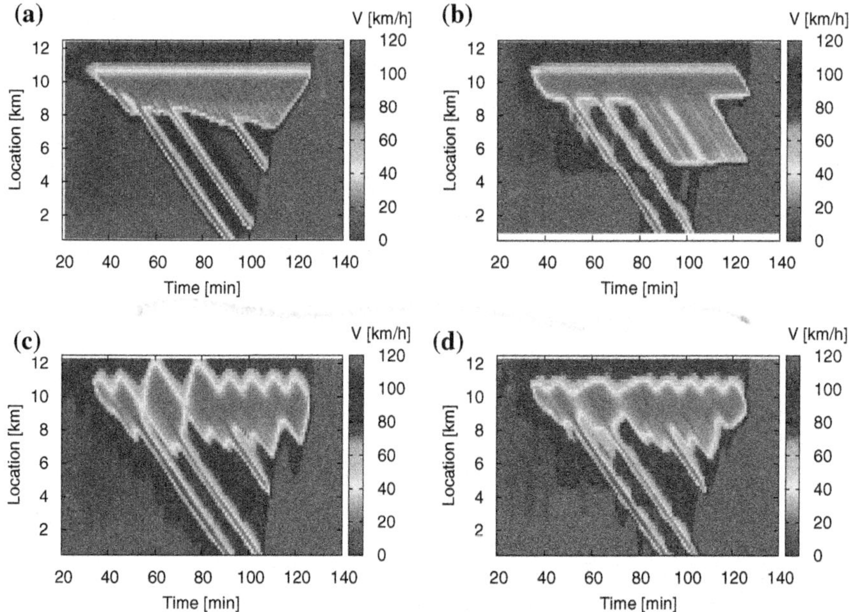

Fig. 5.12 Spatiotemporal speed profiles: **a** ground truth; **b** using stationary detector data (SDD) only; **c** using floating-car data (FCD) only; **d** combining stationary detector and floating-car data. The input data for the Adaptive Smoothing Method resulting in diagrams b–d is shown in Fig. 5.11

5.3.2 Weighting the Data Sources

When using multiple data sources, their relative weighting plays an important role. However, in the Eqs. (5.4 and 5.5), the weights

$$w_i(x,t) = \phi_0\left(x - x_i,\ t - t_i - \frac{x - x_i}{c}\right) \tag{5.8}$$

(with $c = c_{\text{free}}$ or c_{cong}) depend only on the (spatiotemporal) distance between the location in question (x, t) and the data points (x_i, t_i). Consequently, all data points are considered equally important. However, if the data originate from several detector categories m (such as induction-loop detectors, infrared detectors, floating cars) with different magnitude of the associated errors, it is sensible to include an additional weight r_m that represents the *reliability* of the data source to give the more reliable sources a stronger influence on the result. Combining this with the weighting (5.8) according to the spatiotemporal distance between data and interpolation points, all spatiotemporal weights $w_i(x,t)$ in the above formulas have to be replaced by the total weights

$$w_i^{tot}(x,t) = w_i(x,t) r_{m(i)} = \phi_0\left(x - x_i,\ t - t_i - \frac{x - x_i}{c}\right) r_{m(i)}. \tag{5.9}$$

Here, $m(i)$ denotes that the data source i is of type m with reliability weight r_m.

In order to determine the reliability weights r_m, let us assume that (i) the different data sources m bear no systematic errors, (ii) the variance θ_m of the random errors is known, and (iii) the errors of the different sources are uncorrelated.

Now, we assume that, for a given point (x, t), speed estimates v_m from all data types are available such that $\phi_0(x - x_i,\ t - t_i - (x - x_i)/c) = 1$. Then, according to a basic addition rule for a linear combination of independent random variables, the error variance of the weighted arithmetic mean $V(x,t) = \sum_m r_m v_m$ is given by $\theta = \sum_m r_m^2 \theta_m$ where $\sum_m r_m = 1$ must be satisfied. Our objective is to minimize this variance by varying the weights r_m, or the weight vector $\mathbf{r}$. This immediately leads to following *constrained optimization problem*: Minimize

$$\theta(\mathbf{r}) = \sum_m r_m^2 \theta_m, \tag{5.10}$$

subject to

$$\sum_m r_m = 1. \tag{5.11}$$

Constrained optimization problems can be solved using *Lagrange multipliers*. The procedure is as follows:

1. Formulate each constraint n as a constraint function equating to zero, $B_n(\mathbf{r}) = 0$. Here, the only constraint $\sum_m r_m = 1$ results in the function $B_1(\mathbf{r}) = \sum_m r_m - 1$.
2. Define the *Lagrange function* by adding to the objective function to be minimized the constraint functions multiplied by Lagrange multipliers λ_n:

$$L(\mathbf{r}, \boldsymbol{\lambda}) = \theta(\mathbf{r}) - \sum_n \lambda_n B_n(\mathbf{r}), \tag{5.12}$$

where the vector $\boldsymbol{\lambda}$ represents the Lagrange multipliers which are unknown at this stage. In our optimization problem, the Lagrange function is given by $L(\mathbf{r}, \lambda_1) = \sum_m r_m^2 \theta_m - \lambda_1(\sum_m r_m - 1)$.

3. Find a necessary condition for the minimum of the Lagrange function with respect to $\mathbf{r}$:

$$\frac{\partial L(\mathbf{r}, \boldsymbol{\lambda})}{\partial \mathbf{r}} = 0. \tag{5.13}$$

This results in M equations if the weight vector $\mathbf{r}$ consists of M components. In our application, we obtain

$$\frac{\partial L}{\partial r_m} = 2 r_m \theta_m - \lambda_1 \stackrel{!}{=} 0 \quad \Rightarrow \quad r_m = \frac{\lambda_1}{2\theta_m}.$$

4. Determine the unknown Lagrange multipliers by applying the constraints. If there are N constraints, (5.13) and the constraints constitute $M+N$ equations for the $M+N$ unknown components of the vectors $\mathbf{r}$ and $\boldsymbol{\lambda}$. In our optimization problem, we obtain $\lambda_1 = 2/(\sum_{m'} \theta_{m'}^{-1})$ resulting in the final weighting

$$r_m = \frac{\frac{1}{\theta_m}}{\sum\limits_{m'} \frac{1}{\theta'_m}}. \tag{5.14}$$

The weights should be proportional to the inverse of the variance of the errors in the data source.

Problems

5.1 Reconstruction of the traffic situation around an accident

Different data sources provide information about a road segment of length 10 km ($0 \le x \le 10$ km) indicating a road block caused by an accident: (i) At 4.00 p.m., a floating car enters the area and crosses it at 120 km/h. (ii) At 4.19 p.m., another floating car, driving at the same speed, has to stop at the end of a traffic jam at $x = 5$ km. (iii) Two stationary detectors at $x = 4$ and 8 km measure the traffic flow (but not the speed). The detector at $x = 4$ km reports a flow of zero between 4.25 and 4.58 p.m.. The detector at $x = 8$ km reports zero flow between 4.14 and 4.51 p.m. (iv) At 4.40 p.m., a driver reports (via cell phone) that he has been stuck in a traffic jam at $x = 5$ km for a few minutes already. (v) At 4.30 p.m., another caller, driving on the opposite lane, reports an empty road at $x = 7$ km.

1. Visualize the available information in a space-time diagram. Mark all information as one of (i) "free traffic", (ii) "traffic jam", (iii) "empty road", (iv) "do not know; either empty road or stopped traffic".
2. Determine the location and time of the accident, assuming an immediate and total road block causing a traffic jam that propagates upstream with constant velocity. Also, determine the propagation velocity.
3. Determine the time at which the road block clears. (Keep in mind that downstream jam fronts move with a universal propagation velocity of $-15\,\text{km/h}$.)

5.2 Dealing with inconsistent information
When a floating car passes the location x_D of a stationary detector at time t_D, the data for (x_D, t_D) from the two different sources is usually inconsistent. Assume that the floating-car speed data V_2 has a standard deviation of errors σ_2 that is twice as large as those of the stationary detectors (speed V_1, variance $\sigma_1^2 = \frac{1}{4}\sigma_2^2$), and that the errors are independent and not systematic. How do the errors in the fused data improve (or worsen) when using (i) equal or (ii) optimal weights (5.14)?

Further Reading

- Treiber, M., Helbing, D.: Reconstructing the spatio-temporal traffic dynamics from stationary detector data. Cooper@tive Tr@nsport@tion Dyn@mics **1** (2002) 3.1–3.24 (Internet Journal, www.TrafficForum.org/journal)
- Treiber, M., Kesting, A., Wilson, R.E.: Reconstructing the traffic state by fusion of heterogeneous data Computer-Aided Civil and Infrastructure Engineering **26** (2011), 408–419
- van Lint, J., Hoogendoorn, S.P.: A robust and efficient method for fusing heterogeneous data from traffic sensors on freeways. Computer-Aided Civil and Infrastructure Engineering **24** (2009) 1–17

Part II
Traffic Flow Modeling

Chapter 6
General Aspects

Politics is for the present, but an equation is for eternity.
Albert Einstein

Abstract In this chapter we present the general approach to traffic flow modeling and distinguish it from the methods of transportation planning. Furthermore, we introduce model classifications with respect to the aggregation level and with respect to mathematical and conceptional criteria. We also discuss how to model non-motorized traffic.

6.1 History and Scope of Traffic Flow Theory

Traffic flow theory and modeling started in the 1930s, pioneered by the US-American Bruce D. Greenshields (Fig. 6.1). However, since the 1990s, the field has gained considerable attraction as overall traffic demand has increased and more data as well as easy access to computing power has become available.

Both *traffic flow modeling* and *transportation planning* belong to the broader field of *traffic modeling*. However, there are important differences between traffic flow modeling and transportation planning:

- *Temporal aspect:* The timescale in traffic flow dynamics is of the order of minutes to a few hours, while transportation planning covers periods from hours to several days or even years.
- *Objective aspect:* Traffic flow dynamics assumes an externally given traffic demand and fixed infrastructure. Transportation planning models the dynamics of the traffic demand and effects of infrastructure changes.
- *Subjective aspect:* Traffic flow dynamics analyzes human (or automated) *operational* driving behavior (accelerating, braking, lane-changing, turning) while higher-level actions, e.g., activity choice (number and type of trips), destination choice, mode choice, and route choice belong to the realm of transportation planning.

M. Treiber and A. Kesting, *Traffic Flow Dynamics*,
DOI: 10.1007/978-3-642-32460-4_6, © Springer-Verlag Berlin Heidelberg 2013

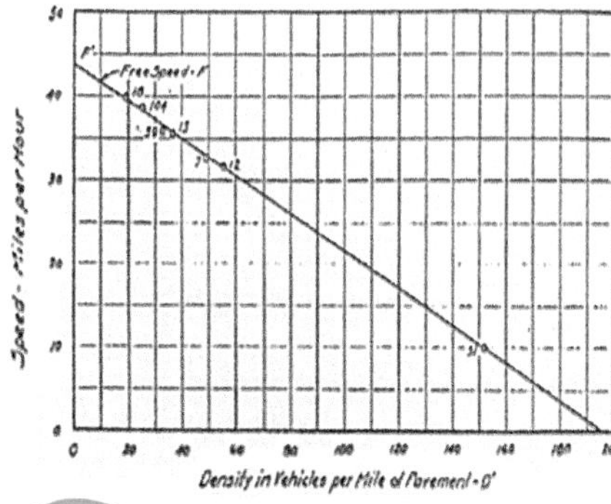

Fig. 6.1 Traffic theory in the 1930s: Historical speed-density diagram and the experiment carried out by Bruce D. Greenshields. [From: Greenshields, B.D., A study of traffic capacity. In: Proceedings of the Highway Research Board, Vol. 14. Highway Research Board, Washington, D.C. (1935)]

Note that "dynamics", i.e., explicit time evolution is not a distinguishing feature by itself: Transportation planning also includes the method of *dynamic traffic assignment* (i.e., route choices that depend on traffic state and time). Furthermore, *supply* (infrastructure) and *demand* are "dynamic" over timescales (years) routinely considered in transportation planning.

The differences between the two fields are reflected in their approaches to tackle a given problem. For example, the probability of traffic jams can be reduced by traffic regulations such as speed limits, on-ramp metering, bans on passing for trucks, or variable-message signs for alternate-route advises. The same effect can be achieved, however, by building, modifying or removing infrastructure elements, creating incentives to use different means of transportation, dispersing rush hours, or reducing overall traffic demand (e.g., by political action). While the former solutions are simulated using traffic flow models, the latter refer to the field of transportation planning. Of course, for a detailed assessment of a measure pertaining to transportation planning (e.g., redesigning a major intersection or building a new bridge), traffic flow simulations come into play.

6.2 Model Classification

Traffic flow models can be categorized with respect to a number of aspects: Aggregation level (the way reality is represented), mathematical structure, and conceptual aspects. This section introduces important classes of models.

6.2.1 Aggregation Level

There are several ways to abstract real-world traffic events and model them, i.e., describe them mathematically (Fig. 6.2):

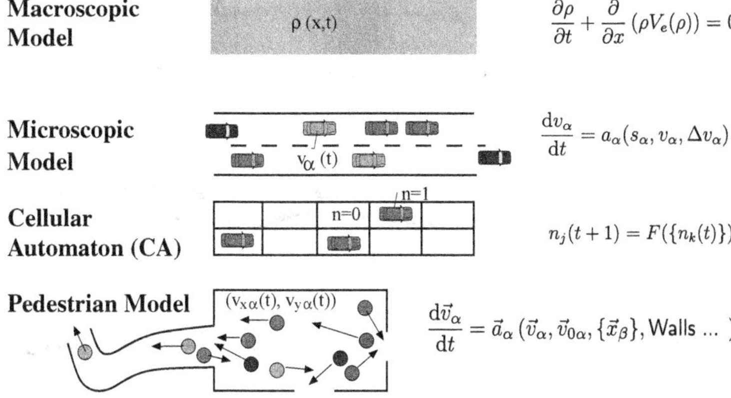

Fig. 6.2 Comparison of various model categories (with respect to the way they represent reality) including typical model equations

Macroscopic models describe traffic flow analogously to liquids or gases in motion. Hence they are sometimes called *hydrodynamic models*. The dynamical variables are *locally aggregated* quantities such as the traffic density $\rho(x,t)$, flow $Q(x,t)$, mean speed $V(x,t)$, or the speed variance $\sigma_V^2(x,t)$. Because the aggregation is local, these quantities generally vary across space and time, i.e., they correspond to dynamic *fields*. Thus, macroscopic models are able to describe *collective phenomena* such as the evolution of congested regions or the propagation velocity of traffic waves. Furthermore, macroscopic model are useful,

- if effects that are difficult to describe macroscopically need not to be considered (e.g., lane changes, several driver-vehicle types),
- if one is interested in macroscopic quantities, only,
- if the computation time of the simulation is critical, e.g., in real-time applications (due to increasing computing power, this aspect is becoming less important), or
- if the available input data come from heterogeneous sources and/or are inconsistent, so data fusion is necessary.

Multiple real-time speed and the capability to incorporate heterogeneous data sources are particularly important for *traffic state estimations* and *predictions*. In this process, the future traffic state is predicted over a time horizon τ and the predictions are updated over smaller time intervals Δt. The predictions are processed such that they can be distributed via traffic message channel, variable-message signs, or serve as input for connected navigation devices.[1]

Microscopic models including *car-following models* and most *cellular automata* describe individual "driver-vehicle particles" α, which collectively form the traffic

[1] Traffic flow modeling and transportation planning are intertwined in these applications: Traffic flow models provide the basis for the route choice.

flow. These models describe the reaction of every driver (accelerating, braking, lane-changing) depending on the surrounding traffic. In a broader context, microscopic traffic flow models are examples of driven multi-particle models. The dynamical variables are vehicle positions $x_\alpha(t)$, speeds $v_\alpha(t)$, and accelerations $\dot{v}_\alpha(t)$. Microscopic models are particularly suited for the following applications:

- Modeling how single vehicles affect traffic: This is becoming more and more important as *advanced driver-assistance systems* (ADAS) such as *adaptive cruise control* (ACC) or infrastructure-to-vehicle (I2V) and vehicle-to-vehicle communication (V2V) as well as other applications of Intelligent Transportation Systems (ITS) see widespread use.
- Situations in which the heterogeneity of the traffic plays an important role, e.g., simulating the effects of speed limits or bans on passing for trucks: As we will see in Chap. 21, this applies to *any* traffic control action since the general objective of all measures for traffic optimization is the homogenization of the traffic.
- Describing human driving behavior, including estimation errors, reaction times, inattentiveness, and anticipation: Microscopic models allow us to assess how different driving styles affect traffic capacity and stability.
- Visualization of interactions between various traffic participants (cars, trucks, buses, cyclists, pedestrians, etc.).
- Generating the surrounding traffic for scientific driving simulators used for physio-psychological studies of human drivers, or even for game simulators.

Mesoscopic models combine microscopic and macroscopic approaches to a hybrid model: In *local-field models*, parameters of a microscopic model may depend on macroscopic quantities such as traffic density or local speed and speed variance. Conversely, in so-called *master equations*, the dynamics of a macroscopic quantity (the number of vehicles in a traffic jam) is described in terms of microscopic stochastic rate equations for in- and out-"flowing" vehicles. *Gas-kinetic traffic models* use idealized "collisions" to describe the dynamics of a quantity called *phase-space density* $\tilde{\rho}(x,t,v)$ which includes traffic density and the local probability distribution of vehicle speed. In the class of *parallel-hybrid models*, critical parts of a traffic network (e.g., intersections and traffic lights) are described microscopically, and the rest macroscopically (see below). Apart from these categories, there is a large spectrum of further mesoscopic models which are beyond the scope of this book.

Aggregation and disaggregation. Macroscopic quantities (density, flow, local speed and speed variance) can be obtained from microscopic quantities (vehicle positions and speeds) by local *aggregation* (cf. Fig. 6.3). This is possible if we can define spatiotemporal regions which are *microscopically large*, such that they contain a significant number of vehicles for averaging (notice that some macroscopic quantities such as traffic density or speed variance are only defined for many vehicles). Simultaneously, these regions must be *macroscopically small*, that is, smaller than the typical lengths and time scales of the traffic patterns of interest (jams, stop-and-go waves, changes in the traffic flow). The standard method for aggregation is *kernel-based moving averaging*, a technique for weighted averaging where vehicles near

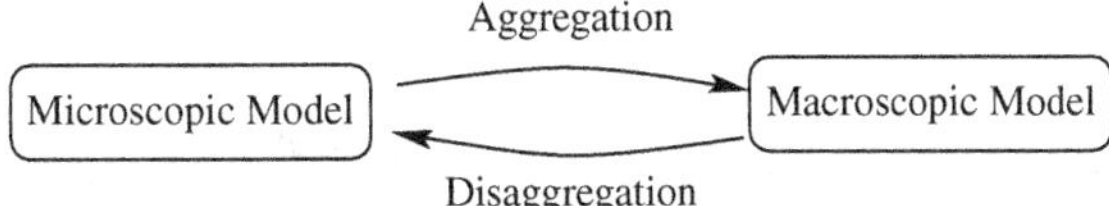

Fig. 6.3 Aggregation and disaggregation connects microscopic and macroscopic models

the spatiotemporal point in question are weighted more than more distant vehicles and the weighting is defined by a localized kernel function.[2] A special method of aggregation is the simulation of the empirical measurement process by *virtual detectors* with subsequent aggregation over time intervals to obtain "virtual one-minute data". The "measurements" of these detectors can be analyzed using the methods introduced in the Chaps. 3 and 4.

The reverse operation, i.e., obtaining single-vehicle information from macroscopic quantities by *disaggregation*, is more difficult. Since the information content of the microscopic configuration is higher than that of the macroscopic fields, this is only possible by using ad-hoc assumptions which generally cannot be well justified.

One application of aggregation and disaggregation are parallel-hybrid models in which, for example, critical road sections are modeled by a microscopic model while the rest is being described by a macroscopic model. For this we need a fit pair of a micro- and a macro-model, which both have the same model parameters. The aggregated results of the microscopic model should correspond to the results of the macro-model. Furthermore, we need a *micro-macro link* for the transition from the microscopic model to the macroscopic model at a given location (aggregation), and a *macro-micro link* for the corresponding disaggregation. One exemplary derivation of a macroscopic model from a car-following model is shown in Sect. 9.4.1.

6.2.2 Mathematical Structure

We can also categorize traffic flow models by their mathematical structure.

Partial differential equations (PDE). In models of this class both location x and time t are continuous and serve as the independent variables of continuous fields such as the local speed $V(x,t)$ or density $\rho(x,t)$. The model equations contain these fields and their derivatives with respect to either of the two variables. This is the distinctive feature of PDEs. This mathematical form is suited to express macroscopic models or gas-kinetic based mesoscopic models. PDE traffic flow models generally allow for analytical steady-state solutions (fundamental diagram), and analytical expressions for propagation velocities of traffic waves and stability properties. Furthermore, in

[2] In this sense, the adaptive smoothing method described in Sect. 5.2 can be considered as a kernel-based aggregation method for data points rather than vehicles.

spite of their inherent mathematical complexity, most PDE traffic flow models allow for a fast numerical solution.

Coupled ordinary differential equations. In this mathematical class, the continuous state variables (e.g., location $x_\alpha(t)$ or speed $v_\alpha(t)$ of vehicle α) depend on only one variable, the time t. The model equations contain the state variables and their time derivatives (the distinctive feature of ordinary differential equations) and are coupled with the equations of the leading vehicle. This is the most natural form to describe time-continuous microscopic models (*car-following models*).

Coupled iterated maps. If the model uses discrete time steps Δt instead of continuous time while the state variables (e.g., speed) remain continuous, the mathematical form is that of a coupled iterated map. The set of state variables at time t are given as a function (the "map") of these variables at time $t - \Delta t$ (and possibly earlier time steps).

Iterated maps are used for both microscopic and macroscopic models. In microscopic models, the continuous state variables are the position, lane and speed of all vehicles. In macroscopic iterated maps, space is discretized into cells and the continuous state variables are traffic density and local speed. The maps are "coupled" since the new state of the vehicles of microscopic models or the cells of macroscopic models depend not only on the old state but on the old state of the neighboring vehicles or cells, respectively.

Formally, iterated maps are identical to differential equations that are numerically solved by an explicit method. *Conceptionally* however, there is a difference: In iterated maps, the duration Δt of one time step is a model parameter and the accuracy of the numerical solution is only restricted by numerical rounding errors. In contrast, the time step used when numerically integrating differential equations is not part of the model, but an auxiliary variable of the numerical method. The mathematically exact solution is obtained in the limit $\Delta t \to 0$ (provided that the integration method is *consistent*), while the numerical solution for finite $\Delta t > 0$ becomes necessarily inaccurate.

Cellular automata. In models of this class, all variables are discrete. *Space* is divided into fixed cells and *time* is updated in fixed intervals. The *state* of each cell is either 0 ("no vehicle") or 1 ("vehicle" or "part of vehicle"). The occupation of the cells is determined at every time step and depends on the occupation at the previous time step. In the traffic context, cellular automata (singular: cellular automaton, CA) are mainly used for microscopic models. However, macroscopic traffic flow models in the form of a CA are conceivable as well.

Discrete state variables, continuous time. Most (sub-)models for lane changes use this mathematical form, even in time-continuous microscopic models: The lane index is an integer, i.e., the lane change is (unrealistically) modeled as an *instantaneous process*. Mesoscopic models using master equations belong to this category as well.

Static models. This class of models, also known as *traffic stream models*, describe pairwise relations between the macroscopic state variables (density, flow, speed or occupation). The speed-density relation $V(\rho)$ and the fundamental diagram $Q(\rho)$ discussed in Part I are examples of these models. The classical route-choice step of transportation planning (or the route calculation of simple navigation devices without live-data feed) uses the speed-flow relation, transformed into a travel time versus flow relation, for each link. This so-called *capacity restraint function* is an increasing function of the traffic demand. Notice that steady-state solutions of dynamic microscopic or macroscopic models can be considered as traffic-stream models as well.

6.2.3 Other Criteria

Depending on the application, traffic flow models can be categorized with respect to several other criteria.

Conceptional foundation. We can distinguish between *heuristic models* and *first-principles models*.[3] Heuristic models use a simple mathematical ansatz (e.g., multivariate-linear or polynomial in the exogenous variables) with the coefficients playing the role of model parameters. They are fitted to the data by, e.g., regression techniques and generally have no intuitive meaning. In contrast, first-principles models are derived from certain postulates. For car-following models, this may be a driving behavior that is determined by desired values for speed, acceleration, deceleration, time gap, and minimum gap (bumper-to-bumper distance). Ideally, each of these postulates is reflected by a model parameter the value of which thereby has an intuitive meaning. Of course, first-principle models are calibrated against empirical data, as well. In "good" first-principles models the calibrated parameter values will assume reasonable values. For example, desired time gaps should be between 1 and 2 s, or accelerations within 0.8–2.5 m/s^2.

Randomness. Random elements can be used to describe aspects of the traffic flow which are unknown, immeasurable, impossible to model, or "genuinely" random.[4] While models without any randomness are called *deterministic models*, those with random elements (also known as noise terms or stochastic terms) are called *stochastic models*. In a computer simulation, the stochastic terms are implemented using (pseudo-)random number generators. Randomness can occur at different points in the model:

- *Acceleration noise* phenomenologically models the unpredictability and irrationality of human driving behavior ("man is not a machine"). Most cellular automata need noise terms to produce meaningful results.
- *Exogeneous noise* added to the input data (gaps and speeds in microscopic models) is a way to model perception and estimation errors of humans (or ACC sensors)

---

3 Since traffic flow models include describing the human behavior, the *first principles* are not as universal and invariant as the first principles in, e.g., physics.

4 Some people say that introducing stochastic elements is tantamount to confessing ignorance.

on a more fundamental level: In contrast to acceleration noise, the acceleration function itself is deterministic while the exogenous variables (input) fluctuate randomly over time.

- *Parameter noise*, i.e., changing the values of model parameters according to a stochastic process is a means to describe unpredictable changes in the mood of drivers leading to changes of the driving behavior.
- *Heterogeneities* in the composition of the vehicles and drivers are described by assigning each driver-vehicle unit a different set of parameter values drawn from given probability distributions (see below). This adds stochastic elements to the initial and boundary conditions (inflow, outflow), and leads, e.g., to stochastically changing road capacities.

Identical versus heterogeneous drivers and vehicles. Traffic models may use identical driver-vehicle units or describe *heterogeneous traffic*. In the latter case, the vehicle pool consists of several vehicle types (motorcycles, cars, trucks, etc.) and the model might incorporate *inter-driver variability* by using several parameter sets for each type (cautious vs. agile drivers, loaded vs. empty trucks, etc.).

Constant versus variable driving behavior. The (usually constant) model parameters determine the driving behavior of a driver-vehicle type. If some of these parameters become time-dependent, we can describe changes in driving behavior. This *intra-driver variability* may be stochastic (see above), or deterministic as a function of the past traffic condition describing, e.g., resignation effects after being stuck in congested traffic for a while. Simulating variable driving behavior is crucial in assessing how adaptations of human drivers or ACC systems to different traffic situations may improve or deteriorate the performance and stability of traffic flow (cf. Sect. 21.5).

Single-lane versus multi-lane models. If the traffic flow model describes several lanes and changes between them, it consists of two components: *Longitudinal dynamics* (acceleration model) and *lateral dynamics* (lane-changing model). Some models intrinsically incorporate lateral dynamics while pure longitudinal models can be extended by a suitable lane-changing model (see Chap. 14).

Which model categories are suited for each of the following applications?

1. Traffic state estimation for traffic reporting or routing applications
2. Modeling human drivers with different driving behaviors
3. Development of an adaptive cruise control systems (ACC)
4. Modeling the impact of ACCs and other driver-assistance systems on traffic
5. Models of large-scale traffic networks (e.g., a state-wide highway network)
6. Models of complex city traffic networks
7. Modeling the effects of traffic regulations such as speed limits or lane-changing restrictions
8. Modeling roadworks or other bottlenecks

6.3 Non-Motorized Traffic

Besides motorized traffic flow, the dynamics of non-motorized traffic is accessible to the model framework developed above. Non-motorized traffic includes pedestrian, bicycle, and mixed traffic (particularly in developing countries, cf. Fig. 6.4). Generally, the microscopic approach is most suitable. However, large events with unidirectional traffic flow (e.g., the pedestrian streams at the yearly pilgrimage event of Mecca) have also been modeled and simulated macroscopically. Similar to models of vehicular traffic, microscopic pedestrian models can be categorized into models with continuous variables, so-called *social-force models*, and cellular automata. Contrary to vehicles, pedestrians can generally move freely in *two* spatial dimensions, i.e., there are two spatial coordinates x and y. In addition to the desired (walking) speed, every pedestrian also has a desired direction. Consequently, the desired *velocity* is a vectorial quantity.

Pedestrian models are used, for example, in the design of airports, public open spaces, shopping malls, and for planning large-scale events, e.g., carnivals or other processions, demonstrations, soccer matches, rock concerts, and other big events. Furthermore, they are used to simulate evacuations from buildings, sports stadiums, airplanes, and ships. A well-known example of pedestrian traffic flow modeling is the simulation of pilgrims at the annual Hajj to Mecca, Saudi Arabia. At this event, mass panic with catastrophic consequences caused by jams occurred frequently in the past. With the help of pedestrian traffic simulations, the infrastructure of the site has been modified and a routing of the pilgrim streams has been introduced which significantly alleviated pedestrian crowding and jamming.

Models for other types of non-motorized traffic such as bicyclists, runners, or inline skaters are nearly nonexistent in the literature, even though there are many possible applications. For example, there is a huge demand for the modeling of mixed traffic consisting of pedestrians, motorcycles, three-wheelers, horse and man-powered carriages, cars, and trucks in developing and emerging countries (cf. Fig. 6.4). Flow models of runners and skaters have the potential to anticipate and optimize the operations of large-scale events such as marathons, skating events, or cross-country ski races. For example, at the annual *Vasaloppet* cross-country ski race in Sweden, significant traffic jams occur during the first kilometers due to its popularity (about 15,000 skiers). As a consequence, the athletes in the last starting groups are delayed by an hour or even more while the clock is ticking (Fig. 6.5). Here, traffic flow models can assist in the redesign and planning process by simulating several scenarios, including a staggered rather than a mass start, changing the organization of the starting field (size and location of the starting groups), changing the infrastructure by detecting and eliminating active bottlenecks or modifying the routing. Finally, simulations may suggest imposing an upper bound for the number of participants.

Fig. 6.4 Mixed traffic in Hyderabad (India) (Courtesy of www.cepolina.com)

Fig. 6.5 Traffic jam of cross-country skiers at the Vasaloppet (Sweden) near the start

Problems

6.1 Speed limit on the German Autobahn?
Some people and organizations regularly ask for the introduction of a general speed limit (of, say, 130 km/h) on the German Autobahn. The reasoning usually includes the following points:

1. A speed limit reduces the number of accidents (*safety effect*).
2. Assuming a given traffic demand, the speed limit of 130 km/h increases the dynamic highway capacity and reduces traffic jams (*traffic effect*).
3. By restricting vehicles to 130 km/h, fuel consumption, CO_2 emissions, and noise pollution are reduced (*environmental effects*).
4. The economic internal and external costs (cost of time and fuel, costs of accidents, costs due to noise-related health problems, etc.) are reduced (*macro-economic effect*).

For which of these effects can we find a quantitative description using traffic flow models? If the answer is positive, which model category would be suitable?

Further Reading

- Greenshields, B.D.: A study of traffic capacity. In: Proceedings of the Highway Research Board, Vol. 14. Highway Research Board, Washington, D.C. (1935) 448–477
- Helbing, D.: Traffic and related self-driven many-particle systems. Reviews of Modern Physics **73** (2001) 1067–1141
- Hoogendoorn, S., Bovy, P.: State-of-the-art of vehicular traffic flow modelling. Proceedings of the Institution of Mechanical Engineers, Part I: Journal of Systems and Control Engineering **215** (2001) 283–303
- Kerner, B.: Introduction to Modern Traffic Flow Theory and Control: The Long Road to Three-Phase Traffic Theory. Springer, Heidelberg (2009)
- Gàbor Orosz, R. Eddie Wilson and Gàbor Stépàn: Traffic jams: dynamics and control. Phil. Trans. R. Soc. A **368** (2010) 4455–4479
- Helbing, D., Farkas, I., Vicsek, T.: Simulating dynamical features of escape panic. Nature **40** (2000) 487–490

Chapter 7
Continuity Equation

Experience without theory is blind, but theory without experience is mere intellectual play.

Immanuel Kant

Abstract The foundations of every macroscopic traffic model are the hydrodynamic relation "flow equals density times speed" and the continuity equation, which describes the temporal evolution of the density as a function of flow differences or gradients. The macroscopic vehicle speed is defined such that it satisfies the hydrodynamic relation, and the continuity equation is directly derived from the conservation of vehicle flows. Thus, both equations are parameter-free and hold for arbitrary macroscopic models. In this chapter, we derive the continuity equation for various road geometries and illustrate it both from the point of view of a stationary observer (Eulerian representation) and a vehicle driver (Lagrangian representation).

7.1 Traffic Density and Hydrodynamic Flow-Density Relation

The continuity equation describes the conservation of vehicles in terms of the *traffic density* and the *hydrodynamic flow-density relation*. Therefore, we will begin with defining these quantities for multi-lane highways.

Traffic density is defined as the number of vehicles per unit length (cf. Sect. 3.3.1). When describing traffic flow on highways with $I > 1$ lanes, we distinguish:

- The *single-lane densities* $\rho_i(x, t)$ on lane $i = 1, \ldots, n$.
- The *total density* $\rho_{\text{tot}}(x, t)$ over all lanes.

M. Treiber and A. Kesting, *Traffic Flow Dynamics*,
DOI: 10.1007/978-3-642-32460-4_7, © Springer-Verlag Berlin Heidelberg 2013

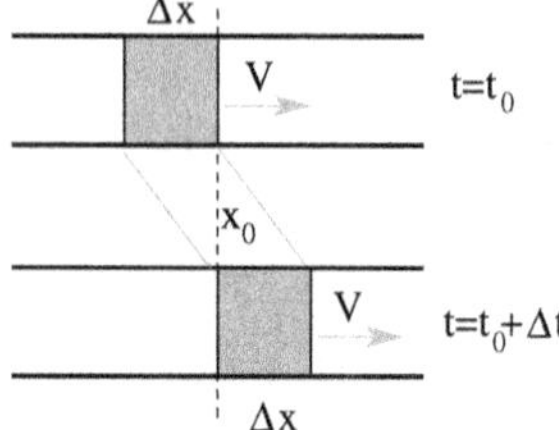

Fig. 7.1 Illustration of the hydrodynamic relation $Q = \rho V$. The colored area contains $\Delta n = \rho \Delta x$ vehicles. Within the time interval $\Delta t = \Delta x / V$, this area completely passes a fixed location x_0. Thus, at this location we have a vehicle flow of $Q = \Delta n / \Delta t = \rho \Delta x / \Delta t = \rho V$

- And the *lane-averaged density* $\rho(x, t)$, also called *effective density* which is defined by $\rho(x, t) = \rho_{\text{tot}}(x, t)/I$.

These definitions are related to each other by

$$\rho_{\text{tot}}(x, t) = \sum_{i=1}^{I} \rho_i(x, t) = I\rho(x, t). \tag{7.1}$$

Notice that the effective density is defined as the simple arithmetic mean of all single-lane densities. While the density definitions ρ and ρ_{tot} are equivalent and interchangeable, one of them may be more useful than the other, depending on the problem at hand. The continuity equation is most conveniently written for the total density ρ_{tot} since vehicles are only conserved on the highway as a whole and not on each lane. However, the speed-density and dynamic speed equations representing the drivers' behavior in first and second-order models, respectively, depend only weakly on the number of lanes.[1] Therefore, the complete macroscopic equations are better formulated in terms of lane-averaged (effective) density and speed fields.

All densities in macroscopic models are to be understood as *real* spatial densities according to the definitions above. Thus, the "hydrodynamic" flow-density relations

$$Q_i(x, t) = \rho_i(x, t) V_i(x, t), \tag{7.2}$$

as illustrated in Fig. 7.1, hold exactly for each individual lane.[2] In this equation, $Q_i(x, t)$ is the flow of lane i at location x and time t, and $V_i(x, t)$ the respective *local speed*.

[1] For example, the capacity per lane of a three-lane highway is a few percent larger than that of a two-lane highway since the obstructing effects of slower vehicles (trucks) decrease with the number of lanes.

[2] If we neglect diffusion, cf. Sect. 8.6.

The local speed $V_i(x, t)$ on lane i, sometimes also denoted as space mean speed, is defined as the arithmetic mean speed of all the vehicles in the interval $[x - \Delta x/2, x+\Delta x/2]$ (or $[x, x+\Delta x]$) at a given time t. The interval must be microscopically large (containing several vehicles) and macroscopically small (see page 56 for details). The same definition applies to other "space mean" quantities such as the lane-averaged speed $V(x, t)$ and the speed variance $\sigma_V(x, t)$.

If we define the *lane-averaged* or *effective* speed $V(x, t)$ using an arithmetic mean that is *weighted* by the lane densities,

$$V(x, t) = \sum_{i=1}^{I} w_i V_i(x, t), \quad w_i = \frac{\rho_i(x, t)}{\rho_{\text{tot}}(x, t)}, \tag{7.3}$$

and the average flow using the *simple* arithmetic mean,[3]

$$Q(x, t) = \frac{1}{I} \sum_{i=1}^{I} Q_i(x, t) = \frac{Q_{\text{tot}}}{I}, \tag{7.4}$$

then the same hydrodynamic relation also holds for the averages and sums over all lanes:

$$\boxed{Q(x, t) = \rho(x, t) V(x, t) \quad \text{Hydrodynamic Flow Relation}} \tag{7.5}$$

and

$$Q_{\text{tot}}(x, t) = \rho_{\text{tot}}(x, t) V(x, t). \tag{7.6}$$

7.2 Continuity Equations for Several Road Profiles

The continuity equation does not depend on the particular macroscopic model being used, but on the geometry of the road infrastructure. We discuss the following cases in order of increasing complexity: (i) homogeneous road section, (ii) highway with on- or off-ramps, (iii) road section in which the number of lanes changes.

---

3 Flow and density are *extensive* quantities, i.e., they depend on the system size (here, the number of lanes) and it is meaningful to use sums of these quantities (e.g., the sum of densities on all lanes). The speed, however, is an *intensive* quantity and it is not meaningful to use sums of such quantities. In general, appropriate averages of extensive and intensive quantities are simple and weighted means, respectively.

7.2.1 Homogeneous Road Section

Let us consider a road section of length Δx without any on- or off-ramps or other geometric inhomogeneities such as changes in the number of lanes (Fig. 7.2 top).[4] The definitions of the local densities and speeds imply that the length Δx must be *microscopically large*, such that it contains sufficiently many vehicles to obtain macroscopic quantities, and *macroscopically small*, such that densities and flow gradients are approximately constant within the road section.[5] Then, the number of vehicles in the road section at time t is given by

$$n(t) = \int\limits_{x}^{x+\Delta x} \rho_{\text{tot}}(x', t)\mathrm{d}x' \approx \rho_{\text{tot}}(x, t)\Delta x \,. \tag{7.7}$$

Since we assumed a homogeneous road section, changes to the number of vehicles can only be caused by inflow Q_{in} or outflow Q_{out} at the section boundaries (cf. Fig. 7.2 top). These boundary flows are given by $Q_{\text{tot}}(x, t)$ and $Q_{\text{tot}}(x + \Delta x, t)$, respectively, resulting in the flow balance

$$\frac{\mathrm{d}n}{\mathrm{d}t} = Q_{\text{in}}(t) - Q_{\text{out}}(t) = Q_{\text{tot}}(x, t) - Q_{\text{tot}}(x + \Delta x, t) \,.$$

Combining this relation with the time-derivative of Eq. (7.7), $\frac{\mathrm{d}n}{\mathrm{d}t} \approx \frac{\partial}{\partial t}(\rho_{\text{tot}}\Delta x) = \Delta x \frac{\partial \rho_{\text{tot}}}{\partial t}$, we obtain

$$\frac{\partial \rho_{\text{tot}}(x, t)}{\partial t} = \frac{1}{\Delta x}\frac{\mathrm{d}n}{\mathrm{d}t} = -\frac{Q_{\text{tot}}(x + \Delta x, t) - Q_{\text{tot}}(x, t)}{\Delta x} \approx -\frac{\partial Q_{\text{tot}}(x, t)}{\partial x},$$

and finally, using the hydrodynamic flow-speed relation $Q_{\text{tot}} = \rho_{\text{tot}} V$ (omitting the function arguments):

$$\boxed{\frac{\partial \rho_{\text{tot}}}{\partial t} + \frac{\partial(\rho_{\text{tot}} V)}{\partial x} = 0 \quad \text{or} \quad \frac{\partial \rho}{\partial t} + \frac{\partial(\rho V)}{\partial x} = 0 \,.} \tag{7.8}$$

Since, for a homogeneous road section, the number I of lanes is constant, the continuity equation for the effective density $\rho = \rho_{\text{tot}}/I$ has the same form.

If the macroscopic model has the form of a coupled iterated map, the road section is divided into several cells k of length Δx_k and the discrete version of the continuity Eq. (7.8) applies:

---

[4] Local changes in driving behavior caused, e.g., by gradients, speed limits, curves, or narrow lanes (without a reduction of the number of lanes), are permitted and do not influence the continuity equation as such. They come into play when closing the equations by speed-flow relations, see Chap. 8.

[5] For highways, both assumptions typically hold for sections of length $\Delta x \approx 100\,\text{m}$.

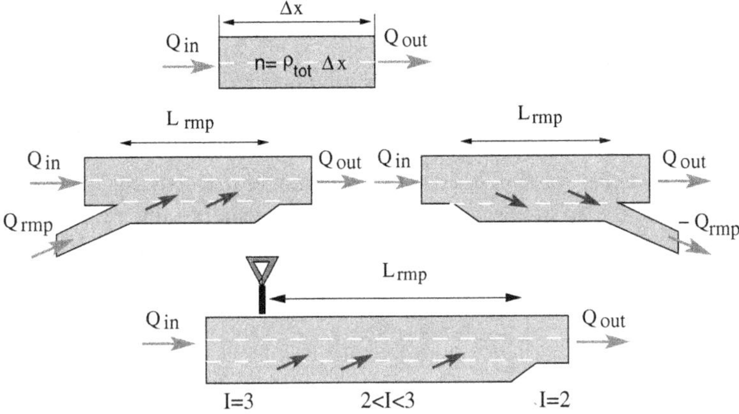

Fig. 7.2 Sketch of the road geometries which yield the continuity Eqs. (7.8), (7.12) and (7.15), respectively: (i) homogeneous road section, (ii) on- or off-ramps, (iii) changes to the number of lanes

$$\boxed{\rho_k(t+\Delta t) = \rho_k(t) + \frac{1}{\Delta x_k}\left(Q_k^{\text{up}} - Q_k^{\text{down}}\right)\Delta t.} \tag{7.9}$$

Here, the inflows Q_k^{up} and outflows Q_k^{down} depend on the respective neighboring cells and are calculated using the *supply-demand method* introduced in Sect. 8.5.7.

7.2.2 Sections with On- and Off-Ramps

On- and off-ramps imply additional in- and outflows $Q_{\text{rmp}}(t)$, which have to be added to those at the section boundaries (cf. Fig. 7.2 center). The balance now reads

$$\frac{\mathrm{d}n}{\mathrm{d}t} = Q_{\text{in}}(t) - Q_{\text{out}}(t) + Q_{\text{rmp}}(t).$$

The ramp flow Q_{rmp} is positive for on-ramps, and negative for off-ramps. If the ramp has more than one lane, Q_{rmp} is the sum of the flow on all lanes of the ramp. Assuming that the in- and outflows are evenly distributed along the length $\Delta x = L_{\text{rmp}}$ of the ramp, we can define a constant flow density $\mathrm{d}Q_{\text{rmp}}/\mathrm{d}x = Q_{\text{rmp}}/L_{\text{rmp}}$. This term is only active within the merging (diverging) sections of the on-ramp (off-ramp). Thus the continuity equation reads

$$\frac{\partial \rho_{\text{tot}}}{\partial t} + \frac{\partial(\rho_{\text{tot}} V)}{\partial x} = \frac{Q_{\text{rmp}}}{L_{\text{rmp}}} = I\,\nu_{\text{rmp}}(x,t)\,. \tag{7.10}$$

Here,

$$\nu_{\text{rmp}}(x,t) = \begin{cases} \frac{Q_{\text{rmp}}(t)}{I L_{\text{rmp}}} & \text{if } x \text{ is within merging or diverging zones,} \\ 0 & \text{otherwise} \end{cases} \tag{7.11}$$

denotes the effective *source density*. By dividing Eq. (7.10) by the number of lanes, we obtain the continuity equation in the presence of ramps for the effective (lane-averaged) density:

$$\boxed{\frac{\partial \rho}{\partial t} + \frac{\partial(\rho V)}{\partial x} = \nu_{\text{rmp}}(x,t)\,.} \tag{7.12}$$

With coupled iterated maps, it is easiest to model the ramp as one cell k whose length is equal to the length of the acceleration/deceleration lane of the ramp. The discrete continuity equation then becomes

$$\rho_k(t+\Delta t) = \rho_k(t) + \frac{1}{\Delta x_k}\left(Q_k^{\text{up}} - Q_k^{\text{down}} + \frac{Q_{k,\text{rmp}}}{I}\right)\Delta t. \tag{7.13}$$

Drivers often change onto the continuous lanes immediately at the beginning of an on-ramp, especially in free traffic. How can this behavior be captured by changing the source term $\nu_{\text{rm}}(x,t)$ of the continuity equation?

On-ramps with very short acceleration lanes force vehicles to change onto the highway at relatively low speeds. Discuss why it is possible to use the continuity equation to describe the perturbations caused by the low speeds?

7.2.3 Changes in the Number of Lanes

When a lane ends, drivers usually merge into the other lane(s) very early, typically 200–1,000 m before the end (or blocking) of the lane on highways and somewhat later in cities.[6] In contrast, when a new lane opens, there are many "early adopters"

[6] This is even the case in case of congested traffic and in countries (e.g., Germany) where, for such situations, traffic regulations require "zipper merging" just before the lane ends. Zipper merging makes full use of the road capacity and minimizes the occurrence of secondary traffic jams caused by gridlock effects (the waiting queue obstructs vehicles driving in other directions).

changing immediately to this lane such that, after a few hundred meters, it is used the same way as the other lanes.

If one were to formulate individual continuity equations for each lane (multi-lane macroscopic models), the equations would be coupled by source terms along the lines of those in Eq. (7.10). In particular, a lane closure would have the same effect on its neighboring lane as an on-ramp, and an opening lane would represent a traffic sink to the through lanes similar to an off-ramp.

However, we are only interested in macroscopic models which describe the dynamics of the effective (lane-averaged) density $\rho(x,t)$ and effective speed $V(x,t)$. The lane changes before a lane closure or after the beginning of an additional lane are modeled by using a *non-integer, location-dependent* number $I(x)$ of lanes (cf. Fig. 7.2 bottom). The averages of all extensive (additive) variables, i.e., flow and density, are related to this continuous number of lanes:

$$Q(x,t) = \frac{Q_{\text{tot}}(x,t)}{I(x)}, \quad \rho(x,t) = \frac{\rho_{\text{tot}}(x,t)}{I(x)}. \tag{7.14}$$

The average speed $V(x,t)$, however, is still given by Eq. (7.3) and the hydrodynamic relation (7.5), $Q = \rho V$ still holds everywhere.

For example, a value of $I = 2.2$ indicates that the third lane is seldom used anymore (or yet), as the flow on this lane is only 0.2 times the average flow on the other lanes. This shows the consistency of the average speed as defined in Eq. (7.3), since (in this example) the local speed on the third lane is weighted by a factor of 0.2. Moreover, with $I(x)$ tending to 2.0, the weighting of the third lane continuously drops to zero, as expected. This also means that the length of the transition zone associated with a non-integer number of lanes should be the same as the length of the typical "merging zone" from or to the non-through lane(s).

The weighted mean speed (7.3) is consistent with continuous changes in the number of lanes, if the upper limit of the sum over all lanes is the smallest *integer* larger than I. Convince yourself that even though the upper limit of the sum is discontinuous (e.g., 3 for $I = 2.01$ vs. 2 for $I = 2$), the lane-averaged effective speed (7.3) is continuous.

The continuity equation for the total density ρ_{tot} is the same as Eq. (7.8), or Eq. (7.10) if ramps are present. However, since the traffic state (free, dense, and congested) and thus the modeled driving dynamics depends on flows and densities *per lane*, we have to express the continuity equation for a changing number of lanes in terms of effective densities, speeds and flows, $\rho = \rho_{\text{tot}}/I(x)$ and $Q_{\text{tot}}/I(x)$, respectively. We insert $\rho_{\text{tot}} = I(x)\rho$ and $Q_{\text{tot}} = I(x)Q$ into Eq. (7.10) and obtain the following continuity equation:

$$\frac{\partial(I\rho)}{\partial t} + \frac{\partial(IQ)}{\partial x} = I\nu_{\text{rmp}}$$
$$I\frac{\partial\rho}{\partial t} + Q\frac{\mathrm{d}I}{\mathrm{d}x} + I\frac{\partial Q}{\partial x} = I\nu_{\text{rmp}}$$
$$\frac{\partial\rho}{\partial t} + \frac{\partial Q}{\partial x} = -\frac{Q}{I}\frac{\mathrm{d}I}{\mathrm{d}x} + \nu_{\text{rmp}}$$

And with $Q = \rho V$:

$$\boxed{\frac{\partial\rho}{\partial t} + \frac{\partial(\rho V)}{\partial x} = -\frac{\rho V}{I}\frac{\mathrm{d}I}{\mathrm{d}x} + \nu_{\text{rmp}}(x) \quad \text{Continuity Equation.}} \tag{7.15}$$

The continuity equation (7.15) describes the most general case including ramps, lane closings, and lane openings. In addition to the ramp term $\nu_{\text{rmp}}(x)$, there is another source density $\nu_I(x) = -\frac{Q}{I}\frac{\mathrm{d}I}{\mathrm{d}x}$ which describes the net flow from ending lanes and to newly opening lanes. Of course, all terms on the right-hand side of the equation are only nonzero within the merging zones of on- and off-ramps, or within the transition zones where vehicles leave lanes that are about to end or enter new lanes.

In the case of coupled iterated maps, the merging zone is modeled similarly to ramps by a cell k of length Δx_k with I_{up} lanes at the upstream end of the cell and I_{down} lanes at the downstream end (cf. Problem 7.6):

$$\rho_k(t+\Delta t) = \rho_k(t) + \frac{1}{\Delta x_k}\left(Q_k^{\text{up}} - Q_k^{\text{down}} + \frac{Q_{k,\text{rmp}}}{I_{\text{down}}} + \frac{I_{\text{up}} - I_{\text{down}}}{I_{\text{down}}} Q_k^{\text{up}}\right)\Delta t. \tag{7.16}$$

7.2.4 Discussion

Let us first stress the fundamental nature of the continuity equation for macroscopic traffic flow models:

> Since the continuity equation is derived solely from the conservation of vehicles, it is a part of *all* macroscopic models. Its form only depends on the modeled road infrastructure and on the mathematical form of the model (partial differential equation, iterated map, or cellular automaton).

Continuity equation without sources. Without on- or off-ramps we have $\partial\rho_{\text{tot}}/\partial t = -\partial Q_{\text{tot}}/\partial x$: The number of vehicles can only change due to in- or outflows at the boundaries of the considered road section. If more vehicles flow out than are flowing in, i.e., $\partial Q_{\text{tot}}/\partial x > 0$, the rate of change in density is negative. If the inflow is larger than the outflow for a sufficiently long time, e.g., due to an accident at the downstream end, then the positive rate of change in the density will eventually lead to a traffic jam. In the absence of ramps and with constant number of lanes, the same continuity equation holds for the lane-averaged effective quantities, $\partial\rho/\partial t = -\partial Q/\partial x$.

Reduction in the number of lanes. If we define density and flow as the average over the *continuous* lanes, the lane changes from the closed lane(s) to the continuous lane(s) cause a net inflow. This is reflected by the source term $-\frac{Q}{I}\frac{\mathrm{d}I}{\mathrm{d}x}$ and causes an increase in density. However, if we use the total density and flow, there will be no "source terms" in the continuity equation.

On-and off-ramps. In addition to the flow gradients, the in- and outflow at ramps also cause a rate of change in density on the highway at the merging or diverging zones. The source terms are proportional to the ramp flows. The effective flow density (source density) $\nu_{\mathrm{rmp}}(x)$ is larger for shorter ramps (since more vehicles have to merge per unit length) and smaller for a larger number of lanes on the highway (since the ramp flow is distributed to more lanes).

7.3 Continuity Equation from the Driver's Perspective

The continuity equation is usually formulated from the perspective of a stationary observer in terms of a partial derivative of the density with respect to time while keeping the location fixed. This is also called the *Eulerian representation*. From the perspective of a driver "drifting" with the traffic, the perceived change in density has an additional *convective* contribution caused by the vehicle motion in the presence of spatial density variations (cf. Fig. 7.3):

$$\Delta\rho \approx \left(\frac{\partial\rho}{\partial t} + V\frac{\partial\rho}{\partial x}\right)\Delta t.$$

In the limit $\Delta t \to 0$ and $\Delta x = V\Delta t \to 0$ (assuming that the density function $\rho(x,t)$ is continuously differentiable), the rate of change in the density perceived by a driver is given by the total time derivative

$$\frac{\mathrm{d}\rho}{\mathrm{d}t} = \frac{\partial\rho}{\partial t} + V(x,t)\frac{\partial\rho}{\partial x}. \tag{7.17}$$

In many publications, the total time derivative is also referred to as *material* derivative, *convective* derivative, or *substantial* derivative. It is composed of the local rate of change $\frac{\partial\rho}{\partial t}$, and the convective rate of change $V\frac{\partial\rho}{\partial x}$ due to spatial changes (see Fig. 7.3).

With $\frac{\partial}{\partial x}(\rho V) = \rho\frac{\partial V}{\partial x} + V\frac{\partial\rho}{\partial x}$ we can rewrite the continuity equation for homogeneous road sections as

$$\frac{\mathrm{d}\rho}{\mathrm{d}t} = \frac{\partial\rho}{\partial t} + V\frac{\partial\rho}{\partial x} = -\rho\frac{\partial V}{\partial x}. \tag{7.18}$$

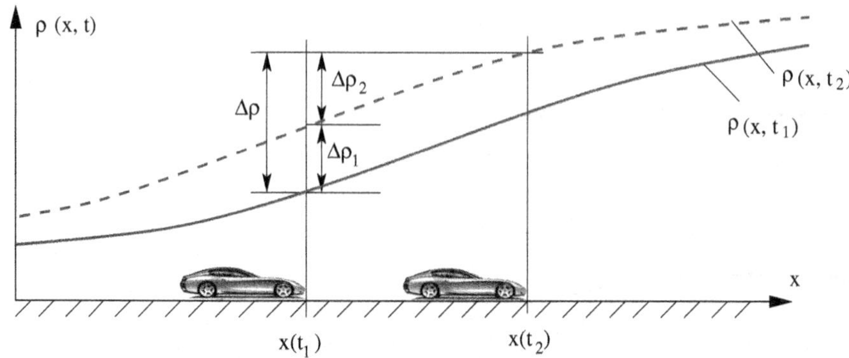

Fig. 7.3 From the perspective of a driver, the change $\Delta\rho = \frac{d\rho}{dt}\Delta t$ in density over time is composed of the local change $\Delta\rho_1 \approx \frac{\partial\rho}{\partial t}(x_1)\Delta t$ at the initial location $x_1 = x(t_1)$ and the convective change $\Delta\rho_2 = \rho(x_2, t_2) - \rho(x_1, t_2) \approx \frac{\partial\rho}{\partial x}\Delta x \approx V\frac{\partial\rho}{\partial x}\Delta t$ due to spatial density variations when moving to a new location x_2

Equation (7.18) states that the density increases if the speed gradient $\frac{\partial V}{\partial x}$ is negative. In the microscopic view, this means that the headway *decreases* when the leading vehicle is driving at a lower speed (which will be made explicit when formulating the Lagrangian view, see Sect. 7.4 below). Furthermore, the density can never be negative, as $\rho(x, t) = 0$ implies $d\rho/dt = 0$ (surely, negative vehicles would be inconsistent).

The two different perspectives are also illustrated in Fig. 7.4: The density profile (different shades) and the speed profiles (the gradients of the five trajectories) describe a stationary downstream jam front, i.e., the density and speed at any given location x are constant, so the local derivatives $\partial\rho(x, t)/\partial t$ and $\partial V(x, t)/\partial t$ are zero. This can also be seen by the stylized time series that would be measured by stationary loop detectors at the positions x_1 and x_2 (Fig. 7.4 right). Each driver, however, perceives a decrease in density since he or she is leaving the traffic jam: $\frac{d\rho}{dt} = V\frac{\partial\rho}{\partial x} < 0$ (cf. Fig. 7.4 bottom). With the (Eulerian) continuity equation for stationary traffic on a homogeneous road being $\frac{\partial}{\partial x}(\rho V) = 0$, the driver will of course observe $\frac{\partial V}{\partial x} = -\frac{V}{\rho}\frac{\partial\rho}{\partial x} = -\frac{1}{\rho}\frac{d\rho}{dt} > 0$.

The relation between local (partial) and substantial (total) derivatives as seen from stationary and comoving observers, respectively, is not only valid for the density but for *arbitrary* continuously differentiable fields $F(x, t)$,

$$\boxed{\frac{dF(x,t)}{dt} = \frac{\partial F(x,t)}{\partial t} + V(x,t)\frac{\partial F(x,t)}{\partial x}.} \tag{7.19}$$

Particularly, this relation holds for the speed $V(x, t)$ itself, so the total speed derivative (rate of change) as seen from the driver's perspective is given by

$$\frac{dV}{dt} = \frac{\partial V}{\partial t} + V\frac{\partial V}{\partial x} = \frac{\partial V}{\partial t} - \frac{V}{\rho}\frac{d\rho}{dt}. \tag{7.20}$$

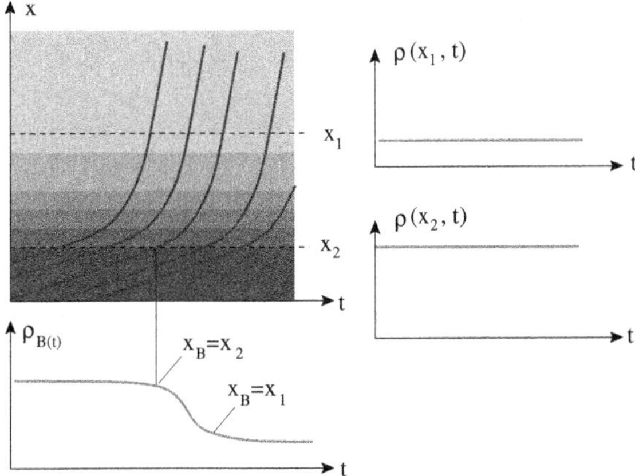

Fig. 7.4 The rate of change in the density, $\frac{\mathrm{d}\rho_B}{\mathrm{d}t} = \frac{\mathrm{d}\rho(x,t)}{\mathrm{d}t} = \frac{\partial \rho}{\partial t} + V \frac{\mathrm{d}\rho}{\mathrm{d}x}$, as perceived by vehicle B (middle trajectory) while driving through the downstream front of a traffic jam, i.e., leaving the jam. Since the jam front shown here is stationary, the local (partial) derivative $\frac{\partial \rho(x,t)}{\partial t}$ is equal to zero

In the situation of Fig. 7.4, the partial time derivatives are zero, so $\frac{\mathrm{d}V}{\mathrm{d}t} = -\frac{V}{\rho}\frac{\mathrm{d}\rho}{\mathrm{d}t}$ is positive which, obviously, is a further signature of leaving the jam.

In Chap. 9, we will use relation (7.19) to formulate the speed equation of second-order macroscopic models.

7.4 Lagrangian Description

In full consequence, the driver's view leads to the *Lagrangian formulation* of the continuity equation. In this view, the independent variable x is expressed in terms of the *vehicle index* n. Assuming no sources and sinks, the transformation can be expressed by[7]

$$x \to n(x,t) = -\int_0^x \rho(x', 0)\mathrm{d}x' + \int_0^t Q(x, t')\mathrm{d}t'. \tag{7.21}$$

In this equation, we assume $n(0,0) = 0$ and a vehicle numbering consistent with that in Chap. 3, i.e., the first vehicle (with the largest x value) has the lowest index.

Furthermore, the dependent variable traffic density is given in terms of the distance headway $h = s + l_{\mathrm{veh}}$ from front bumper to front bumper (cf. Chap. 3),[8]

$$\rho(x,t) = \frac{1}{h(n(x,t),t)}. \tag{7.22}$$

[7] To distinguish the formulation from microscopic equations, we do not use the microscopic vehicle index α.

[8] In order to avoid confusion with differential operators, we denote the distance headway in this section by h instead of the symbol d used in Chap. 3.

As a result, the fields relevant for the Lagrangian description are the distance headway field $h(n,t)$ and the Lagrangian speed field $v(n,t)$ defined by

$$V(x,t) = v(n(x,t),t). \tag{7.23}$$

With these definitions, the chain rule of differentiation allows us to transform the continuity equation (7.18) into the Lagrangian form. The total time derivative of this equation transforms as follows:

$$\begin{aligned}\frac{\mathrm{d}\rho}{\mathrm{d}t} = \frac{\partial \rho}{\partial t} + V\frac{\partial \rho}{\partial x} &= \left(\frac{\partial}{\partial t} + V\frac{\partial}{\partial x}\right)\left[\frac{1}{h\big(n(x,t),t\big)}\right] \\ &= -\frac{1}{h^2}\left(\frac{\partial h}{\partial n}\frac{\partial n}{\partial t} + \frac{\partial h}{\partial t} + V\frac{\partial h}{\partial n}\frac{\partial n}{\partial x}\right) \\ &= -\frac{1}{h^2}\left(\rho V\frac{\partial h}{\partial n} + \frac{\partial h}{\partial t} - \rho V\frac{\partial h}{\partial n}\right) \\ &= -\frac{1}{h^2}\frac{\partial h}{\partial t}. \end{aligned} \tag{7.24}$$

Here, we made use of the relations

$$\frac{\partial n}{\partial t} = Q = \rho V, \quad \frac{\partial n}{\partial x} = -\rho \tag{7.25}$$

derived from Eq. (7.21). With $\mathrm{d}x = -1/\rho\,\mathrm{d}n$ (again obtained from Eq. (7.21)), we obtain for the second term of Eq. (7.18) the transformation

$$\rho\frac{\partial V}{\partial x} = -\rho^2\frac{\partial v}{\partial n} = -\frac{1}{h^2}\frac{\partial v}{\partial n}$$

and hence the continuity equation for homogeneous road sections in Lagrangian form,

$$\frac{\partial h}{\partial t} + \frac{\partial v}{\partial n} = 0. \tag{7.26}$$

Considering also ramps and a variable number $I(x)$ of lanes, we obtain from Eq. (7.15) the general continuity equation in Lagrangian formulation:

$$\boxed{\frac{\partial h}{\partial t} + \frac{\partial v}{\partial n} = -h^2\left[\nu_{\text{rmp}}(x(n,t),t) - \frac{v}{Ih}I'(x)\right]} \tag{7.27}$$

where $I = I(x(n,t))$ and $I'(x) = \frac{\mathrm{d}I}{\mathrm{d}x}$ must be expressed in terms of n and t. To this end, we express x as a function of n using the relation $\frac{\partial x}{\partial n} = -h(n,t)$. Assuming that the vehicle numbers are defined such that vehicle $n = 0$ crosses $x = 0$ at time $t = 0$ and that this reference vehicle is connected with the independent coordinate

$n = 0$ for all times,[9] we obtain

$$x(n,t) = \int_0^t v(0,t')\mathrm{d}t' - \int_0^n h(n',t)\mathrm{d}n'. \tag{7.28}$$

Generally, the advantage of the Lagrangian description lies in the existence of simpler and less nonlinear numerical integration schemes for the homogeneous part (left-hand side) enabling a faster model calibration. However, this comes with the disadvantage of more complicated source terms (right-hand side): Since the fixed space coordinate x is replaced by the moving vehicle number coordinate n, all infrastructure inhomogeneities are no longer stationary but move backwards in the direction of increasing n according to Eq. (7.28). For illustration, the relation for steady-state homogeneous flow ($v(n,t) = \text{const.}$ and $h(n,t) = \text{const.}$) reads $x(n,t) = -hn + tv$.

Problems

7.1 Flow-density-speed relations
Prove that the hydrodynamics relations (7.5) and (7.6) hold. Furthermore, show that they do *not* hold for per-lane speed averages, regardless of whether they are unweighted or weighted by the flows of the lanes.

7.2 Conservation of vehicles
Using the continuity equation, show that the total number of vehicles on a closed ring road with varying number of lanes $I(x)$ (but no on- or off-ramps) never changes.

7.3 Continuity equation I
Consider a two-lane highway with an on-ramp of length $L = 300\,\text{m}$, beginning at $x = 0$. The inflow is 600 vehicles per hour. Write down the continuity equation for the total traffic density for $0 \le x \le L$ as well as for $x > L$. (i) Assume that the inflow of the on-ramp is evenly distributed across the full length L. (ii) How can we model the common behavior of drivers merging early onto the highway if there is free traffic and merging late (near the end of the ramp) in congested conditions?

7.4 Continuity equation II
Use the continuity equation to determine the traffic flow $Q(x)$ in a stationary state, i.e., $\partial\rho/\partial t = 0$ and constant average per-lane demand $Q(x,t) = Q_0$ at the upstream boundary at $x = 0$. Distinguish the two following cases: (i) The road section has no on- or off-ramps but a variable number of lanes, $I(x)$. (ii) The road section has

---

[9] In general, this is not true for the other vehicles. For example, a vehicle entering the highway upstream of the reference vehicle between the vehicles $n = n_1$ and $n_1 - 1 \ge 0$ will get the coordinate n_1, so the coordinates of all vehicles further upstream ($n \ge n_1$) need to be incremented by one to avoid ambiguities. Likewise, any vehicle entering the highway downstream of the reference vehicle will decrement the coordinate n of all vehicles further downstream by one.

a constant number of lanes I, and an off-ramp between $x = 300$ and $500\,\text{m}$ with constant outflow Q_{out} as well as an on-ramp between $x = 700$ and 1,000m with constant inflow Q_{in}. All ramps have constant differential entering (exiting) rates over the length of the merging (diverging) lanes.

7.5 Continuity equation III

Consider a three-lane highway with constant traffic demand $Q_{\text{tot}} = 3600$ veh/h. One of the lanes is blocked due to roadworks and the merging zone is between $x = 0$ and $x = L = 500\,\text{m}$.

1. Find the average per-lane density ρ and the average flow Q with respect to the two continuous lanes. Assume a uniform, density-independent vehicle speed of 108 km/h.
2. Compare the effects of the lane closure in the previous part to the effects of an on-ramp of length $L = 500\,\text{m}$ on a two-lane highway. Find a ramp flow Q_{rmp} and a ramp term $\nu_{\text{rmp}}(x)$ (which may be variable within the $0 \le x \le 500\,\text{m}$) such that the continuity equation is identical to the one found in part 1 of this problem.

7.6 Continuity equation for coupled maps

Show that the steady-state condition $\rho_k(t + \Delta t) = \rho_k(t)$ for the coupled map (7.16) leads to the flow balance

$$Q_{k,\text{rmp}} = Q_k^{\text{down}} I_k^{\text{down}} - Q_k^{\text{up}} I_k^{\text{up}}.$$

Show that this implies that the coupled map (7.16) is consistently defined even if ramps and changes of the number of lanes occur *simultaneously* in a road cell.

7.7 Parabolic fundamental diagram

Consider the speed-density relation $V(\rho) = V_0(1 - \rho/\rho_{\text{max}})$ where V_0 is the desired speed and ρ_{max} the maximum density.

1. Write down the equation for the fundamental diagram $Q(\rho)$.
2. Determine the maximum possible flow and the density at which it is obtained, as a function of V_0 and ρ_{max}.

Chapter 8
The Lighthill–Whitham–Richards Model

There is nothing more powerful than an idea whose time has come.

Victor Hugo

Abstract The continuity equation, which holds for all macroscopic models, describes the rate of change of the density in terms of gradients (or differences) of the flow. The model is closed by specifying flow or local speed. In this chapter we discuss the simpler approach in which the flow is given as a static function of the density, i.e., by a fundamental diagram. The models of this class of *first-order models* which are also called *Lighthill–Whitham–Richards models* differ only in the functional form of the fundamental diagram and in their mathematical representation.

8.1 Model Equations

The continuity equation is a partial differential or difference equation for the macroscopic quantities ρ (density) and V (speed) or Q (flow). Due to the hydrodynamic relation "flow equals density times speed" these two options are equivalent. While the parameterless continuity equation is always valid, we need an additional equation for the flow or speed to complete the model.

> Since the continuity equation is completely determined by the geometry of the road infrastructure, the macroscopic models differ in their modeling of speed or flow, only.

In 1955 and 1956, Lighthill and Whitham, and independently also Richards, proposed the following *static relation* to complement the continuity equation:

$$Q(x,t) = Q_e(\rho(x,t)) \quad \text{or} \quad V(x,t) = V_e(\rho(x,t)). \tag{8.1}$$

M. Treiber and A. Kesting, *Traffic Flow Dynamics*,
DOI: 10.1007/978-3-642-32460-4_8, © Springer-Verlag Berlin Heidelberg 2013

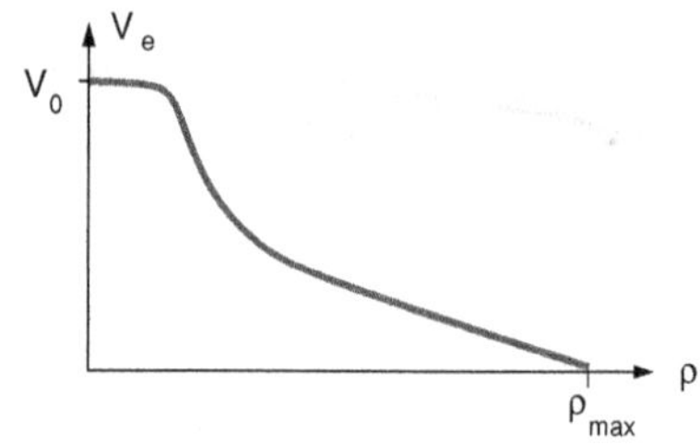

Fig. 8.1 Schematic example of a static speed-density relation for Lighthill–Whitham–Richards (LWR) models (see also the empirical data in Figs. 4.4 and 4.12)

This relation assumes that traffic flow $Q(x,t) = \rho(x,t)\,V(x,t)$ or speed $V(x,t)$ is always in local equilibrium with respect to the actual density: Traffic flow and local speed *instantaneously* follow the density, not only for steady-state traffic but in *all* situations. The precise form of the speed-density relation $V_e(\rho)$ (cf. Fig. 8.1) or the *fundamental diagram* $Q_e(\rho) = \rho V_e(\rho)$ is usually determined by fitting against empirical speed-density or flow-density data (see Fig. 4.12).[1]

Inserting the assumption (8.1) of local equilibrium into the continuity equation (7.8) for homogeneous road sections and applying the chain rule $\frac{\partial Q_e}{\partial x} = \frac{\mathrm{d}Q_e(\rho)}{\mathrm{d}\rho}\frac{\partial \rho}{\partial x}$ yield the simplest form of a *Lighthill–Whitham–Richards model*:

$$\boxed{\frac{\partial \rho}{\partial t} + \frac{\mathrm{d}Q_e(\rho)}{\mathrm{d}\rho}\frac{\partial \rho}{\partial x} = 0 \quad \text{LWR Model.}} \tag{8.2}$$

This equation can also be written as

$$\frac{\partial \rho}{\partial t} + \left(V_e + \rho \frac{\mathrm{d}V_e}{\mathrm{d}\rho}\right)\frac{\partial \rho}{\partial x} = 0. \tag{8.3}$$

On- and off-ramps as well as changes in the number of lanes are described by the corresponding additional terms in the generic continuity equation (7.15) assuming local speed equilibrium $V(x,t) = V_e(\rho(x,t))$ wherever applicable. Since Eq. (8.2) does not specify the functional form of the fundamental diagram $Q_e(\rho)$, and many (more or less realistic) specific functions have been proposed, LWR refers to a whole *class of models*. Thus the common usage of the plural, LWR *models*. All models in this class only have one dynamic equation, the continuity equation. Therefore, they are also referred to as *first-order models*. In contrast, the second-order models discussed in Chap. 9 assume that the local speed is an independent dynamic quantity which, consequently, is modeled by an additional *dynamic* equation.

---

[1] Speed-density and flow-density plots are one of the most important visualizations of aggregated traffic data and have already been discussed in Sects. 4.2 and 4.4. Strictly speaking, the fundamental diagram describes the one-dimensional manifold of steady states parameterized as a function of the density. However, in several publications, the (scattered) flow-density data itself is often incorrectly referred to as the fundamental diagram as well.

8.2 Propagation of Density Variations

The partial differential equations (8.2) and (8.3) are *nonlinear wave equations*, describing the propagation of *kinematic waves*. In the following, we derive the propagation velocity $\tilde{c}$ of such waves, or smooth density variations in general, by using the traveling-wave ansatz

$$\rho(x,t) = \rho_0(x - \tilde{c}t). \tag{8.4}$$

The function $\rho_0(x) = \rho(x,0)$ defines the initial density distribution, which, according to Eq. (8.4), uniformly moves with velocity $\tilde{c}$. Let $\rho_0'(x)$ be the derivative of ρ_0 with respect to its (only) argument. By invoking the chain rule, we obtain

$$\frac{\partial \rho}{\partial t} = -\tilde{c}\rho_0'(x - \tilde{c}t) \quad \text{and} \quad \frac{\partial \rho}{\partial x} = \rho_0'(x - \tilde{c}t).$$

Substituting these partial derivatives into the LWR model equation (8.2) yields the condition

$$-\tilde{c}\rho_0'(x - \tilde{c}t) + \frac{\mathrm{d}Q_e}{\mathrm{d}\rho}\rho_0'(x - \tilde{c}t) = 0$$

which should hold for all x and t. This is only possible if the propagation velocity $\tilde{c}$ depends on the density according to

$$\boxed{\tilde{c}(\rho) = \frac{\mathrm{d}Q_e}{\mathrm{d}\rho} = \frac{\mathrm{d}(\rho V_e(\rho))}{\mathrm{d}\rho},} \tag{8.5}$$

or, again with the notation $f'(x) = \frac{\mathrm{d}f}{\mathrm{d}x}$,

$$\tilde{c}(\rho) = Q_e'(\rho) = V_e(\rho) + \rho V_e'(\rho). \tag{8.6}$$

Equation (8.5) states that the propagation velocity $\tilde{c}(\rho)$ of density variations in a fixed reference frame is proportional to the gradient of the steady-state flow-density relation (fundamental diagram). The density variations may propagate either *in driving direction* (free traffic; left part of the fundamental diagram, cf. Fig. 8.2) or *against* the driving direction (congested traffic; right part of the fundamental diagram).

To find the relationship between the propagation velocity and the vehicle speed v, we define a *relative* propagation velocity from the point of view of a driver (comoving coordinates, Lagrangian view) and insert Eqs. (8.5) and (8.6):

$$\tilde{c}_{\text{rel}}(\rho) = \tilde{c}(\rho) - V = \tilde{c}(\rho) - V_e(\rho) = \rho V_e'(\rho).$$

Since $V_e'(\rho)$ is non-positive for all correctly specified models (cf. Figs. 8.1 and 4.12) we have $\tilde{c}_{\text{rel}} \leq 0$. Thus, from the perspective of a driver, density variations always propagate *backwards* (upstream), or are, at most, stationary if traffic is completely free

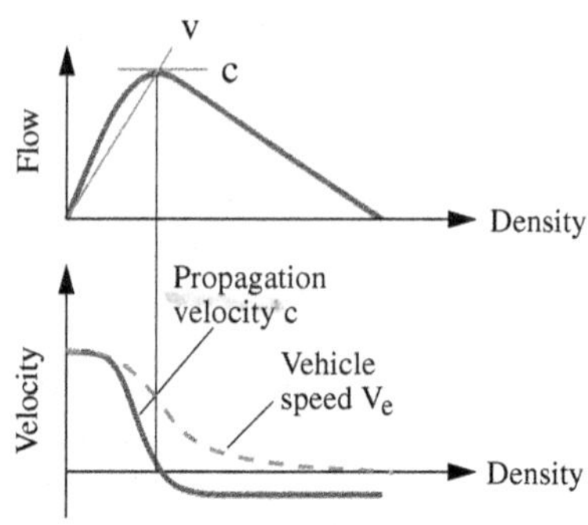

Fig. 8.2 Propagation velocity $\tilde{c} = Q_e'(\rho)$ of density and speed variations in the LWR model in comparison with the local vehicle speed $V_e(\rho)$. In the fundamental diagram (*top*), $\tilde{c}$ is given by the slope of the tangent while V is given by the slope of the secant through the origin

and no interactions between drivers are present. This is reflected in most microscopic models by the fact that the (modeled) drivers only observe and react to the *leading* vehicle and not to the *following* vehicle (see Chap. 10).

8.3 Shock Waves

8.3.1 Formation

Continuous LWR models of the form (8.2) describe density variations of constant amplitudes but with varying local propagation velocities: the lower the local density, the higher the propagation velocity. For illustration purposes, we can think of the density profile (the plot of density vs. location for a given time instant) as a stack of thin horizontal layers which move independently with a velocity given by evaluating Eq. (8.5) for the corresponding density (proportional to the "vertical location" of this layer). Thus, the "top" layers move more slowly (possibly even backwards) compared to the "bottom" layers (cf. Fig. 8.3 top, see also Fig. 8.4). For a "stop-and-go wave", this means that the upstream front becomes steeper and the downstream front disperses (Fig. 8.3 middle). Thus, from the driver's perspective, the transition free → congested traffic becomes more and more abrupt while the vehicles at the transition congested → free traffic accelerate more and more slowly over time.[2]

Eventually, the gradient $\partial\rho/\partial x$ will tend to infinity at the upstream front. At this point, Eq. (8.5) figuratively predicting "breaking waves", breaks down (Fig. 8.3 bottom). After all, there can only be one unique density value at any given time and location, so "breaking" waves are physically absurd. Instead, we will observe a discontinuous transition indicated by the vertical line in Fig. 8.3 (bottom panel) which is the defining feature of a *shock wave* or *shock front*. This is confirmed by simulation: Fig. 8.4 shows how the gradient of the upstream front gradually increases

[2] In the mathematical literature, this widening of the downstream transition zone is called a *dispersion fan*.

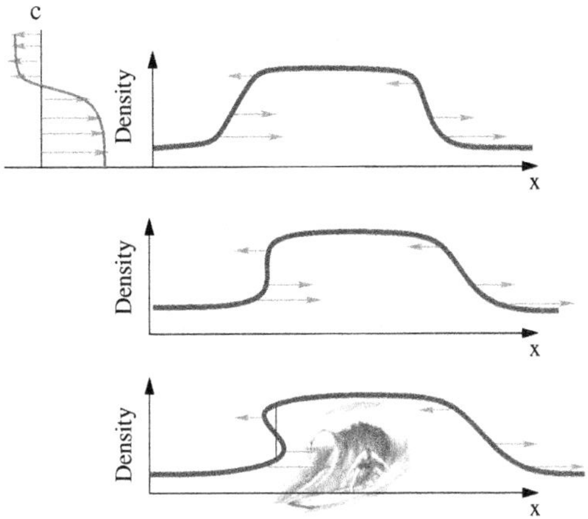

Fig. 8.3 Emergence of shock waves due to the density-dependent local propagation velocities in the LWR model

until a discontinuity emerges at about $t = 10$ min and $x = 4$ km. Since shocks are associated with infinite accelerations or decelerations, they do not reflect real-world traffic, so LWR models are unrealistic in this respect.

In general, the evolution of the transition regions depend on mathematical properties of the flow-density relation (fundamental diagram) $Q_e(\rho)$:

- If the fundamental diagram is concave (the second derivative $Q_e''(\rho) < 0$), the transition lower → higher density steepens and develops to (or remains a) discontinuous shock while the transition higher → lower density disperses over time.
- If the fundamental diagram is locally convex in the considered density range, a transition higher → lower density eventually becomes a shock while the transition lower → higher density disperses.
- If the fundamental diagram has no curvature in the density range in question, all transitions (whether continuous or shocks) propagate at constant velocity $Q_e'(\rho)$ while the shape of the transition remains unchanged.

This means, the qualitative dynamics shown in Fig. 8.4 is valid for concave fundamental diagrams, only.

Explain why density transitions evolve as described in the box above with the help of Fig. 8.3.

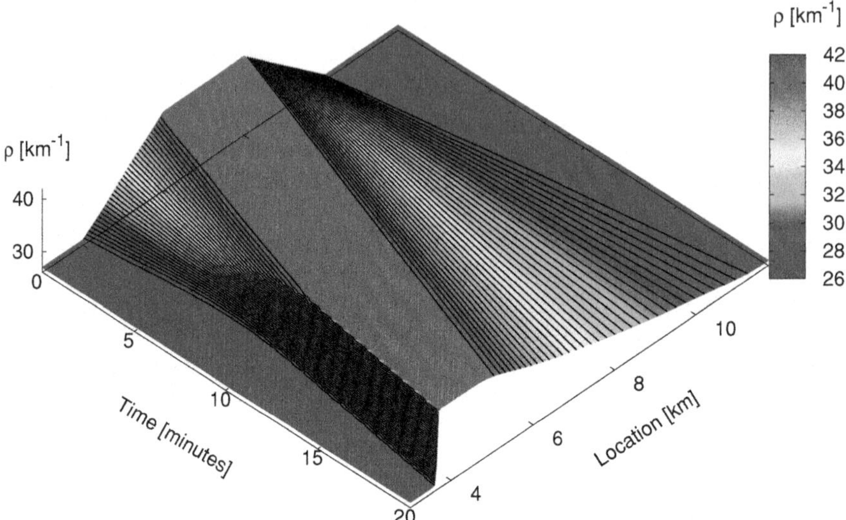

Fig. 8.4 Numerical solution to the LWR model with a continuous speed-density relation and trapezoid-like initial density distribution (*left upper boundary* of the plot). The situation corresponds to a localized region of congested traffic (*red*) surrounded by free traffic (*blue*) with initially continuous transitions. The transition free → congested traffic evolves into a shock wave while the transition congested → free traffic disperses. Outside of the actual shock forming at time $t \approx 10$ min and location $x \approx 4$ km, the contours of equal density are *straight lines*. This corresponds to a uniform motion of each layer of constant density

8.3.2 Derivation of the Propagation Velocity

While the details of the transitions free → congested and congested → free in the LWR models are unrealistic,[3] the propagation of the wave positions *as a whole*, and also the motion of the transition zones to and from extended congested traffic, is described realistically.

In order to derive the propagation velocities, we consider a discontinuous transition from state 1 (free traffic) to state 2 (congested traffic) as depicted in Fig. 8.5. Without loss of generality, we consider a single-lane road.[4] Within a sufficiently small road section $0 \leq x \leq L$ fixed in the stationary coordinate system around the instantaneous location of the shock front $x_{12}(t)$, we can assume constant flow and density at both sides of the front, i.e,

---

[3] Personal experience from the authors tells us that there is considerable dispersion in the downstream front of the "mega-jam" that is formed by the participants of marathon, inline-skating or cross-country skiing events after the starter's gun. However, in vehicular traffic, this dispersion is very limited.

[4] In order to obtain the same result for $I > 1$ we would substitute n with n/I in the following equations. Furthermore, we would replace all densities and speeds by their respective effective values as defined in Sect. 7.1

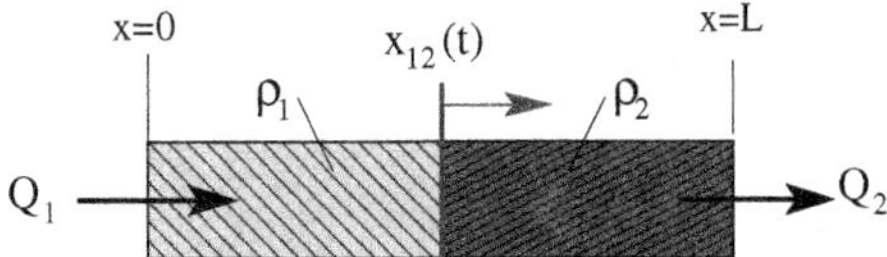

Fig. 8.5 A shock front at location $x_{12}(t)$ with constant flow and density within small road sections on either side

$$\rho(x,t) = \begin{cases} \rho_1 & \text{for } x \le x_{12}(t) \\ \rho_2 & \text{for } x > x_{12}(t) \end{cases}, \quad Q(x,t) = \begin{cases} Q_1 = Q_e(\rho_1) & \text{for } x \le x_{12}(t) \\ Q_2 = Q_e(\rho_2) & \text{for } x > x_{12}(t) \end{cases}.$$

The location $x_{12}(t)$ of the front itself, however, is time-dependent. To find the velocity $c_{12} = \frac{\mathrm{d}x_{12}}{\mathrm{d}t}$, we will express the rate of change in the number of vehicles, $\frac{\mathrm{d}n}{\mathrm{d}t}$, in two different ways. From the conservation of vehicles, we get the balance equation

$$\frac{\mathrm{d}n}{\mathrm{d}t} = Q_1 - Q_2. \tag{8.7}$$

With the definition of the density, we can also write the number of vehicles as

$$n = \rho_1 x_{12} + \rho_2 (L - x_{12}). \tag{8.8}$$

Taking the time derivative yields

$$\begin{aligned} \frac{\mathrm{d}n}{\mathrm{d}t} &= \frac{\mathrm{d}}{\mathrm{d}t} \left(\rho_1 x_{12} + \rho_2 (L - x_{12})\right) \\ &= (\rho_1 - \rho_2) \frac{\mathrm{d}x_{12}}{\mathrm{d}t} \\ &= (\rho_1 - \rho_2) c_{12}. \end{aligned}$$

Comparing both expressions for $\frac{\mathrm{d}n}{\mathrm{d}t}$ gives us

$$\boxed{c_{12} = \frac{Q_2 - Q_1}{\rho_2 - \rho_1} = \frac{Q_e(\rho_1) - Q_e(\rho_2)}{\rho_2 - \rho_1} \quad \text{Propagation of Shock Waves.}} \tag{8.9}$$

Notice that we did not make use of flow-density relations in deriving the first equal sign in Eq. (8.9). Therefore, the motion of sharp transitions is given by $c_{12} = \frac{Q_2 - Q_1}{\rho_2 - \rho_1}$ in *any* first-order or second-order macroscopic model.

8.3.3 Vehicle Speed Versus Propagation Velocities

The LWR model allows us to extract all relevant velocities directly from the fundamental diagram (Fig. 8.6):

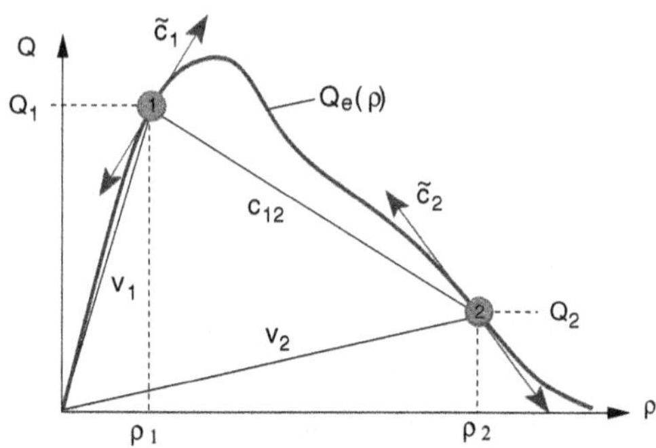

Fig. 8.6 Visualization of how to obtain vehicle speeds and propagation velocities from the fundamental diagram: Shown are the propagation velocity $\tilde{c}(\rho) = \frac{dQ_e}{d\rho}$ of density variations and vehicle speed $V_e = Q_e/\rho$ for the two states ① and ②, and the propagation velocity c_{12} of a shock front separating these states

1. The *propagation velocity of density variations* $\tilde{c}(\rho) = Q'_e(\rho)$ is given by the slope of the fundamental diagram.
2. The *propagation velocity of shock fronts* c_{12} is given by the slope of the secant connecting points of the fundamental diagram corresponding to traffic on either side of the front.
3. The *vehicle speed* $V_e = Q_e(\rho)/\rho$ is given by the slope of the secant connecting the origin with the corresponding point on the fundamental diagram.

Using Eq. (8.5), we can use these relations to distinguish free and congested traffic by the sign of the propagation velocity $\tilde{c}$ of small density and speed variations:

- *Free traffic* is characterized by the left-hand side of the fundamental diagram, i.e., by densities below the critical value ρ_C at static capacity C (state of stationary flow). The propagation velocity is positive.
- *Congested traffic* is characterized by densities at the right-hand side of the fundamental diagram, $\rho > \rho_C$, i.e., the propagation velocity is negative.

In a more complex situation such as that shown in Fig. 8.7, the various speeds and velocities can be read off as the slopes of tangents, secants or lines through the origin at (or along) the traffic states marked by the symbols ①–⑦ in Fig. 8.8. Particularly, we can distinguish following situations:

- The temporary bottleneck A restricts the flow to its capacity $C_A = Q_2^{\text{tot}}$. Since $Q_{\text{in}} > C_A$, the traffic breaks down. Free traffic ② emerges downstream of the bottleneck, while a traffic jam ⑥ forms upstream. The congestion grows due to its propagating upstream front (the secant between ③ and ⑥ has slope $c_{36} < 0$) and stationary downstream front (slope $c_{62} = 0$). However, the vehicle speed

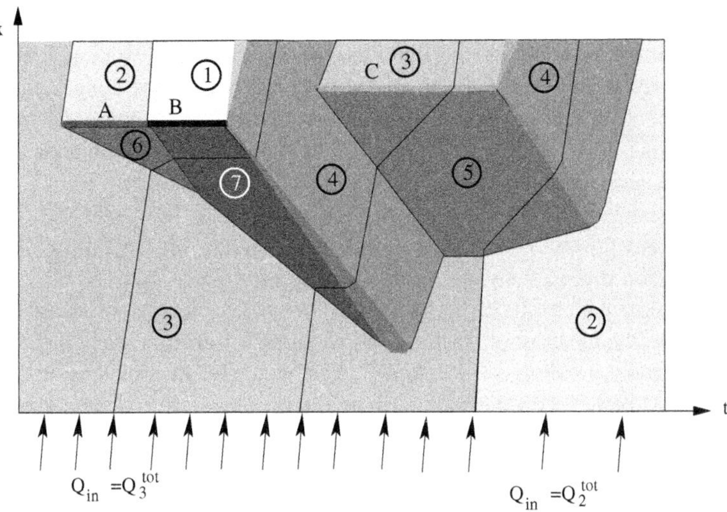

Fig. 8.7 Spatiotemporal traffic dynamics of an LWR model with fundamental diagram as shown in Fig. 8.8. The influx Q_{in} corresponds to state ③ in the fundamental diagram, but decreases after some time and then corresponds to state ②. Furthermore, there are three temporary bottlenecks: Bottleneck A (e.g., a traffic accident) has capacity $C_A = Q_2^{\text{tot}}$, bottleneck B corresponds to a temporary full road closure (e.g., to tow away vehicles involved in the accident), and bottleneck C is a less severe obstruction with capacity $C_C = Q_3^{\text{tot}}$. The slopes of the three trajectories (*black*) indicate the local vehicle speed. The transitions from high to low density "soften" over time while the others remain discontinuous, i.e., shocks

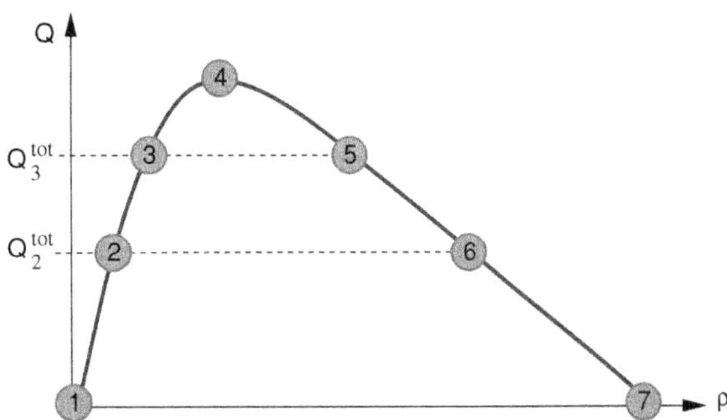

Fig. 8.8 Schematic fundamental diagram. The circles correspond to the traffic states illustrated in Fig. 8.7

$V_6 = Q_6/\rho_6$ inside the congested area is positive. The transition between ② and ③ (in the top-left corner of Fig. 8.7) has a propagation velocity c_{23} that is only slightly less than the local vehicle speeds V_2 and V_3.

- The full road closure (bottleneck B) reduces the flow to zero on either side of the bottleneck. However, the road is empty on the downstream side (state ①) while a traffic jam with maximum density forms on the upstream side (state ⑦). A number of fronts emerge with velocities $c_{67} < c_{37} < c_{36} < 0$ and $c_{71} = 0$.

- Re-opening the road creates a maximum-flow state ④ as the vehicles start moving again. The transition between ④ and the maximum-density state ⑦ propagates with velocity $c_{74} \approx c_{67} < 0$. Note that only the transitions congested → free traffic disperse (to an unrealistic degree) while the others remain discontinuous shocks. This is a consequence of the concave fundamental diagram used here (i.e., the second derivative of $Q_e(\rho)$ is non-positive for all densities).
- Finally, the weakest temporary bottleneck C causes a flow $Q_3^{\text{tot}} = Q_5^{\text{tot}}$ both upstream and downstream, thus $c_{53} = 0$. Depending on the inflow, the slow-moving traffic state ⑤ may grow ($c_{45} < 0$), shrink ($c_{25} > 0$), or be bounded by a stationary upstream front ($c_{35} = 0$). When this bottleneck is removed, the situation is similar to the clearing of the full road closure: A transition from slow-moving traffic ⑤ to the maximum-flow state ④ propagates backwards ($c_{54} < 0$) until it reaches the upstream front of ⑤, marking the full resolution of the congestion.

8.4 Numerical Solution

With the exception of the section-based model (see Sect. 8.5) applied to very simple situations, the LWR models, i.e., the continuity equation (7.8) or (7.10) with a steady-state speed-density relation $V_e(\rho)$, needs to be solved ("integrated") numerically. This is generally done by *finite-difference* methods: Space is divided into cells of generally constant length Δx (although this is not required), and time in the index k increasing in the downstream direction.[5] All the dynamics at scales below Δx and Δt is ignored. So, the density inside each cell k at time t can be characterized by a single value $\rho_k(t)$ (and by the speed $V_k(t) = V_e(\rho_k(t)))$. Furthermore, the flow $Q_{k,k+1}(t)$ between neighboring cells is constant during each time interval Δt. The equations for the LWR models have the form of a so-called *conservation law* for which many specialized explicit solution methods are available.[6] In the simplest case, they take on the form (7.9).

The most common integration method for LWR models is the *Godunov scheme*. This method is based on an exact solution of the continuity equation for one time step assuming stepwise initial conditions given by the actual densities $\{\rho_k\}$ of the cells. Such exact solutions exist if we make sure that neither information (carried by the vehicles or by the propagation velocities) propagates over more than

[5] This is in contrast to the vehicle index where the first (must downstream) vehicle has the lowest index.

[6] In explicit integration schemes, the new state, i.e., all densities $\rho_k(t + \Delta t)$, are given in terms of the old state $\{\rho_k(t)\}$. Implicit methods are characterized by relations between the old and new states that cannot be easily solved for the new state. In traffic-flow models, only explicit methods play a role.

one cell during one time step. Since the vehicle speed is given by the gradient of the secant connecting the origin with a point on the fundamental diagram, the maximum vehicle speed is smaller than, or at most equal to, the maximum propagation speed $|Q_e'(\rho)|$. Thus, one arrives at the first *Courant-Friedrichs-Lévy condition* (CFL condition) for LWR models,

$$\Delta t \max_{\rho \in [0, \rho_{\max}]} \left(|Q_e'(\rho)| \right) < \Delta x. \tag{8.10}$$

The CFL condition restricts the time step to a value which is proportional to the cell size, i.e., the numerical complexity increases with the inverse of the *square* of the cell size (see Sect. 9.5 below for details).[7] Although the Godunov method is based on exact analytical solutions, it entails discretization errors since, after each time step, the density structure inside each cell resulting from the analytical solution is "flattened" to obtain the stepwise initial conditions for the analytical solution of the next time step. These discretization errors lead to the phenomenon of *numerical diffusion* which increases with the cell size (see Sect. 9.5 for details).

8.5 LWR Models with Triangular Fundamental Diagram

The simplest of the Lighthill–Whitham–Richards models uses a "triangular" fundamental diagram (cf. Fig. 8.9):

$$\boxed{Q_e(\rho) = \begin{cases} V_0 \rho & \text{if } \rho \le \rho_C = \frac{1}{V_0 T + l_{\text{eff}}} \quad \text{(free traffic)}, \\ \frac{1}{T}\left(1 - \rho l_{\text{eff}}\right) & \text{if } \rho_C < \rho \le \rho_{\max} = \frac{1}{l_{\text{eff}}} \quad \text{(congested traffic)}. \end{cases}} \tag{8.11}$$

As with the other LWR models, this model can be formulated in continuous and discrete variables:

- The continuous version (8.2) is called *section-based model.*
- The discrete version is formulated as an iterated coupled map with time and space discretized into time steps and cells, respectively, and supplemented by a special "supply-demand" update rule. This model is known as *cell-transmission model* (CTM), see Sect. 8.5.7.

Among the class of LWR models, the section-based model is the most efficient in simulations. In particular, there is no need to numerically solve the hyperbolic partial differential equation (8.2) with Eq. (8.11) defining this model. Due to the specific properties of the triangular fundamental diagram (only two distinct propagation

---

[7] The numerical complexity $\mathscr{C}$ indicates the number of multiplications or other operations on a computer which are necessary to obtain a certain approximate solution. Typically, the absolute value is irrelevant and the numerical complexity is given in terms of a scaling relation, here $\mathscr{C} \propto \Delta x^{-2}$.

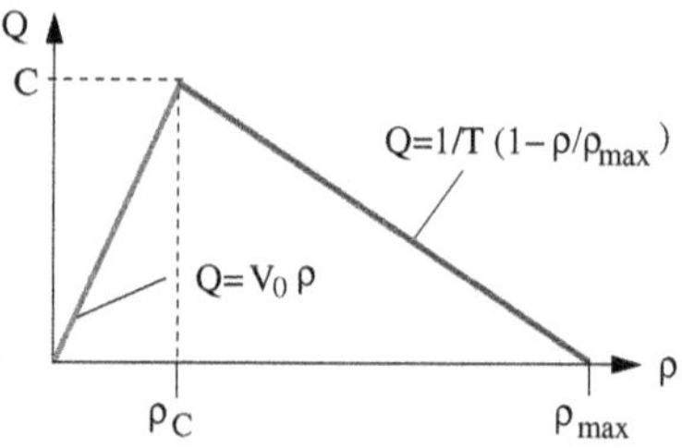

Fig. 8.9 Triangular fundamental diagram, as used in the cell-transmission model and the section-based model

velocities of density variations; one for free traffic and one for congested traffic), it is possible to break down the road into road sections defined such that each section has an inhomogeneity or bottleneck at the downstream end. In each road section, there is at most one jam front. Then, instead of solving the model equation (8.2) for all locations x, it is sufficient to solve a single integral for the motion of the jam front. The inflow at the upstream end of each section is given either by the outflow of an adjacent section or by the source boundary conditions of the simulated system.

The special shape of the fundamental diagram allows for efficient numeric update rules for the cell-transmission model (CTM) as well. Moreover, it is straightforward to generalize the CTM to road networks. The CTM is widely used in model-supported traffic state estimation. Furthermore, it is the only macroscopic model for pedestrians in common use.

Another implication of the fixed propagation velocities is the absence of dispersion at the transitions from high to low density (Fig. 8.3, see also Figs. 8.7 and 8.11).[8] Sometimes a weak dispersion is observed in reality; however, the absence of such dispersion is certainly more realistic than the very strong dispersion caused by most of the other fundamental diagrams.

8.5.1 Model Parameters

The three model parameters of the section-based and cell-transmission models are shown in Table 8.1 with typical values for different situations including the large-scale modeling of pedestrian flows.[9] Since pedestrian flows are two-dimensional, the units of T and ρ_{max} and their meanings are different from those of vehicular traffic. Specifically, the flow and the inverse of the time headway (time gap plus occupancy time) is to be interpreted as *flow density* Q^* (flow per unit width of the way) with the unit 1 $(\text{ms})^{-1}$, so T has the unit second times meter, and can no longer be interpreted as time gap (cf. Table 8.1).

---

[8] With the exception of *numerical diffusion* caused by the discretization of the CTM, cf. Sect. 9.5.

[9] A prerequisite for using LWR models for pedestrian flow are unidirectional pedestrian streams. This is satisfied in the arguably most prominent application example, namely the model-assisted planning and organization of the flow of pilgrims at the *Hajj* in Mecca, Saudi Arabia.

Table 8.1 Model parameters of the section-based model and the discrete cell-transmission model (CTM), and their typical values for highway, city, and pedestrian traffic

Parameter	Highway	City traffic	Pedestrian traffic
Desired speed V_0	110 km/h	50 km/h	1.3 m/s
Time gap T	1.4 s	1.2 s	0.5 ms
Maximum density ρ_{max}	120 vehicles/km	120 vehicles/km	4 pedestrians/m^2

As in the case in all macroscopic models, the parameters are *averages* over the individual vehicles (or pedestrians). For vehicular traffic, the maximum density corresponds to the inverse of the minimum distance headway l_{eff}, which is the average vehicle length plus the average minimum gap s_0 in stopped traffic:

$$l_{\text{eff}} = s_0 + l = \frac{1}{\rho_{\text{max}}}. \tag{8.12}$$

8.5.2 Characteristic Properties

The numerical efficiency (and even analytical solvability of the section-based model, for some cases) stems from a number of properties which make the LWR models with a triangular fundamental diagram stand out from the rest of the LWR models.

Analytical inverse function of the fundamental diagram. The most important input and control variable in real-world (i.e., open) systems is the traffic flow, while local speed and density are dependent variables of the flow and the traffic state. Thus, the density is described by the inverse function of the fundamental diagram, i.e., the density-flow relation $\rho(Q)$. However, this function is not unique since, for a given flow Q, there are two possible density values: one from the so-called *branch for free flow*, $\rho_{\text{free}}(Q)$ and one from the *branch for congested traffic*, $\rho_{\text{cong}}(Q)$. In the triangular fundamental diagram, these two density-flow relations are given by the simple relations

$$\rho_{\text{free}}(Q) = \frac{Q}{V_0}, \tag{8.13}$$

$$\rho_{\text{cong}}(Q) = \frac{1 - QT}{l_{\text{eff}}} = \rho_{\text{max}}(1 - QT). \tag{8.14}$$

Analytical expression for the capacity. The maximum flow Q_{max}, i.e., the effective capacity C per lane on a homogeneous road section is given by the intersection of the two branches,

$$\frac{C}{I} = Q_{\text{max}} = \frac{1}{T + \frac{l_{\text{eff}}}{V_0}} = \frac{1}{T\left(1 + \frac{|c|}{V_0}\right)} \tag{8.15}$$

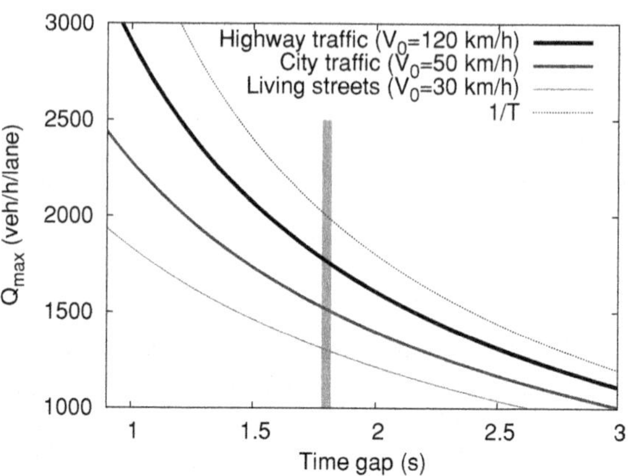

Fig. 8.10 Maximum effective capacity $Q_{\max}$ per lane for the triangular fundamental diagram assuming an effective vehicle length $l_{\text{eff}} = 8\,\text{m}$ ($\rho_{\max} = 125$ vehicles/km). The vertical bar at $T = 1.8\,\text{s}$ corresponds to a rule of thumb taught in German driving schools: "Don't come closer (in meters) than half of your speedometer reading (in km/h)"

where c is the propagation velocity of density changes in congested traffic (see below). Notice that, as in other LWR models, there is no hysteresis, so the maximum flow is unique.

Equation (8.15) shows that the capacity is *always smaller than the inverse of the mean time gap* T, see Fig. 8.10.[10] Furthermore, it can be shown that, for a given time gap, the section-based model is the model with the largest capacity. The corresponding density ρ_C at capacity is given by

$$\rho_C = \frac{1}{V_0 T + l_{\text{eff}}}. \tag{8.16}$$

Constant propagation velocity of density variations in free traffic. The propagation velocities of flow, density, and speed variations in free traffic are given by the rising slope of the fundamental diagram,

$$c_{\text{free}} = \left. \frac{\mathrm{d}Q_e}{\mathrm{d}\rho} \right|_{\rho < \rho_C} = V_0. \tag{8.17}$$

Thus, the variations propagate along the vehicles at the desired speed V_0. Consequently, there are no interactions between the vehicles: Otherwise, drivers would react to the behavior of the leading vehicles and perturbations would propagate upstream relative to the vehicle motion, i.e., $c_{\text{free}} < v_0$. This constant propagation velocity is also true for larger perturbations and discontinuous jumps $\rho_1 \to \rho_2$ as long as the densities ρ_1 and ρ_2 are both smaller than the density ρ_C at capacity: $c_{12} = V_0$ if $\rho_1 \le \rho_C$ and $\rho_2 \le \rho_C$.

---

[10] The parameter T is not to be confused with the reaction time of microscopic models (see Sect. 12.2 below) even though the microscopic equivalent of the section-based model exhibits this reaction time in congested situations (see Sect. 8.5.4).

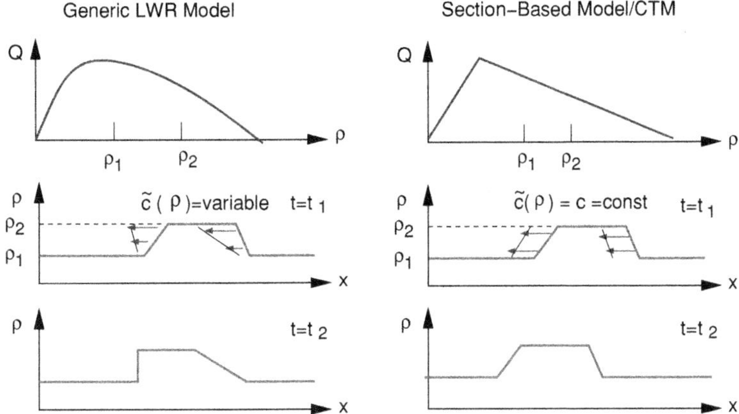

Fig. 8.11 Propagation of perturbations in a generic LWR model (*left*) and the special case of a section-based model/CTM (*right*). Each column shows the fundamental diagram (*top*), an initial density profile in congested traffic at time t_1 (*middle*), and the density profile at a later time $t_2 > t_1$ (*bottom*). The perturbations propagate with constant velocity leading to an immutable form of the transitions in the section-based model/CTM, only

Constant propagation velocity of density variations in congested traffic. In congested traffic, the propagation velocities of flow, density, or speed variations are given by the falling slope of the fundamental diagram:

$$c_{\text{cong}} = c = \left.\frac{\mathrm{d}Q_e}{\mathrm{d}\rho}\right|_{\rho>\rho_C} = -\frac{l_{\text{eff}}}{T} = -\frac{1}{\rho_{\max}T}. \tag{8.18}$$

The same is true for discontinuities if the density on each side of the jump is larger than ρ_C: $c_{12} = c$ if $\rho_1 > \rho_C$ and $\rho_2 > \rho_C$. The negative value of c means that speed and density variations propagate backwards, not only relative to the other vehicles but also for a stationary observer (cf. Fig. 8.11).

Equation (8.18) explains the main influencing factors of the propagation velocity: Perturbations propagate by one effective vehicle length per time gap The absolute value $|c|$ is about 14–16 km/h in Europe, and 18–20 km/h in the United States.

Why is the propagation velocity c of density variations in congested traffic slightly more negative in the U.S. than it is in Europe? Consider differences between typical vehicles used in these regions.

Propagation velocities of transitions from free to congested traffic. Let us denote the free traffic state upstream of the transition by the index 1 and the congested traffic state by index 2. Then, the *upstream jam front* propagates with the velocity

$$c_{\rm up} = c_{12} = \frac{Q_2 - Q_1}{\rho_2 - \rho_1} = \frac{Q_2 - Q_1}{\rho_{\rm max}(1 - Q_2 T) - Q_1/V_0}. \tag{8.19}$$

This velocity can take on any value between c and V_0 and thus the front may move in either direction (growing or shrinking traffic jams).

Constant propagation velocity of downstream jam fronts. First we notice that, for all LWR models, the outflow at the downstream end of a moving traffic jam is characterized by the *maximum-flow state* $(\rho_{\rm C}, Q_{\rm max})$. For a triangular fundamental diagram, the maximum-flow state corresponds to the top of the triangle, so the downstream front propagates at the *same* velocity as small variations within the jam,

$$c_{\rm down} = c = -\frac{1}{\rho_{\rm max} T} = -\frac{l_{\rm eff}}{T}. \tag{8.20}$$

If there is a bottleneck, the downstream front gets stationary and pinned at the bottleneck, and the outflow is reduced to the capacity of the bottleneck (see Sect. 8.5.6). We may summarize the propagation properties of LWR models with a triangular fundamental diagram by following statements:

> There are only two distinct propagation velocities of small perturbations which are given by $c_{\rm free} = V_0$ and $c_{\rm cong} = c = -l_{\rm eff}/T$ for free and congested traffic, respectively. The propagation velocities of discontinuous upstream or downstream fronts can be given analytically and lie in the range $[c, V_0]$. Downstream fronts are either stationary (at bottlenecks), or move upstream at velocity c (homogeneous roads). In contrast to all other first-order models, neither front shows dispersion.

8.5.3 Model Formulation with Measurable Quantities

In Eq. (8.11), we formulated the triangular fundamental diagram in terms of parameters directly describing properties of the vehicles and aspects of driving behavior: Mean desired speed V_0 in free traffic, mean desired time gap T in car-following situations, and mean effective vehicle length $l_{\rm eff} = 1/\rho_{\rm max}$. However, we cannot directly determine the time gap and the effective length with the most commonly available aggregated stationary detector data. On the other hand, as discussed in Sects. 3.3.1 and 4.4, indirectly estimating T and $l_{\rm eff}$ by linear regression techniques of the congested flow-density data leads to large and uncontrollable systematic errors. In order to get a formulation in which the model can be better calibrated by stationary detector data, i.e., its parameters can be estimated, we make use of the relations (8.18) and (8.15) to express T and $l_{\rm eff}$ by the more easily accessible quantities effective capacity

$Q_{\max}$ per lane, and propagation velocity c in congested traffic:

$$\frac{1}{l_{\text{eff}}} = \rho_{\max} = Q_{\max}\left(\frac{1}{V_0} - \frac{1}{c}\right), \quad T = \frac{1}{Q_{\max}\left(1 - \frac{c}{V_0}\right)}. \tag{8.21}$$

We can approximatively estimate $Q_{\max}$ by the outflow of moving traffic waves.[11] Notice that we cannot use the outflow at stationary fronts since these are always connected with a bottleneck, and the outflow indicates the capacity of the bottleneck rather than the capacity at homogeneous sections. Alternatively, $Q_{\max}$ can be estimated by the maximum average flow observed over sufficiently long periods (10–30 min).[12] To estimate the propagation velocity c, we determine the time interval which is needed by the downstream front of a moving traffic wave or another identifiable structure to pass two consecutive stationary detectors.

Using the new parameters, we can express the triangular fundamental diagram by

$$\boxed{Q_e(\rho) = \begin{cases} V_0\rho & \text{if } \rho \le \frac{Q_{\max}}{V_0} \text{ (free flow)}, \\ Q_{\max}\left[1 - \frac{c}{V_0}\right] + c\rho & \text{if } \rho > \frac{Q_{\max}}{V_0} \text{ (congested flow)}. \end{cases}} \tag{8.22}$$

When applied to model-based traffic-state estimation, the CTM or the section-based model will be used in this formulation.

8.5.4 Relation to Car-Following Models

By using the definitions of the density (inverse distance headway between two vehicles) and the vehicle speed (slope of the tangent of the trajectory), we can generate trajectories $x_i(t)$ of individual vehicles i reflecting the macroscopic density and local speed fields (cf. Fig. 8.12). Assuming that the trajectory of vehicle i crosses the location x_0 at time t_0, the trajectories are defined by

$$x_i(t_0) = x_0, \quad \int\limits_{x_{i+1}(t_0)}^{x_i(t_0)} \rho(x, t_0)\mathrm{d}x = 1, \quad v_i(t) = \frac{\mathrm{d}x_i}{\mathrm{d}t} = V(x_i(t), t). \tag{8.23}$$

This *macro-micro relation* is valid for arbitrary time-continuous macroscopic models. So, providing a single point (x_0, t_0) crossed by a trajectory defines *all* trajectories

[11] Strictly speaking, this determines the dynamic capacity which differs from the static capacity by the capacity drop. However, LWR models do not describe the phenomenon of a capacity drop.

[12] Since the evolution of a traffic breakdown takes 10 min and more, higher flows are possible over shorter time periods. However, LWR models cannot describe traffic instabilities, so these flow peaks are irrelevant for this model class.

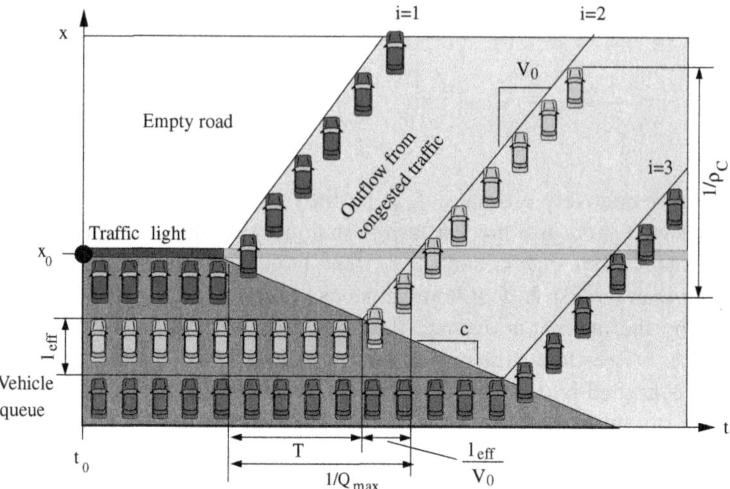

Fig. 8.12 Visualization of the macro-micro relation (8.23) for the section-based model and a situation where a waiting queue behind a *red* traffic light dissolves as the light turns *green*. Shown are the density field as solution of the section-based model (*shaded*; *the darker*, the higher the density), and the trajectories generated for the "initial condition" $x_1(t_0) = x_0$ (*black bullet*)

for *all* locations at *all* times (including the past) within the spatiotemporal interval of the macroscopic data.

In the case of the section-based model and congested traffic ($\rho(x,t) \geq \rho_C$), these generated trajectories are the same as those produced by *Newell's model*, a car-following model discussed in Sect. 10.8:

$$v_i(t+T) = v_{i-1}(t). \tag{8.24}$$

This equation describes a behavior where the driver of vehicle i exactly copies the speed profile of his or her leader $i-1$ at a constant delay T which is equal to the local time gap (cf. Fig. 8.12). In this sense, the time gap parameter T may be interpreted as a reaction time. However, this is only true within congested traffic or for the maximum-flow state. For example, in Fig. 8.12, the first vehicle instantly accelerates to its desired speed without any reaction time, after the traffic signal switches to green.

Equation (8.24) can be derived by keeping in mind that, in congested traffic, the section-based model describes density and speed variations propagating with constant velocity c, so $V(x,t) = V(x-ct,0)$ for all times t. According to Eq. (8.23), the speed profile of the leading vehicle $i-1$ is given by the local speed field $V(x,t)$ along the trajectory:

$$v_{i-1}(t) = V(x_{i-1}(t), t) = V(x_{i-1}(t) - ct,\ 0). \tag{8.25}$$

Furthermore, by means of the triangular fundamental diagram, the time gap of vehicle i to the leading vehicle $i-1$ is given by T for congested conditions ($\rho > \rho_C$) as required here. Thus, the position of vehicle i at time t is $x_i(t) = x_{i-1}(t) - v_{i-1}T - l_{\text{eff}} = x_{i-1}(t-T) - l_{\text{eff}}$ for any time. Then, its speed profile is given by

$$\begin{aligned} v_i(t) &= V(x_i(t), t) = V\,(x_i(t) - ct,\ 0) \\ &= V(x_{i-1}\,(t-T) - l_{\text{eff}} - ct,\ 0) \\ &= V\left(x_{i-1}(t-T) - c\left(t + \frac{l_{\text{eff}}}{c}\right),\ 0\right) \\ &= V\,(x_{i-1}(t-T) - c(t-T),\ 0) = V\,(x_{i-1}(t-T),\ t-T) \\ &= v_{i-1}(t-T), \end{aligned} \tag{8.26}$$

i.e., by Eq. (8.24). In the previous-to-last line, we made use of the relation $c = -l_{\text{eff}}/T$ for the propagation velocity in congested situations.

8.5.5 Definition of Road Sections

Both in the cell-transmission and section-based models, a unidirectional road stretch is divided into several sections or cells k. We assume that the index k increases in downstream direction.

8.5.5.1 Partitioning for the Section-Based Model

When defining the sections of the section-based model, we keep in mind that, with the exception of the propagation of the upstream front, *all* propagation velocities of continuous or discontinuous changes in free or congested traffic assume one of only three values, namely c, V_0, or zero. Thus, the only propagation requiring a numerical update is that of the upstream jam fronts. We define the sections such that, in each section, there is at most one upstream jam front:

> A *road section* k of the section-based model is a homogeneous segment of length $L^{(k)}$ and capacity $C^{(k)}$ (over all lanes) with an inhomogeneity or *bottleneck* of capacity $C_B^{(k)}(t) \le C^{(k)}$ at its downstream end. A bottleneck is characterized by a transition to a lower capacity. A general road inhomogeneity can also represent a transition to higher capacity. In this case, two homogeneous road segments are just joined together.

We emphasize that we define a bottleneck by the *transition* zone to a lower capacity, not by the region of lower local capacity itself. For example, if the bottleneck is caused by road construction, the bottleneck is not the construction zone itself (which constitutes the next homogeneous road segment) but it is located at the *beginning* of the construction zone. By this partitioning, stationary downstream jam fronts can only be found at the downstream boundary of a road section.[13]

8.5.5.2 Partitioning for the Cell-Transmission Model

When simulating with the *a priori* discretized CTM, one generally uses cells of equal length Δx which are generally shorter (100–500 m) than the segments of the section-based model (whose length may be 5 km, or more). There is no structure in each cell of the CTM, i.e., the traffic variables ρ, V, and Q take on unique values. In determining the appropriate cell size, one must take into account that the update time step is limited by the first CFL condition (8.10) which, for the triangular fundamental diagram, takes on the form

$$\Delta t < \frac{\Delta x}{V_0}. \tag{8.27}$$

8.5.6 *Modeling Bottlenecks*

The most important types of bottlenecks are sketched in Fig. 8.13. The bottlenecks $k = 1, 2, 5$, and 6 are so-called *flow-conserving bottlenecks* while the on-ramp and off-ramp bottlenecks contain additional sources and sinks.

- When passing flow-conserving bottlenecks all vehicles pass from road section k to $k+1$, no vehicle leaves section k or enters section $k+1$ differently. From the viewpoint of an economist, flow-conserving bottlenecks represent a change in the *supply* (local road capacity).
- Non-flow-conserving bottlenecks contain sources and sinks constituted by on-ramps, off-ramps, junctions, interchanges, intersections, and similar. From the viewpoint of an economist, non-flow-conserving bottlenecks represent a change in the traffic *demand*.

We can also distinguish the bottlenecks whether they are permanent on the time scales of typical simulations (bottlenecks 1–4 of Fig. 8.13), or temporary such as obstructions caused by accidents, or, on even shorter time scales, by traffic lights.[14]

Regardless of the type of bottleneck, the influence on traffic flow can be characterized by a single property, at least within the class of LWR models: The bottleneck strength (cf. Chap. 18). We state:

---

[13] This agrees with most observations: Jams are observed at the begin or upstream of a bottleneck but rarely within the zone of reduced capacity.

[14] In a more coarse-grained picture, one would model a signalized intersection by a permanent bottleneck whose capacity is given by the maximum number of passing vehicles per signal cycle.

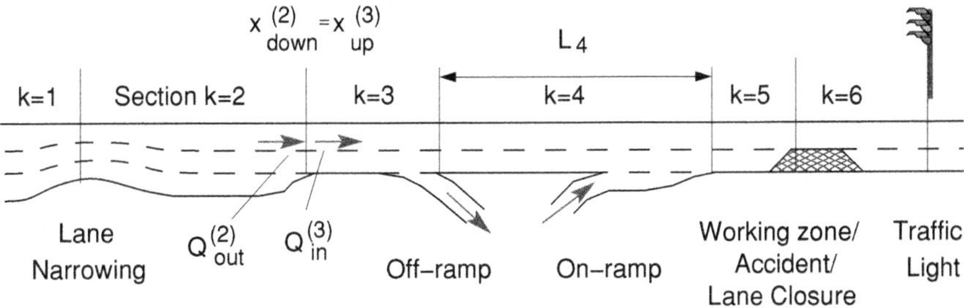

Fig. 8.13 Definition of the bottleneck types and the road sections of the section-based model. Each section consists of a homogeneous road stretch with a bottleneck attached downstream

> In LWR or other first-order macroscopic traffic flow models, bottlenecks are characterized by the reduction of local capacity with respect to the homogeneous upstream road section. Only a single property is relevant for traffic flow dynamics: The *effective bottleneck strength* $\Delta Q = \Delta C/I$, i.e., the capacity reduction $\Delta C = C - C_{\text{bottl}}$ divided by the number I of lanes in this section.

In the following, we describe different types of bottlenecks and determine the associated reduction ΔC of the total capacity, and the bottleneck strength $\Delta Q = \Delta C/I$.

8.5.6.1 Classical Flow-Conserving Bottlenecks

This includes all flow-conserving bottlenecks that are caused by *local changes of the road attributes* except changes in the number of lanes. Examples are uphill and downhill gradients, curves, changes of the speed limits, entering or exiting the city limits, and local narrowings, e.g., at road construction sites. Also included are local capacity changes induced by a locally changed driving behavior, e.g., at entrances and exits of tunnels, or even when passing a spectacular accident in the opposite driving direction (*behaviorally-induced bottleneck*).[15]

The defining property of this type of bottleneck is a local reduction ΔQ of the capacity per lane. In all microscopic and macroscopic traffic flow models, this type of flow-conserving bottleneck is modeled by local changes of the model parameters, i.e., by changing the average driving behavior.

In the LWR models with a triangular fundamental diagram, changing the drivers' behavior implies changing the mean desired speed V and/or the time gap T while the average vehicle length $l_{\text{eff}} = 1/\rho_{\text{max}}$, obviously, should remain unchanged.

[15] In Germany, there is anecdotic evidence that behaviorally-induced bottlenecks have caused traffic breakdowns on otherwise completely homogeneous road sections.

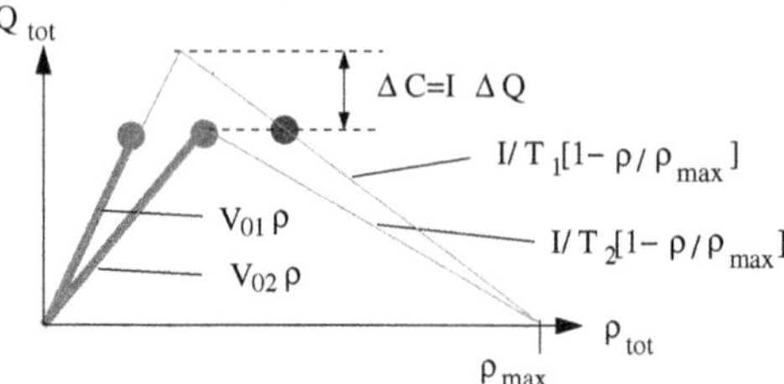

Fig. 8.14 Describing a classical flow-conserving bottleneck by changing the parameters T and V_0 of the triangular fundamental diagram. The points indicate steady-state traffic situations after breakdown (see the main text)

When describing the situation upstream and downstream of the local change with the indices 1 and 2, respectively, the bottleneck strength is given by

$$\Delta Q_{\text{cl}} = \frac{\Delta C_{\text{cl}}}{I} = \frac{V_{01}}{V_{01}T_1 + l_{\text{eff}}} - \frac{V_{02}}{V_{02}T_2 + l_{\text{eff}}}. \tag{8.28}$$

Notice that the bottleneck strength (8.28) can be positive (a true bottleneck), or negative (transition to a road segment with increased capacity). However, in the framework of first-order models, only true bottlenecks can lead to a traffic breakdown.[16]

Figure 8.14 shows the situation when simultaneously reducing the desired speed V_0 (caused, e.g., by a change of the speed limit) and increasing the mean time gap T (which may be caused by locally reduced visibility conditions near curves or inside tunnels leading to a more defensive driving style). The thin gray lines show the fundamental diagram, i.e., possible steady states of infinite homogeneous road sections corresponding to the sections inside and outside the bottleneck. The thick lines shows possible steady states of free traffic at and upstream of the bottleneck. The three bullets indicate the situation after a breakdown:

- Free traffic downstream of the zone of capacity restriction is characterized by T_1 and V_{01} (left bullet).
- Traffic in the range of the bottleneck or in downstream sections characterized by T_2 and V_{02} is characterized by the maximum-flow state (middle).
- Jammed or congested traffic upstream of the bottleneck is characterized by the right-hand side of the fundamental diagram at the bottleneck capacity (right bullet).

Notice that the fundamental diagram $Q_e^{\text{tot}}(\rho_{\text{tot}})$ for the *total* flow and density (rather than that for lane-averaged effective quantities) is most useful to describe the criteria for the breakdown while the dynamics after a breakdown has occurred is best described by effective equations, i.e., by the effective (lane-averaged) fundamental diagram. While this does not play a role here, it will be relevant for the next two types of bottlenecks.

[16] Remarkably, in models with dynamic speed (second-order macroscopic models and most microscopic models), even a local *increase* of capacity can lead to a breakdown ("more is less") which is mediated by the speed perturbations associated with any local change together with traffic flow instabilities.

8.5.6.2 Lane-Drop Bottlenecks

This type of bottleneck plays an intermediate role. From a global viewpoint over all lanes, it is a flow-conserving bottleneck (since there are no off- or on-ramps or other sources and sinks). When formulating the effective continuity equation (7.15) in terms of the lane-averaged effective flow Q and density ρ, however, the result is mathematically equivalent to the effective continuity equation in the presence of ramps.

Letting aside local changes of the driving behavior as described by the classical flow-conserving bottlenecks, the fundamental diagram for the total flow and density will be congruently inflated or deflated whenever the number I of lanes changes (cf. Fig. 8.15). Particularly, when reducing the number of lanes from I_1 to $I_2 < I_1$, the total and effective capacities are reduced by

$$\Delta C_{\text{lane drop}} = (I_1 - I_2) Q_{\text{max}}, \quad \Delta Q_{\text{lane drop}} = \frac{I_1 - I_2}{I_1} Q_{\text{max}}, \tag{8.29}$$

respectively. Again, the congestion resulting from a traffic breakdown at the bottleneck is characterized by the bottleneck capacity, only, while other features of the bottleneck (such as the length of the transition zone) are irrelevant for LWR models. Notice that, consistent with the general definition, the bottleneck is located at the *beginning* of the section with a reduced number of lanes. Particularly, the LWR models will not produce any congestion *inside* the section with a reduced number of lanes (unless a congestion caused by a bottleneck further downstream propagates into this region). This is in agreement with most observations: For example, road constructions often produce congestions upstream of the construction site while the actual roadworks zone is less congested.

Since most lane drops are connected with new speed limits and behavioral changes, one typically combines lane-drop bottlenecks with classical flow-conserving bottlenecks.

Notice that, in certain situations, the concept of lane-drop bottlenecks may be extended to *fractional* lane drops. For example, many work zones include narrow lanes with no physical separation to the opposite traffic flow which drivers avoid if possible, and which are even forbidden for certain vehicles. If such a lane can take, say, 70 % of the traffic of normal lanes, a transition from a normal to such a narrow lane would be modeled by a lane drop by 0.3 lanes.

8.5.6.3 On-Ramps

In contrast to the previous bottleneck types, on-ramp bottlenecks are caused by the increasing demand of the merging vehicles (additional flow Q_{rmp}) rather than by a decreasing supply (local capacity reduction by ΔC). If we assume that all on-ramp vehicles can merge to the main-road regardless of traffic conditions, the main-road traffic flow upstream of the on-ramp is restricted by the capacity C of all the

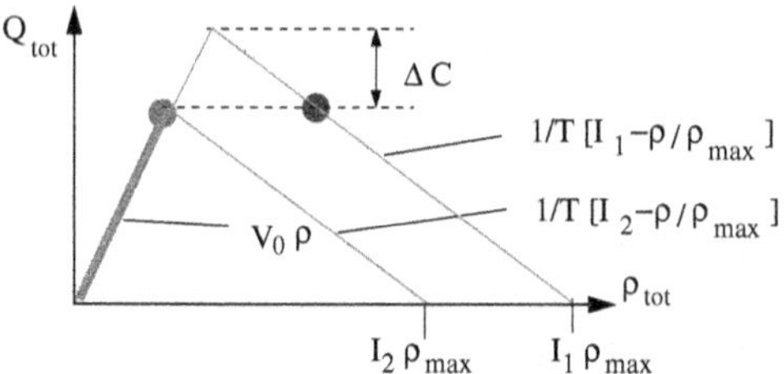

Fig. 8.15 Change of the fundamental diagram when reducing the number of lanes from I_1 to I_2. The bullets denote the steady-state situation downstream (*left*) and upstream (*right*) the lane drop after the congestion has formed

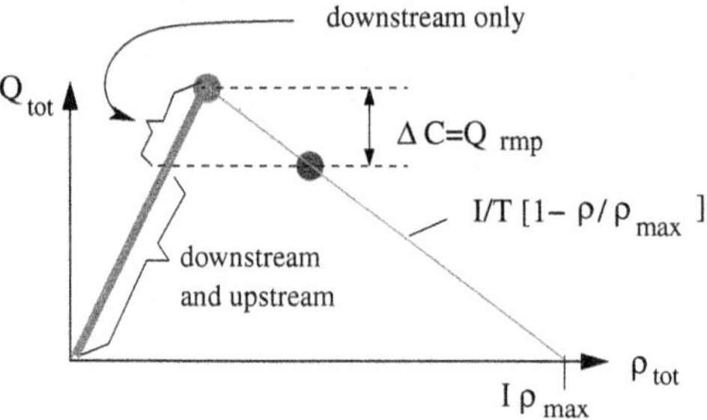

Fig. 8.16 Modeling an on-ramp bottleneck. Upstream the ramp, the sustainable main-road traffic flow is restricted by $C - Q_{\rm rmp}$

main-road lanes at the location of the on-ramp minus the ramp flow $Q_{\rm rmp}$. Thus, we can describe the effective reduction of the capacity ΔC and the bottleneck strength ΔQ by

$$\Delta C_{\rm rmp} = Q_{\rm rmp}, \quad \Delta Q_{\rm rmp} = \frac{Q_{\rm rmp}}{I} \tag{8.30}$$

where I is the number of main-road lanes at and upstream of the on-ramp (cf. Fig. 8.16). If the on-ramp consists of more than one lane, the on-ramp flow is to be understood as the total flow over all ramp lanes. Notice that, in contrast to the flow-conserving bottlenecks, the bottleneck strength of on-ramp bottlenecks is generally time-dependent.

As for the other bottleneck types, the traffic flow immediately downstream of the bottleneck is given by the maximum-flow state at this location (left bullet in Fig. 8.16). Similarly to the situation at a lane-drop bottleneck, an on-ramp bottleneck will often be combined with a classical flow-conserving bottleneck representing a change in the driving behavior.

8.5.6.4 Off-Ramps

At first sight, off-ramps (or lane additions) do not constitute bottlenecks. After all, off-ramps or additional lanes reduce the traffic demand per lane on the main-road, i.e., the bottleneck strength is negative. However, near these inhomogeneities, the lane-changing frequency and the speed variance between the vehicles, and thus perturbations, increase. Moreover, lanes are used less efficiently. So, even off-ramps or lane additions can constitute a bottleneck. This explains why one regularly observes congestions behind off-ramps.[17]

This situation, however, brings us to the limits of first-order macroscopic models: Capacity reductions at off-ramps or near lane additions cannot be described by LWR models without resorting to additional *ad-hoc* assumptions. A straightforward approach would be to increase the time-gap parameter T near the off-ramp (in regions with increased speed variances, the non-equilibrium time gaps are, on average, greater than that at steady state), or to introduce a fractional lane drop (less effective lane usage). Thus, off-ramps are modeled by effective flow-conservative bottlenecks immediately upstream of the diverge.

8.5.6.5 Traffic Lights

In the framework of macroscopic traffic flow models, traffic lights are modeled by a time-dependent flow-conserving bottleneck: During the red (and part of the yellow) phases, the bottleneck capacity is equal to zero while the bottleneck is nonexistent, otherwise. This means, the time-dependent bottleneck strength of a lane controlled by a traffic light is given by

$$\Delta Q_{\text{TL}} = \begin{cases} Q_{\max} & \text{red/yellow,} \\ 0 & \text{green.} \end{cases} \tag{8.31}$$

8.5.7 Numerical Solution of the Cell-Transmission Model

8.5.7.1 Single Roads

As for the other LWR models, the appropriate numerical integration method for the cell-transmission model (CTM) is the explicit Godunov scheme (see Sect. 8.4). Because of the special properties of this model (constant propagation velocities and no dispersion), the exact solution of the Godunov scheme in each update step is given simply by displacing the stepwise initial conditions. Depending on the density of each cell, the initial density jump between the cells propagates into the upstream cell (propagation velocity $c_{12} < 0$) or downstream ($c_{12} > 0$), so we need a case distinction. Summing up all steps of the Godunov scheme for the triangular fundamental diagram leads to the simple and intuitive *supply-demand method*: If supply is the limiting factor, information travels upstream. If demand limits the traffic flow, information travels downstream.

---

[17] Part of off-ramp induced congestions, however, have a more trivial reason: If the off-ramp *itself* is congested (caused, e.g., by an insufficient capacity on the secondary road network at junctions or by congestion on the target highway at interchanges), the off-ramp queue can *spill over* and obstruct a lane of the original road.

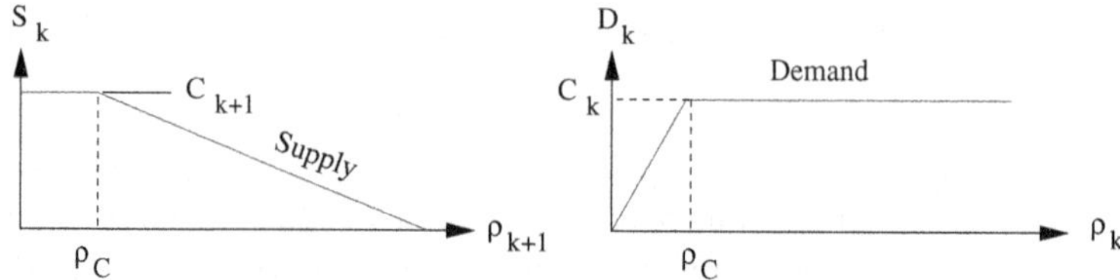

Fig. 8.17 Supply and demand functions as a function of the cell densities

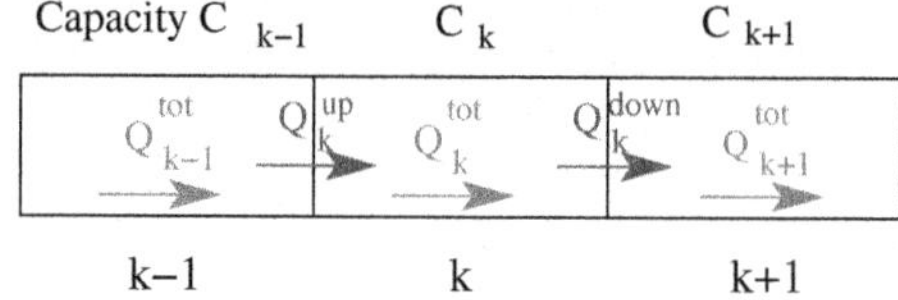

Fig. 8.18 Cells of the CTM for a simple straight road and definition of the relevant quantities for the supply-demand method

The steps for the update rule of the supply-demand method for cell k are described in the following (cf. the Figs. 8.17 and 8.18).

1. Determine the supply and demand at the cell boundaries. We first consider the downstream boundary, i.e., the transition from cell k to $k+1$. The *supply* S_{k+1} is given by the maximum flow the receiving cell $k+1$ can accommodate. In any case, it is restricted by the capacity C_{k+1} of this cell. However, in case of congested traffic, it is further restricted to the actual flow of this cell. This gives rise to the supply function (cf. Fig. 8.17)

$$S_{k+1}(t) = \begin{cases} Q_{k+1}(t)^{\text{tot}} & \text{cell } k+1 \text{ congested}, \rho_{k+1}(t) > \rho_C, \\ C_{k+1} & \text{otherwise.} \end{cases} \tag{8.32}$$

The *demand* D_k is given by the (potential) outflow from cell k at its downstream boundary if there are no restrictions on the supply side. Obviously, for free traffic, it is equal to the actual flow in this cell. If cell k is congested, we use the fact that, in LWR models, the outflow from congested zones always takes on the maximum value:

$$D_k = \begin{cases} Q_k^{\text{tot}} & \text{free traffic in cell } k, \rho_k \le \rho_C, \\ C_k & \text{otherwise.} \end{cases} \tag{8.33}$$

The supply S_k and demand D_{k-1} at the upstream boundary of cell k are defined analogously by replacing k with $k-1$.

2. Determine the flows through the cell boundaries. The mechanism is identical to that in trading: Supply must meet demand, i.e., the number of traded goods cannot be greater than either demand or supply. At the downstream boundary, the supply (receiving potential of cell $k+1$) must meet the demand (sending potential of cell k) while, at the upstream boundary, cell k provides the supply and cell $k-1$ the demand. This gives rise to following boundary flows:

$$Q_k^{\text{up}} = Q_{k-1}^{\text{down}} = \min\left(S_k, D_{k-1}\right), \tag{8.34}$$

$$Q_k^{\text{down}} = Q_{k+1}^{\text{up}} = \min\left(S_{k+1}, D_k\right). \tag{8.35}$$

3. Update the cells. The explicit, first-order update step is already given by Eq. (7.9):

$$\rho_k^{\text{tot}}(t+\Delta t) = \rho_k^{\text{tot}}(t) + \frac{1}{\Delta x_k}\left(Q_k^{\text{up}} - Q_k^{\text{down}}\right)\Delta t, \tag{8.36}$$

$$Q_k^{\text{tot}}(t+\Delta t) = I\,Q_e\left(\rho_k^{\text{tot}}(t+\Delta t)/I\right). \tag{8.37}$$

By applying the supply-demand method to the total instead of the effective lane-averaged quantities, $\rho_k^{\text{tot}} = I\rho_k$, $Q_k^{\text{tot}} = I\,Q_k$, all sorts of flow-conserving bottlenecks are taken care of by the associated capacity reductions. Non-conserving bottlenecks are discussed in the following.

8.5.7.2 Road Networks

The true potential of the supply-demand method comes into effect when modeling the road infrastructure in a whole region by a *directed network* consisting of nodes and unidirectional links. The links correspond to one direction of single (possibly inhomogeneous) roads which are treated as described above. In the following, we describe how to apply the supply-demand method to the simplest node types, namely two-to-one merges (e.g., on-ramps) and one-to-two diverges (off-ramps).

Merges. For the merging cell 3 of the geometry depicted in Fig. 8.19, we generalize Eqs. (8.32)–(8.36) as follows:

$$S_{12,3} = \begin{cases} Q_3^{\text{tot}} & \text{if } \rho_3 > \rho_C, \\ C_3 & \text{otherwise}, \end{cases} \tag{8.38}$$

$$D_{12,3} = \begin{cases} Q_1^{\text{tot}} & \text{if } \rho_1 \le \rho_C, \\ C_1 & \text{otherwise}, \end{cases} + \begin{cases} Q_2^{\text{tot}} & \text{if } \rho_2 \le \rho_C, \\ C_2 & \text{otherwise}, \end{cases} \tag{8.39}$$

$$Q_3^{\text{up}} = \min(S_{12,3}, D_{12,3}), \tag{8.40}$$

$$\rho_3^{\text{tot}}(t+\Delta t) = \rho_3^{\text{tot}} + \frac{1}{\Delta x_3}\left(Q_3^{\text{up}} - Q_3^{\text{down}}\right)\Delta t. \tag{8.41}$$

The flow Q_3^{down} at the downstream boundary of the merging cell is given by the normal supply-demand rule (8.34) for the cells 3 and 4 of Fig. 8.19.

Notice that condition (8.40) only defines the *sum* $Q_1^{\text{tot}} + Q_2^{\text{tot}} = Q_3^{\text{up}}$ through the boundaries $1 \to 2$ and $1 \to 3$ while the flows Q_1^{tot} and Q_2^{tot} themselves are not uniquely defined, at least, if the supply $S_{12,3}$ rather than the demands restrict the throughput such that jams propagate into at least one of the road segments 1 or 2. In this case, we need additional assumptions such as cutting off the excess demand from the outflow of road segment 1 (road segment 2 has priority or traffic from road

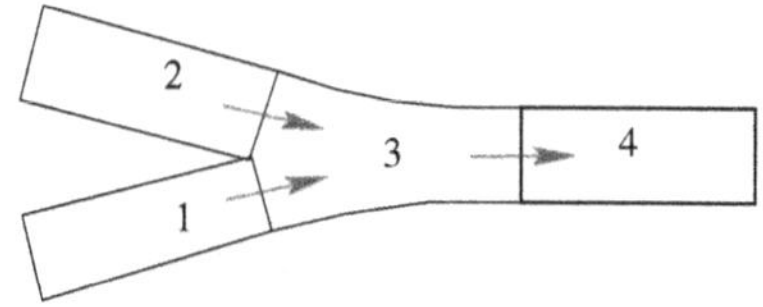

Fig. 8.19 Simple cell geometry for a merge as described by Eqs. (8.38)–(8.41)

2 forces entry), or splitting the excess demand proportionally to the capacities C_1 and C_2 (both roads have same priority).

Diverges. To describe one-to-two diverges, we need to prescribe the fraction of the traffic flow (splitting probability) entering each downstream road. As for merges, we need additional information if one or both of the downstream links are congested. This situation is nontrivial if only one of the diverging links, say, road segment 2, is congested. Drivers who want to diverge onto road segment 2 generally are not willing to use the uncongested alternative. As a consequence, they obstruct the road segment upstream of the merge and reduce the flow to the uncongested branch as well. This is also known as *gridlock phenomenon*.

Sources and sinks. The supply-demand method is applied to all cell boundaries that are connected to other cells. In addition, we need external boundary conditions in the form of *sources* and *sinks* for all but the most trivial networks (closed ring roads). Typically, these boundary conditions are given in terms of inflows, e.g., obtained from detector data or from the specification of the scenarios to be investigated. The boundary flows are integrated into the supply-demand scheme with the additional condition that the sources always represent virtual cells with free traffic. The sinks are generally represented by a virtual cell of infinite capacity just sucking out the complete demand of the last real cell. However, sinks may also be driven by external flow data which, then, represent the time-dependent capacity of the virtual sink cell, so they can cause congestions.

8.5.8 Solving the Section-Based Model

Following properties are the basis for the efficiency when analytically solving or numerically simulating the section-based model:

1. Inside regions of free and congested traffic, density and speed variations propagate uniformly at velocities V_0 and $c = -l_{\text{eff}}/T$, respectively (cf. Fig. 8.11). Since $V_0 > 0$ and $c < 0$, free and congested traffic are exclusively controlled by the upstream and downstream boundaries of a road section, respectively.
2. A possibly existing transition free → congested traffic is always discontinuous, i.e., a shock front propagating according to the shock-wave formula (8.9). The propagation velocity can assume values in the range between c and V_0, i.e., the upstream jam front can propagate in both the upstream and downstream direction.

3. The transition congested $\rightarrow$ free is either stationary and fixed at a bottleneck, i.e., a downstream boundary of a road section, or it propagates uniformly upstream at velocity c. However, since the outflow from moving jams of LWR models always represents the maximum-flow state (which, in our case, is given by the top of the triangular fundamental diagram), we can formally attribute the outflow to a form of congested traffic.

This can be summarized by following rule for the internal structure of road sections:

There is at most a single transition free $\rightarrow$ congested traffic (*upstream jam front*) in each road section. Transitions from congested $\rightarrow$ free traffic (*downstream fronts*) are always located at the downstream boundaries of the road sections. Thus, each road section can assume exactly one of three qualitative states: (i) completely free, (ii) completely congested, (iii) partly congested with free (congested) traffic in the upstream (downstream) parts and a transition between these states at $x_k^*(t)$. Other options (such as several jams or transitions congested $\rightarrow$ free inside of sections) *are nonexistent.*

8.5.8.1 Dynamics Within a Road Section

From the above considerations it follows that we can describe the complete spatiotemporal dynamics inside each partly congested road section by an ordinary integral for the upstream jam front while completely free or jammed sections are updated by simply translating the initial state. Explicitly solving the continuity equation (i.e., a *partial* differential equation) is not necessary. In formulating update rules, we assume that the inflow $Q_{\text{in}}(t)$ at the upstream boundary (located at x_{up}) and the outflow $Q_{\text{out}}(t)$ at the downstream boundary (at x_{down}) are externally given in this stage (cf. Fig. 8.13). For ease of notation, we omit the cell index k.

Qualitative state 1: Free traffic on the whole road section. Only the upstream boundary flow is relevant in this situation. The flow field and the associated density (8.13) propagate with the vehicles:

$$\begin{aligned} Q_{\text{free}}(x,t) &= \frac{1}{I} Q_{\text{in}}(t'), \quad t' = t - \frac{x - x_{\text{up}}}{V_0}, \\ \rho_{\text{free}}(x,t) &= \rho_{\text{free}}\left(Q_{\text{free}}(x,t)\right). \end{aligned} \tag{8.42}$$

Qualitative state 2: Congested traffic. We treat this case analogously but take the downstream boundary, the propagation velocity c and the congested branch (8.14) of the fundamental diagram for calculating the density:

$$Q_{\text{cong}}(x,t) = \frac{1}{I} Q_{\text{out}}(t') = \frac{1}{I} C_{\text{B}}(t'), \quad t' = t - \frac{x_{\text{down}} - x}{|c|},$$
$$\rho_{\text{cong}}(x,t) = \rho_{\text{cong}}\left(Q_{\text{cong}}(x,t)\right). \tag{8.43}$$

Qualitative state 3: Partly congested traffic. Here, both boundary conditions are relevant, and the density and flow fields of the free and congested regions of the cell are calculated by Eqs. (8.42) and (8.43), respectively. Additionally, we need an equation for the time-dependent location $x^*(t)$ of the upstream jam front. With Eq. (8.19), we obtain

$$\boxed{\frac{\mathrm{d}x^*}{\mathrm{d}t} = \frac{Q_2^* - Q_1^*}{\rho_{\text{cong}}(Q_2^*) - \rho_{\text{free}}(Q_1^*)} = \frac{Q_2^* - Q_1^*}{\rho_{\max}(1 - Q_2^* T) - Q_1^*/V_0}.} \tag{8.44}$$

Notice that, by means of the propagation rules (8.42) and (8.43), the equation of motion (8.44) for the location of the front only depends on *past* values of the boundary flows:

$$Q_1^* = Q_{\text{free}}(x^*,t) = \frac{Q_{\text{in}}}{I}\left(t - \frac{x^* - x_{\text{up}}}{V_0}\right), \tag{8.45}$$
$$Q_2^* = Q_{\text{cong}}(x^*,t) = \frac{C_{\text{B}}}{I}\left(t - \frac{x_{\text{down}} - x^*}{|c|}\right)$$
$$= \frac{C_{\text{B}}}{I}\left(t - \rho_{\max} T \left(x_{\text{down}} - x^*\right)\right). \tag{8.46}$$

Since the boundary conditions can be taken from stationary detector measurements and only past values are required, we can use the equation of motion (8.44) for traffic-state estimation and short-term prediction.

In many cases, the detectors not only provide flow but also aggregated speed values. Why should one, nevertheless, only use the flows in calculating Eq. (8.44) (leaving the speed information only for the qualitative distinction of free or congested) although using the speed would eliminate the necessity to estimate the parameters of the fundamental diagram?

Figure 8.20 gives an example of the mechanism behind Eq. (8.44) for a road section defined by the boundaries $x_{\text{up}} = 0$ and $x_{\text{down}} = L$ with a variable demand $Q_{\text{in}}(t)$ (which may represent a rush hour) and a supply represented by the bottleneck capacity $C_{\text{B}} = C - \Delta C$ at the downstream boundary. We add an additional time dependence

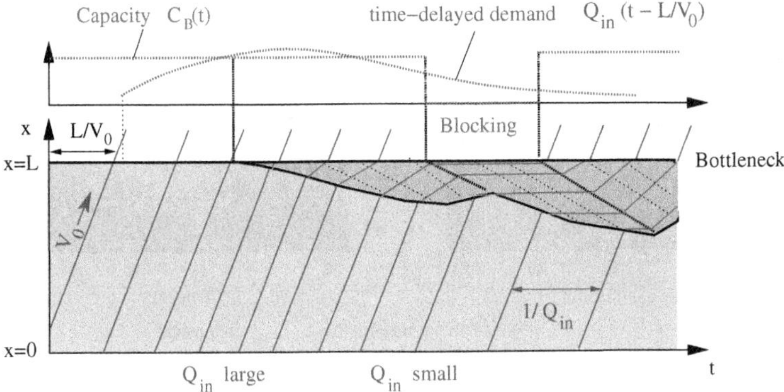

Fig. 8.20 Schematic traffic flow dynamics inside a partly congested road section of length L as described by the section-based model. Free and congested traffic is painted in *blue* and *red*, respectively. Both the upstream and downstream boundary flows (*dotted lines* in the upper diagram) are time-dependent. For visualization, the graphics also shows some trajectories (*blue* and *red solid lines* in the lower diagram) and the propagation of density changes (*dotted lines*). A jam forms where the *blue line* of the upper diagram intersects the *red line* from below

by temporarily blocking the passage ($C_B = 0$) which may represent an obstruction caused by an accident.

As soon as the delayed demand (delay time L/V_0) exceeds the bottleneck capacity C_B, a congestion with associated moving front $x^*(t)$ forms (represented by the boundary between the blue and red regions in Fig. 8.20). The propagation of the jam front is then calculated by Eq. (8.44). After some time, the delayed demand decreases while the supply C_B is constant so that the jam front reverts its direction. However, as soon as the information about the total blockage ($C_B = 0$) has propagated to the location x^* of the upstream front, the jam increases once more. Finally, after the blocking has been removed and this information has propagated to the jam front, the latter reverts its direction for the last time and the jam dissolves at the time where $x^*(t) = L$.

Figure 8.21 demonstrates how to apply Eq. (8.44) to model-based traffic-state estimation or short-term prediction (see Sect. 5.2 for an offline method to reconstruct the spatiotemporal speed fields). We first notice that the parallel structures of the local speed fields in both the free and congested regions indicate constant propagation velocities in these regions agreeing with the implications of a triangular fundamental diagram. In this figure, we see how the propagation of the jam fronts (white lines) according to Eq. (8.44) reflects the actual situation, at least, if the three parameters of the model are correctly estimated.

It is crucial for potential real-time applications that this approach only requires detector information from the past. For example, in the situation depicted in the top left graphics of Fig. 8.21, the necessary downstream information from the detector at $x = 477.5$ km lies in the past by up to 40 min, and the required upstream data lie in the past as well. In the situations depicted in the top left graphics of Fig. 8.21, we also

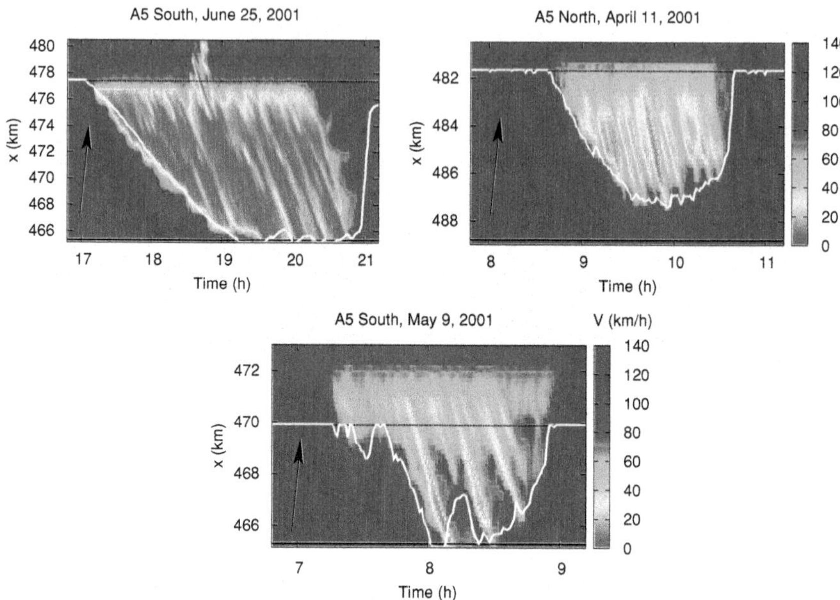

Fig. 8.21 Simulation of the jam front dynamics in Eq. (8.44) (*white lines*) for real data from the German Autobahn A5. Shown is a situation with a temporary accident-induced bottleneck (*top left*) and two situations with a permanent bottleneck (*top right* and *bottom*). The upstream and downstream boundary conditions have been taken from stationary detectors positioned at the *black horizontal lines*. The local speed field only serves for visualization and calibration purposes but has not been used directly. The calibrated parameters of the triangular fundamental diagram are $V_0 = 100\,\text{km/h}$, $T = 2\,\text{s}$ and $\rho_{\text{max}} = 100$ vehicles/h (*upper row*), and $V_0 = 100\,\text{km/h}$, $T = 1.5\,\text{s}$ and $\rho_{\text{max}} = 80$ vehicles/h lower graphics)

see that the propagation of the jam front as described by Eqs. (8.44) and (8.45) works for the maximum-flow state (around 21:00 h or 9:00 pm, respectively) as well. This justifies the formal attribution of the maximum-flow state to the regime of congested traffic.

8.5.8.2 Dynamics at the Boundaries

The boundary conditions of each road section mediate the coupling between neighboring road sections and provide the boundary flows $Q_{\text{in}}(t)$ and $Q_{\text{out}}(t)$ assumed as given up to now.

The basic dynamics is the same as for the cells of the CTM, i.e., given by the supply-demand method with additional data-driven sources and sinks. Particularly, the outflow $Q_{\text{out}}^{(k)}$ from road section k is equal to the inflow $Q_{\text{in}}^{(k+1)}$ of the downstream road section.

However, in contrast to the CTM, the elements of the section-based model are generally structured and can assume a third state "partially congested" in addition to the states "free" and "congested" of the CTM cells. This leads to following additions to the supply-demand method:

1. If $D_k(t) > S_{k+1}(t)$ and section k is in the state "free", the new state of this section changes to "partially congested". The position of the jam front x_k^* is given by integrating Eq. (8.44) with the initial condition x_k^{down}.[18]
2. If $D_{k-1}(t) < S_k(t)$ and section k is in the state "congested", the new state of this section changes to "partially congested". The position of the jam front x_k^* is given by integrating Eq. (8.44) with the initial condition x_k^{up}.
3. If section k is "partially congested" and the jam front x^* reaches the upstream boundary, this cell becomes "congested".
4. If section k is "partially congested" and the jam front x^* reaches the downstream boundary, this cell becomes "free".

8.5.9 Examples

The compact formulation (8.42)–(8.44) of the solutions to the section-based model allows for an instructive insight into the dynamics of traffic jams in realistic open systems in the presence of bottlenecks.

The general situation of Fig. 8.22 shows three road sections. The beginning of road section 2 represents the bottleneck. The reduced local capacity $C_2 = C_\text{B} = C_1 - \Delta C$ of this road section can figuratively be interpreted as a *capacity hole*. Shown is the situation where road section 1 is in the partially congested state, section 2 is in the maximum-flow state, and section 3 is completely free. This means, there are *four* traffic states denoted by ①–④ in this figure. Furthermore, we show a situation in this figure where the demand Q_1 exceeds the supply Q_2 which is equal to the jam flow,[19] so the gradient of the secant connecting the states ① and ② is negative corresponding to an upstream moving jam front x_{12}^*. Traffic state ③ corresponds to the maximum-flow state of road section 2, and state ④ to the free flow of section 3. Since, by the supply-demand conditions, the flows $Q_2 = Q_3 = Q_4$ of the states ② to ④ are equal, the locations of the corresponding transitions are stationary corresponding to horizontal secants in the figure. In the following, we discuss the dynamics by examples of increasing complexity.

---

[18] In order to arrive at this rule, we must require the first CFL condition (8.27) to be satisfied. Since the road sections generally are several kilometer long, this rarely is an issue.

[19] Here and in the whole subsection, the indices of flows and densities refer to the *traffic states*, and not to the road sections. Specifically, in Fig. 8.22, we have three road sections and four traffic states.

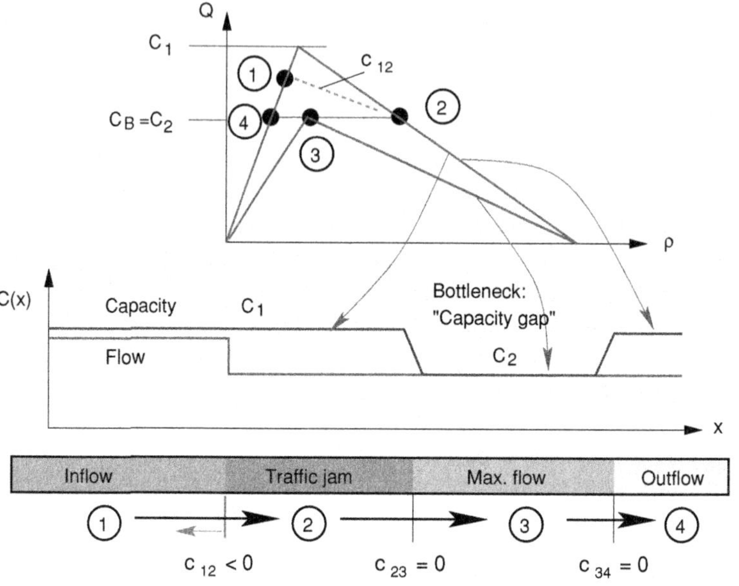

Fig. 8.22 Evolution of the traffic state for the section-based model for three road sections. The capacity of section 2 is assumed to be lower than that of the other sections, so the beginning of this section constitutes a flow-conserving bottleneck. Vehicles pass, in this order, the states ① to ④

8.5.9.1 Congestion at a Traffic Light

We consider a situation with a constant inflow $Q(0,t) = Q_{\text{in}} < C$ at the upstream boundary of a road section. The downstream boundary at $x = L$ is controlled by a traffic light. The traffic light is red in the period $t \in [t_1, t_2]$ and green, otherwise.[20] The situation is modeled by two road sections 1 and 2 upstream and downstream of the traffic light, respectively. Initially, we assume free traffic everywhere. Furthermore, we assume free traffic in section 2 at any time, i.e. no queues from traffic lights further downstream "spill over" to the location of the considered traffic light. This means, the supply $S(t)$ at the transition between the sections 1 and 2 is determined solely by the time-dependent maximum throughput $C_{\text{TL}}(t)$ allowed by the traffic light (cf. Fig. 8.23):

$$S(t) = C_{\text{TL}}(t) = \begin{cases} 0 & \text{if } t_1 < t < t_2 \text{ (light is red)}, \\ C & \text{otherwise.} \end{cases} \tag{8.47}$$

[20] Yellow/amber phases are associated to the red or green phases as appropriate for the respective country and average driving behavior.

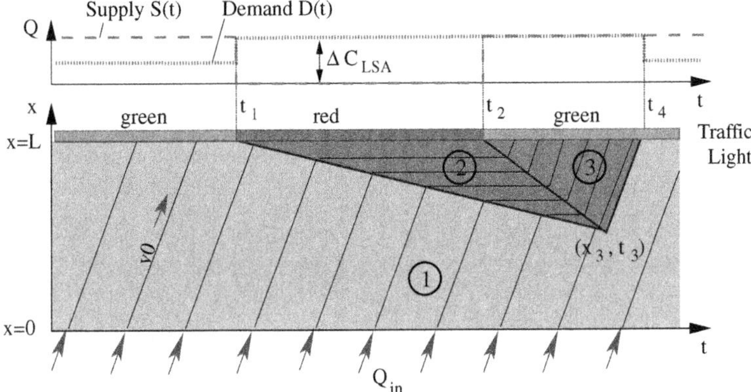

Fig. 8.23 Modeling a queue behind a single traffic light. The *upper part* shows the time-dependent supply and demand at the section boundary formed by the traffic light. The *lower part* shows the spatiotemporal evolution of free traffic (region ①), the waiting queue (region ②), and the maximum-flow state (region ③). Some vehicle trajectories (*black lines*) are drawn for purposes of visualization

Phase 1: Free traffic. For $t < t_1$ the traffic light is green and the supply $S = C$ is greater than the demand $D = Q_{\text{in}}$. Consequently, free traffic persists.

Phase 2: Formation of a queue. For $t_1 \le t < t_2$, the traffic light is red, i.e., the supply $S = C_{\text{B}} = 0$. Consequently, section 1 gets partially congested. The motion (8.44) of the upstream boundary of the queue (transition between the traffic states ① and ②) is given by

$$\frac{\mathrm{d}x^*}{\mathrm{d}t} = c_{12} = \frac{-Q_{\text{in}}}{\rho_{\max} - \frac{Q_{\text{in}}}{V_0}},$$

resulting, after integration, in $x^*(t) = L + c_{12}(t - t_1)$. In this phase, the propagation velocity can take on values between $c_{12} = -l_{\text{eff}}/T = c$ (for the maximum possible inflow $Q_{\text{in}} = C$), and $c_{12} = 0$ (no inflow). At maximum inflow, the velocity of the upstream front is equal to that of moving downstream fronts. Consequently, the section-based model (and LWR models in general) can only describe sustained traffic waves at maximum inflow.

Phase 3: Dissolution of the queue. For $t \ge t_2$, the traffic light is green and the supply $S(t) = C$ is given again by the capacity. Because of the queue (section 1 is partly congested), the demand is equal to C as well. So, the maximum-flow state ③ develops. With Eq. (8.44), we calculate the propagation velocity of the transition from state ② (standing queue) and state ② (maximum flow),

$$\frac{\mathrm{d}x_{23}}{\mathrm{d}t} = c_{23} = \frac{Q_3 - Q_2}{\rho_3 - \rho_2} = \frac{Q_{\max} - 0}{\rho_{\text{cong}} - \rho_{\max}} = -\frac{1}{\rho_{\max} T} = c. \tag{8.48}$$

In summary, we have two transitions in this phase:

- The transition ① → ② with the transition front given by $x_{12}(t) = L + c_{12}(t - t_1)$ as in phase 2.
- The transition ② → ③ with the transition front obtained by integrating Eq. (8.48) resulting in $x_{23}(t) = L + c(t - t_2)$. The queue dissolves completely at time t_3 where the two fronts intersect:

$$t_3 = \frac{c_{12}t_1 - ct_2}{c_{12} - c} = t_1 + \left(\frac{-c}{c_{12} - c}\right)(t_2 - t_1).$$

Microscopically, this means that the last vehicle being stopped by the traffic light sets into motion at time t_3 and at location $x_3 = L + c(t_3 - t_2)$.

Phase 4: Dissolution of the maximum-flow state and free traffic again. With the intersection of the transitions ① → ② and ② → ③, we obtain a new transition from state ① to state ③. Again we make use of Eq. (8.44) to arrive at a propagation velocity

$$c_{13} = \frac{Q_{\text{max}} - Q_{\text{in}}}{\rho_{\text{cong}}(Q_{\text{max}}) - \rho_{\text{free}}(Q_{\text{in}})} = V_0,$$

i.e., the propagation velocity is equal to the vehicle speed V_0. Hence, this transition passes the stopping line $x = L$ of the traffic light at time

$$t_4 = t_3 + \frac{x_3}{V_0} = t_2 + (t_2 - t_1)\frac{Q_{\text{in}}}{Q_{\text{max}} - Q_{\text{in}}}$$

resulting in free traffic, everywhere. Notice that a periodic situation is only possible if the next red phase of the traffic light begins at some time $t_5 \geq t_4$. Otherwise, traffic flow becomes over-saturated and the queue grows without bounds.

Verify that the limit for a cyclic (not oversaturated) situation at an isolated traffic light is characterized by the relation

$$Q_{\text{in}} = C\left(\frac{t_4 - t_2}{t_4 - t_1}\right) = C\left(\frac{\text{duration of green}}{\text{duration of total cycle}}\right), \tag{8.49}$$

i.e., the inflow Q_{in} is equal to the mean capacity $\overline{C}$ averaged over one cycle.

8.5.9.2 Several Signalized Intersections: Progressive Signal System

For the sake of simplicity, we ignore the traffic flow of all secondary roads and only consider one direction of the main-road which we model as a homogeneous

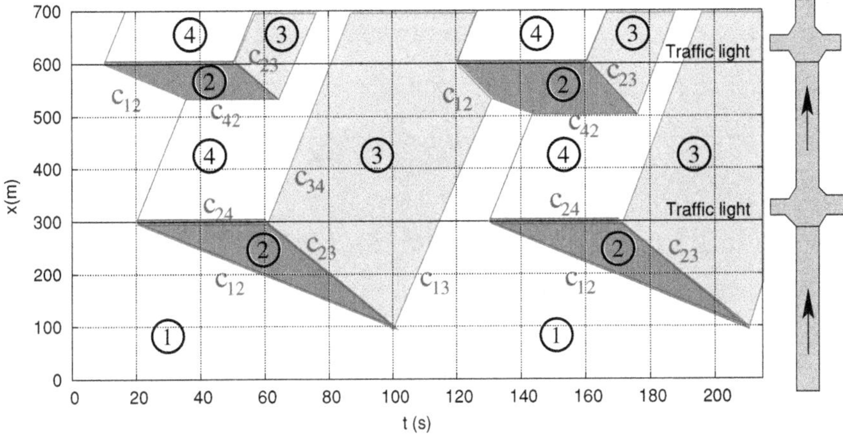

Fig. 8.24 Simulating the traffic flow across two signalized intersections with the section-based model. The duration of the *red* and *green* phases of the traffic lights is given by $\tau_r = 40$ and $\tau_g = 70\,\text{s}$, respectively, and the relative phase shift $\Delta\tau = 10\,\text{s}$. The spatiotemporal traffic states are characterized by the inflow region ①, the queues ②, the outflows ③, and the empty regions ④ downstream of *red lights* (assuming no vehicles from secondary roads enter)

road of capacity C (if the lights are green). The traffic lights at each intersection all have the same cycle time $\tau_{\text{cyc}} = \tau_r + \tau_g$ and the same durations τ_r and τ_g of the red and green phases. By varying the relative phase shifts $\Delta\tau_k = t_r^{(k)} - t_r^{(k-1)}$ of the beginning red phases at intersection k, we can optimize the traffic lights to a progressive signal system, i.e., try creating "green waves" with platoons of moving vehicles in sync. While the basic optimization is simple (at least, in this idealized case), we need simulations to (i) check the stability of the resulting synchronization against variations of the demand, (ii) assess the robustness against perturbations of the signalization arising, e.g., by giving priority to the buses and trams of the public transport system, or introducing an isolated pedestrian traffic light that is not connected to the system,[21] (iii) estimate the maximum demand at which the synchronization breaks down, and (iv) determine how fast the system recovers from over-saturation.

Figure 8.24 shows an example with two signalized intersections assuming a constant inflow at the source (which would be a good approximation if this example represents the first traffic light of a passage through a town or city). Since the limit (8.49) is not yet reached, one could optimize this system such that queues only form at the most upstream traffic light. However, if the relative phases are not optimized, queues form further downstream as well. This is demonstrated in Fig. 8.24 showing the result for relative phases $\Delta\tau = t_r^{(k)} - t_r^{(k-1)} = -10\,\text{s}$ while the optimized phase shift of adjacent traffic lights would be $\tau(x_2 - x_1)/V_0 = 20\,\text{s}$. Generalizations to less idealized situations are easy to implement and simulate.

[21] This really happens as is observed by one of the authors in his home city.

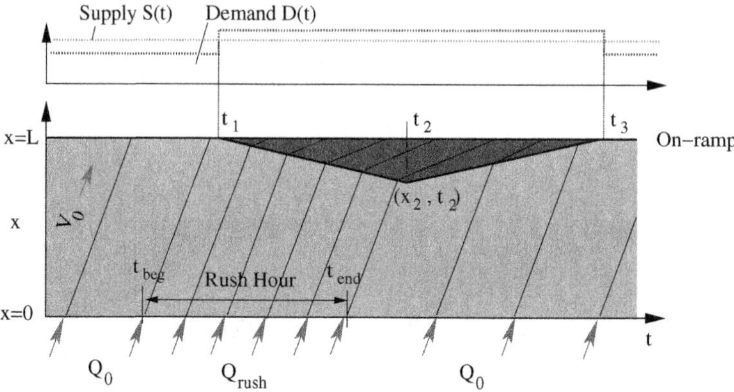

Fig. 8.25 On-ramp bottleneck during rush-hour conditions. The traffic breakdown is provoked by the traffic peak during the rush hour (see the main text)

8.5.9.3 On-Ramp Bottleneck During Rush-Hour Conditions

We consider a road section of capacity C with I lanes to which an on-ramp (merging length L_{rmp}) leads at the main-road position $x = L$. The traffic demand (inflow) of the ramp is constant and given by Q_{rmp} while traffic inflow on the main-road temporarily increases for a time interval $t \in [t_{\text{beg}}, t_{\text{end}}]$ representing the rush hour (Fig. 8.25):

$$Q_{\text{in}} = \begin{cases} Q_{\text{rush}} & t_{\text{beg}} < t < t_{\text{end}}, \\ Q_0 & \text{otherwise.} \end{cases}$$

Furthermore, we assume a situation where the inflow Q_0 allows for sustained free traffic while Q_{rush} exceeds the capacity of the on-ramp bottleneck:

$$Q_0 < C_{\text{B,rmp}} = C - Q_{\text{rmp}} < Q_{\text{rush}}.$$

Here, traffic evolves in following phases:

Phase 1: Free traffic, $t < t_{\text{beg}}$. Since we have demanded $Q_{\text{in}} = Q_0 < C_{\text{B,rmp}}$ in this phase, the supply (the bottleneck capacity) is greater than the demand (inflow), so there is free traffic everywhere. Since the resulting steady-state flow does not contain any explicit time dependence, we can integrate the relevant continuity equation (7.12) (multiplied by the number I of lanes) without resorting to any specification for the speed. This results in

$$Q^{\text{tot}}(x) = \begin{cases} Q_0 & x < L - L_{\text{rmp}}/2, \\ Q_0 + Q_{\text{rmp}} & x > L + L_{\text{rmp}}/2, \\ Q_0 + Q_{\text{rmp}}\left(x - L + L_{\text{rmp}}/2\right) & \text{otherwise,} \end{cases}$$

and with the free-flow branch of the fundamental diagram

$$Q(x) = \frac{Q^{\text{tot}}(x)}{I}, \quad \rho(x) = \frac{Q(x)}{V_0}, \quad V(x) = V_0.$$

Phase 2: Rush hour and traffic breakdown, $t_1 \le t < t_2$. Traffic breaks down at the on-ramp bottleneck as soon as the increased demand reaches the position of the on-ramp. Since the "rush-hour" information travels downstream with the vehicles (propagation velocity $c_{\text{free}} = V_0$), this occurs at time

$$t_1 = t_{\text{beg}} + \frac{L}{V_0},$$

at least, if we neglect the physical length L_{rmp} of the merging lane. The resulting congested traffic upstream of the ramp is characterized by the effective bottleneck capacity (bottleneck capacity per lane) C_B/I:

$$Q_{\text{cong}} = \frac{C_B}{I} = Q_{\max} - \frac{Q_{\text{rmp}}}{I}, \quad \rho_{\text{cong}} = \rho_{\max}(1 - Q_{\text{cong}}T), \quad V_{\text{cong}} = \frac{Q_{\text{cong}}}{\rho_{\text{cong}}}.$$

The region of free traffic downstream of the ramp is characterized by the maximum-flow state: $Q = C/I = Q_{\max}$, $\rho = Q/V_0$, $V = V_0$. Finally, the region of free traffic further upstream can accommodate the demand during rush hour, at least for some time until the congestion reaches the considered location:

$$Q_{\text{free}} = Q_{\text{rush}}, \quad \rho_{\text{free}} = \frac{Q_{\text{rush}}}{V_0}, \quad V_{\text{free}} = V_0.$$

The propagation of the transition free → congested is, again, given by the shock-front formula (8.44):

$$c_{12} = \frac{Q_2 - Q_1}{\rho_2 - \rho_1} = \frac{Q_{\text{cong}} - Q_{\text{rush}}}{\rho_{\max}(1 - Q_{\text{cong}}T) - \frac{Q_{\text{rush}}}{V_0}}. \tag{8.50}$$

Phase 3: Rush hour is over and jam dissolves, $t_2 < t \le t_3$. First, we determine the time t_2 where the upstream front of the congested zone reverts its propagation direction, i.e., the congestion begins to shrink. To this end, we determine the intersection of the line $x^*(t) = L + c_{12}(t - t_1)$ giving the spatiotemporal positions of the upstream front with the line $x_{\text{end}}(t) = x^*(t)$ which describes how the information ("rush hour is over") propagates downstream. After some simple arithmetic, the intersection condition $x^*(t) = x_{\text{end}}(t)$ provides the time

$$t_2 = \frac{L - c_{12}t_1 + V_0 t_{\text{end}}}{V_0 + c_{12}},$$

and the maximum length of the congested region (cf. Fig. 8.22)

$$L_{\text{max}}^{\text{cong}} = L - x_2 = L - x^*(t_2) = -c_{12}(t_2 - t_1).$$

After time t_2, the propagation velocity of the jam front becomes positive:

$$\tilde{c}_{12} = \frac{Q_{\text{cong}} - Q_0}{\rho_{\text{max}}(1 - Q_{\text{cong}}T) - \frac{Q_0}{V_0}},$$

and, eventually, the jam front $\tilde{x}^*(t) = x_2 + \tilde{c}_{12}(t - t_2)$ reaches the ramp position at time

$$t_3 = t_2 + \frac{L_{\text{max}}^{\text{cong}}}{\tilde{c}_{12}} = t_2 - \frac{c_{12}(t_2 - t_1)}{\tilde{c}_{12}}$$

which coincides with the dissolution of the jam. Since $c_{12} < 0$ and $\tilde{c}_{12} > 0$, we have $t_3 > t_2$ as required for reasons of consistency.

8.5.9.4 Reduction of Number of Lanes

When reducing the number of lanes from I_1 to $I_2 < I_1$, the road capacity (8.29) decreases by the amount $\Delta C = (I_2 - I_1)Q_{\text{max}}$ (cf. Fig. 8.15). As soon as the demand Q_{in} exceeds the new capacity

$$C_{\text{B}} = C - \Delta C = I_2 Q_{\text{max}}, \tag{8.51}$$

traffic breaks down at the location of the lane drop resulting in a jam which is characterized by the flow variables (cf. Fig. 8.15)

$$Q_{\text{cong}} = \frac{C_{\text{B}}}{I_1} = \frac{I_2}{I_1} Q_{\text{max}}, \tag{8.52}$$

$$\rho_{\text{cong}} = \rho_{\text{max}}(1 - Q_{\text{cong}}T) = \rho_{\text{max}}\left(1 - \frac{I_2}{I_1} Q_{\text{max}} T\right), \tag{8.53}$$

$$V_{\text{cong}} = \frac{Q_{\text{cong}}}{\rho_{\text{cong}}} = \frac{Q_{\text{cong}}}{\rho_{\text{max}}(1 - Q_{\text{cong}}T)} = \frac{I_2 V_0}{I_1 + (I_1 - I_2)\rho_{\text{max}} V_0 T}. \tag{8.54}$$

Relation (8.54) explains the observation that even a relatively mild bottleneck (e.g. a lane drop from 3 to 2 lanes) leads to drastically reduced speeds upstream as soon as there is congestion (while in the actual region of a reduced lane number the speed remains at V_0). We make this explicit by following calculation.
Example: For the model parameters $V_0 = 144$ km/h, $T = 1.5$ s and $l_{\text{eff}} = 1/\rho_{\text{max}} = 7$ m, we obtain following speed values.

$$\text{Drop from 3 to 2 lanes: } V_{\text{cong}} = 0.17\,V_0 = 24.9\,\text{km/h},$$
$$\text{Drop from 3 lanes to 1 lane: } V_{\text{cong}} = 0.05\,V_0 = 7.1\,\text{km/h},$$
$$\text{Drop from 2 lanes to 1 lane: } V_{\text{cong}} = 0.09\,V_0 = 13.6\,\text{km/h}.$$

Since these values depend only weakly on the model parameters (for $V_0 = 72\,\text{km/h}$ we obtain instead of the above results the values 19.8, 6.2, and 11.5 km/h, respectively), the plain information of the average speed in jammed regions allows one to conclude the kind of bottleneck (at least when knowing or assuming that it is a lane-drop bottleneck).

Justify the observation that, when driving in congested traffic caused by a successive lane drop from three lanes to two lanes to one lane, one can drive more quickly after the first drop to 2 lanes although the local road capacity is reduced.

8.6 Diffusion and Burgers' Equation

Shock waves are not very realistic in describing traffic flow. After all, they imply unbounded accelerations in the microscopic picture of Eq. (8.23). Furthermore, the associated discontinuities turn out to be problematic for a numerical solution—at least for non-triangular fundamental diagrams. As a simple phenomenological solution, one may introduce diffusion to the continuity equation by adding a diffusion term $D\partial^2\rho/\partial x^2$ with the diffusion constant $D > 0$:

$$\boxed{\frac{\partial\rho}{\partial t} + \left[V_e(\rho) + \rho\frac{\mathrm{d}V_e}{\mathrm{d}\rho}\right]\frac{\partial\rho}{\partial x} = D\frac{\partial^2\rho}{\partial x^2} \qquad \text{LWR model with diffusion.}} \tag{8.55}$$

Notice that, generally, this nonlinear partial differential equation is solved numerically (cf. Sect. 9.5) while analytical solutions are only feasible for the simplest situations as described below.

For the case of a linear speed-flow relation (i.e., a parabolic fundamental diagram, see Problem 7.7) we can transform Eq. (8.55), which is then called *Burgers' equation* by the so-called *Cole-Hopf transformation* to a linear diffusion equation which can be solved analytically.

Verify that Burgers' equation is consistent with vehicle conservation.

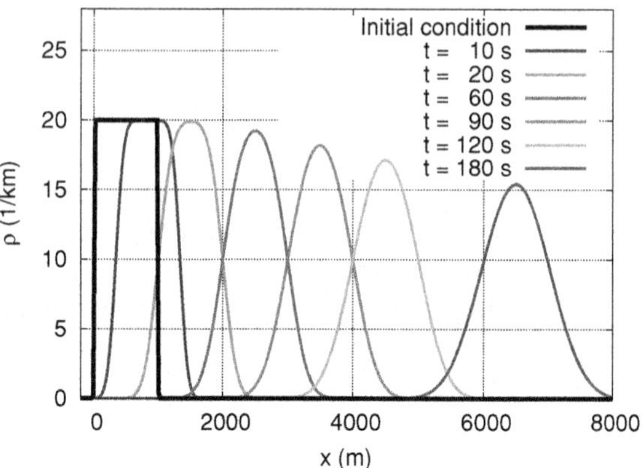

Fig. 8.26 Solution to the diffusions-transport equation (8.56) in an infinite system for an initial density profile $\rho_0(x)$ depicted by the *thick black lines*

Linearizing Eq. (8.55) for small density intervals or assuming a triangular fundamental diagram (8.11) and keeping to the free or congested region, we obtain a linear *transport-diffusion equation*

$$\frac{\partial \rho}{\partial t} + \tilde{c}\frac{\partial \rho}{\partial x} = D\frac{\partial^2 \rho}{\partial x^2}, \tag{8.56}$$

where the propagation velocity $\tilde{c} = Q'_e(\rho)$ is the same as in the section-based model. Particularly, we obtain $\tilde{c} = c_{\text{free}}$ and c_{cong} for the triangular fundamental diagram in the free and congested regimes, respectively.

Figure 8.26 shows a solution to Eq. (8.56) corresponding to free traffic with initially 20 vehicles uniformly distributed in the region $x \in [0, 1\text{ km}]$, and no traffic outside. Notice that the density profile for $D = 0$ corresponds to the solution of the section-based model, i.e., it is given by uniformly translating the rectangular initial profile. This means, diffusion just "smears out" the solution to the original LWR models without diffusion.

We can express the solution to the diffusion-transport equation (8.56) for general initial density profiles $\rho_0(x) = \rho(x, 0)$ by an integral, at least, if the boundaries are sufficiently far away so as to not influence the dynamics:

$$\rho(x,t) = \int \rho(x', t_0) g(x - x', t - t_0)\, \mathrm{d}x'. \tag{8.57}$$

Here, the so-called *Green's function*

$$g(x,t) = f_N^{(\mu,\sigma^2)}(x) = \frac{1}{\sqrt{4\pi D t}} \exp\left[-\frac{(x - \tilde{c}t)^2}{4Dt}\right] \tag{8.58}$$

corresponds to the solution for an ideally (point-like) localized initial density distribution satisfying $\int \rho(x,0)\,dx = 1$, i.e., representing a single vehicle on an otherwise empty road. The Green's function is given by a Gaussian with time-dependent expectation $\mu(t) = \tilde{c}t$ and variance $\sigma^2(t) = 2Dt$.[22]

By inspecting Eq. (8.57), one sees immediately that the evolution of the rectangular initial density profile shown in Fig. 8.26 can be expressed in terms of two cumulative Gaussians (cf. Problem 8.7).

Why must we require D to be nonnegative for any sensible traffic flow model? Discuss with the help of the solution (8.57).

Verify that Eq. (8.57) is a solution to the diffusion-transport equation (8.56) by directly inserting the solution into Eq. (8.56).

Problems

8.1 Propagation velocity of a shock wave free → congested
Justify why the propagation velocity (8.9) takes on values between c and V_0 for LWR models with a triangular fundamental diagram. What are the conditions for realizing the extreme values? What is the range of shock propagation velocities in the parabolic fundamental diagram of Problem 7.7?

8.2 Driver interactions in free traffic
In the LWR with a triangular fundamental diagram, all continuous and discontinuous density changes within free traffic propagate at velocity V_0. Draw conclusions what this means for the interactions between the drivers.

8.3 Dissolving queues at a traffic light
In the triangular fundamental diagram, the velocity c of moving downstream fronts of congested traffic is given by $c = -l_{\text{eff}}/T$. Try to intuitively understand this relation for a queue of standing vehicles behind a traffic light after it turns green.

8.4 Total waiting time during one red phase of a traffic light
Calculate the total waiting time of all vehicles caused by one red phase of duration τ_r for the LWR model with a triangular fundamental diagram. Assume the conditions

[22] Here, the limits of a collective macroscopic description become obvious: A single vehicle cannot "smear out" by diffusion effects.

of the example in Sect. 8.5.9.1, i.e., a constant inflow Q_{in}, and sufficient capacity to avoid over-saturation, i.e., the queue completely dissolves before the next red phase. Express the solution as a function of τ_r, the maximum density, the propagation velocities c^{up} of the upstream boundary, and the universal propagation velocity c^{cong} of congested traffic.

8.5 Jam propagation on a highway I: Accident
Consider traffic flow on a two-lane highway between the road kilometers 0 and 10 during and after a closure of one lane at kilometer 10 at 15:00 h. In the considered time period, the traffic demand (inflow at $x = 0$) is constant and given by 3024 vehicles/h. The lane closure is effective for half an hour until 15:30 h.

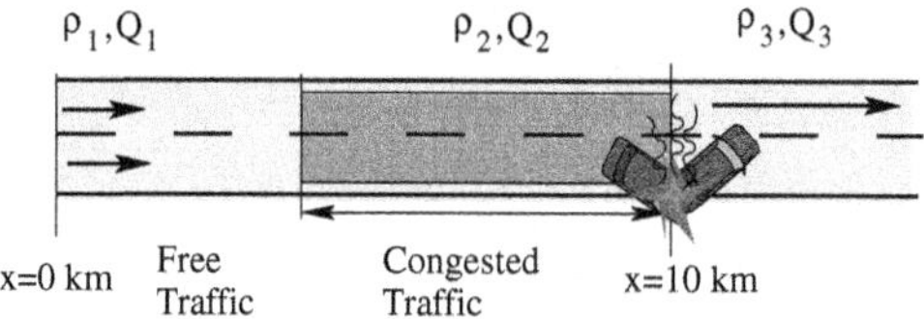

Solve the following questions using the section-based model with the parameters $l_{\text{eff}} = 8$ m, $T = 1.5$ s, and $V_0 = 28$ m/s.

1. Calculate the total road capacity and the effective capacity (per lane) prior to the lane closure. Does the capacity satisfy the demand? Calculate the traffic density and the traveling time to traverse the 10 km long road stretch.
2. Show that, after the lane closure is active, the bottleneck capacity of the remaining lane does not satisfy the demand. Calculate the effective and total traffic density of the forming jam. Assume that the drivers symmetrically use both lanes (i.e., consider locations upstream of the transfer zone where people change lanes to the through lane).
3. Calculate the growth rate of the jam by determining the velocity of its upstream front. *Hint:* Distinguish carefully between total and effective (lane-averaged) densities and flows.
4. After the lane closure has been lifted, the downstream front of the jam sets into motion. Since it moves faster than the upstream front, the jam eventually dissolves. Calculate the velocity of the moving downstream front and the time for complete dissolution. Also calculate the position of the last vehicle to be obstructed by the jam at obstruction time.
5. Visualize the spatiotemporal dynamics of the jam by drawing its boundaries in a space-time diagram.
6. How much time does a vehicle need to traverse the 10 km long road section if it enters at 15:30 h?

8.6 Jam propagation on a highway II: Uphill grade and lane drop Consider following highway section containing a three-to-two lane drop and an uphill grade on parts of the two-lane region:

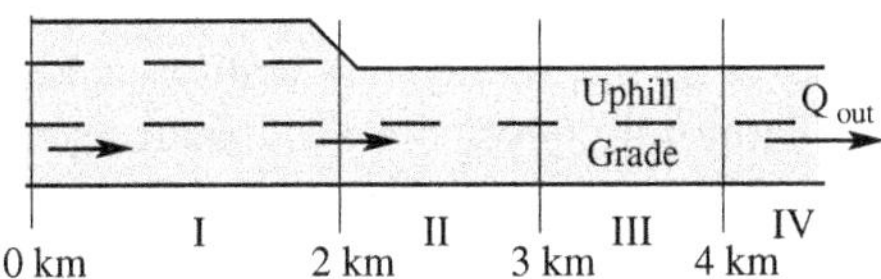

The effective traffic flow is modeled with a triangular fundamental diagram. Because of a high percentage of trucks, the mean free speed in the uphill region III is only $V_{03} = 60$ km/h while $V_0 = 120$ km/h applies elsewhere. Furthermore, assume an increase of the time gap from $T = 1.5$ to $T_3 = 1.9$ s in the gradient region III, and an effective vehicle length $l_{\text{eff}} = 10$ m everywhere.

1. Show that the capacity per lane is given by 2,000 vehicles/h outside the uphill region, and 1,440 vehicles/h in the uphill region.
2. Before the onset of the rush hour at 4:00 pm, assume free steady-state traffic everywhere and a total inflow of 2,000 vehicles/h. Calculate the effective and total densities and the speeds in all regions I–IV.
3. At 4:00 pm, the total traffic demand at $x = 0$ increases abruptly to 3,600 vehicles/h. Does this cause a breakdown? If so, at which time and where?
4. Assume now a breakdown at $x = 3$ km and consider two stages of the developing congestion: (i) The upstream jam front is in region I at $x = 1$ km, (ii) the front is in region II at $x = 2.5$ km. Calculate, for both stages, flows and densities of all four regions of the considered road stretch.
5. At which velocity does the upstream jam front propagate in stages (i) and (ii) of Part (4) of this problem?

8.7 Diffusion-transport equation
Solve the diffusion-transport equation (8.56) for a constant propagation velocity $\tilde{c}$ on an infinite homogeneous road for the initial conditions (cf. Fig. 8.26)

$$\rho(x,0) = \begin{cases} \rho_0 & 0 \le x \le L, \\ 0 & \text{otherwise.} \end{cases}$$

Hint: Express the solution in terms of the (cumulative) distribution function $\Phi(z)$ of the standard normal distribution.

Further Reading

- Lighthill, M.J., Whitham, G.B.: On kinematic waves: II. A theory of traffic on long crowded roads. Proceedings of the Royal Society A **229** (1955) 317–345
- Richards, P.: Shock waves on the highway. Operations Research **4** (1956) 42–51
- Daganzo, C.F.: The cell transmission model: A dynamic representation of highway traffic consistent with the hydrodynamic theory. Transportation Research Part B: Methodological **28** (1994) 269–287
- Daganzo, C.F.: The cell transmission model part II: Network traffic. Transportation Research Part B: Methodological **29** (1995) 79–93
- Daganzo, C.F.: Urban gridlock: Macroscopic modeling and mitigation approaches. Transportation Research Part B: Methodological **41** (2007) 49–62
- Helbing, D.: A section-based queueing-theoretical traffic model for congestion and travel time analysis in networks. Journal of Physics A: Mathematical and General **36** (2003) L593–L598

Chapter 9
Macroscopic Models with Dynamic Velocity

Real knowledge is to know the extent of one's ignorance.
Confucius

Abstract In the macroscopic first-order (Lighthill-Whitham-Richards, LWR) models presented in the previous two chapters, the local speed and flow are statically coupled to the density by the fundamental relation. This implies instantaneous adaptation to new circumstances and leads to unbounded accelerations and other unrealistic consequences such as the lack of hysteresis effects or traffic instabilities. In the *second-order models* considered in this chapter, the local speed possesses its own dynamical acceleration equation describing speed changes as a function of density, local speed, their gradients, and possibly other exogenous factors. Second-order models are the models of choice to macroscopically describe traffic-flow instabilities leading to traffic waves, the capacity drop phenomenon, or scattered flow-density data. Besides discussing representative models, this chapter describes approximative numerical integration techniques which are more demanding than that of the LWR models.

9.1 Macroscopic Acceleration Function

The first-order LWR models considered in the previous chapters are characterized by a single dynamical partial differential equation (or iterated coupled map) for the density which, in essence, is a consequence of the conservation of the number of vehicles. The local speed $V(x, t)$ of these models does not possess any independent dynamics since it is statically coupled to the density by a speed-density relation. Such models can describe traffic breakdowns at bottlenecks due to insufficient capacity and the propagation of the resulting congested regions. From a microscopic point of view, the associated instantaneous speed adaptations imply unbounded accelerations which, clearly, is unrealistic. Moreover, finite speed adaptation times and reaction times are the main factors leading to growing *traffic waves* and capacity-drop

M. Treiber and A. Kesting, *Traffic Flow Dynamics*,
DOI: 10.1007/978-3-642-32460-4_9, © Springer-Verlag Berlin Heidelberg 2013

phenomena (see Chap. 18), or to traffic flow *instabilities* in general (see Chap. 15). Consequently, LWR models cannot describe these observations.[1]

If such phenomena are required, one needs models where the local speed is treated as a second independent field which is governed by a second dynamical *acceleration equation*. Such an equation describes the local acceleration (in Lagrangian coordinates, i.e., in a system comoving with the drivers) as a function of density, speed, gradients thereof, and possibly other exogenous factors. Hence, this class of models is also known as the class of *second-order models*, in contrast to the LWR models which are also termed *first-order models*.

In time-continuous second-order models,[2] the acceleration equation is a second partial differential equation of the general form

$$\frac{\mathrm{d}V(x,t)}{\mathrm{d}t} = \left(\frac{\partial}{\partial t} + V(x,t)\frac{\partial}{\partial x}\right) V(x,t) = A[\rho(x,t), V(x,t)]. \tag{9.1}$$

This equation implies that the rate of change of the local speed $\frac{\mathrm{d}V(x,t)}{\mathrm{d}t} = \frac{\partial V}{\partial t} + V\frac{\partial V}{\partial x}$ in (Lagrangian) coordinates is equal to an *acceleration function* $A(x,t) = A[\rho(x,t), V(x,t)]$.[3] The total time derivative on the left-hand side is also referred to as the *material* derivative, *convective* derivative, *Lagrangian* derivative, or *substantial* derivative. It is composed of the local rate of speed change $\frac{\partial V}{\partial t}$ that a stationary detector would measure, and the convective rate of change $V\frac{\partial V}{\partial x}$ due to moving to a new location (Fig. 9.1).[4]

The different second-order models are solely distinguished by their acceleration function characterizing the macroscopic acceleration as a function of the density and speed field of the neighborhood. However, just a function $A(\rho(x,t), V(x,t))$ is not sufficient since this would correspond to very short-sighted drivers not taking into account what happens in front of them. In fact, we will show in Sect. 15.4.2 that such a model would be *unconditionally* unstable. Consequently, we introduce spatial *anticipation*, either by allowing density and speed gradients $\frac{\partial \rho}{\partial x}$, $\frac{\partial V}{\partial x}$ (and possibly higher-order gradients), or nonlocalities. To implement nonlocalities, we take the density and speed fields at the spatial coordinate $x_\mathrm{a} > x$ in front of the actual position x (Fig. 9.2),

$$\rho_\mathrm{a}(x,t) = \rho(x_\mathrm{a},t), \quad V_\mathrm{a}(x,t) = V(x_\mathrm{a},t), \quad x_\mathrm{a} > x. \tag{9.2}$$

---

[1] One can also add reaction times *ad hoc* to LWR models. This alone, however, does not lead to traffic waves of growing amplitude or hysteresis effects.

[2] We will not explicitly consider second-order models formulated as iterated coupled maps. Time-continuous models assume this form as an approximation anyway at time of numerical integration (Sect. 9.5).

[3] The arguments of the acceleration function are enclosed in brackets instead of parentheses to indicate a *functional* dependence including gradients and nonlocalities as described below.

[4] For a further discussion, see Sect. 7.3.

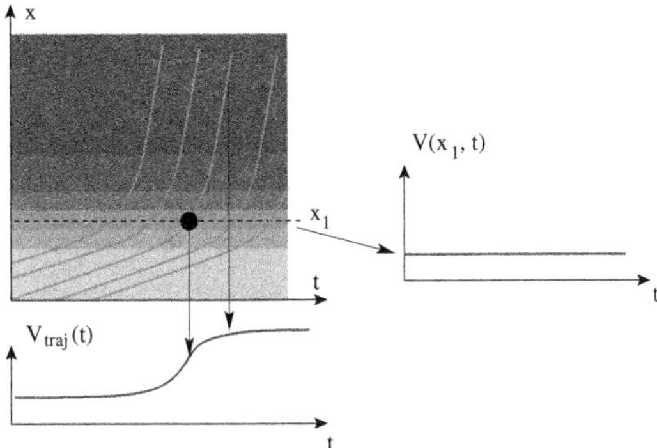

Fig. 9.1 Visualization of the total (Lagrangian) time derivative along a trajectory (driver's point of view) in contrast to the partial (Eulerian) time derivative taken in a stationary coordinate system. Along a trajectory, the rate of change is given by $\frac{\mathrm{d}V(x(t),t)}{\mathrm{d}t} = \frac{\partial V}{\partial x}\frac{\mathrm{d}x}{\mathrm{d}t} + \frac{\partial V}{\partial t} = V\frac{\partial V}{\partial x} + \frac{\partial V}{\partial t}$. Here, we show the situation of steady-state flow with a stationary downstream front $\left(\frac{\partial V}{\partial t} = 0\right)$ with accelerating vehicles $\left(\frac{\mathrm{d}V}{\mathrm{d}t} > 0, \frac{\partial V}{\partial x} > 0\right)$. This situation corresponds to that of Fig. 7.4

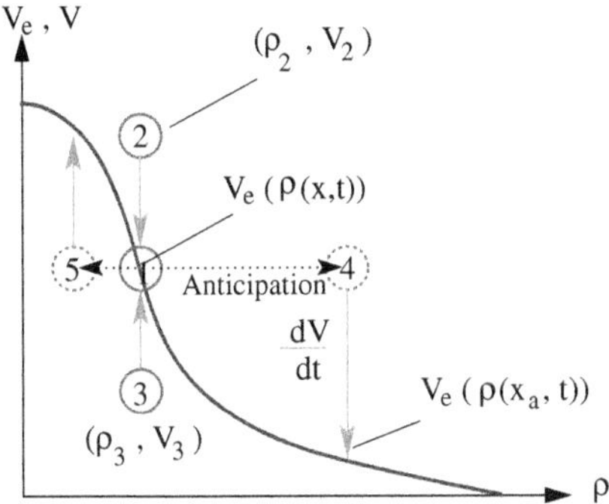

Fig. 9.2 Effect of the speed adaptation and anticipation terms of the second-order model (9.11). The points ① to ③ characterize local situations (ρ, V) in steady state (point ①) and in non-equilibrium (② and ③). The arrows from the states ② and ③ to ① describe local speed adaptation, while the *arrows* from ④ and ⑤ describe anticipative adaptation

Since both gradients and nonlocalities describe the same aspect (spatial anticipation), they are interchangeable. Therefore, *local* macroscopic models can be written in the general form

$$\frac{\mathrm{d}V}{\mathrm{d}t} = \left(\frac{\partial}{\partial t} + V\frac{\partial}{\partial x}\right)V = A_{\mathrm{loc}}\left(\rho, V, \frac{\partial \rho}{\partial x}, \frac{\partial V}{\partial x}\right), \tag{9.3}$$

while nonlocal second-order macroscopic models take on the general form

$$\frac{\mathrm{d}V}{\mathrm{d}t} = \left(\frac{\partial}{\partial t} + V\frac{\partial}{\partial x}\right)V = A_{\mathrm{nonloc}}(\rho, V, \rho_{\mathrm{a}}, V_{\mathrm{a}}). \tag{9.4}$$

At first sight, nonlocal macroscopic models look more complicated. After all, we have to solve a set of two nonlocal partial differential equations. While this seems to be a formidable task, approximative numerical solutions to nonlocal models are, in fact, computationally faster and numerically more stable than local models (see Sect. 9.5).

9.2 Properties of the Acceleration Function

9.2.1 Steady-State Flow

As is the case for LWR models, most of the second-order models possess a unique steady state on homogeneous roads described by the steady-state speed $V_e(\rho)$ for a given density. With the hydrodynamic relation $Q = \rho V$, this results in a fundamental diagram $Q_e(\rho)$.[5]

Since the left-hand sides of Eqs. (9.1), (9.3) or (9.4) are equal to zero for stationary-homogeneous conditions, the same must apply to the acceleration functions. Considering local and nonlocal acceleration functions as defined by Eqs. (9.3) and (9.4), respectively, and taking into account stationarity $\left(\frac{\partial}{\partial t} = 0\right)$ and homogeneity $\left(\frac{\partial}{\partial x} = 0, x_{\text{a}} = x\right)$, we arrive at the steady-state conditions

$$A_{\text{loc}}\,(\rho, V_e, 0, 0) = 0, \quad A_{\text{nonloc}}\,(\rho, V_e, \rho, V_e) = 0. \tag{9.5}$$

These are implicit equations for the steady-state speed-density relation $V_e(\rho)$ and the fundamental diagram $Q_e(\rho) = \rho V_e(\rho)$. The conditions for plausible speed-density relations are the same as for the LWR models, namely $V_e'(\rho) \leq 0$, $V_e(0) = V_0$ and $V_e(\rho_{\max}) = 0$ (cf. Fig. 8.1).

9.2.2 Plausibility Conditions

In contrast to the LWR models, the local speed $V(x,t)$ is generally not equal to the steady-state speed $V_e(\rho(x,t))$. Nevertheless, the acceleration function should model the desire of the drivers to approach the steady state. In homogeneous situations (no speed or density gradients) this leads to following criteria for plausible acceleration functions, i.e., plausible models:

$$A(\rho, V, \ldots) < 0, \quad \text{if } V > V_e(\rho), \quad \frac{\partial \rho(x,t)}{\partial x} = 0, \text{ and } \frac{\partial V(x,t)}{\partial x} = 0. \tag{9.6}$$

---

[5] There are a few models *without* unique steady states on homogeneous roads. Such models are controversial and will not considered here.

Analogously, $A(\rho, V, \dots) > 0$ if $V < V_e(\rho)$. We can summarize both conditions by demanding that the partial derivative of the acceleration functions with respect to the local speed should be strictly negative:

$$\frac{\partial A_{\text{loc}}\left(\rho, V, \frac{\partial\rho}{\partial x}, \frac{\partial V}{\partial x}\right)}{\partial V} < 0, \qquad \frac{\partial A_{\text{nonloc}}\left(\rho, V, \rho_{\text{a}}, V_{\text{a}}\right)}{\partial V} < 0. \tag{9.7}$$

This condition states that, when increasing the local speed and leaving everything else unchanged (*ceteris paribus*), the acceleration decreases. Notice that Eq. (9.7) is valid for non-steady-state conditions as well, i.e., in the presence of gradients and nonlocalities, so the constraints of Eq. (9.6) are no longer needed.

However, the existence of a steady-state, (Eq. 9.5) and drivers attempting to reach it (Eq. 9.7) is not sufficient since this alone would result in unconditionally unstable traffic flow: Models without additional negative feedback mechanisms would always produce traffic waves, even in completely free traffic, which is even more unrealistic than the total absence of such waves in the LWR models. In real traffic, instabilities are only observed in a density range corresponding to congested traffic. It turns out that *anticipation*, either in the form of gradients or by nonlocalities, provides sufficient stabilization to restrict flow instabilities to realistic density ranges. In terms of gradients, denser or more congested traffic ahead is characterized by a positive density gradient. So, in anticipation, one would reduce the acceleration or increase the braking deceleration. In contrast, positive speed gradients imply that the cars ahead drive faster, so it is appropriate to accelerate. This leads us to the third set of plausibility conditions:

$$\frac{\partial A\left(\rho, V, \frac{\partial\rho}{\partial x}, \frac{\partial V}{\partial x}\right)}{\partial\left(\frac{\partial\rho}{\partial x}\right)} \le 0, \qquad \frac{\partial A\left(\rho, V, \frac{\partial\rho}{\partial x}, \frac{\partial V}{\partial x}\right)}{\partial\left(\frac{\partial V}{\partial x}\right)} \ge 0. \tag{9.8}$$

When modeling anticipation by nonlocalities, a plausible driver's behavior is even more explicit (cf. Fig. 9.2): Reduce speed if the density ρ_{a} ahead is higher than the local density ρ, increase speed if the speed V_{a} ahead is higher than the local speed V. Moreover, the degree of reaction should increase monotonically with the differences, so we require

$$\frac{\partial A(\rho, V, \rho_{\text{a}}, V_{\text{a}})}{\partial\rho_{\text{a}}} \le 0, \qquad \frac{\partial A(\rho, V, \rho_{\text{a}}, V_{\text{a}})}{\partial V_{\text{a}}} \ge 0. \tag{9.9}$$

Notice that a single of the four conditions summarized in Eqs. (9.8) and (9.9) is enough to provide sufficient stability. Therefore, we did not formulate these conditions as strict inequalities (in contrast to Eq. 9.7). One of these conditions, however, must be satisfied as a strict inequality.

9.3 General Form of the Model Equations

Here, we will formulate general forms for effectively single-lane and single-class second-order macroscopic models.[6] Since vehicle conservation is always valid, second-order models obey the same *continuity equations* as LWR models, i.e., Eq. (7.15) in the most general case with ramps and changes of the number of lanes. The *acceleration equation* is characterized by the exogenous factors (speed, density, and gradients and nonlocalities thereof) imposed by the plausibility conditions discussed in Sect. 9.2.2. In order to arrive at a compact formulation, we express the speed and density gradients in terms of a gradient of a *traffic pressure* P while the nonlocalities can be expressed in terms of a generalized targeted speed $V_e^*(\rho, V, \rho_a, V_a)$ that can be seen as a generalization of the steady-state speed $V_e(\rho)$. Furthermore, the acceleration equation may also contain diffusion terms—second-order spatial derivatives—analogously to the LWR models with diffusion described in Sect. 8.6. Finally, on road sections parallel to the acceleration and deceleration lanes of ramps, the acceleration equation generally contains additional ramp terms describing the influence of entering or exiting vehicles on the local speed of the main-road (cf. Problem 9.1).

In summary, most local and nonlocal second-order models formulated by continuous variables can be represented by following generic continuity and acceleration equations:

$$\frac{\partial \rho}{\partial t} + \frac{\partial(\rho V)}{\partial x} = \frac{\partial}{\partial x}\left(D\frac{\partial \rho}{\partial x}\right) - \frac{\rho V}{I}\frac{\mathrm{d}I}{\mathrm{d}x} + \nu_{\text{rmp}}(x,t) \tag{9.10}$$

$$\boxed{\frac{\partial V}{\partial t} + V\frac{\partial V}{\partial x} = \frac{V_e^*(\rho, V, \rho_a, V_a) - V}{\tau} - \frac{1}{\rho}\frac{\partial P}{\partial x} + \frac{1}{\rho}\frac{\partial}{\partial x}\left(\eta\frac{\partial V}{\partial x}\right) + A_{\text{rmp}}(x,t).} \tag{9.11}$$

In the following subsections, we describe the different terms of this set of equations.

9.3.1 Local Speed Adaptation

In local models, the generalized targeted speed V_e^* is equal to the local steady-state speed $V_e(\rho)$ and the first term on the right-hand side of Eq. (9.11), also denoted as *speed adaptation term* or *relaxation term* describes the mean acceleration of the vehicles in the local neighborhood in order to reach the steady-state speed corresponding to the local density. In the simplest case, the *speed adaptation time* is constant (does not depend on density or speed), and the acceleration is proportional to the difference

[6] As this qualifying statement implies, there are also explicit multi-lane models with separate density and speed fields for each lane, and multi-class models with partial densities and local speed fields for each vehicle class (e.g., cars and trucks). Such models are beyond the scope of this book and we refer to the references at the end of this chapter.

of the local speed V to the steady-state speed $V_e(\rho)$. Then, τ represents a characteristic *speed adaptation time* in which the distance to the steady-state is $1/e$ times the original distance. The speed-adaptation term ensures that the consistency conditions (9.5) and (9.7) are satisfied. The effect of this term can be visualized by Fig. 9.2: If one is outside of steady-state (local equilibrium), the adaptation term represents an acceleration (vertical arrows) towards the steady-state speed. Depending of the traffic context (city streets, minor and major roads, highways), the speed adaptation τ is of the order of few seconds (city streets) and up to 20–30 s (highways).

9.3.2 Nonlocal Anticipation

In nonlocal models, the generalized targeted speed V_e^* of the speed adaptation term depends not only on the local density but on the density $\rho_a = \rho(x_a, t)$ at an anticipated location *ahead* of the actual position, $x_a > x$, and possibly on the actual and anticipated speeds V and V_a, respectively. In the simplest case, $V_e^* = V_e(\rho_a)$ is directly given by the steady-state speed at the anticipated location resulting in the nonlocal adaptation term. So, the relaxation and nonlocal anticipation terms can be expressed by

$$A_{\text{relax+antic}}(x, t) = \frac{V_e(\rho(x_a, t)) - V(x, t)}{\tau}. \tag{9.12}$$

Thus, an anticipatory driving style is simply described by adapting the speed to the steady-state speed as in local models, but taking the steady-state speed at a position ahead. This satisfies the first plausibility condition of the set (9.9) as a strict inequality.

We can visualize the effect of this anticipation term again by Fig. 9.2: If denser traffic or a congestion is ahead ($\rho_a > \rho$), a driver who already is in *local* steady-state at point ①, would nevertheless adapt his or her speed to the lower value of the steady-state speed at the anticipated position, i.e., he or she reacts as if the traffic state is given by the virtual point ④. Conversely, if less congested traffic is ahead (downstream jam front), the driver would react according to the virtual point ⑤ resulting in a positive acceleration. Such anticipation terms are the direct equivalent of anticipative local pressure terms (cf. Problem 9.3).

It is straightforward to generalize this anticipation concept. For example, to model a direct reaction to faster traffic ahead (rather than taking the indirect route over the density), one would make V_e^* explicitly depend on $V_a = V(x_a, t)$ thereby satisfying the second plausibility condition of Eq. (9.9) as a strict inequality. In local models, this corresponds to making the traffic pressure $P(\rho, V)$ dependent on the speed.

9.3.3 Limiting Case of Zero Adaptation Time

When multiplying the acceleration equation (9.11) with τ and taking the limit $\tau \to 0$, one observes immediately that this equation reduces to the LWR condition $V(x, t) = V_e(\rho(x, t))$. This is plausible since the limit $\tau \to 0$ signifies that the speed is rigidly coupled to the density which is a defining feature of the class of first-order (LWR) models.

9.3.4 Pressure Term

The pressure term $-\frac{1}{\rho}\frac{\partial P}{\partial x}$ introducing the *traffic pressure* P describes a response of the local ensemble of vehicles or drivers on density gradients and, in some models, speed gradients. The resulting dependence of the acceleration function on density and speed gradients should satisfy the plausibility conditions (9.8).

Notice that we deliberately speak of the behavior of a "local ensemble of vehicles or drivers" instead of directly referring to driver reactions: The name "pressure term" has its origins in macroscopic models derived from gas-kinetic considerations such as the GKT model (Sect. 9.4.3) where this term describes a purely kinematic (statistical) effect of speed variance *without a single vehicle accelerating or braking*.[7] Besides this first-principles interpretation, traffic pressure terms are also used purely phenomenologically and then, indeed, describe the anticipation of the drivers (cf. Problem 9.3). Therefore, the pressure term generally has both a behavioral and a kinematic component:

$$-\frac{1}{\rho}\frac{\partial P}{\partial x} = -\frac{1}{\rho}\frac{\partial P_{\text{antic}}(\rho, V)}{\partial x} - \frac{1}{\rho}\frac{\partial P_{\text{kin}}(\rho, V)}{\partial x}. \tag{9.13}$$

In most models using the behavioral interpretation of the traffic pressure, we can write this term as

$$A_{\text{antic}} = -\beta_1\frac{\partial \rho}{\partial x} + \beta_2\frac{\partial V}{\partial x}, \quad \beta_1 = \frac{1}{\rho}\frac{\partial P_{\text{antic}}}{\partial \rho}, \quad \beta_2 = -\frac{1}{\rho}\frac{\partial P_{\text{antic}}}{\partial V} \tag{9.14}$$

with non-negative sensitivities β_1 and β_2 that may depend on ρ and V themselves. The sensitivity β_1 with respect to density gradients states that one accelerates less (or brakes harder) when the density gradient is positive, i.e., denser traffic is ahead (cf. Fig. 9.3). The sensitivity β_2 describes that drivers tend to accelerate more (brake less) when traffic flow ahead is faster. Notice that the behavioral pressure terms of local models are equivalent to the nonlocal contributions of the targeted speed V_e^* of the nonlocal models.

In contrast to the behavioral part of traffic pressure representing anticipative driver reactions, the *kinematic* part

$$A_{\text{kin}} = -\frac{1}{\rho}\frac{\partial P_{\text{kin}}}{\partial x}, \quad P_{\text{kin}} = \rho\sigma_V^2(x,t), \tag{9.15}$$

[7] Notice the analogy to the physical pressure as defined by statistical physics. There, pressure is proportional to the velocity variance of the molecules, and forces are proportional to pressure gradients.

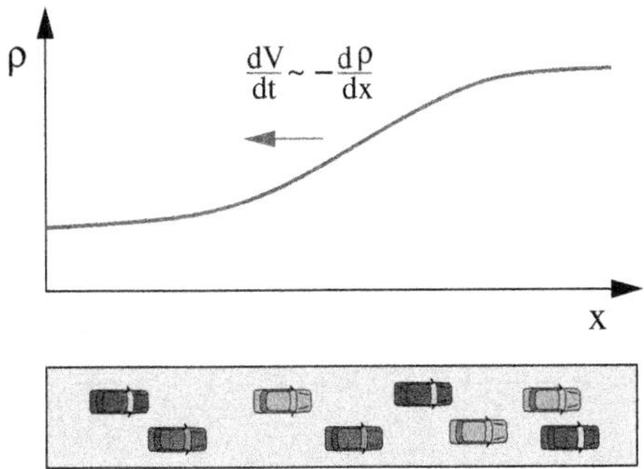

Fig. 9.3 Sensitivity to density gradients when entering a traffic jam. In addition to the adaptation to the local steady-state speed, there is an additional negative contribution $A_{\text{antic}} = -\beta_1 \frac{\partial \rho}{\partial x}$ to the acceleration

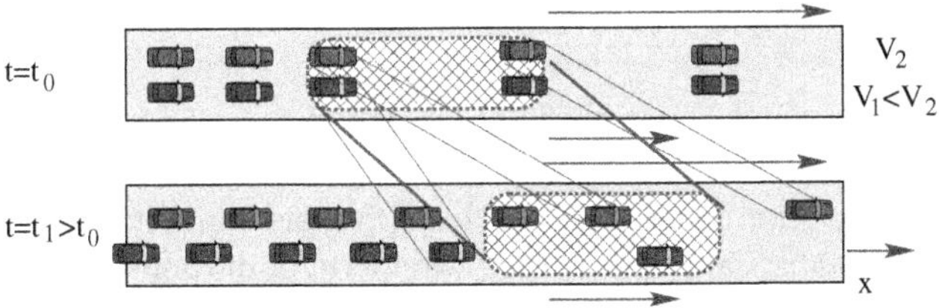

Fig. 9.4 Effects of kinematic dispersion at a transition from higher to lower density assuming, for simplicity, constant speed and speed variance everywhere. The convective speed change $\frac{dV}{dt} = \frac{\partial V}{\partial t} + V\frac{\partial V}{\partial x}$ in the reference region moving with the average speed (*hatched box*) is determined by the average of the slower vehicles downstream (moving more slowly than the reference) and the faster vehicles upstream (moving more quickly). Since there are fewer vehicles downstream, the fraction of faster vehicles inside the reference region, and thereby the local macroscopic speed, increases without any acceleration of the vehicles

is a sole consequence of a finite speed variance

$$\sigma_V^2(x,t) = \left\langle (v_i - V(x,t))^2 \right\rangle \tag{9.16}$$

of the vehicles i in the microscopically large and macroscopically small local neighborhood. The vehicles contributing to the variance are exactly the same that contribute to the local speed, $V(x,t) = \langle v_i \rangle$. The kinematic pressure term is only effective if (i) the speed variance is nonzero, and (ii) there are density gradients ($\partial P_{\text{kin}}/\partial x \neq 0$). Then, it leads to a macroscopic speed change *even if not a single vehicle accelerates*. As an illustrative example, we consider the downstream front of a region of decreased density where $\frac{\partial \rho}{\partial x} < 0$ (hatched region of Fig. 9.4). Since a finite variance implies finite speed differences, the faster vehicles leave the region of increased density more quickly than the slower vehicles. This has two consequences: (i) The width of the transition between higher and lower density at the downstream front grows. This is represented by a diffusion term in the continuity equation (see Sect. 9.3.5) below.[8]

[8] The mechanism is different from the dispersion already present in the LWR models with a concave fundamental diagram, although the effect is the same.

(ii) The vehicles separate at the downstream transition according to their speed: The faster vehicles move downstream relative to the center of the moving downstream front while the slower vehicles move upstream. Since, because of the density gradient, there are fewer vehicles downstream than upstream, the proportion of fast vehicles increases at the transition zone. This is modeled exactly by the kinematic macroscopic acceleration

$$A_{\text{kin}} = -\frac{1}{\rho}\frac{\partial}{\partial x}\left(\rho(x,t)\sigma_V^2(x,t)\right) \approx -\frac{\sigma_V^2}{\rho}\frac{\partial \rho}{\partial x} > 0$$

assuming that the speed variance is constant.[9]

9.3.5 Diffusion Terms

Some macroscopic models contain diffusion terms, i.e., second-order derivatives with respect to space, in the continuity or acceleration equations. Generally, they are phenomenologically introduced to smooth sharp transitions and shocks. From the point of view of statistical physics, a diffusion term in the continuity equation is a consequence of erratic microscopic motion components (*random walk* of particles described by their velocity variance which is proportional to the temperature of physical systems). Moreover, "speed diffusion", i.e., a diffusion term in the acceleration equation, is the consequence of a finite *viscosity*.[10]

In vehicular traffic, finite speed variances are caused by the drivers' heterogeneity (there are faster and slower drivers), and by the unsystematic erratic components of the driver's acceleration, e.g., caused by estimation errors or lack of attention. However, the order of magnitude of the diffusion constant η[11] exceeds the magnitude caused by these effects, so speed diffusion terms do not reflect properties of single drivers at a microscopic level. Nevertheless, speed diffusion may be useful (i) to improve the numerical properties of a model, (ii) to eliminate shock waves, and (iii) to investigate the effects of *numerical* diffusion which are unavoidable when numerically integrating macroscopic models (cf. Sect. 9.5).

---

[9] This is not exactly true since the variance decreases with the density which leads to an opposite contribution. However, the direct effect of the density gradient prevails.

[10] In one of his groundbreaking papers of the year 1905, Albert Einstein explained the old puzzle of *Brownian motion* of particles in a fluid in terms of the fluid temperature. Microscopically, the temperature is proportional to the velocity variance of the particles which is defined similarly as in Eq. (9.16) for the vehicles. In this work, A. Einstein also uncovered the microscopic origin of viscosity.

[11] In physical systems, η is known as *dynamic viscosity*, and the speed diffusion coefficient $\nu = \eta/\rho$ corresponds to the *kinematic viscosity*. Notice, however, that the viscosities in physical systems are generally *transverse viscosities* (mediated by shear flows) while that of traffic flow models are of a longitudinal nature.

Show that a density diffusion term in the continuity equation does not violate vehicle conservation by considering the dynamics of the total number of vehicles $\int \rho(x,t)\mathrm{d}x$ in a closed system ("ring road") of length L.

9.3.6 On- and Off-Ramp Terms

While, in the continuity equation, ramps and lane drops give rise to the same source terms $-\frac{\rho V}{I}\frac{\partial I}{\partial x}$ and $\nu_{\text{rmp}}(x,t)$ as in the LWR models (Sect. 7.2), additional contributions A_{rmp} may appear in the acceleration equation. As is the case for the kinematic pressure term, A_{rmp} describes changes of the macroscopic local speed by changes in the vehicle composition rather than by microscopic accelerations. In order to derive A_{rmp} one takes into consideration that the right-hand side of the acceleration equation describes the rate of change of the mean speed in microscopically large and macroscopically small road elements *comoving* with the vehicles (hatched regions in Fig. 9.5). We calculate this rate of change assuming that on-ramp vehicles merge to the main-road at speed $V_{\text{rmp}} < V$. Conversely, we assume that drivers about to leave the main-road reduce their speed, on average, to V_{rmp} *before* they diverge to the off-ramp. For the case of on-ramps, new vehicles enter the hatched regions at a negative relative speed. This leads to (i) an increase of the density as described by the source term $\nu_{\text{rmp}}(x,t)$ of the continuity equation, and (ii) to a reduced mean speed as described by A_{rmp}. In Problem 9.1 we show that both on-ramps and off-ramps lead to following equation for the rate of change of the mean speed:

$$A_{\text{rmp}}(x,t) = \frac{(V_{\text{rmp}} - V)}{\rho}\left|\nu_{\text{rmp}}(x,t)\right| = \frac{(V_{\text{rmp}} - V)|Q_{\text{rmp}}|}{\rho I L}. \tag{9.17}$$

Analogously to the source term of the continuity equation, this term is only nonzero at main-road locations parallel to the merging or diverging sections of ramps. Finally, we notice that lane drops do *not* give rise to a term in the acceleration equation, at least, if there are no microscopic accelerations or decelerations.

9.4 Overview of Second-Order Models

In this section, we present three well-known representatives of second-order models.

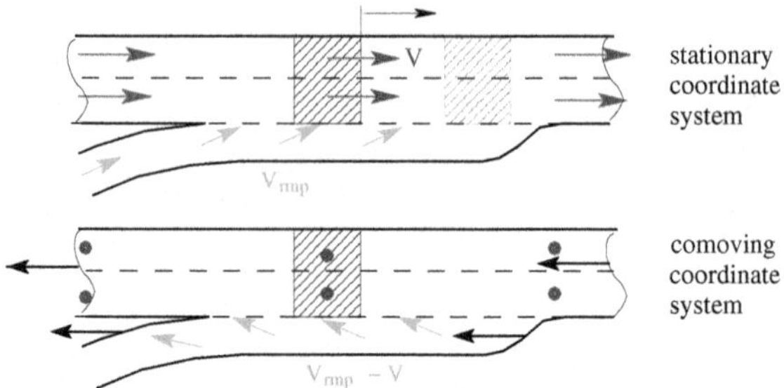

Fig. 9.5 Origin of the ramp term (9.17) of the acceleration equation. The main-road vehicles are stationary relative to the hatched region in a coordinate system comoving with the vehicles while the on-ramp vehicles enter the hatched region at a relative speed $V_{\text{rmp}} - V$

9.4.1 Payne's Model

A simple representative of a local second-order macroscopic model is *Payne's model*, sometimes also called the Payne-Whitham model. Its acceleration equation is

$$\boxed{\frac{\partial V}{\partial t} + V\frac{\partial V}{\partial x} = \frac{V_e(\rho) - V}{\tau} + \frac{V_e'(\rho)}{2\rho\tau}\frac{\partial \rho}{\partial x} \qquad \text{Payne's model.}} \tag{9.18}$$

Payne's model is a special case of the general macroscopic acceleration equation (9.11) for a constant speed relaxation time τ, zero diffusion η, and a density dependent traffic pressure $P = -V_e(\rho)/2\tau$. As in the LWR models, different equations for the steady-state speed-density relation $V_e(\rho)$ (satisfying $V_e(0) = V_0$, $V_e'(\rho) \leq 0$, and $V_e(\rho_{\text{max}}) = 0$) characterize a whole class of Payne-Whitham models.

Traffic pressure. The connection between the traffic pressure and driver interactions can be made more explicit by augmenting the traffic pressure by the constant $V_0/(2\tau)$ resulting in

$$P_{\text{Payne}}(x,t) = \frac{V_0 - V_e(\rho(x,t))}{2\tau}. \tag{9.19}$$

While this does not change the acceleration equation (where only the gradient of the pressure is relevant), this form of the pressure term is directly proportional to the difference between the desired and steady-state speeds, i.e., proportional to the driver-driver interactions and tends to zero for $\rho \to 0$, i.e., $V = V_0$.

On the other hand, when interpreting the pressure as a purely kinematic effect, we can relate the pressure P_{Payne} to the speed variance according to Eq. (9.15) resulting in

$$\sigma_V^2 = \frac{P_{\text{Payne}}}{\rho} = \frac{V_0 - V_e}{2\tau\rho}.$$

When formulating the pressure term as

$$-\frac{1}{\rho}\frac{\partial P}{\partial x} = -\frac{1}{\rho}\frac{\partial}{\partial x}\left(\frac{V_0 - V_e(\rho(x,t))}{2\tau}\right) = -\frac{1}{\rho}\frac{V_e'(\rho)}{2\tau}\frac{\partial \rho}{\partial x} = \frac{c_0^2(\rho)}{\rho}\frac{\partial \rho}{\partial x},$$

Payne's model is formally identical to the equations describing a compressible one-dimensional gas with generally density dependent sonic velocities $\pm c_0 = \pm\sqrt{V_e'/2\tau}$.[12]

Relation to a microscopic model. In Sect. 10.8, we will derive Payne's model from a simple car-following model, namely *Newell's model.*

Parameters and simulation. Apart from the parameters of the steady-state relation $V_e(\rho)$, the only additional parameter is the speed relaxation time τ. From the microscopic derivation it follows that τ is identical to the reaction time T_r while, in reality, these characteristic times have different orders of magnitude: The speed adaptation time is of the order of 10s (more in highway traffic, less in city traffic) while the reaction time is about 1 s. Typical values for τ adopted in the simulation are between 1 and 5 s. This corresponds to sonic velocities c_0 of the order of $\pm 10\,\mathrm{m/s^2}$.

Payne's model is difficult to simulate and prone to physical and numerical instabilities (cf. Chap.15). Dedicated numerical methods are necessary.

Limiting case of the adaptation time tending to zero. Since the speed relaxation time enters not only the relaxation term of Payne's model but also the pressure term, we will not simply obtain $V = V_e(\rho)$—the LWR model—when multiplying the acceleration equation by τ and setting $\tau = 0$. Instead, we obtain

$$V = V_e(\rho) + \frac{V_e'(\rho)}{2\rho}\frac{\partial \rho}{\partial x}.$$

This relation for the static local speed depends on the density and additionally on density gradients. Inserting this in the applicable continuity equation, e.g., in Eq. (7.8) for homogeneous roads, we obtain

$$\frac{\partial \rho}{\partial t} + \frac{\partial}{\partial x}\left(\rho V_e(\rho) + \frac{V_e'(\rho)}{2}\frac{\partial \rho}{\partial x}\right) = 0,$$

i.e., the LWR model with diffusion. We can make this more explicit by writing this equation as

$$\frac{\partial \rho}{\partial t} + \frac{\partial(\rho V_e(\rho))}{\partial x} = \frac{\partial}{\partial x}\left(D(\rho)\frac{\partial \rho}{\partial x}\right) \tag{9.20}$$

[12] This is a purely mathematical analogy, see Sect. 9.3.4.

with the density dependent diffusion coefficient

$$D = D(\rho) = -\frac{V_e'(\rho)}{2}.$$

9.4.2 Kerner–Konhäuser Model

The Kerner–Konhäuser model (KK model) is another well-known local second-order model which is related to Payne's model. Its acceleration equation is

$$\boxed{\frac{\partial V}{\partial t} + V\frac{\partial V}{\partial x} = \frac{V_e(\rho) - V}{\tau} - \frac{c_0^2}{\rho}\frac{\partial \rho}{\partial x} + \frac{\eta}{\rho}\frac{\partial^2 V}{\partial x^2} \quad \text{KK model.}} \tag{9.21}$$

This model is purely phenomenological, i.e., it is not based on a microscopic model. Instead, its equations are formulated analogously to that of a one-dimensional compressible gas with sonic velocities $\pm c_0$ and a variable speed diffusion coefficient $D_v = \eta/\rho$. In contrast to Payne's model, the sonic velocities $\pm c_0$ (or, in another interpretation, the speed variance c_0^2) are constant, and there is an additional speed diffusion term corresponding to a constant dynamic compression viscosity η. However, both terms have no microscopic foundation. As for Payne's model, a whole class of KK models can be formulated depending on the specific form of the steady-state speed density relation $V_e(\rho)$. Typical values for the three dynamic model parameters are $\tau = 10\,\text{s}$, $c_0^2 = 200\,\text{m}^2/\text{s}^2$, and $\eta = 150\,\text{m/s}$. This set of parameters is supplemented by the parameters of the steady-state speed density relation.

The diffusion term has been introduced to prevent unrealistically sharp transitions, particularly shocks. However, diffusion also impairs the numerical efficiency of simulations since it favors *numerical instabilities*. If such instabilities are present, the dynamical quantities oscillate wildly and grow beyond all bounds until the simulating program eventually crashes.[13] Therefore, they must be avoided at all costs, in contrast to *physical* instabilities which are desirable in a certain density range of congested traffic — after all, physical instabilities are the cause of the observed traffic waves. In Sect. 9.5 we show that diffusion restricts the numerical update time step Δt by the so-called second *Courant–Friedrichs–Lévy* (CFL) condition $\Delta t < (\Delta x)^2/(2D_v)$, i.e., the numerical complexity[14] increases inversely proportional to the *third* power of the cell size of the numerical grid indicating the resolution.

Example: Assume we want to simulate traffic of density $\rho = 15$ vehicles/km $=$ 0.015 vehicles/m (free traffic) with the KK model using a spatial discretization (cell size) of $\Delta x = 50\,\text{m}$. Then, for a typical value of $\eta = 150\,\text{m/s}$, we obtain a speed

---

[13] Do not confuse this with a simulated crash of vehicles. The latter is characterized by densities exceeding the maximum density ρ_{max}.

[14] The numerical complexity C indicates the number of multiplications or other operations on a computer which are necessary to obtain a certain approximate solution.

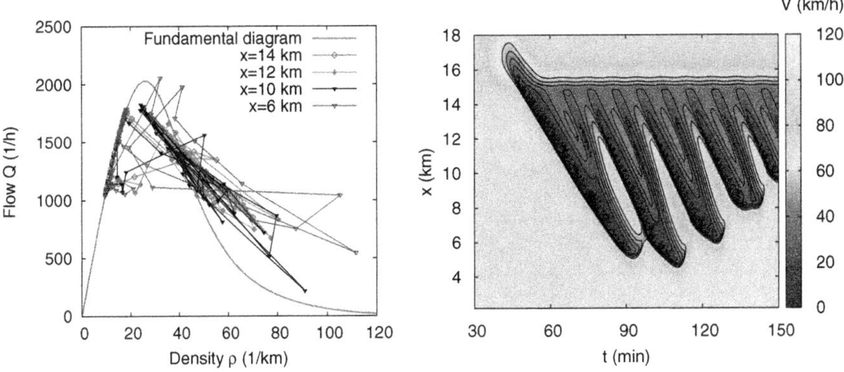

Fig. 9.6 Simulation of a highway section with an on-ramp (details in the main text). *Left* fundamental diagram (*gray smooth line*) and flow-density one-minute data of virtual detectors (*jagged lines*); *right* spatiotemporal dynamics of the local speed

diffusion $D_v = 10{,}000\,\mathrm{m}^2/\mathrm{s}$. Thus, the second CFL condition implies the condition $\Delta t < 0.125\mathrm{s}$. In comparison, without diffusion, the first CFL condition (which is discussed in Sect. 9.5) allows time steps of up to one second. Moreover, since $D_v \to \infty$ for $\rho \to 0$, we have to introduce an additional cap for D_v.

Figure 9.6 shows a simulation of a highway section with the KK model. We have assumed an on-ramp flow of 400 vehicles/h per main-road lane, and a main-road flow initially increasing from 1,100 vehicles/h per lane to 1,800 vehicles/h/lane (0:20 h), then linearly decreasing to 1,100 vehicles/h/lane at 2:00 h and keeping this inflow afterwards. The parameters are $\tau = 30\,\mathrm{s}$, $c_0 = 15\,\mathrm{m/s}$, $\eta = 150\,\mathrm{m/s}$. Furthermore, we have assumed the steady-state speed-density relation

$$V_e(\rho) = V_0 \frac{1 - \rho/\rho_{\max}}{1 + 200(\rho/\rho_{\max})^4}$$

with $v_0 = 120\,\mathrm{km/h}$, which is often used for this model. The left hand diagram f Fig. 9.6 shows the theoretical fundamental diagram $Q_e(\rho) = \rho V_e(\rho)$ together with flow-density data obtained from one-minute averages of *virtual detectors* at various locations. As the name implies, virtual detectors simulate the measuring and data aggregation process by recording the speed and passage times of all passing vehicles and aggregating them by averaging the microscopic data (see Chap. 3). Notice that the flow-density data do not lie on the fundamental diagram, not even on average.[15] This is a signature of the speed being a dynamical variable rather than coupled rigidly to the density.

[15] Flow-density data of real traffic scatter even more distinctly in the congested regime. To simulate this, we need to include further factors leading to scattered data points, particularly, a heterogeneous traffic composition. To simulate this macroscopically, we can make the model parameters time dependent. However, it is better to simulate heterogeneity microscopically (cf. Chap. 12).

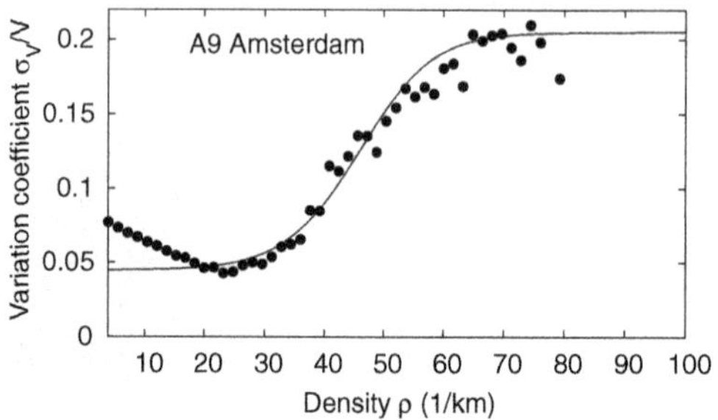

Fig. 9.7 Observed coefficient of variation σ_V / V of the speed (*bullets*) and theoretical GKT curve $\sqrt{\alpha(\rho)}$ according to Eq. (9.22) (*curved line*)

The spatiotemporal local speed profile of the right hand diagram of Fig. 9.6 shows growing traffic waves propagating against the direction of the traffic stream in the stationary system. This corresponds to the observations, at least qualitatively (cf. Fig. 5.1).

9.4.3 Gas-Kinetic-Based Traffic Model

The *gas-kinetic based traffic model* (GKT model) is one of few second-order macroscopic models that are derived from a microscopic model with explicit consideration of vehicle-driver heterogeneities. The GKT model characterizes this heterogeneity in terms of the empirically measurable speed variance depending on the density according to (cf. Fig. 9.7)

$$\sigma_V^2(\rho) = \alpha(\rho)[V_e(\rho)]^2 \tag{9.22}$$

with

$$\alpha(\rho) = \alpha_{\text{free}} + \frac{\alpha_{\text{cong}} - \alpha_{\text{free}}}{2} \left[1 + \tanh\left(\frac{\rho - \rho_{\text{cr}}}{\Delta\rho} \right) \right]. \tag{9.23}$$

The relative speed standard deviation $\sqrt{\alpha(\rho)} = \sigma_V / V_e$ (i.e., the coefficient of variation, cf. Sect. 3.2) has typical values between $\sqrt{\alpha_{\text{free}}} = 5\text{–}10\,\%$ (free traffic) and $\sqrt{\alpha_{\text{cong}}} = 20\,\%$ (congested traffic).

As its name implies, the microscopic model underlying the GKT model has properties of an idealized one-dimensional gas consisting of rigid particles. At any time, the stochastic velocity components of the particle (vehicle) speeds are assumed to be uncorrelated (*molecular chaos*) with a prescribed variance. Specific traffic-related properties enter into this model as follows: (i) the particles are self-driven with a driving force on particle i corresponding to the acceleration $\dot{v}_i = (v_{0i} - v_i)/\tau$ where v_{0i} is the *desired* (not the steady-state) speed, and τ is the speed adaptation time, (ii) the interaction range (effective length) of the rigid particles increases with their speed according to $l_i^{\text{eff}} = l_i + v_i T$ with T corresponding to a safety time gap, (iii) several lanes are assumed, and the probability p of passing a slower vehicle without necessity to brake is proportional to the fraction of free space on the road section (i.e., space which is not occupied by the effective vehicle lengths) in relation to its total length, (iv) the velocity variance depends on the local density according to Eq. (9.22), and (v) the rigid-body interactions are anisotropic violating conservation of momentum: Whenever the interaction range of a particle is about to intersect the range of the slower particle ahead and no passing is possible (probability $1 - p$), the particle decelerates instantaneously to the speed of the slower particle ahead.

These specifications allow to derive a macroscopic model using the standard methods of gas-kinetics. The resulting acceleration equation is

$$\boxed{\frac{\partial V}{\partial t} + V\frac{\partial V}{\partial x} = \frac{V_e^*(\rho, V, \rho_{\text{a}}, V_{\text{a}}) - V}{\tau} - \frac{1}{\rho}\frac{\partial P}{\partial x} \quad \text{GKT model.}} \tag{9.24}$$

The pressure term

$$P(x,t) = \rho\sigma_V^2(\rho) \tag{9.25}$$

is the sole consequence of the speed variance, i.e., a purely kinematic effect implying no microscopic accelerations. Moreover, there is no speed diffusion. All acceleration components corresponding to accelerations of single drivers (in order to reach the desired speed or as a consequence of interactions) are contained in the *generalized targeted speed* V_e^*. This quantity depends on the local traffic state (ρ, V) and on the traffic state $(\rho_{\text{a}}, V_{\text{a}})$ at the anticipated location $x_{\text{a}} > x$, where

$$\rho_{\text{a}} = \rho(x_{\text{a}}, t), \quad V_{\text{a}} = V(x_{\text{a}}, t), \quad x_{\text{a}} = x + \gamma V(x,t)T. \tag{9.26}$$

This means, the GKT model is a *nonlocal* model. The anticipation distance $s_{\text{a}} = \gamma VT$ is a multiple γ (typical values are 1–1.5) of the safety gap VT. The generalized targeted speed V_e^* itself reads

$$V_{\text{e}}^*(\rho, V, \rho_{\text{a}}, V_{\text{a}}) = V_0\left[1 - \frac{\alpha(\rho)}{\alpha(\rho_{\max})}\left(\frac{\rho_{\text{a}} VT}{1 - \rho_{\text{a}}/\rho_{\max}}\right)^2 B\left(\frac{V - V_{\text{a}}}{\sigma_V}\right)\right]. \tag{9.27}$$

In this formula, the *Boltzmann factor*

$$B(x) = 2\left[x f_N(x) + (1 + x^2)\Phi(x)\right] \tag{9.28}$$

increases monotonically with the normalized speed difference $x = (V - V_{\text{a}})/\sigma_V$ and depends on the standard normal distribution $\Phi(x) = \int_{-\infty}^x f_N(x')\mathrm{d}x'$ and its density $f_N(x) = 1/\sqrt{2\pi}\exp(-x^2/2)$. In contrast to most other macroscopic traffic flow models, the steady-state speed-density relation $V_e(\rho)$ is not explicitly given but

Table 9.1 Parameters of the gas-kinetic based traffic model (GKT model) with typical values

Parameter	Typical value highway	Typical value city traffic
Desired speed v_0	120 km/h	50 km/h
Time gap T	1.2 s	1.2 s
Maximum density ρ_{max}	160 vehicles/km	160 vehicles/km
Speed adaptation time τ	20 s	8 s
Anticipation factor γ	1.2	1.0

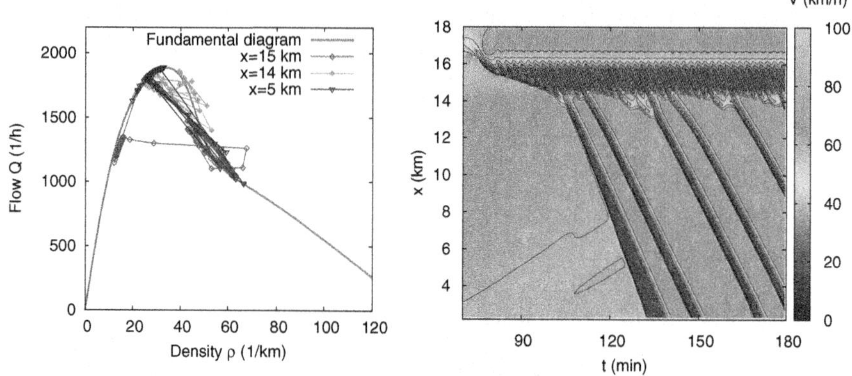

Fig. 9.8 Simulation of a highway section with an intersection consisting of an off-ramp and an on-ramp with the GKT model. *Left* fundamental diagram and flow-density one-minute data of virtual detectors; *right* spatiotemporal profile of the local speed

results from the implicit steady-state condition (9.5), i.e., $V_e^*(\rho, V_e, \rho, V_e) = V_e$ (cf. Problem 9.4).

Table 9.1 lists typical values of the five GKT model parameters (excluding the parameters of the empirically determined variance-density relation (9.23)). The first three parameters are the same as that of Payne's model and the KK model when assuming a triangular fundamental diagram 8.11 in these models.

In the simulations. the GKT model is *robust* in the sense that small parameter changes generally lead to small changes in the simulation result.[16] Moreover, this model contains *intuitive parameters* which can be simulated with *realistic values* (cf. Table 9.1).

In spite of its complex mathematical form, the GKT model is easier and more effective to simulate than Payne's model or the KK model. Figure 9.8 shows a simulation of a highway section with an intersection consisting of an off-ramp located 1.5 km upstream of an on-ramp. Both ramps have merging/diverging lengths of 500 m and flows of 500 vehicles per hour and main-road lane. The simulated

[16] Obvious and realistic exceptions include simulations on the verge of a traffic breakdown. In this case, reducing the simulated capacity by increasing T will trigger congestion, i.e., the output is discontinuous with respect to the input.

traffic waves display a similar spatiotemporal dynamics to the real traffic waves emerging from the intersection "Bad Homburg" on the German Autobahn A5-South (Fig. 5.1).

9.5 Numerical Solution

9.5.1 Overview

Apart from very simple special cases, second-order macroscopic models can only be solved approximatively by numerical integration.[17] The methods of choice are explicit *finite differences* which are applied nearly exclusively. In this method, the highway stretch is subdivided into cells of generally equal length Δx, and time is discretized into time steps with generally constant update time intervals Δt. When simulating a whole network, all roads are updated simultaneously by the same global time step. In each simulation time step, spatial derivatives are approximated by suitable difference quotients (hence the name of this class of methods), and the new traffic state at time $t+\Delta t$ is estimated based on the old state at time t by approximating the time derivatives by finite differences and solving for the new state. This is iterated until the end of the simulated time is reached. In this sense, the cell-transmission model (cf. Sect. 8.5) can be interpreted as a time-continuous model (the section-based model) with a dedicated finite-difference integration method.

When numerically solving a model with a finite-difference method, we need to distinguish between *explicit* and *implicit* methods. In the latter, the state and the spatial derivatives are calculated using both the old and the new (yet unknown) state, so, from the finite-difference approximation of the time derivative, we obtain a coupled system of equations for the new state which needs to be solved separately. In contrast, explicit methods give the new state explicitly in terms of the old state, i.e., we arrive at an iterated coupled map. Generally, explicit methods are easier to implement and numerically faster but they are also prone to numerical instabilities. Implicit methods are not practical for all but the simplest road networks (i.e., simple homogeneous roads), so explicit methods are applied nearly exclusively.[18] In the following, we will take a closer look at them.

To determine efficient integration schemes, we need to keep in mind that all macroscopic traffic flow models include the continuity equation which is derived from the conservation of the number of vehicles. Furthermore, without single-vehicle accelerations (i.e., the vehicles represent passive particles instead of driven particles),

---

[17] We will only discuss aspects of numerical integration that are directly relevant for application. For a deeper insight, we refer to the literature at the end of this chapter.

[18] This is valid more generally for the numerical integration of *hyperbolic partial differential equations*, to which the continuous macroscopic models belong.

global momentum is conserved as well.[19] Therefore, it is essential that the schemes are constructed in a way that these conservation conditions are satisfied exactly. To this purpose, we reformulate the equations of the traffic flow models in terms of *conservation laws*. In one dimension, pure conservation laws have the form

$$\frac{\partial u}{\partial t} + \frac{\partial f}{\partial x} = 0$$

where u is the density of the conserved quantity, and f the associated flux. Because momentum conservation is not valid for traffic flow and because ramps may be present, macroscopic traffic flow models cannot be formulated as pure conservation laws. Therefore, source terms s are added if necessary.

For the conservation of the number of vehicles, the associated density is simply the vehicle density ρ, and for the conservation of momentum, it is the flow $Q = \rho V$. When expressing the general continuity and acceleration equations (9.10) and (9.11) in terms of ρ and Q by eliminating the local speed via the relation $\rho = QV$ (cf. Problem 9.5), we obtain following conservation laws with sources:

$$\frac{\partial \rho}{\partial t} + \frac{\partial Q}{\partial x} = \nu_{\text{rmp}} - \frac{Q}{I}\frac{\mathrm{d}I}{\mathrm{d}x}, \tag{9.29}$$

$$\frac{\partial Q}{\partial t} + \frac{\partial}{\partial x}\left[\frac{Q^2}{\rho} + P - \eta\frac{\partial}{\partial x}\left(\frac{Q}{\rho}\right)\right] = \frac{\rho V_e^* - Q}{\tau} + S_{\text{inh}}, \tag{9.30}$$

where the source term of the flow equation associated with road inhomogeneities reads

$$S_{\text{inh}} = \frac{Q^2}{\rho I}\frac{\mathrm{d}I}{\mathrm{d}x} - \frac{Q\nu_{\text{rmp}}}{\rho} + \rho A_{\text{rmp}}. \tag{9.31}$$

All vehicle accelerations not depending on gradients (the latter are part of the pressure term) are contained in the generalized relaxation source term $(\rho V_e^* - Q)/\tau$. Notice that variable quantities outside of gradients (such as the advective term $V\frac{\partial V}{\partial x}$ in the original acceleration equation) violate the conservation property and are not allowed.[20] In vector notation, the above equations can be written more compactly as

$$\frac{\partial \mathbf{u}}{\partial t} + \frac{\partial \mathbf{f}(\mathbf{u})}{\partial x} = \mathbf{s}(\mathbf{u}) \tag{9.32}$$

---

[19] With driving forces, momentum is no longer conserved. However, it can be shown that the *source terms* resulting from the individual accelerations are similarly innocuous as the source terms $\eta(x,t)$ of the continuity equation originating from ramps or lane drops, or the ramp source term A_{rmp} of the acceleration equation.

[20] In contrast, nonlocalities are allowed.

where

$$\mathbf{u} = \begin{pmatrix} \rho \\ Q \end{pmatrix}, \quad \mathbf{f} = \begin{pmatrix} Q \\ \frac{Q^2}{\rho} + P - \eta \frac{\partial}{\partial x}\left(\frac{Q}{\rho}\right) \end{pmatrix}, \quad \mathbf{s} = \begin{pmatrix} \nu_{\text{rmp}} - \frac{Q}{I}\frac{\mathrm{d}I}{\mathrm{d}x} \\ \frac{\rho V_e^* - Q}{\tau} + S_{\text{inh}} \end{pmatrix}. \tag{9.33}$$

Here, $\mathbf{u}$ denotes the components of the traffic state, and $\mathbf{f}$ and $\mathbf{s}$ the associated fluxes and sources. For the Payne and KK models, the relaxation source term of the flow equation reads $(V_e(\rho) - V)/\tau$ while, for the GKT model, this term contains additional nonlocalities.

9.5.2 Upwind and McCormack Scheme

Two explicit integration methods turned out to be effective and useful for most second-order macroscopic models. The simple first-order *upwind method* calculates gradients as asymmetric upwind differences, i.e., it takes over the information coming from the upstream direction. It is suitable for nonlocal models such as the GKT model (9.24) since its nonlocalities handle downstream information propagating upstream which is relevant for congested traffic (see Eq. (9.35) below). When applying this method to local methods, one has to determine the local traffic state (free or congested), and switch to downwind finite differences for the case of congested traffic (see below). However, the *McCormack-Method* is more suited for this model class.

The McCormack method includes two steps: (i) calculating a "predictor" using upwind finite differences, (ii) calculating a refinement, the "corrector" by using the arithmetic means of the old and new traffic states for the temporal update with downwind finite differences. While this is, strictly speaking, an implicit scheme (*Crank–Nicholson method*) it becomes explicit when approximating the new state with the predictor.

In order to formulate the integration schemes, we subdivide the road section into cells of length Δx and time into time intervals Δt. Denoting the traffic state $\mathbf{u}(x,t)$ at location $j\Delta x$ and time $n\Delta t$ by $\mathbf{u}_j^n = \mathbf{u}(j\Delta x, n\Delta t)$ and defining $\mathbf{f}_j^n$ and $\mathbf{s}_j^n$ analogously, the update from time t to time $t + \Delta t$ (from n to $n+1$) is specified by

$$\begin{aligned} \mathbf{u}_j^{n+1} &= \mathbf{u}_j^n - \frac{\Delta t}{\Delta x}(\mathbf{f}_j^n - \mathbf{f}_{j-1}^n) + \Delta t\, \mathbf{s}_j^n && \text{Upwind method,} \\ \tilde{\mathbf{u}}_j^{n+1} &= \mathbf{u}_j^n - \frac{\Delta t}{\Delta x}(\mathbf{f}_j^n - \mathbf{f}_{j-1}^n) + \Delta t\, \mathbf{s}_j^n && \text{McCormack predictor,} \\ \mathbf{u}_j^{n+1} &= \frac{1}{2}\left(\tilde{\mathbf{u}}_j^{n+1} + \mathbf{u}_j^n - \frac{\Delta t}{\Delta x}(\tilde{\mathbf{f}}_{j+1}^{n+1} - \tilde{\mathbf{f}}_j^{n+1}) + \Delta t\, \tilde{\mathbf{s}}_j^{n+1}\right) && \text{McCormack corrector.} \end{aligned} \tag{9.34}$$

If the fluxes $\mathbf{f}$ contain gradients (which is true if the model contains diffusion terms), these gradients are approximated by finite differences with the opposite asymmetry (downwind differences for calculating the upwind scheme or the McCormack predictor, upwind for calculating the McCormack corrector).

9.5.3 Approximating Nonlocalities

Generally, the anticipated position x_a of the anticipated quantities $Q_a = Q(x_a, t)$ and $\rho_a = \rho(x_a, t)$ will not be a integer multiple j' of the cell width, so a numerical approximation by interpolation is necessary. For the two methods above, a piecewise linear interpolation has the same numerical consistency order as the other discretizations, so we will adopt it. If the model defines the anticipation distance $s_a = x_a - x$ at position $x = j\Delta x$, the piecewise linear interpolation is realized by following simple yet efficient scheme:

$$(\mathbf{u}_a)_j^n = \mathbf{u}_{j+k}^n + \left(\mathbf{u}_{j+k+1}^n - \mathbf{u}_{j+k}^n\right)\left(\frac{s_a}{\Delta x} - k\right), \quad k = \left\lfloor \frac{s_a}{\Delta x} \right\rfloor. \tag{9.35}$$

Here, $\mathbf{u} = (\rho, Q)^T$ are the variables of the flow-conservative formulation of the model equations as above, and the floor function $\lfloor x \rfloor$ denotes the largest integer not greater than x. In most cases, the cell size is greater than the anticipation distance, so $k = 0$, and Eq. (9.35) corresponds to a weighted arithmetic average between the values of the actual and the neighboring downstream cell.

9.5.4 Criteria for Selecting a Numerical Integration Scheme

There are many other integration methods which may be more efficient for specific models or applications. To give a guide for selecting them, we will now discuss some relevant selection criteria.

Information flow. Due to its asymmetric finite-difference approximation for gradients, the upwind method considers only information of the actually considered (local) cell and the neighboring cell in the upstream direction. Therefore, it is only suited for local models whose *velocity!characteristic* always propagate in the downstream direction (conversely, the "downwind" method would be suitable for information flow propagating upstream). The characteristic velocities are generalizations of the propagation velocity $\tilde{c} = Q_e'(\rho)$ of first-order models and will be considered in the Sect. 9.5.5. In the GKT model, the characteristic velocities are always positive while, for Payne's model and the KK model, they become negative under congested conditions. Therefore, the upwind method is the method of choice for the GKT model.[21]

If one intends to use simple asymmetric finite differences for the local models, it is crucial to dynamically switch between upwind and downwind finite differences depending on the local characteristic velocities. Exactly this switching is realized by the *supply-demand method* for integrating the cell-transmission model described in Sect. 8.5.7: If demand rules (free traffic), this corresponds to using upwind finite differences while in supply-dominated regimes (congested traffic) downwind differences are selected. Generally, methods applying asymmetric finite differences with a dynamical event-oriented switching are called *Godunov schemes*. For local second-order models, the actual method is more complicated than this schematic description: There are two characteristic velocities that may have different propagation directions, and the corresponding eigenmodes have to be separated. So, for this model class, it is better to apply schemes taking into account both upstream and downstream cells, such as the McCormack method.

Consistency and convergence order. A numerical method for integrating ordinary differential equations is consistent if the local discretization error tends to zero in the limit $\Delta t \to 0$. A consistent method has the consistency order p if the discretization error is proportional to $(\Delta t)^p$ for sufficiently small time steps Δt. These definitions can also be applied to partial differential equations by demanding that the spatial discretization Δx changes with Δt such that the quotient $\Delta x / \Delta t$ is kept constant (which is consistent with the first Courant–Friedrichs–Lévy condition (9.39) described below). For smooth density and speed profiles, the upwind method has the consistency order $p = 1$, and the McCormack scheme $p = 2$. This means, by halving Δx and Δt simultaneously, the local error is reduced by a factor of two and four in the upwind and McCormack methods, respectively (at least if Δt is sufficiently small where the criteria for "sufficiently" depend on the method).

Discretization errors. In the simulations, discretization errors typically result in an artificial smoothing (numerical diffusion), or spurious high-frequency oscillations (numerical dispersion), see Sect. 9.5.6. The consistency order specifies how the numerical errors scale with Δt. However, nothing is said about the *prefactors* of this scaling relation. For realistic update time intervals of, say, 0.5 s, a method of consistency order 1 may result in smaller discretization errors than one with order 2. Moreover, the consistency order is defined for a very fine discretization, and the scaling may be different for realistic update time intervals (if there is a scaling at all). So, it boils down to empirical tests to determine which method is most efficient for actual simulations. Ultimately, such tests lead to the recommendations given above (upwind for nonlocal second-order models with positive characteristic velocities, McCormack for the rest).

---

[21] The numerical approximation (9.35) of the nonlocalities of this model implies taking information from downstream cells thereby ensuring the upstream information transport.

Numerical instability. Besides discretization errors, explicit integration methods also imply *numerical instabilities* when certain limits of the time step Δt are exceeded. Since this topic is crucial, we discuss it in its own subsection.

9.5.5 Numerical Instabilities

Numerical instabilities typically result in wild oscillations growing beyond all bounds and eventually leading to a crash of the simulation. Therefore, they must be avoided at all costs. The artificial oscillations caused by *numerical* instabilities have to be distinguished from real traffic waves caused by *physical* instabilities that good second-order models are able to reproduce. While numerical instabilities are clearly a bug of the simulator, physical instabilities emerging under appropriate conditions are a feature.[22] Besides boundless growth, the signature of numerical instabilities are oscillations whose spatial and temporal periods are two times (or a low multiple of) the corresponding space and time discretizations. This can serve as a criterion to distinguish them from physical instabilities which have much larger periods. For traffic flow models, following categories of numerical instabilities are relevant.

(1) **Convective instability**. Numerical instabilities of this class appear if, in the exact model, flow and density changes mediated by first-order spatial derivatives can enter cells which have not been considered in the numerical update. Therefore, this kind of instability is termed a convective instability.[23] In asymmetric first-order methods, this is already the case if the integration scheme uses the "wrong" kind of asymmetric spatial finite differences, e.g., upwind finite differences when the information flow has components pointing upstream (i.e., at least one characteristic velocity is negative). However, there are also restrictions on Δt when choosing the correct method. To derive a quantitative criterion, we define the *characteristic velocities* in the limit of small perturbations by linearizing Eq. (9.32) for a homogeneous road and without diffusion terms around the steady state $\rho(x,t) = \rho_0$, $Q(x,t) = Q_0 = Q_e(\rho_0)$ and express the result in terms of the perturbation vector $\mathbf{w} = (\rho(x,t) - \rho_0,\ Q(x,t) - Q_0)^T$:

$$\frac{\partial \mathbf{w}}{\partial t} + \underline{\mathbf{C}} \cdot \frac{\partial \mathbf{w}}{\partial x} = \underline{\mathbf{L}} \cdot \mathbf{w} \tag{9.36}$$

where

$$\underline{\mathbf{C}} = \begin{pmatrix} 0 & 1 \\ -V^2 + \frac{\partial P}{\partial \rho} & 2V + \frac{\partial P}{\partial Q} \end{pmatrix}, \tag{9.37}$$

---

[22] For a model parameterization corresponding to unrealistically unstable traffic flow, the simulation may lead to densities above the maximum density, i.e., to simulated *physical* rather than *numerical* crashes.

[23] This kind of numerical instability may not be confused with real physical convective instabilities (traffic flow instabilities grow but propagate in only one direction) which play a significant role in traffic flow dynamics, see Sect. 15.5.

$$\underline{\mathbf{L}} = \begin{pmatrix} 0 & 0 \\ \frac{1}{\tau}\left(\tilde{V}_e + \rho\frac{\partial \tilde{V}_e}{\partial \rho}\right) & \frac{1}{\tau}\left(\rho\frac{\partial \tilde{V}_e}{\partial Q} - 1\right) \end{pmatrix}. \tag{9.38}$$

Here, $\tilde{V}_e = V_e(\rho)$ for Payne's model and the KK model, and $\tilde{V}_e(\rho, Q) = V_e^*(\rho, Q, \rho, Q)$ for the GKT model. The characteristic propagation velocities can be calculated as the eigenvalues of the matrix $\underline{\mathbf{C}}$:

- For Payne's model, we obtain $c_{1,2} = V \pm \sqrt{-V_e'(\rho)/2\tau}$,
- for the KK model $c_{1,2} = V \pm \sqrt{\theta}$,
- and for the GKT model $c_{1,2} = V(1 \pm \sqrt{3\alpha})$ plus negligible contributions proportional to $\alpha(\rho)V$.

If only neighboring cells are considered by the numerical update, the characteristic velocities imply following stability condition which is also called the first *Courant–Friedrichs–Lévy* (CFL) condition,

$$\Delta t < \frac{\Delta x}{\max |c|} \tag{9.39}$$

where $\max |c| \approx V_0$ for the Payne and GKT models, and $\max |c| = V_0 + \sqrt{\theta_0}$ for the KK model.

(2) **Diffusive instability**. The following consideration shows that diffusion terms may lead to numerical instabilities as well: Assume a diffusion term in the continuity equation of the form $\frac{\partial \rho}{\partial t} = \cdots + D\frac{\partial^2 \rho}{\partial x^2}$. The simplest way to approximate the second-order derivative by finite differences reads

$$\frac{\partial^2 \rho}{\partial x^2} \approx \frac{\rho_{j+1}^n - 2\rho_j^n + \rho_{j-1}^n}{\Delta x^2}.$$

In the simplest explicit integration scheme (*Euler* update), one calculates the new density by calculating the changing rate based on the old state:

$$\rho_j^{n+1} \approx \rho_j^n + \cdots + D\frac{\mathrm{d}\rho_j^n}{\mathrm{d}t}\Delta t \approx \rho_j^n + \cdots + D\Delta t\,\frac{\rho_{j+1}^n - 2\rho_j^n + \rho_{j-1}^n}{\Delta x^2}. \tag{9.40}$$

Now we consider high-frequency oscillations of period $2\Delta x$ by setting $\rho_j^n = \rho_e + A_n(-1)^j$. Inserting this in Eq. (9.40) results in

$$\rho_j^{n+1} = \rho_e + A\left(1 - \frac{4D\Delta t}{(\Delta x)^2}\right)(-1)^j.$$

This means that these high-frequency oscillations grow beyond any bounds if $\Delta t > (\Delta x)^2/(2D)$. Both the upwind and McCormack methods apply the same explicit time update as this simple example. So, we must require for these methods

$$\Delta t < \frac{(\Delta x)^2}{2D_v}, \quad \Delta t < \frac{(\Delta x)^2}{2D}. \tag{9.41}$$

The conditions (9.39) and (9.41) are known as the first and second *Courant–Friedrichs–Lévy* (CFL) conditions. Notice that other explicit integration schemes may have different CFL conditions.

(3) **Relaxation instability.** If partial (or ordinary) differential equations have relaxation terms (which is true for the second-order models), explicit integration methods will "overshoot" when approximating this relaxation unless

$$\Delta t < \frac{1}{\max(|\lambda_1|, |\lambda_2|)}, \tag{9.42}$$

where λ_1 and λ_2 are the eigenvalues of the matrix $\underline{\mathbf{L}}$, Eq. (9.38).[24] This is the "classical" instability mechanism for feedback-control systems: If the feedback response time (here, the update time step Δt) is greater than the smallest intrinsic time scale of the system to be controlled (here, the minimum of $|1/\lambda_1|$ and $|1/\lambda_2|$), an oscillating numerical instability arises.

For Payne's model and the KK model, condition (9.42) results in

$$\Delta t < \tau\,. \tag{9.43}$$

For the GKT model, the condition is more restrictive. For general densities, we obtain (cf. Problem 9.6)

$$\Delta t < \frac{\tau}{1 + \frac{2\alpha(\rho) V_0 \rho Q_e}{\alpha_{\max}} \left(\frac{T}{1-\rho/\rho_{\max}}\right)^2}\,. \tag{9.44}$$

For densities near the maximum density $\rho_{\max}$, this simplifies to

$$\Delta t \left(1 + 2\frac{V_0}{V_e}\right) < \tau. \tag{9.45}$$

S, the condition becomes most restrictive for densities near the maximum density ($V_e \to 0$).

(4) **Nonlinear instabilities**. We have derived all previous instability sources for small perturbations of density and flow, i.e., in the linear regime. For nonlinear amplitudes, particularly when fully developed (physical) traffic waves are present, further *nonlinear instabilities* may arise. Under most conditions, they cannot be characterized or derived. Only *Trial and Error* helps.

---

24 Strictly speaking, this is valid for homogeneous roads, only. However, no significant changes occur if sources, sinks, or lane drops are present.

9.5.6 Numerical Diffusion

The most conspicuous consequence of discretization errors is the so-called *numerical diffusion*: To second order $\mathcal{O}(\Delta t)^2$ in the time evolution, the numerical solution is equivalent to the *exact* solution of a *modified equation* which contains additional diffusion. Let us denote by $\mathbf{u}(x,t') = (\rho_0, Q_0)^{\mathrm{T}} + \mathbf{w}(x,t')$ the exact solution to the linearized form (9.36) of the original model for $t' \geq t$ where the initial profile $\mathbf{u}(x,t)$ at time t is equal to the discrete values at the grid points, $\mathbf{u}(j\Delta x, n\Delta t) = \mathbf{u}_j^n$, and linearly interpolated, elsewhere. Then, the numerical diffusion terms are characterized by a diffusion matrix $\underline{\mathbf{D}}$ defined by

$$\frac{1}{\Delta t}\left[\mathbf{u}_j^{n+1} - \mathbf{u}(j\Delta x, t+\Delta t)\right] = \underline{\mathbf{D}}\frac{\partial^2 \mathbf{u}}{\partial x^2} + \mathcal{O}(\Delta t)^2 . \tag{9.46}$$

This means, at time $t+\Delta t$, the difference between the numerical approximation $\mathbf{u}_j^{n+1}$ and the exact solution $\mathbf{u}(j\Delta x,\ t+\Delta t)$ at the same location is, to first order in Δt, given by integrating additional diffusion terms specified by $\underline{\mathbf{D}}$. The numerical diffusion terms depend on the model, on the state (traffic density), and on the discretizations Δx and Δt. For the two considered integration schemes, evaluating Eq. (9.46) yields

$$\begin{array}{ll} \underline{\mathbf{D}}_{\text{num}} = \frac{\Delta x}{2}\underline{\mathbf{C}} \cdot \left(\underline{\mathbf{1}} - \frac{\Delta t}{\Delta x}\underline{\mathbf{C}}\right) & \text{Upwind-Method,} \\ \underline{\mathbf{D}}_{\text{num}} = 0 & \text{McCormack-Method.} \end{array} \tag{9.47}$$

When integrating the GKT model with the upwind method, we essentially obtain the same scalar diffusions

$$D = V\frac{\Delta x}{2}\left(1 - V\frac{\Delta t}{\Delta x}\right) \tag{9.48}$$

in both the density and flow equations. Notice that the diffusions become negative (or more precisely, at least one eigenvalue of the diffusion matrix is negative) if and only if the upwind method becomes convectively unstable (the first CFL criterion (9.39) is violated).

The McCormack method has no numerical diffusion. Here, the discretization errors lead to *numerical dispersion* and other errors of higher order. While a signature of numerical diffusion is unnaturally smooth density and speed profiles, numerical dispersion leads to spurious artificial high-frequency waves near high density or speed gradients.

Problems

9.1 Ramp term of the acceleration equation

Derive the ramp term (9.17) for an on-ramp. Assume that the entering positions of the merging vehicles are uniformly distributed over the whole length L_{rmp} of the merge section, and none of the vehicles accelerates.

9.2 Kinematic dispersion

Consider a two-lane road with the same initial density profile on both lanes,

$$\rho(x,0) = \begin{cases} 15 & x < 0, \\ 15 - 100x & 0 \le x \le 0.1, \\ 5 & x > 0.1. \end{cases}$$

Here, ρ is given in units of vehicles/km (per lane), and x in kilometers. All vehicles on the right lane drive at 72 km/h and those on the left lane at 144 km/h, i.e., the initial lane-averaged velocity is $V(x,0) = 108$ km/h = const. Assume furthermore, that no vehicles accelerate or brake.

1. Determine the local speed variance $\sigma_V^2(x)$ at a given cross section of the road.
2. The speed variance corresponds to a kinematic pressure term $P_\theta = \rho\theta$. Which macroscopic acceleration $A(x,0)$ results from it at time $t = 0$?
3. Discuss using this example how a nonzero macroscopic acceleration may arise even if no vehicles accelerate or brake.
4. In the presence of a heterogeneous traffic composition, different actual speeds (as in this example) may be a consequence of distributed desired speeds. Find one of the principle limits of modeling heterogeneous traffic with the single-class macroscopic models (i.e., models having only one density field representing the whole population) by discussing the qualitative traffic flow dynamics of the above example for times $t > 0$.

9.3 Modeling anticipation by traffic pressure

We can model an anticipative driving style by evaluating the speed adaptation term $(\mathrm{d}V/\mathrm{d}t)_{\text{relax}} = (V_e - V)/\tau$ at a position $x_\mathrm{a} = x + d$ one distance headway d ahead of the actual position:

$$\left(\frac{\mathrm{d}V}{\mathrm{d}t}\right)_{\text{relax+antic}} = \frac{V_e(\rho(x_\mathrm{a},t)) - V(x,t)}{\tau}. \tag{9.49}$$

1. Express d by a macroscopic quantity.
2. Show that the anticipative part of Eq. (9.49) can be approximated by the pressure term $-\frac{1}{\rho}\frac{\partial P_\mathrm{a}}{\partial x}$ with the "anticipative" pressure component

   $$P_\mathrm{a} = -\frac{V_e(\rho(x,t))}{\tau}.$$

   Hint: Expand the adaptation term in a Taylor series to first order around x and assume a constant density.
3. Assume a situation where the lane-averaged density increases, in a distance of 200 m, from 20 to 40 vehicles/km. Furthermore, assume the steady-state speed-density relation

   $$V_e(\rho) = V_0\left(1 - \frac{\rho}{\rho_{\max}}\right).$$

Calculate the anticipative component of Eq. (9.49) when describing the anticipation (i) by the nonlocal part of the adaptation term, (ii) by the pressure term P_{a}. *Hint*: The anticipative component is characterized by the full acceleration (9.49) minus the local contribution obtained by setting $x_a = x$ in this equation.

9.4 Steady-state speed of the GKT model
Calculate the GKT steady-state speed-density relation for homogeneous roads. *Hint*: "Steady-state" means stationary traffic flow, i.e., $\frac{\partial}{\partial t} = 0$. Furthermore, homogeneous traffic flow implies $\frac{\partial}{\partial x} = 0$.

9.5 Flow-conserving form of second-order macroscopic models
Derive the conservation laws (9.29) and (9.30) resulting from the general second-order model (9.10), (9.11) for $D = 0$. *Hint*: Eliminate V in the continuity equation with the help of the definition $Q = \rho V$ for the flow. Multiply the acceleration equation by ρ and substitute $\rho \frac{\partial V}{\partial t} = \frac{\partial Q}{\partial t} - V \frac{\partial \rho}{\partial t}$. Now, use the continuity equation to eliminate $\frac{\partial \rho}{\partial t}$ and consolidate the resulting terms.

9.6 Numerics of the GKT model
Consider the numerical stability thresholds when integrating the GKT model with the upwind method for a cell size of 50 m. Assume the parameters of Table 9.1 and a constant speed variation coefficient $\sqrt{\alpha(\rho)} = 10\,\%$. Furthermore, assume that, in the simulation, the density is always below 100 vehicles/km. Determine the maximum time step Δt to avoid all sources of linear numerical instability. Calculate the numerical diffusions at a local speed of 72 km/h and a time step $\Delta t = 1$ s.

Further Reading

- Payne, H.: Models of freeway traffic and control. In: Bekey, G.A. (ed.) Mathematical Models of Public Systems, vol. 1, pp. 51–61. Simulation Council, La Jolla, CA (1971)
- Treiber, M., Hennecke, A., Helbing, D.: Derivation, properties, and simulation of a gas-kinetic based, non-local traffic model. Physical Review E **59** (1999) 239–253
- Kerner, B., Konhäuser, P.: Structure and parameters of clusters in traffic flow. Physical Review E **50** (1994) 54–83
- LeVeque, R.: Numerical Methods for Conservation Laws. Birkhäuser, Basel (1992)
- Hoogendoorn, S.: Continuum modeling of multiclass traffic flow. Transportation Research Part B **34** (2000) 123–146

Chapter 10
Elementary Car-Following Models

Progress is the realization of Utopias.
Oscar Wilde

Abstract Microscopic models describe traffic flow dynamics in terms of single vehicles. The mathematical formulations include car-following models, the topic of this and the next two chapters, and cellular automata, which are described in Chap. 13. This chapter begins with a discussion of general principles that apply to all microscopic models of traffic flow, such as the microscopic steady-state equilibrium, the micro-macro transition to the fundamental diagram, and heterogeneous traffic. The next sections discuss Newell's car-following model and the Optimal Velocity Model (and variants thereof) as the generic examples of simple car-following models.

10.1 General Remarks

Car following models are the most important representatives of microscopic traffic flow models (cf. Sect. 10.6). They describe traffic dynamics from the perspective of individual *driver-vehicle units*.[1] In a strict sense, car-following models describe the driver's behavior only in the presence of interactions with other vehicles while free traffic flow is described by a separate model. In a more general sense, car-following

[1] This technical term expresses the fact that the driving behavior depends not only on the driver but also on the acceleration and braking capabilities of the vehicle.

M. Treiber and A. Kesting, *Traffic Flow Dynamics*,

DOI: 10.1007/978-3-642-32460-4_10,

models include all traffic situations such as car-following situations, free traffic, and also stationary traffic. In this case we say that the microscopic models is complete:

> A car-following model is *complete* if it is able to describe all situations including acceleration and cruising in free traffic, following other vehicles in stationary and non-stationary situations, and approaching slow or standing vehicles, and red traffic lights.

Depending on the actions modeled, one distinguishes between *acceleration models* models for longitudinal movement, *lane-changing models* for lateral movement, and *decision models* for other discrete-choice situations such as entering a priority road. The latter two classes will be described in Chap. 14. Furthermore, more complex and realistic models describe the complex and often subtle interactions between acceleration and discrete-choice situations such as acceleration to the prevailing speed on the target lane in preparation of a lane change. This implies that the longitudinal and transversal dynamics can no longer be separated but must be part of a single complex model. Such models often form the simulation core of commercial traffic simulation software.

The first car-following models were proposed more than fifty years ago by Reuschel (1950), and Pipes (1953). These two models already contained one essential element of modern microscopic modeling: The minimum bumper-to-bumper distance to the leading vehicle (also known as the "safety distance") should be proportional to the speed. This can be expressed equivalently by requiring that the time gap should not be below a fixed *safe time gap*. We emphasize that, for obvious reasons, the relevant spatial or temporal distances are the net, i.e., rear-bumper-to-front-bumper, distances. In contrast, the commonly used term *time headway* generally refers to the time interval between the passage times of the front bumpers of two consecutive vehicles, i.e., including the *occupancy time interval* needed for a vehicle to move forward its own length. Unfortunately, this distinction (which is essential for vehicular traffic) is often ignored.[2] To avoid confusion and in order to be consistent with Sect. 2.2, we will refer to "gaps" if net quantities are meant and define gaps and headways as follows (the modifiers in parentheses will be omitted if the meaning is clear from the context):

> distance headway = (distance) gap + length of the leading vehicle,
> (time) headway = time gap + occupancy time interval of the leading vehicle.

[2] This has essentially historic reasons. The term "time headway" originates from rail transport indicating the succession time interval between two trains. Since this interval is measured in terms of minutes or hours, a distinction between gross and net quantities is irrelevant, in this context.

In this chapter, we will describe *minimal models* for the longitudinal dynamics that do not describe realistic driving behavior. Particularly, they yield unrealistic acceleration values. Nevertheless, they capture many essential features at a qualitative level and can be implemented and simulated easily (sometimes even allowing an analytical solution). Therefore, they are suited to introduce the essential concepts. Real applications require more refined models which will be presented in the subsequent chapters.

Regarding the classification of Sect. 6.2, the minimal models, as well as the cellular automata treated in Chap. 13, belong to the class of *heuristic models*. In contrast, the strategy-based models of the Chaps. 11 and 12 can be considered as *first-principles* models.

Examples of minimal models include the first ever car-following models of Reuschel and Pipes in which the speed is varied instantaneously as a function of the actual distance to the leading vehicle. Another class of minimal models are the *General Motors* (GM) based car-following models in which the acceleration depends on the speed difference and the distance gap according to a power law while the driver's own speed is not considered as an influencing factor. These models are not complete since they cannot describe either free traffic or approaches to standing obstacles. In this chapter, we will therefore focus on other models.

10.2 Mathematical Description

Each driver-vehicle combination α is described by the state variables location $x_\alpha(t)$ (position of the front bumper along the arc length of the road, increasing in driving direction), and speed $v_\alpha(t)$[3] as a function of the time t, and by the attribute "vehicle length" l_α. Depending on the model, additional state variables are required, for example, the acceleration $\dot{v}_\alpha = \mathrm{d}v/\mathrm{d}t$, or binary activation-state variables for brake lights or indicators. We define the vehicle index α such that vehicles pass a stationary observer (or detector) in ascending order, i.e., the first vehicle has the lowest index (cf. Fig. 10.1). Notice that this implies that the vehicles are numbered in *descending* order with respect to their location x.[4]

From the vehicle locations and lengths, we obtain the (bumper-to-bumper) *distance gaps*

$$s_\alpha = x_{\alpha-1} - l_{\alpha-1} - x_\alpha = x_l - l_l - x_\alpha \tag{10.1}$$

[3] To distinguish the vehicle speed in microscopic models from the local speed in macroscopic models, we denote the speed of individual vehicles in lowercase (in analogy to the notation for single-vehicle data). Generally, the relation $V = \langle v_\alpha \rangle$ applies.

[4] There is no generally accepted convention for the vehicle numbering in the literature. The converse numbering scheme (ascending in space, descending in time) is used as well, particularly in the literature of traffic engineers.

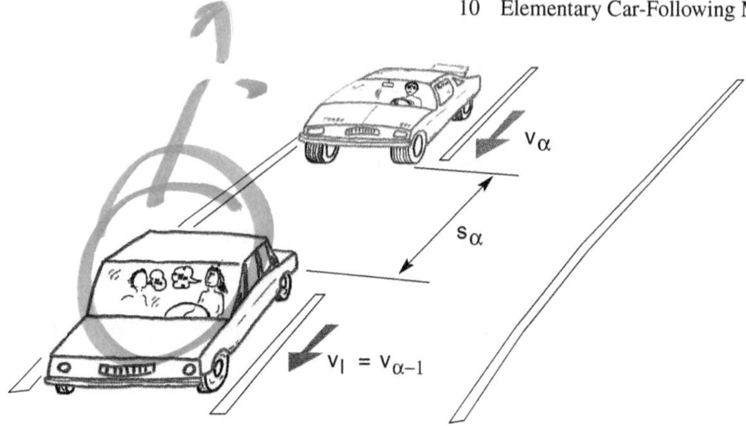

Fig. 10.1 Defining the state variables of car-following models

which (together with the vehicle speeds) constitute the main input of the microscopic models. For ease of notation, we sometimes denote the index $\alpha - 1$ of the leading vehicle with the symbol l (see Fig. 10.1).

The minimal models (and many of the more realistic models of the subsequent chapters) describe the response of the driver as a function of the gap s_α to the lead vehicle, the driver's speed v_α, and the speed v_l of the leader. In *continuous-time models*, the driver's response is directly given in terms of an *acceleration function* $a_{\text{mic}}(s, v, v_l)$ leading to a set of *coupled ordinary differential equations* of the form

$$\dot{x}_\alpha(t) = \frac{\mathrm{d}x_\alpha(t)}{\mathrm{d}t} = v_\alpha(t), \tag{10.2}$$

$$\dot{v}_\alpha(t) = \frac{\mathrm{d}v_\alpha(t)}{\mathrm{d}t} = a_{\text{mic}}(s_\alpha, v_\alpha, v_l) = \tilde{a}_{\text{mic}}(s_\alpha, v_\alpha, \Delta v_\alpha). \tag{10.3}$$

In most acceleration functions, the speed v_l of the leader enters only in form of the speed difference (approaching rate)[5]

$$\Delta v_\alpha = v_\alpha - v_{\alpha-1} = v_\alpha - v_l. \tag{10.4}$$

The corresponding models can be formulated more concisely in terms of the alternative acceleration function

$$\tilde{a}_{\text{mic}}(s, v, \Delta v) = a_{\text{mic}}(s, v, v - \Delta v). \tag{10.5}$$

Taking the time derivative of Eq. (10.1), one can reformulate Eq. (10.2) by

$$\dot{s}_\alpha(t) = \frac{\mathrm{d}s_\alpha(t)}{\mathrm{d}t} = v_l(t) - v_\alpha(t) = -\Delta v_\alpha(t). \tag{10.6}$$

[5] Again, there is no common consensus about the definition. Sometimes, the speed difference is defined as $v_l - v$, i.e., as the negative approaching rate.

The set of Eqs. (10.3) and (10.6) can be considered as the generic formulation of most time-continuous car-following models. In this formulation, the coupling between the gap s_α and the speed v_α as well as the coupling between the speed v_α and the speed v_l of the leader becomes explicit.

There are also *discrete-time* car-following models, where time is not modeled as a continuous variable but discretized into finite and generally constant time steps. Instead of differential equations, one obtains *iterated coupled maps* of the general form

$$v_\alpha(t + \Delta t) = v_{\text{mic}}\left(s_\alpha(t), v_\alpha(t), v_l(t)\right), \tag{10.7}$$

$$x_\alpha(t + \Delta t) = x_\alpha(t) + \frac{v_\alpha(t) + v_\alpha(t + \Delta t)}{2} \Delta t. \tag{10.8}$$

The driver's response is no longer modeled by an acceleration function but by a *speed function* $v_{\text{mic}}(s, v, v_l)$ indicating the speed that will be reached at the end of the next time step.

Compared to continuous models, discrete-time car-following models are generally less realistic and less flexible but require less computing power for their numerical integration. Most discrete-time car-following models have been proposed at times where computing was more expensive. Nowadays, hundreds of thousand vehicles can be simulated with time-continuous models on a PC in real-time, so this numerical advantage becomes less relevant. Most commercial traffic simulation software uses time-continuous models.

We emphasize that the Eqs. (10.2) and (10.6) represent kinematic facts that are valid *a priori*—in analogy to the continuity equations of the macroscopic models. Therefore, a specific time-continuous model is uniquely characterized by its acceleration function a_{mic}. Similarly, a specific discrete-time model is completely characterized by its speed function v_{mic}. When simulating heterogeneous traffic consisting of a variety of driving styles and vehicle classes (such as cars and trucks), each driver-vehicle combination is described by different acceleration functions $a^\alpha_{\text{mic}}(s, v, v_l)$ or speed functions $v^\alpha_{\text{mic}}(s, v, v_l)$, respectively.

Numerical integration. In general, time-continuous models cannot be solved analytically and an integration scheme is necessary for an approximate numerical solution of the system of Eqs. (10.3) and (10.6). For traffic flow applications, only explicit update schemes with a fixed time step are practical. Furthermore, the performance of the standard fourth order *Runge-Kutta scheme* is generally inferior to simpler lower-order update methods.[6]

---

[6] For typical single-lane simulations, the mixed first-order-second-order scheme (10.9), (10.10) is more efficient by a factor of about two compared to the Runge-Kutta scheme, and by a factor of about three with respect to the simple first-order *Euler update*, Eq. (10.9) combined with the positional update $x_\alpha(t + \Delta t) = x_\alpha(t) + v(t)\Delta t$. For multi-lane simulations, the difference is less pronounced. Ignoring bookkeeping costs, numerical efficiency is defined by the inverse of the number of evaluations of the acceleration function for obtaining a numerical solution with given accuracy.

While this may seem surprising at first sight, it can be understood when considering the standard principles of numerical mathematics: Higher-order schemes are guaranteed to be more effective than lower-order schemes only if the exact solution is sufficiently smooth *and* the update time is sufficiently small: Specifically, the standard Runge-Kutta scheme has its nominal consistency order of four only if the acceleration function is at least three times continuously differentiable with respect to all independent variables and with respect to time. While this is satisfied for single-lane simulations of some car-following models, it always breaks down for multi-lane simulations: When lane changes occur, the acceleration function of the subject vehicle (active lane change) and the old and new followers is not even continuous in time. In this case, the accuracy of higher-order schemes may even be *worse* than that of lower-order schemes for the same time step. Furthermore, the proposed scheme (10.9), (10.10) has an intuitive meaning in the context of car-following models: It corresponds to drivers that act only at the beginning of each time step but do nothing in between (see Sect. 12.2 for details).

Assuming a constant update time step Δt, a simple but efficient explicit update method is given by the "ballistic" assumption of constant accelerations during each time step,

$$v_\alpha(t+\Delta t) = v_\alpha(t) + a_{\text{mic}}\left(s_\alpha(t), v_\alpha(t), v_l(t)\right)\Delta t, \tag{10.9}$$

$$x_\alpha(t+\Delta t) = x_\alpha(t) + \frac{v_\alpha(t) + v_\alpha(t+\Delta t)}{2}\,\Delta t. \tag{10.10}$$

Consequently, the combination of a continuous-time model with the ballistic update scheme (10.9), (10.10) is *mathematically* equivalent to discrete-time models if one sets

$$a_{\text{mic}}(s, v, v_l) = \frac{v_{\text{mic}}(s, v, v_l) - v}{\Delta t}. \tag{10.11}$$

However, there is a *conceptual* difference: For discrete-time models, the time step Δt plays the role of a model parameter typically describing the reaction time, the time headway, or the speed adaptation time. For time-continuous models, the update time Δt is an auxiliary variable of the approximate numerical solution which preferably should be small as the true solution is obtained in the limit $\Delta t \to 0$ (at least if the numerical method is consistent and stable).

10.3 Steady State Equilibrium and the Fundamental Diagram

Since the driver-vehicle units of microscopic models are equivalent to *driven particles* of physical systems, there is no equilibrium in the strict sense. Instead, there is a stationary state where the forces and the entering and exiting energy fluxes are balanced. Strictly physically, this can be interpreted in terms of a balance of the forces (the sum of friction, wind drag, and engine driving force equates to zero) or energy fluxes (engine power equals change in potential and kinetic energy plus energy dissipation rate by friction and wind drag). More relevant for traffic flow, however, is the concept of balancing the *social forces*: The desire to go ahead generates a positive (accelerating) social force while the interactions with other vehicles generally lead to negative social forces in order to avoid critical situations and crashes.[7] In any case, such a balanced state is denoted as *steady-state equilibrium*. For microscopic models, a consistent description of the steady-state equilibrium requires identical driver-vehicle units on a homogeneous road.[8] Technically, this implies that the model parameters are the same for all drivers and vehicles, i.e., the acceleration or speed functions characterizing the respective model do not depend on the vehicle index, $a^{\alpha}_{\text{mic}}(s, v, v_l) = a_{\text{mic}}(s, v, v_l)$, and $v^{\alpha}_{\text{mic}}(s, v, v_l) = v_{\text{mic}}(s, v, v_l)$, respectively. From the modeling point of view, the steady-state equilibrium is characterized by the following two conditions:

- *Homogeneous traffic*: All vehicles drive at the same speed ($v_\alpha = v$) and keep the same gap behind their respective leaders ($s_\alpha = s$).
- *No accelerations*: $\dot{v}_\alpha = 0$ or $v_\alpha(t + \Delta t) = v_\alpha(t)$ for all vehicles α.

For time-continuous models with acceleration functions of the form a_{mic} or $\tilde{a}_{\text{mic}}$ this implies

$$a_{\text{mic}}(s, v, v) = 0, \quad \text{or} \quad \tilde{a}_{\text{mic}}(s, v, 0) = 0, \tag{10.12}$$

respectively, while the condition

$$v_{\text{mic}}(s, v, v) = v \tag{10.13}$$

is valid for discrete-time models with the speed function (10.7). Depending on the model, the microscopic steady-state relations (10.12) or (10.13) can be solved for

- the equilibrium speed $v_e(s)$ as a function of the gap (microscopic fundamental diagram, see below),
- the equilibrium gap $s_e(v)$ for a given speed.

Microscopic fundamental diagram. The Eqs. (10.12) and (10.13) allow for a one-dimensional manifold of possible steady states that can be parameterized by the

---

[7] The physical interpretation of forces becomes more relevant when simulating driven particles where the available driving power is a major limiting factor. Examples include trucks going up a steep hill, bicycle traffic, inline skating, and cross-country-skiing mass events.

[8] One can conceive of microscopic models having no unique steady state even in this case. Such models "without a fundamental diagram" have been postulated by some researchers. However, they are controversial (see the KKW model in Sect. 13.3.2 for an example).

distance gap s and described by the equilibrium speed function $v_e(s)$ which is also termed the *microscopic fundamental diagram.*[9]

Transition to macroscopic relations. In order to obtain a *micro–macro relation* between the distance gap s and the density ρ we directly apply the definition of the density as number of vehicles per road length. For a given vehicle length l, we obtain

$$s_\alpha = s = \frac{1}{\rho} - l. \tag{10.14}$$

Furthermore, the steady-state equilibrium implies that the speed of all vehicles is the same and equal to the macroscopic speed

$$V(x,t) = \langle v_\alpha(t) \rangle = v_e(s). \tag{10.15}$$

With these relations, we can derive the macroscopic steady-state speed-density diagram and the macroscopic fundamental diagram:

$$V_e(\rho) = v_e\left(\frac{1}{\rho} - l\right), \quad Q_e(\rho) = \rho v_e\left(\frac{1}{\rho} - l\right). \tag{10.16}$$

10.4 Heterogeneous Traffic

Microscopic models play out their advantages when describing different drivers and vehicles, i.e., *heterogeneous traffic*. Including different drivers and vehicles is crucial when modeling the effects of active traffic management such as variable message signs, speed limits, or ramp metering, or when simulating traffic-related effects of new driver-assistance systems as discussed in Chap. 21 of this book. Heterogeneous traffic can be microscopically modeled in two ways:

1. All driver-vehicle units are described by the same model using different parameter values. The heterogeneity can be applied on the level of vehicle classes (e.g., different parameters for cars and trucks), individually (distributed parameters), or both (different parameter distributions for cars and trucks). The last combined approach has the advantage that it automatically leads to realistic correlations between the parameters.[10]
2. Different driver-vehicle classes can also be described with different *models*. This allows us to directly represent qualitatively different driving characteristics between, e.g., cars and trucks or between human driving and semi-automated driving with the help of adaptive cruise control (ACC) systems.

[9] Not to be confused with the data related microscopic flow-density diagram, i.e., a scatter plot of the points $(1/d_\alpha, 1/\Delta t_\alpha)$ that can be derived from single-vehicle detector data, see Chap. 3.

[10] For example, trucks tend to drive more slowly than cars and they tend to accelerate more slowly as well. Thus, speed and acceleration parameters are positively correlated.

We emphasize that simulating heterogeneous traffic is only sensible in the context of multi-lane traffic models. Otherwise, a single long queue will eventually form behind the slowest vehicle, which is unrealistic. Finally, when parameterizing heterogeneous traffic, it is favorable if the model parameters have an intuitive meaning (such as that of the models presented in Chap. 11).

10.5 Fact Sheet of Dynamical Model Characteristics

In order to compare the properties of the different microscopic models introduced in this chapter and the following Chaps. 11 and 12, we will provide, for each model, simulation results for two example scenarios in form of a *fact sheet*. We consider these characteristics for the different microscopic models as one of the core elements of this book. Using the simulation tool on the book's website[11] the reader can interactively reproduce the characteristics, and can also produce new simulation results by changing the model parameters (see Fig. 10.2 for a screenshot).

In the following, we will describe the simulation scenarios with the help of Fig. 10.3 displaying the essential characteristics of a given model in terms of a *fact sheet*. The model itself (the Optimal Velocity Model) will be discussed in Sect. 10.6 below.

10.5.1 Highway Scenario

The left column of Fig. 10.3 shows the simulation of a highway with an on-ramp during rush-hour conditions. In order to display the "pure" characteristics of the acceleration models, we simulate a single freeway lane. Lane changes (see Chap. 14 for details) take place in the 1 km long merging zone of the on-ramp centered at $x = 0$, only. The traffic demand during the rush hour is modeled by prescribing a time-varying inflow at the upstream end of the simulated main-road section with a peak value of 2,200 vehicles per hour.[12] The on-ramp inflow of 550 vehicles/h is assumed to be constant. Furthermore, "free" boundary conditions are assumed at the downstream end of the road: As soon as a vehicle crosses the downstream boundary, it is taken out of the system. The next vehicle behaves as though there is an empty road of infinite length until it leaves the system as well.

Due to the high traffic demand, traffic breaks down near the merging zone around $x = 0$ at about $t = 10$ min. Fig. 10.3a depicts the spatiotemporal evolution of

[11] see: www.traffic-flow-dynamics.org

[12] The speed of the entering vehicles must be given as well. However, since the speed will reach equilibrium during the first few hundred meters, its initial value is irrelevant as long as it allows the vehicles to be inserted at the rate prescribed by the flow. Specifically, we determine the gap at insertion time and set the speed to the equilibrium speed given by the microscopic fundamental diagram for this gap.

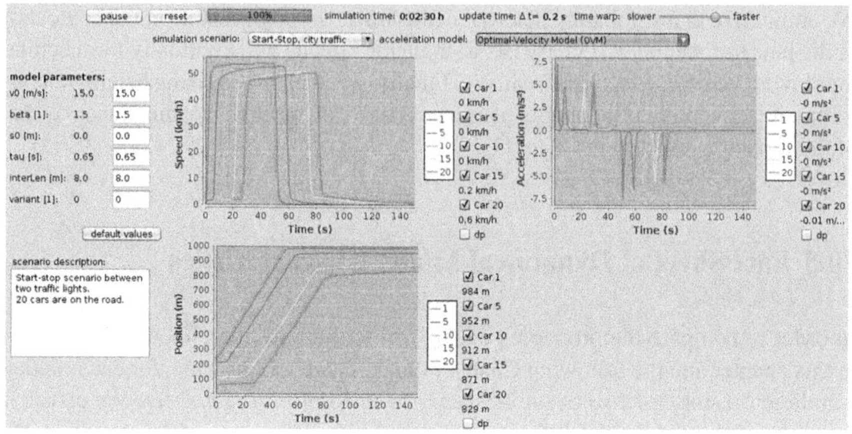

Fig. 10.2 Screenshot of the interactive simulation tool on the book's website. Shown is the city scenario for the Optimal Velocity Model simulated with the standard parameters

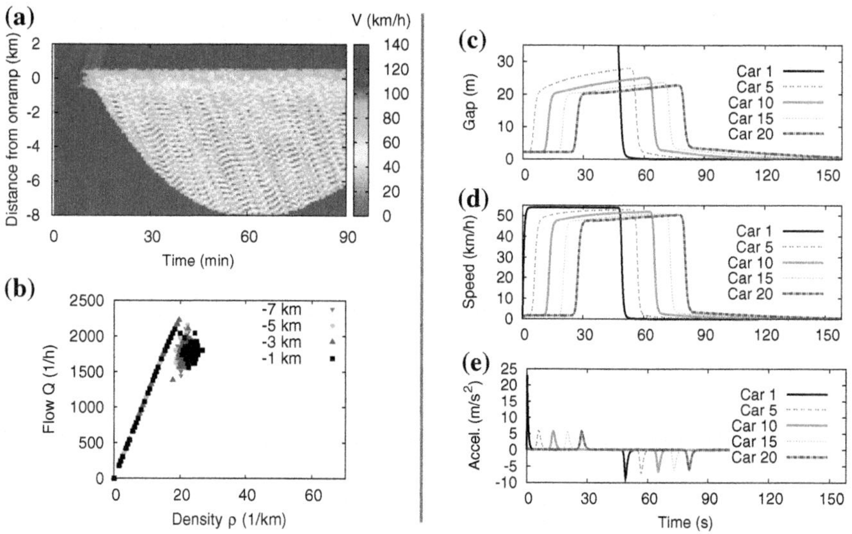

Fig. 10.3 Fact sheet of the Optimal Velocity Model with the OV function (10.21) and model parameters according to Table 10.1. The *left column* displays results of the highway scenario with an on-ramp: **a** Local speed, and **b** flow-density data of virtual detectors. The *right column* depicts the city start–stop scenario where vehicles move from one traffic-light controlled intersection to the next. See Sect. 10.5 for a detailed description of the simulation scenarios

the congestion following the traffic breakdown by plotting the *macroscopic local speed* $V(x, t)$. The local speed is derived from the set $\{x_\alpha(t)\}$ of simulated trajectories by a linear interpolation over space at fixed time according to

$$V(x, t) = \frac{x - x_\alpha(t)}{x_{\alpha-1}(t) - x_\alpha(t)} v_{\alpha-1}(t) + \frac{x_{\alpha-1}(t) - x}{x_{\alpha-1}(t) - x_\alpha(t)} v_\alpha(t). \tag{10.17}$$

For a given spatiotemporal point (x, t), the vehicle index α is determined such that $x_\alpha(t) \leq x < x_{\alpha-1}(t)$. This plot shows that the congested region caused by the traffic breakdown grows for $t < 60$ min. Later on, it shrinks due to the decreasing demand. The following aspects of the simulated traffic dynamics can be compared with real-world traffic:

- Driving in free traffic.
- Braking characteristics when approaching the upstream jam front (in real traffic flow, the deceleration profile is smooth and the deceleration rarely exceeds 2 m/s^2).
- Acceleration characteristics when exiting the stationary downstream jam front near the on-ramp bottleneck (again, the real acceleration is smooth and the accelerations generally do not exceed 2 m/s^2).
- Capacity of the *activated* bottleneck, i.e., the flow downstream of the bottleneck after breakdown (observed values are of the order of 2,000 vehicles/h per lane).
- Stability of traffic flow in free and congested conditions (free traffic should be stable and congested traffic unstable with respect to oscillations and stop-and-go traffic, see Chap. 15 for details).

Finally, the simulated spatiotemporal dynamics of regions with traffic oscillations can be compared with following observed facts (see Sect. 18.3 for details): All oscillations and waves of a congested area propagate upstream at the same velocity (-20 to -15 km/h, depending only weakly on the country and the traffic composition). Furthermore, the wavelength of contiguous stop-and-go waves varies between 1 and 3 km, and the wave amplitude grows during the upstream propagation. Finally, the decelerations and accelerations of the vehicles entering and leaving the waves rarely exceed 2 m/s^2.

Figure 10.3b shows flow-density data at several locations. To enable a direct comparison with real data, we simulate the data capturing process as well by means of *virtual detectors* positioned at several locations. To this end, we create *virtual single-vehicle data* by recording the passing times and speeds of each vehicle trajectory when it passes one of the detectors. Furthermore, we generate aggregated virtual data by determining, for each one-minute aggregation interval, the number of passing vehicles (or, equivalently, the traffic flow), and the arithmetic mean speed according to Sect. 3.2. Finally, flow and density is calculated and plotted as in the Sects. 3.3.1 and 4.4, respectively. Following characteristics can be compared with real observations (for example, that displayed in Sect. 4.4):

- The free branch of the flow-density data shows comparatively little scattering and a negative curvature near capacity.

- The congested flow-density data shows wide scattering.[13]
- In many cases, real flow-density data are distributed according to an inverse-λ shape corresponding to a capacity drop of typically 10–20 % and associated hysteresis effects.

10.5.2 City Scenario

The right column of the fact sheet 10.3 shows a typical inner-city situation with intersections and traffic lights. The simulation is initialized by a queue of 20 vehicles waiting behind a red traffic light which turns green at $t = 0$. During the simulation, the vehicle queue moves in single file to the next red traffic light 740 m further downstream. Red traffic lights are simulated by a standing virtual vehicle of zero length positioned at the stopping line which is removed when the light turns green. Following driving characteristics can be compared with reality:

- Starting phase: The acceleration ranges between 1 and 2.5 m/s^2 and smoothly decreases to zero as cruising speed is approached. The first vehicle takes 3–4 s to pass the stopping line of the traffic light. Afterwards, the cars pass at a rate of one vehicle every 1.5–2 s.
- Cruising phase: In this phase, all vehicles should travel at close to the desired speed. The vehicles should move as a platoon with time gaps of the order of 1–2 s corresponding to distance gaps of about 15–30 m.
- Approaching phase: All vehicles should decelerate smoothly such that the values for the *jerk*

$$J = \left|\frac{\mathrm{d}\dot{v}}{\mathrm{d}t}\right| = \left|\frac{\mathrm{d}^2 v}{\mathrm{d}t^2}\right| \tag{10.18}$$

  remain below 2 m/s^3. The braking decelerations themselves generally do not exceed 2 m/s^2 and the braking becomes less pronounced for vehicles more at the back of the platoon.

10.6 Optimal Velocity Model

The *Optimal Velocity Model* (OVM) is a time-continuous model whose acceleration function is of the form $a_{\text{mic}}(s, v)$, i.e., the speed difference exogenous variable is missing. The acceleration equation is given by

---

[13] We point to the fact that this difference is partly caused by the way of presentation. When plotting flow-density, speed-density or speed-flow diagrams from the same data, only the flow-density data show a distinct discrepancy of the amount of scattering on the free and congested branches, see Fig. 4.12.

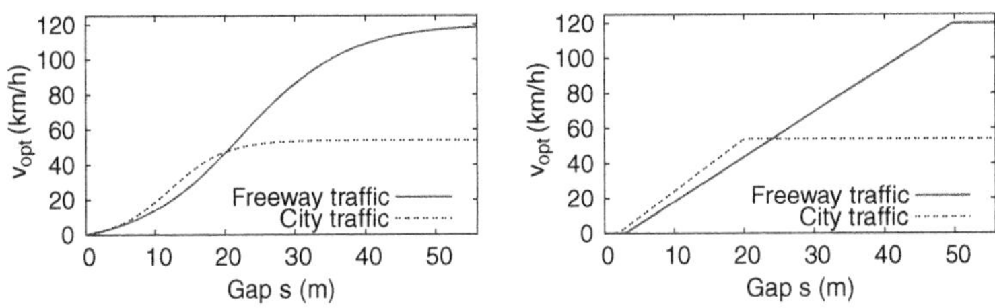

Fig. 10.4 Optimal velocity functions (10.21) (*left*) and (10.22) (*right*) for the parameter values of Table 10.1

$$\boxed{\dot{v} = \frac{v_{\text{opt}}(s) - v}{\tau} \quad \text{Optimal Velocity Model.}} \tag{10.19}$$

This equation describes the adaption of the actual speed $v = v_\alpha$ to the *optimal velocity* $v_{\text{opt}}(s)$ on a time scale given by the *adaptation time* τ. Comparing the acceleration equation (10.19) with the steady-state condition (10.12) it becomes evident that the optimal velocity (OV) function[14] $v_{\text{opt}}(s)$ is equivalent to the microscopic fundamental diagram $v_e(s)$. It should obey the plausibility conditions

$$v'_{\text{opt}}(s) \geq 0, \quad v_{\text{opt}}(0) = 0, \quad \lim_{s\to\infty} v_{\text{opt}}(s) = v_0, \tag{10.20}$$

but is arbitrary, otherwise. Thus, the acceleration equation (10.19) defines a whole class of models whose members are distinguished by their respective optimal velocity functions. The OV function originally proposed by Bando et al.,

$$v_{\text{opt}}(s) = v_0 \frac{\tanh\left(\frac{s}{\Delta s} - \beta\right) + \tanh\beta}{1 + \tanh\beta}, \tag{10.21}$$

uses a hyperbolic tangent.[15] Besides the parameter τ which is relevant for all optimal velocity models,[16] the OVM of Bando et al. has three additional parameters, the desired speed v_0, the transition width Δs, and the form factor β (see Fig. 10.4 and Table 10.1).

A more intuitive OV function can be derived by characterizing free traffic by the desired speed v_0, congested traffic by the time gap T in car-following mode under stationary conditions, and standing traffic by the minimum gap s_0. In analogy to the Section-Based Model of Sect. 8.5, we obtain

---

[14] Strictly speaking, this is an "optimal speed function". In order to be consistent with the literature (and the model name), we will nevertheless stick to velocity instead of speed, in this context.

[15] We have adapted the notation with respect to the original publication such that $v_0 = v_{\text{opt}}(0)$ strictly has the meaning of the desired speed if no other vehicles are present.

[16] Sometimes, τ is replaced by the sensitivity parameter $a = 1/\tau$.

Table 10.1 Parameter of two variants of the Optimal Velocity Model (OVM)

Parameter	Typical value highway	Typical value city traffic
Adaptation time τ	0.65 s	0.65 s
Desired speed v_0	120 km/h	54 km/h
Transition width Δs [v_{opt} according to Eq. (10.21)]	15 m	8 m
Form factor β [v_{opt} according to Eq. (10.21)]	1.5	1.5
Time gap T [v_{opt} according to Eq. (10.22)]	1.4 s	1.2 s
Minimum distance gap s_0 [v_{opt} according to Eq. (10.22)]	3 m	2 m

$$v_{opt}(s) = \max\left[0,\ \min\left(v_0, \frac{s - s_0}{T}\right)\right]. \tag{10.22}$$

This relation is the microscopic equivalent to the triangular fundamental diagrams of the macroscopic Section-Based and Cell-Transmission Models (cf. Fig. 10.4). The simulation results are similar to that of the hyperbolic tangent OV function. Again, typical parameter values are given in Table 10.1.

Model properties. For the simulations of the model fact sheet in Fig. 10.3, we have assumed a speed adaptation time $\tau = 0.65$ s. Obviously this is an unrealistically low value since typical time scales for reaching a desired speed are of the order of 10 s. As a consequence, the periods of the stop-and-go waves (about 1–2 min) are too low. Furthermore, in the simulation of the city scenario, the accelerations and decelerations of the vehicle platoon (up to $22\,\mathrm{m/s^2}$ and down to $-10\,\mathrm{m/s^2}$, respectively) differ from real accelerations by a factor of about ten.[17] However, when increasing the adaptation time τ by only 5 %, the simulations eventually lead to negative distance gaps s_α corresponding to *accidents*. On the other hand, when decreasing τ by 5 %, we obtain absolute string stability even for congested freeway traffic. While this agrees with the theoretical stability limits (see Chap. 15) it is at variance with the observations. We conclude that the simulation outcome is qualitatively correct. However:

- On a quantitative level, the OVM results are unrealistic.
- On a qualitative level, the simulation outcome has a strong dependency on the fine tuning of the model parameters, i.e., the OVM is not *robust*.

These deficiencies are mainly due to the fact that the OVM acceleration function does not contain the speed difference as exogenous variable, i.e., the simulated driver reaction depends only on the gap but is the same whether the leading vehicle is slower or faster than the subject vehicle. This corresponds to an extremely short-sighted driving style.

[17] In reality, accelerations above $4\,\mathrm{m/s^2}$ (corresponding to 7 s for accelerating from zero to 100 km/h) and below $-9\,\mathrm{m/s^2}$ (corresponding to an emergency braking maneuver on a dry road) are physically impossible. Furthermore, everyday accelerations (which the simulations should reproduce) generally are only a fraction of the accelerations at these limits.

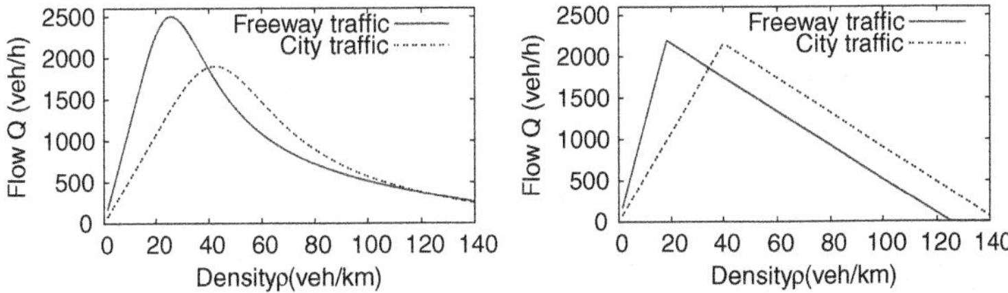

Fig. 10.5 Fundamental diagrams of the OVM for the OV functions (10.21) (*left*) and (10.22) (*right*) for the parameter values of Table 10.1

Steady-state equilibrium. As already mentioned, the steady-state condition (10.12) for the OVM leads to $v_e(s) = v_{\text{opt}}(s)$, i.e., the microscopic fundamental diagram Fig. 10.4 is given by the OV function. The macroscopic fundamental diagram (Fig. 10.5) is obtained from the microscopic one using Eq. (10.16). In particular, the OV function (10.22) results in the triangular macroscopic fundamental diagram

$$Q_e(\rho) = \min\left(v_0\rho, \frac{1-\rho(l+s_0)}{T}\right).$$

10.7 Full Velocity Difference Model

By extending the OVM with an additional linear stimulus for the speed difference, one obtains the *Full Velocity Difference Model* (FVDM):

$$\boxed{\dot{v} = \frac{v_{\text{opt}}(s) - v}{\tau} - \gamma \Delta v \quad \text{Full Velocity Difference Model.}} \tag{10.23}$$

As in the OVM, the steady-state equilibrium is directly given by the optimal velocity function v_{opt}. When assuming suitable values for the speed difference sensitivity γ of the order of $0.6\,\text{s}^{-1}$, the FVDM remains accident-free for speed adaptation times of the order of several seconds. Furthermore, the fact sheet (Fig. 10.6) shows that the waves in the congested region of the freeway scenario are more realistic than that of the OVM, although the wavelengths remain too short. Furthermore, the accelerations remain in a realistic range.

However, in contrast to the OVM, the Full Velocity Difference Model is not *complete* in the sense defined at the beginning of Sect. 10.1, i.e., it is not able to describe all traffic situations. The reason is that the term $\gamma \Delta v$ describing the sensitivity to speed difference in Eq. (10.23) does not depend on the gap. Consequently, a slow vehicle (or a red traffic light corresponding to a standing virtual vehicle) leads to a significant decelerating contribution even if it is miles away. Thus, simulated vehicles do not reach their desired speed even on a long road with no other vehicles.

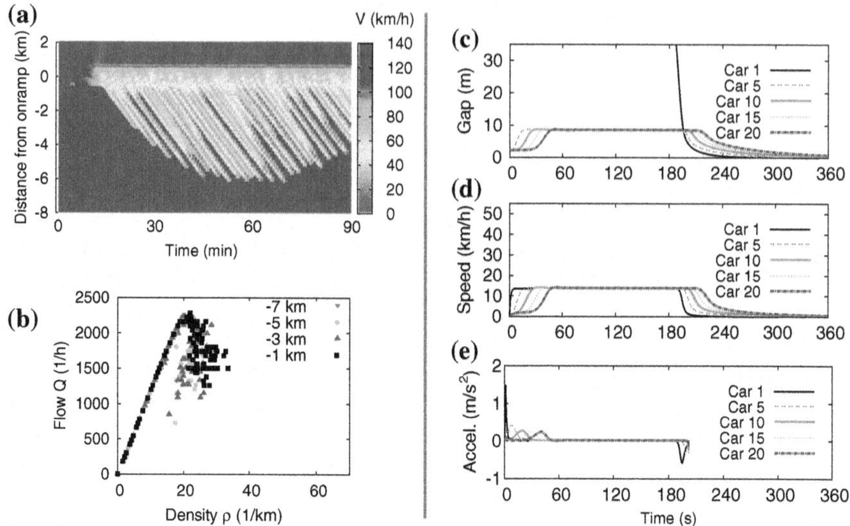

Fig. 10.6 Fact sheet of the Full Velocity Difference Models (FVDM), Eq. (10.23), with the OV function (10.21), the sensitivity $\gamma = 0.6\,\mathrm{s}^{-1}$, and the speed adaptation time $\tau = 5\,\mathrm{s}$. The vehicle length and the parameters v_0, β and Δs of the OV function are given by Table 10.1. The simulation scenarios are discussed in detail in Sect. 10.5

In fact, the maximum speed in the city scenario of the model fact Sheet of Fig. 10.6 is less than 15 km/h (see Problem 10.4 for a quantitative analysis).

Improved Full Velocity Difference Model. In the following, we show how a model developer would proceed to resolve this problem. Obviously, the sensitivity to speed differences must decrease with the gap s and tend to zero as $s \to \infty$. This can be realized by replacing the contribution $-\gamma \Delta v$ of Eq. (10.23) by a multiplicative term $-\tilde{\gamma} \Delta v / s$. However, now the sensitivity diverges for $s \to 0$ which is unrealistic. Furthermore, $\tilde{\gamma}$ has a different unit compared to γ (and, consequently, a different numerical value) which should be avoided if possible.[18] The arguably simplest approach to resolve these new problems consists in applying the inverse proportionality only if the gap is larger than the interaction length $v_0 T$. Hence, the resulting acceleration equation of the new "complete" variant of the FVDM is given by

$$\dot{v} = \frac{v_{\mathrm{opt}}(s) - v}{\tau} - \frac{\gamma \, \Delta v}{\max[1, s/(v_0 T)]}. \tag{10.24}$$

Figure 10.7 displays the city scenario for this model variant. It turns out that model (10.24) is able to realistically simulate the cruising phase, in contrast to the

[18] Three principles for improving existing models are the following: (i) Introduce as few new parameters as possible (ideally zero), (ii) do not change the meaning of existing parameters, (iii) keep it as simple as possible, but not simpler.

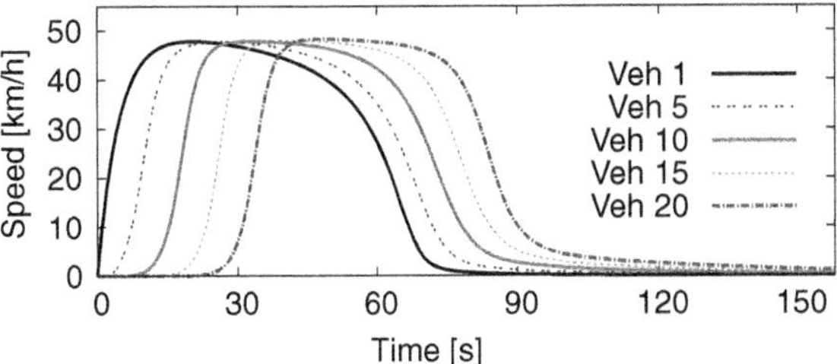

Fig. 10.7 The city scenario of Fig. 10.6 for the "complete" Full Velocity Difference Model (10.24) with the OV function (10.21). The model parameters are the same as in the original simulation

original model (10.23), and produces realistic accelerations, in contrast to the OVM. However, the robustness problem is not resolved.[19]

10.8 Newell's Car-Following Model

Newell's car-following model is the arguably simplest representative of time-discrete models of the type (10.7). Its speed function is directly given by the optimal speed function (10.22) corresponding to the triangular fundamental diagram (10.22) with $s_0 = 0$,

$$\boxed{v(t+T) = v_{\text{opt}}(s(t)), \quad v_{\text{opt}}(s) = \min\left(v_0, \frac{s}{T}\right) \quad \text{Newell's Model.}} \tag{10.25}$$

When restricting to the car-following regime, Newell's model has two parameters: The time gap or reaction time T, and the (effective) vehicle length l_{eff}. Since in this regime the kinematic wave velocity is constant and given by

$$w = c_{\text{cong}} = -l_{\text{eff}}/T,$$

the set of model parameters can alternatively be expressed by $\{T, w\}$ or by $\{l_{\text{eff}}, w\}$. The standard value for the time gap is $T = 1$ s while the wave speed should be within the observed range $w \in [-20\,\text{km/h}, -15\,\text{km/h}]$ corresponding to a plausible effective vehicle length l_{eff} of about 5 m. The minimum condition of the optimal velocity function makes the model complete by defining a free-flow regime and introducing the desired speed v_0 as a third model parameter. It is straightforward to generalize Newell's model by replacing Eq. (10.22) with other microscopic fundamental diagrams.

Newell's model can also be considered as a continuous-in-time model with a time delay assuming that the drivers have a constant *reaction time* $T_r = T$. In this interpretation, Eq. (10.25) has the mathematical form of a *delay-differential equation*

[19] The reader can verify this by simulating the improved FVDM on the book's website www.traffic-flow-dynamics.org and changing the sensitivity γ.

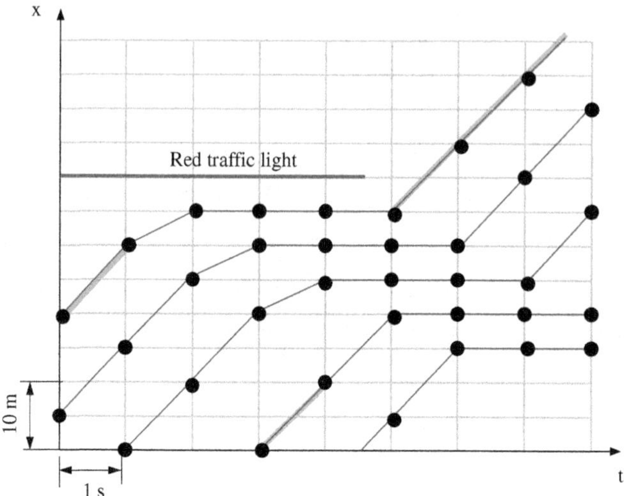

Fig. 10.8 Trajectory plot of the OVM with the triangular fundamental diagram ($l_{\text{eff}} = 5$ m, $v_0 = 10$ m/s, $T = \tau = 1$ s) with an update time $\Delta t = 1$ s, and using Eq. (10.26) for the positional update. Shown are vehicles approaching a traffic light (*red line*) turning *green* at $t = 4.5$ s (*end of the line*). The vehicle trajectories correspond to free traffic if drawn *green*, and to bound traffic, otherwise

which will be discussed in Chap. 12 in more detail. This model has several interesting properties which we now investigate further.

Relation to the Optimal Velocity Model. According to Eq. (10.11), Newell's model is mathematically equivalent to the OVM (10.19) in the car-following regime (bound traffic) if one sets $\tau = T$ and updates the OVM speed according to the explicit integration scheme (10.9) and the vehicle positions by the simple *Euler scheme*[20]

$$x_\alpha(t + \Delta t) = x_\alpha(t) + v_\alpha(t + \Delta t)\, \Delta t. \tag{10.26}$$

As a consequence, the parameter T of Newell's model has the additional meaning of a speed adaptation time τ.

Figure 10.8 shows that this equivalence only applies for the triangular fundamental diagram and only in the bound traffic regime, i.e., for gaps s satisfying $v_e(s) < v_0$ or $s < s_0 + v_0 T$. Otherwise, discretization errors are present.

Generally, the OVM is updated with time steps significantly smaller than the adaptation time. However, this does not invalidate the reasoning above, at least, qualitatively. In any case, the steady-state equilibria of the two models are equivalent.

[20] We emphasize that the usual second-order "ballistic" update scheme (10.8) may not be applied, here.

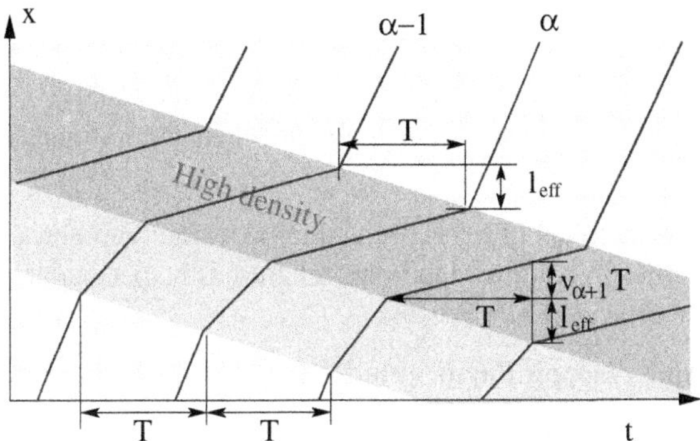

Fig. 10.9 Relation between Newell's model and the Section-Based Model: The *shaded regions* represent the evolution of the macroscopic local density. The gradient $v_e(\rho) = (1/\rho - l_{\text{eff}})/T$ corresponds to the local speed

Relation to the macroscopic Section-Based Model. When disaggregating the solutions of the Section-Based Model (8.2) with the function (8.11) by generating trajectories from the density and speed fields of congested traffic using the macro-micro relation (8.23), these trajectories are simultaneously solutions of Newell's model (as illustrated by Fig. 10.9).

Interpretation from the driver's point of view. The trajectories corresponding to the solution of Newell's model for congested traffic shown in Fig. 10.9 are given by the recursive relations

$$\begin{aligned} x_\alpha(t+T) &= x_{\alpha-1}(t) + wT = x_{\alpha-1}(t) - l_{\text{eff}}, \\ v_\alpha(t+T) &= v_{\alpha-1}(t). \end{aligned} \tag{10.27}$$

This means that the trajectory of the follower is completely determined by the trajectory of the leading vehicle.

In Newell's car-following model, the position of a vehicle following another vehicle at time $t + T$ is given by the position of the leader at time t minus the (effective) vehicle length l_{eff}. As a corollary, the speed profile of a vehicle *exactly* reproduces that of its leader with a time delay T.

The different meanings of the parameter T. From the above considerations we conclude that the parameter T of Newell's model can be interpreted in four different ways:

1. As the *reaction time* when interpreting Eq. (10.25) as a time-delay differential equation or when considering the trajectories (10.27).

2. As the *time gap* of the microscopic fundamental diagram (10.22).
3. As the *speed adaptation time* following from the equivalence between Newell's model and the OVM combined with speed update rule (10.9).
4. And as the *numerical update time* $T = \Delta t$ when interpreting Eq. (10.25) as a discrete-time model.

The interpretation in terms of a reaction time or a time gap can only be applied for congested traffic. In contrast, the interpretation as a speed adaptation time or a numerical update time is generally valid.

Relation to the macroscopic Payne's model. Besides illustrating another interesting property of Newell's model, this paragraph shows how to derive macroscopic from microscopic traffic flow models by the micro–macro relations between the vehicle speed and the local speed field and between distance and local density, respectively, and by first-order Taylor expansions.

Left-hand side of Newell's model equation. In deriving a macroscopic equivalent of $v_\alpha(t+T)$, we start with the expression (10.17) for the macroscopic local speed. For points (x, t) lying on the trajectory of vehicle α, we have

$$v_\alpha(t) = V(x_\alpha(t), t) = V(x, t). \tag{10.28}$$

Furthermore, the change in position and speed during one update time step $\Delta t = T$ is expressed in terms of local macroscopic fields by a Taylor expansion up to first order,

$$\begin{aligned} v_\alpha(t+T) &= V(x_\alpha + v_\alpha T, t+T) \\ &= V(x_\alpha, t) + \frac{\partial V(x,t)}{\partial x} v_\alpha T + \frac{\partial V(x,t)}{\partial t} t \\ &= V(x,t) + \left(V(x,t) \frac{\partial V}{\partial x} + \frac{\partial V}{\partial t} \right) T. \end{aligned} \tag{10.29}$$

Right-hand side of Newell's car-following model equation. First, we apply the micro–macro relation between the optimal velocity function and the macroscopic speed-density relation, $v_{\text{opt}}(s) = v_e(s) = V_e(\rho)$. In a second step, the local density ρ in the argument is determined such that the approximation error is minimal. Since $d_\alpha = s_\alpha + l_{\alpha-1} = 1/\rho$ denotes the distance *between* the vehicles α and $\alpha - 1$, we evaluate ρ at the intermediate location $x_\alpha + d_\alpha/2 = x + d_\alpha/2$ (Fig. 10.10) and consistently express everything up to first order by macroscopic local quantities,

$$\begin{aligned} v_{\text{opt}}(s_\alpha(t)) &= V_e\left(\rho\left(x + d_\alpha/2, t\right)\right) \\ &= V_e(\rho(x,t)) + V_e'(\rho) \frac{\partial \rho}{\partial x} \frac{d_\alpha}{2} \\ &= V_e(\rho(x,t)) + \frac{V_e'(\rho)}{2\rho} \frac{\partial \rho}{\partial x}. \end{aligned} \tag{10.30}$$

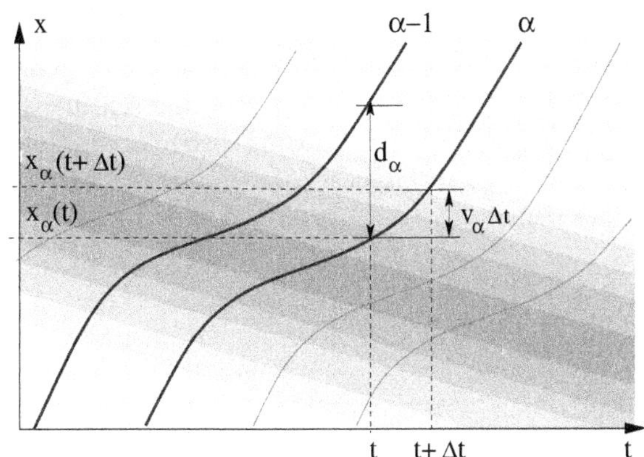

Fig. 10.10 Illustration of the derivation of Payne's model from the (*generalized*) Newell's model. Shown are the trajectories $x_\alpha(t)$ and $x_{\alpha-1}(t)$ of the subject and leading vehicles, respectively, and the associated local density (*shaded*). The error of the micro–macro transition is minimal when defining the local density at as $x_\alpha + d_\alpha/2$ as the inverse of the distance Δx_α

In the second line we have applied the first-order Taylor expansion, and the chain rule. In the third line, we have expressed $d_\alpha/2$ by $\frac{1}{2\rho(x,t)}$.[21] Equating (10.29) with Eq. (10.30) leads to

$$\frac{\partial V}{\partial t} + V\frac{\partial V}{\partial x} = \frac{V_e(\rho) - V}{T} + \frac{V_e'(\rho)}{2\rho T}\frac{\partial \rho}{\partial x}. \tag{10.31}$$

When identifying $T = \tau$ this corresponds to Payne's model (9.18). We conclude that Newell's model and its extensions to other fundamental diagrams are always *approximatively* equivalent to Payne's model. Simultaneously, Newell's model is exactly equivalent to the Section-Based Model if restricting conditions apply (traffic in the car-following regime, triangular fundamental diagram).

Relation of Newell's model with anticipation to the FVDM. In order to compensate for at least part of the reaction time delay described by Newell's model, a driver would try to predict the distance gap (the only exogenous stimulus of Newell's model) by a certain time interval T_a into the future. Using the rate of change $\dot{s} = -\Delta v$ for an estimate of the gap at this time, $\hat{s}(t + T_a) = s(t) - T_a \Delta v$, this results in the *generalized Newell's model*

$$v(t+T) = v_{\text{opt}}\left(s(t) - T_a \Delta v\right) \approx v_{\text{opt}}(s(t)) - v_{\text{opt}}'(s) T_a \Delta v. \tag{10.32}$$

According to Eq. (10.11) this is equivalent to a time-continuous model given by

$$\frac{\mathrm{d}v}{\mathrm{d}t} = \frac{v_{\text{opt}}(s) - v}{T} - \frac{T_a v_{\text{opt}}'(s)}{T}\Delta v.$$

[21] The displacement d_α appears only in terms that are already of first order. Therefore, we only need the zeroth order relation $d_\alpha(t) = 1/\rho(x,t)$.

This corresponds to a Full Velocity Difference Model with a gap dependent sensitivity $\gamma(s) = T_a v'_{\text{opt}}(s)/T$. From the OV plausibility conditions (10.20) it follows that $\lim_{s\to\infty} v'_{\text{opt}}(s) = 0$, i.e., the sensitivity tends to zero when the gap becomes sufficiently large (for the triangular fundamental diagram it is exactly zero for $s > s_0 + v_0 T$). This means, the resulting FVDM-like model is complete, similarly to the "improved" FVDM presented in Sect. 10.7.

Problems

10.1 Dynamics of a single vehicle approaching a red traffic light

A single car in city-traffic conditions can be described by the following time-continuous acceleration model:

$$\frac{\mathrm{d}v}{\mathrm{d}t} = \begin{cases} \frac{v_0 - v}{\tau} & \text{if } \Delta v \le \sqrt{2b(s - s_0)}, \\ -b & \text{otherwise.} \end{cases}$$

Here, s denotes the distance to the next car or the next traffic light (whichever is nearer), and Δv is the approaching rate. A red traffic light is modeled by a virtual standing vehicle of zero dimension at the stopping line which is removed when the light turns green.

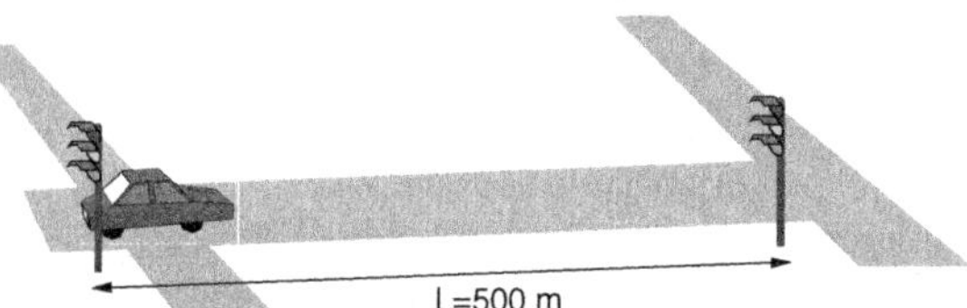

1. What is the meaning of the model parameters v_0, τ, s_0, and b? Describe the qualitative acceleration profile after the initially standing car starts moving, and the deceleration profile when approaching a red traffic light. Which essential human property is not taken care of by this model?
2. The first traffic light turns green at $t = 0$ s. Calculate the speed and the acceleration as a function of time for general model parameters assuming that the second traffic light is always green.
3. Consider now a situation where the subject car is approaching a red traffic light with cruising speed $v_0 = 50$ km/h assuming $s_0 = 2$ m and $b = 2$ m/s^2. At which distance to the traffic light does the driver initiate his or her braking maneuver? What is the braking deceleration and the final distance of the standing car to the stopping line?
4. Calculate the trajectory and the speed profile of the car during the complete start-stop cycle for a distance of 500 m between the stopping lines of the two traffic lights assuming $\tau = 5$ s and values for the other model parameters as above.

Hint: There are two phases: The acceleration phase eventually going into cruising mode, and the braking phase. As an essential step, you have to determine the location and the time where the braking maneuver begins. You can assume that the car has already reached its cruising speed at this time.

10.2 OVM acceleration on an empty road
Consider a single vehicle on an empty road whose acceleration is described by the *Optimal Velocity Model*

$$\frac{dv}{dt} = \frac{v_0 - v}{\tau}$$

assuming the initial conditions $x(0) = 0$, and $v(0) = 0$. (i) At which time does the vehicle reach its maximum acceleration? What is its value? (ii) Determine the parameter τ if the desired speed is given by 120 km/h and the maximum acceleration by 2 m/s^2. (iii) At which time does the vehicle reach a speed of 100 km/h?

10.3 Optimal Velocity Model on a ring road
Consider a closed ring road with identical vehicles. Initially ($t = 0$ s), all vehicles are motionless and evenly positioned with a gap of 20 m between each other. Calculate the speed profile of all vehicles for the OVM with the optimal velocity function (10.22) assuming $\tau = 1$ s, $v_0 = 72$ km/h, $s_0 = 2$ m, and $T = 1.8$ s.

10.4 Full Velocity Difference Model
When modeling the city scenario of the fact sheet with the FVDM (Fig. 10.6), neither vehicle reaches, or at least approaches, its cruising speed $v_0 = 54$ km/h although the distance between the traffic light would allow for the cruising speed. In order to find the underlying mechanism for this, calculate the stationary speed if there is a red traffic light (modeled by a standing virtual vehicle) at an arbitrarily large distance (i) for general parameters, (ii) for the values $v_0 = 54$ km/h, $\tau = 5$ s, and $\gamma = 0.6\,\mathrm{s}^{-1}$. Compare the result with the right column of Fig. 10.6.

10.5 A simple model for emergency braking maneuvers
Critical situations requiring emergency braking maneuvers can be described by following microscopic model:

$$\frac{dv}{dt} = \begin{cases} 0 & \text{if } t < T_r, \\ -b_{\text{max}} & \text{otherwise.} \end{cases}$$

1. Give an intuitive meaning of the parameters T_r and b_{max}.
2. Calculate the braking distance and the overall stopping distance for initial speeds of 50 and 70 km/h assuming $b_{\text{max}} = 8$ m/s^2 and $T_r = 1$ s.
   Hint: The overall stopping distance is composed of the *braking distance*, i.e., the vehicle displacement during the actual braking phase, and the *reaction distance* the vehicle travels during the reaction time of the driver.
3. Imagine a situation where a child suddenly runs into the road from a hidden position behind a vehicle. A driver driving according to the above model just

manages to stop if his or her initial speed is 50 km/h. At what speed would this driver collide with the child if the initial speed is 70 km/h and the situation is otherwise unchanged?

Further Reading

- Reuschel, A.: Fahrzeugbewegungen in der Kolonne. Österreichisches Ingenieur-Archiv **4** (1950) 193–215
- Pipes, L.A.: An operational analysis of traffic dynamics. Journal of Applied Physics **24** (1953) 274–281
- Chandler, R.E., Herman, R., Montrol, E.W.: Traffic dynamics: Studies in car-following. Operation Research **6** (1958) 165–184
- Gazis, D.C., Herman, R., Rothery, R.W.: Nonlinear follow-the-leader models of traffic flow. Operation Research **9** (1961) 545–567
- Bando, M., Hasebe, K., Nakayama, A., Shibata, A., Sugiyama, Y.: Dynamical model of traffic congestion and numerical simulation. Physical Review E **51** (1995) 1035–1042
- Jiang, R., Wu, Q., Zhu, Z.: Full velocity difference model for a car-following theory. Physical Review E **64** (2001) 017101
- Newell, G.F.: A simplified car-following theory: a lower order model. Transportation Research Part B: Methodological **36** (2002) 195–205

Chapter 11
Car-Following Models Based on Driving Strategies

Ideas are like children: you always love your own the most.
Lothar Schmidt

Abstract The models introduced in this chapter are derived from assumptions about real driving behavior such as keeping a "safe distance" from the leading vehicle, driving at a desired speed, or preferring accelerations to be within a comfortable range. Additionally, kinematical aspects are taken into account, such as the quadratic relation between braking distance and speed. We introduce two examples: The simplified Gipps model, and the Intelligent Driver Model. Both models use the same input variables as the sensors of *adaptive cruise control* (ACC) systems, and produce a similar driving behavior. Characteristics that are specific to the human nature, like erroneous judgement, reaction time, and multi-anticipation, are discussed in the next chapter.

11.1 Model Criteria

The models introduced in this chapter are formally identical to the minimal models presented in the previous chapter. They are defined by an acceleration function a_{mic} (see Eq. (10.3)) or a speed function v_{mic} (see Eq. (10.7)). In contrast to the minimal models, the acceleration or speed functions encoding the driving behavior should at least model the following aspects:

1. The acceleration is a strictly decreasing function of the speed. Moreover, the vehicle accelerates towards a *desired speed* v_0 if not constrained by other vehicles or obstacles:

$$\frac{\partial a_{\text{mic}}(s, v, v_l)}{\partial v} < 0, \quad \lim_{s \to \infty} a_{\text{mic}}(s, v_0, v_l) = 0 \quad \text{for all } v_l. \tag{11.1}$$

DOI: 10.1007/978-3-642-32460-4_11,

2. The acceleration is an increasing function of the distance s to the leading vehicle:

$$\frac{\partial a_{\text{mic}}(s, v, v_l)}{\partial s} \geq 0, \quad \lim_{s\to\infty} \frac{\partial a_{\text{mic}}(s, v, v_l)}{\partial s} = 0 \quad \text{for all } v_l. \tag{11.2}$$

   The inequality becomes an equality if other vehicles or obstacles (including "virtual" obstacles such as the stopping line at a red traffic light) are outside the interaction range and therefore do not influence the driving behavior. This defines the *free-flow acceleration*

$$a_{\text{free}}(v) = \lim_{s\to\infty} a_{\text{mic}}(s, v, v_l) = \geq a_{\text{mic}}(s, v, v_l). \tag{11.3}$$

3. The acceleration is an increasing function of the speed of the leading vehicle. Together with requirement (1), this also means that the acceleration decreases (the deceleration increases) with the speed of approach to the lead vehicle (or obstacle):

$$\frac{\partial \tilde{a}_{\text{mic}}(s, v, \Delta v)}{\partial \Delta v} \leq 0 \text{ or } \frac{\partial a_{\text{mic}}(s, v, v_l)}{\partial v_l} \geq 0, \quad \lim_{s\to\infty} \frac{\partial a_{\text{mic}}(s, v, v_l)}{\partial v_l} = 0. \tag{11.4}$$

   Again, the equality holds if other vehicles (or obstacles) are outside the interaction range.
4. A minimum gap (bumper-to-bumper distance) s_0 to the leading vehicle is maintained (also during a standstill). However, there is no backwards movement if the gap has become smaller than s_0 by past events:

$$a_{\text{mic}}(s, 0, v_l) = 0 \quad \text{for all } v_l \geq 0, \quad s \leq s_0. \tag{11.5}$$

By virtue of relation (10.11), these requirements (or *plausibility conditions*) for the acceleration function naturally imply conditions for the speed function v_{mic} of models formulated in terms of coupled maps.

A car-following model meeting these requirements is *complete* in the sense that it can consistently describe all situations that may arise in single-lane traffic. Particularly, it follows that (i) all vehicle interactions are of finite reach, (ii) following vehicles are not "dragged along",

$$a_{\text{mic}}(s, v, v_l) \leq a_{\text{mic}}(\infty, v, v_l') = a_{\text{free}}(v) \quad \text{for all} \quad s, v, v_l, \text{ and } v_l', \tag{11.6}$$

and (iii) an equilibrium speed $v_e(s)$ exists, which has the properties already postulated for the optimal-speed function (10.20):

$$v_e'(s) \geq 0, \quad v_e(0) = 0, \quad \lim_{s\to\infty} v_e(s) = v_0. \tag{11.7}$$

This means that the model possesses a unique steady-state flow-density relation, i.e., a fundamental diagram.[1]

These conditions are necessary but not sufficient. For example, when in the car-following regime (steady-state congested traffic), the time gap to the leader has to remain within reasonable bounds (say, between 0.5 and 3 s). Furthermore, the acceleration has to be constrained to a "comfortable" range (e.g., $\pm 2\,\mathrm{m/s^2}$), or at least, to physically possible values. Particularly, when approaching the leading vehicle, the quadratic relation between braking distance and speed has to be taken into account. Finally, any car-following model should allow instabilities and thus the emergence of "stop-and-go" traffic waves, but should not produce accidents, i.e., negative bumper-to-bumper gaps $s < 0$.[2]

Which of the car-following models introduced in Chap. 10 satisfy the conditions (11.1)–(11.5)?

11.2 Gipps' Model

Gipps' model presented here is a modified version of the one described in his original publication. It is simplified, but conceptually unchanged. Although it produces an unrealistic acceleration profile, this model is probably the simplest complete and accident-free model that leads to accelerations within a realistic range.

11.2.1 Safe Speed

Accidents are prevented in the model by introducing a "safe speed" $v_{\text{safe}}(s, v_l)$, which depends on the distance to and speed of the leading vehicle. It is based on the following assumptions:

1. Braking maneuvers are always executed with constant deceleration b. There is no distinction between comfortable and (physically possible) maximum deceleration.
2. There is a constant "reaction time" Δt.

---

[1] If one were to weaken condition (11.1) to $\partial a_{\text{mic}}/\partial v \leq 0$, it is possible to formulate models that do *not* have a fundamental diagram. Such models are proposed in the context of B. Kerner's *three-phase theory*.

[2] Traffic-flow models are meant to describe *normal* conditions, while accidents are almost always caused by *exceptional* driving mistakes that are not part of normal driving behavior and thus not part of the intended scope of the model.

3. Even if the leading vehicle suddenly decelerates to a complete stop (worst case scenario), the distance gap to the leading vehicle should not become smaller than a minimum gap s_0.[3]

Condition 1 implies that the *braking distance* that the leading vehicle needs to come to a complete stop is given by

$$\Delta x_l = \frac{v_l^2}{2b}.$$

From condition 2 it follows that, in order to come to a complete stop, the driver of the considered vehicle needs not only his or her braking distance $v^2/(2b)$, but also an additional *reaction distance* $v\,\Delta t$ travelled during the reaction time.[4] Consequently, the *stopping distance* is given by

$$\Delta x = v\Delta t + \frac{v^2}{2b}. \tag{11.8}$$

Finally, condition 3 is satisfied if the gap s exceeds the required minimum final value s_0 by the difference $\Delta x - \Delta x_l$ between the stopping distance of the considered vehicle and the breaking distance of the leader:

$$s \geq s_0 + v\Delta t + \frac{v^2}{2b} - \frac{v_l^2}{2b}. \tag{11.9}$$

The speed v for which the equal sign holds (the highest possible speed) defines the "safe speed"

$$v_{\text{safe}}(s, v_l) = -b\Delta t + \sqrt{b^2 \Delta t^2 + v_l^2 + 2b(s - s_0)}. \tag{11.10}$$

11.2.2 Model Equation

The simplified Gipps' model is defined as an iterated map with the "safe speed" (11.10) as its main component:

$$\boxed{v(t + \Delta t) = \min\left[v + a\Delta t,\, v_0,\, v_{\text{safe}}(s, v_l)\right] \quad \text{Gipps' model.}} \tag{11.11}$$

This model equation reflects the following properties:

- The simulation update time step is equal to the reaction time Δt.

---

[3] This condition is not present in the original paper, but is necessary to ensure an accident-free model in the presence of numerical errors arising from discretization.

[4] In contrast to the original publication, we assume the speed to be constant within the reaction time.

- If the current speed is greater than $v_{\text{safe}} - a\Delta t$ or $v_0 - a\Delta t$, the vehicle will reach the minimum of v_0 and v_{safe} during the next time step.[5]
- Otherwise the vehicle accelerates with constant acceleration a until either the safe speed or the desired speed is reached.

11.2.3 Steady-State Equilibrium

The homogeneous steady state implies $v(t+\Delta t) = v_l = v$, thus

$$v = \min(v_0, v_{\text{safe}}) = \min\left(v_0, -b\Delta t + \sqrt{b^2\Delta t^2 + v^2 + 2b(s-s_0)}\right),$$

which yields the steady-state speed-gap relation

$$v_e(s) = \max\left[0, \min\left(v_0, \frac{s-s_0}{\Delta t}\right)\right] \tag{11.12}$$

and, assuming constant vehicle lengths l, the familiar "triangular" fundamental diagram

$$Q_e(\rho) = \min\left(v_0\rho, \frac{1-\rho l_{\text{eff}}}{\Delta t}\right), \tag{11.13}$$

where $l_{\text{eff}} = (l + s_0)$. As in the Newell model, the parameter Δt can be interpreted in four different ways: (i) As the reaction time introduced in the derivation of v_{safe}, (ii) as the numerical update time step of the actual model equation (11.11), (iii) as a speed adaption time in Eq. (11.11) (at least, if $v(t+\Delta t)$ is restricted by v_{safe} or v_0), or (iv) as the "safety time gap" $(s - s_0)/v_e$ in congested traffic as deduced from the fundamental diagram (11.12).

11.2.4 Model Characteristics

Unlike the minimal models described in the previous chapter, the Gipps' model is transparently derived from a few basic assumptions and uses parameters that are easy to interpret and assign realistic values (Table 11.1). Furthermore, Gipps' model is—again, in contrast to the minimal models—*robust* in the sense that meaningful results can be produced from a comparatively wide range of parameter values.
Highway traffic. The simulation of the highway scenario (Fig. 11.1, left) produces more realistic results than the OVM or the Newell model: The speed field in panel

[5] Strictly speaking, this means that deceleration $(v - v_{\text{safe}})/\Delta t$ is not restricted to b. In multi-lane simulations, it can be greater if another vehicle "cuts in" in front of the considered vehicle.

Table 11.1 Parameters of the simplified Gipps' model and typical values in different scenarios

Parameter	Typical value Highway	Typical value City traffic
Desired speed v_0	120 km/h	54 km/h
Adaption/reaction time Δt	1.1 s	1.1 s
Acceleration a	1.5 m/s^2	1.5 m/s^2
Deceleration b	1.0 m/s^2	1.0 m/s^2
Minimum distance s_0	3 m	2 m

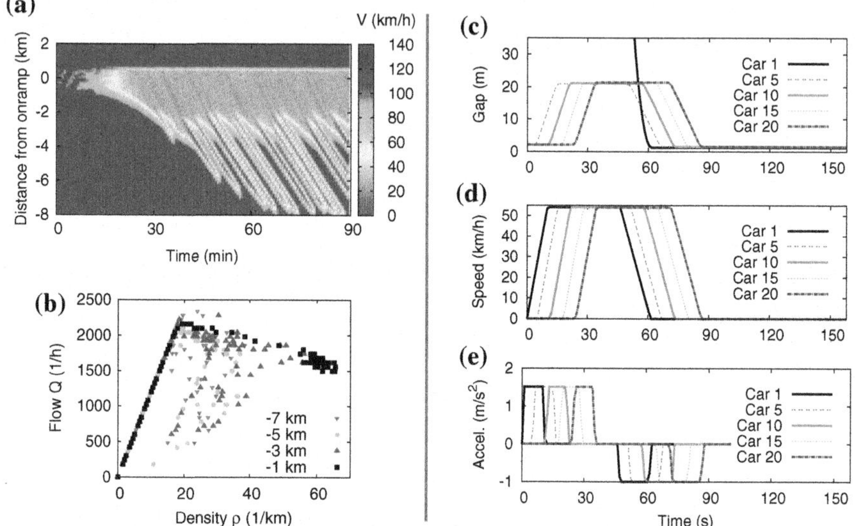

Fig. 11.1 Fact sheet of Gipps' model (11.11), (11.10). Simulation of the two standard scenarios "highway" (*left*) and "city traffic" (*right*) with the parameter values listed in Table 11.1. See Chap. 10.5 for a detailed description of the scenarios

(a) exhibits small perturbations which are caused by vehicles merging from the on-ramp and grow into stop-and-go waves while propagating upstream. The propagation velocity $c_{\text{cong}} = -l_{\text{eff}}/\Delta t$ is constant and of the order of the empirical value (≈ -15 km/h). Furthermore, the wave length (of the order of 1–1.5 km) is not too far away from the empirical values (1.5–3 km).

The flow-density diagram in Fig. 11.1b, obtained from virtual detectors, shows a strongly scattered cloud of data points in the region of congested traffic, i.e., everywhere to the right of the straight line indicating free traffic. Such a wide scattering is in agreement with empirical data (cf. Figs. 4.11 and 4.12). By looking at scatter plots of individual detectors, one observes that detectors that are closer to the bottleneck produce data points that are shifted towards greater densities and closer to the fundamental diagram of steady-state traffic. Moreover, the data points of virtual detectors positioned inside the region of stationary traffic immediately upstream of the bottleneck (solid black squares) lie on the fundamental diagram itself. This apparent density increase near the outflow region of a congestion, also known as *pinch effect*, can be observed empirically. However, the systematic density underestimation, which conspicuously increases with the degree of the scattering of the data points, suggests that the *real* density increase is smaller, or even nonexistent. This means that the pinch effect is essentially a result of data misinterpretation, or, more specifically, by estimating the density with the time mean speed instead of the space mean speed (cf. Sect. 3.3.1). This interpretation is confirmed by simulation as will be shown in Fig. 11.5b. We draw an important conclusion that is not restricted to Gipps' model (and not even to traffic flow models):

> When using empirical data to assert the accuracy and predictive power of models, one has to simulate both the actual traffic dynamics *and* the process of data capture and analysis.

City traffic. Compared to the simple models of the previous chapter, the city-traffic scenario (Fig. 11.1, right column) is closer to reality as well. However, the acceleration time-series is unrealistic. By definition, there are only three values for the acceleration: Zero, a, and $-b$ (cf. Panel (e)). The resulting driving behavior is excessively "robotic" and the abrupt transitions are unrealistic.

Moreover, Gipps' model does not differentiate between comfortable and maximum deceleration: Assuming that b in Eq. (11.10) denotes the maximum deceleration, the model is accident-free but every braking maneuver is performed *very* uncomfortably with full brakes. On the other hand, when interpreting b as the comfortable deceleration and allowing for heterogeneous and/or multi-lane traffic the model possibly produces accidents if leading vehicles (which might be simulated using different parameters or even different models) brake harder than b.

In summary, Gipps' model produces good results in view of its simplicity. Modified versions of this model are used in several commercial traffic simulators. One example of such a modification is *Krauss' model* which essentially is a stochastic version of the Gipps model.

11.3 Intelligent Driver Model

The time-continuous *Intelligent Driver Model* (IDM) is probably the simplest complete and accident-free model producing realistic acceleration profiles and a plausible behavior in essentially all single-lane traffic situations.

11.3.1 Required Model Properties

As Gipps' model, the IDM is derived from a list of basic assumptions (*first-principles model*). It is characterized by the following requirements:

1. The acceleration fulfills the general conditions (11.1)–(11.5) for a complete model.
2. The equilibrium bumper-to-bumper distance to the leading vehicle is not less than a "safe distance" $s_0 + vT$ where s_0 is a minimum (bumper-to-bumper) gap, and T the (bumper-to-bumper) time gap to the leading vehicle.
3. An *braking strategy!intelligent* controls how slower vehicles (or obstacles or red traffic lights) are approached:
   - Under normal conditions, the braking maneuver is "soft", i.e., the deceleration increases gradually to a comfortable value b, and decreases smoothly to zero just before arriving at a steady-state car-following situation or coming to a complete stop.
   - In a critical situation, the deceleration exceeds the comfortable value until the danger is averted. The remaining braking maneuver (if applicable) will be continued with the regular comfortable deceleration b.
4. Transitions between different driving modes (e.g., from the acceleration to the car-following mode) are smooth. In other words, the time derivative of the acceleration function, i.e., the *jerk* J, is finite at all times.[6] This is equivalent to postulating that the acceleration function $a_{\text{mic}}(s, v, v_l)$ (or $\tilde{a}_{\text{mic}}(s, v, \Delta v)$) is continuously differentiable in all three variables. Notice that this postulate is in contrast to the action-point models such as the *Wiedemann Model* where acceleration changes are modeled as a series of discrete jumps.
5. The model should be as parsimonious as possible. Each model parameter should describe only one aspect of the driving behavior (which is favorable for model calibration). Furthermore, the parameters should correspond to an intuitive interpretation and assume plausible values.

11.3.2 Mathematical Description

The required properties are realized by the following acceleration equation:

$$\boxed{\dot{v} = a\left[1 - \left(\frac{v}{v_0}\right)^{\delta} - \left(\frac{s^*(v, \Delta v)}{s}\right)^2\right] \quad \text{IDM.}} \tag{11.14}$$

[6] Typical values of a "comfortable" jerk are $|J| \leq 1.5\,\text{m/s}^3$.

The acceleration of the Intelligent Driver Model is given in the form $\tilde{a}_{\text{mic}}(s, v, \Delta v)$ and consists of two parts, one comparing the current speed v to the desired speed v_0, and one comparing the current distance s to the desired distance s^*. The desired distance

$$s^*(v, \Delta v) = s_0 + \max\left(0, vT + \frac{v\Delta v}{2\sqrt{ab}}\right) \tag{11.15}$$

has an equilibrium term $s_0 + vT$ and a *dynamical term* $v\Delta v/(2\sqrt{ab})$ that implements the "intelligent" braking strategy (see Sect. 11.3.4).[7]

11.3.3 Parameters

We can easily interpret the model parameters by considering the following three standard situations:

- When *accelerating on a free road from a standstill*, the vehicle starts with the maximum acceleration a. The acceleration decreases with increasing speed and goes to zero as the speed approaches the desired speed v_0. The exponent δ controls this reduction: The greater its value, the later the reduction of the acceleration when approaching the desired speed. The limit $\delta \to \infty$ corresponds to the acceleration profile of Gipps' model while $\delta = 1$ reproduces the overly smooth acceleration behavior of the Optimal Velocity Model (10.19).
- When *following a leading vehicle*, the distance gap is approximatively given by the *safety distance* $s_0 + vT$ already introduced in Sect. 11.3.1. The safety distance is determined by the time gap T plus the minimum distance gap s_0.
- When *approaching slower or stopped vehicles*, the deceleration usually does not exceed the comfortable deceleration b. The acceleration function is smooth during transitions between these situations.

Each parameter describes a well-defined property (Fig. 11.2). For example, transitions between highway and city traffic, can be modeled by solely changing the desired speed (Table 11.2). All other parameters can be kept constant, modeling that somebody who drives aggressively (or defensively) on a highway presumably does so in city traffic as well.

Since the IDM has no explicit reaction time and its driving behavior is given in term of a continuously differentiable acceleration function, the IDM describes more closely the characteristics of semi-automated driving by adaptive cruise control

[7] The maximum condition in Eq. (11.15) ensures that the conditions (11.1)–(11.5) for model completeness hold for all situations. Strictly speaking, this condition violates the postulate of a smooth acceleration function. However, it comes into effect only in two situations: (i) For finite speeds if the leading car is much faster, (ii) for stopped queued vehicles when the queue starts to move. The first situation may arise after a cut-in maneuver of a faster vehicle. Since $s \gg s_0$ for this case, the resulting discontinuity is small. In the second case, the maximum condition prevents an overly sluggish start and the associated discontinuous acceleration profile may even be realistic.

Fig. 11.2 By using intuitive model parameters like those of Gipps' model or the Intelligent Driver Model (IDM) we can easily model different aspects of the driving behavior (or physical limitations of the vehicle) with corresponding parameter values

Table 11.2 Model parameters of the Intelligent Driver Model (IDM) and typical values in different scenarios (vehicle length 5 m unless stated otherwise)

Parameter	Typical value Highway	Typical value City traffic
Desired speed v_0	120 km/h	54 km/h
Time gap T	1.0 s	1.0 s
Minimum gap s_0	2 m	2 m
Acceleration exponent δ	4	4
Acceleration a	1.0 m/s^2	1.0 m/s^2
Comfortable deceleration b	1.5 m/s^2	1.5 m/s^2

(ACC) than that of a human driver. However, it can easily be extended to capture human aspects like estimation errors, reaction times, or looking several vehicles ahead (see Chap. 12).

In contrast to the models discussed previously, the IDM explicitly distinguishes between the safe time gap T, the speed adaptation time $\tau = v_0/a$, and the reaction time T_r (zero in the IDM, nonzero in the extension described in Chap. 12). This allows us not only to reflect the conceptual difference between ACCs and human drivers in the model, but also to differentiate between more nuanced driving styles such as "sluggish, yet tailgating" (high value of $\tau = v_0/a$, low value for T) or "agile, yet safe driving" (low value of $\tau = v_0/a$, normal value for T, low value for b).[8] Furthermore, all these driving styles can be adopted independently by ACC systems (reaction time $T_r \approx 0$, original IDM), by attentive drivers (T_r

[8] Obviously, the first behavior promotes instabilities which will be confirmed by the stability analysis in Chap. 15.

comparatively small, extended IDM), and by sleepy drivers (T_r comparatively large, extended IDM).

11.3.4 Intelligent Braking Strategy

The term $v\Delta v/(2\sqrt{ab})$ in the desired distance s^* (11.15) of the IDM models the dynamical behavior when approaching the leading vehicle. The equilibrium terms $s_0 + vT$ always affect s^* due to the required continuous transitions from and to the equilibrium state. Nevertheless, to study the braking strategy itself, we will set these terms to zero, together with the free acceleration term $a[1 - (v/v_0)^\delta]$ of the IDM acceleration equation. When approaching a standing vehicle or a red traffic light ($\Delta v = v$), we then find

$$\dot{v} = -a\left(\frac{s^*}{s}\right)^2 = -\frac{av^2(\Delta v)^2}{4abs^2} = -\left(\frac{v^2}{2s}\right)^2\frac{1}{b}. \tag{11.16}$$

With the *kinematic deceleration* defined as

$$b_{\text{kin}} = \frac{v^2}{2s}, \tag{11.17}$$

this part of the acceleration can be written as

$$\dot{v} = -\frac{b_{\text{kin}}^2}{b}. \tag{11.18}$$

When braking with deceleration b_{kin}, the braking distance is exactly the distance to the leading vehicle, thus b_{kin} is the minimum deceleration required for preventing a collision. With Eq. (11.18), we now understand the self-regulating braking strategy of the IDM:

- A "critical situation" is defined by b_{kin} being greater than the comfortable deceleration b. In such a situation, the actual deceleration is *even stronger* than necessary, $|\dot{v}| = b_{\text{kin}}^2/b > b_{\text{kin}}$. This overcompensation decreases b_{kin} and thus helps to "regain control" over the situation.
- In a non-critical situation ($b_{\text{kin}} < b$), the actual deceleration is less than the kinematic deceleration, $b_{\text{kin}}^2/b < b_{\text{kin}}$. Thus, b_{kin} increases in the course of time and approaches the comfortable deceleration.

Hence, the braking strategy is *dynamically self-regulating* towards a situation in which the kinematic deceleration equals the comfortable deceleration. One can show (see Problem 11.4) that this self-regulation is explicitly given by the differential equation

$$\frac{\mathrm{d}b_{\text{kin}}}{\mathrm{d}t} = \frac{v\,b_{\text{kin}}}{s\,b}\big(b - b_{\text{kin}}\big). \tag{11.19}$$

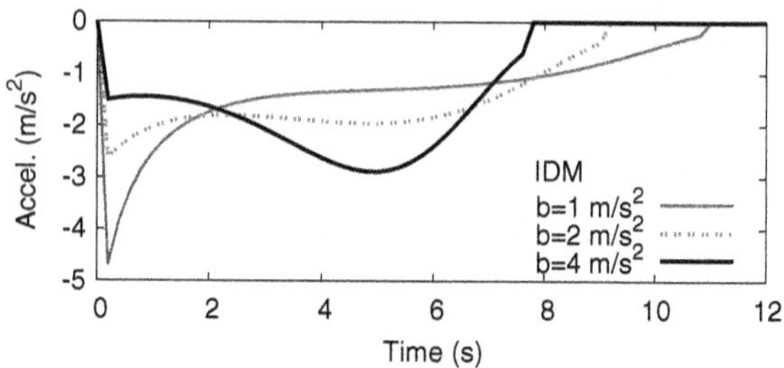

Fig. 11.3 Acceleration time-series of approaching the stop line of a *red* traffic light for different values of the comfortable deceleration. The initial speed is $v = 54$ km/h. The traffic light switches to *red* (at time $t = 0$) when the vehicle is 60 m away

Thus, the kinematic deceleration drifts towards the comfortable deceleration in *any* situation.

In the above considerations, we have ignored parts of the IDM acceleration function. To estimate their effects, the time series of Fig. 11.4e display the complete IDM dynamics when approaching an initially very distant, standing obstacle ($b_{\text{kin}} \ll b$): First, the deceleration increases towards the comfortable deceleration according to Eq. (11.19). However, due to the defensive nature of the neglected terms, the comfortable value is never realized, at least for the first vehicle. Eventually, the deceleration smoothly reduces until the vehicle stops with exactly the minimum gap s_0 left between itself and the obstacle. The following vehicles experience slightly larger decelerations than the comfortable ones, but without having to perform any emergency braking or being in danger of a collision.

Figure 11.3 shows the effects of the self-regulatory braking strategy in a situation where the vehicle is suddenly forced to stop. Drivers with $b = 1\,\text{m/s}^2$ will perceive this situation as "critical" ($b_{\text{kin}} = v^2/(2s) = 1.9\,\text{m/s}^2$) and overcompensate with even stronger deceleration. In contrast, if the comfortable deceleration is given by $b = 4\,\text{m/s}^2$, the comfortable deceleration is initially well above the kinematic deceleration and the simulated driver will brake only weakly, so that b_{kin} increases. Again, due to the other terms in the acceleration function, the actual deceleration will not reach the value of comfortable deceleration.

Why do "IDM drivers" act in a more anticipatory manner for smaller values of b? Yet why are very small values of b (less than about 1 m/s^2) not meaningful?

Consider the situation of approaching a standing obstacle as described above and convince yourself that the effect of the dynamical part of s^* on the acceleration prevails against all other terms. Furthermore, show that these other terms are negative in nearly all situations, thus making the driving behavior more defensive.

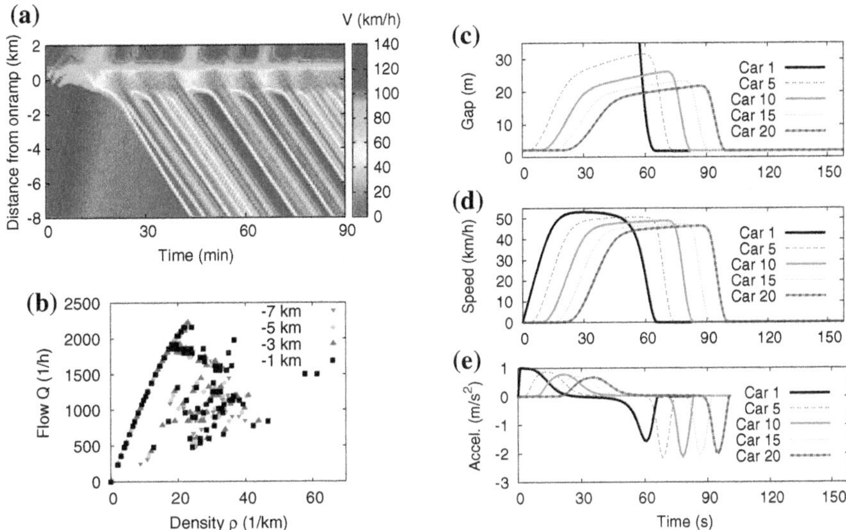

Fig. 11.4 Fact sheet of the Intelligent Driver Model (11.14). The two standard scenarios "highway" (*left*) and "city traffic" (*right*) are simulated with parameters as listed in Table 11.2. See Sect. 10.5 for a detailed description of the scenarios

11.3.5 Dynamical Properties

The *fact sheet* of the IDM, Fig. 11.4, shows IDM simulations of the two standard scenarios "traffic breakdown at a highway on-ramp" and "acceleration and stopping of a vehicle platoon in city traffic".

Highway traffic. The speed field in the highway scenario (Fig. 11.4a) exhibits dynamics similar to the one found in Gipps' model (cf. Fig. 11.1): Stationary congested traffic is found close to the bottleneck, while, further upstream, stop-and-go waves emerge and travel upstream with a velocity of approximately -15 km/h. The wavelength tends to be smaller than in real stop-and-go traffic, but the empirical spatiotemporal dynamics are otherwise reproduced very well. The growing stop-and-go waves in the simulations are caused by a collective instability called *string instability* which will be discussed in more detail in Chap. 15. As we will see in this chapter, the IDM is either unstable with respect to stop-and-go waves (string-unstable) or absolutely stable, depending on the parameters and traffic density. The model is free of accidents, however, except for very unrealistic parameters under specific circumstances.

The flow-density diagram of virtual loop detectors in Fig. 11.4b reproduces typical aspects of empirical flow-density data:

- Data points representing free traffic fall on a line, while data points from congested traffic are widely scattered.

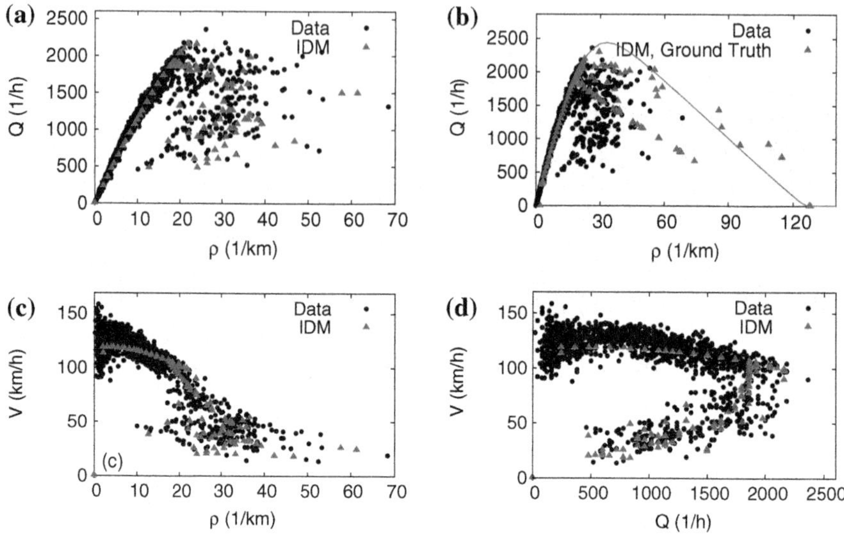

Fig. 11.5 **a** Fundamental diagram, **c** speed-density diagram, and **d** speed-flow diagram showing data from a virtual detector in the highway simulation shown in Fig. 11.4 (positioned 1 km upstream of the ramp). For comparison, empirical data from a real detector on the Autobahn A5 near Frankfurt, Germany, is shown. Velocities have been calculated using arithmetic means in both the real data and the simulation data. **b** Flow-density diagram with the same empirical data but using the real (local) density for the IDM simulation rather than the density derived from the virtual detectors

- The free-traffic branch is not a perfectly straight line but is slightly curved, especially towards the maximum flow.
- Near the maximum flow, the points are arranged in a pattern that looks like a mirror image of the Greek letter λ (*inverse-λ form*), meaning that for a range of densities (here ≈18 − 25 veh/h), both free and congested traffic states are possible. Thus, the IDM reproduces the empirically observed bistability and the resulting hysteresis effects like the *capacity drop* (about 300 veh/h or 15 % in the present example).

Comparing the virtual detector data with the real data in Fig. 11.5a, c, and d, we find almost quantitative agreement of the flow-density, speed-density, and speed-flow diagrams. Contrary to Gipps' model, the IDM also reproduces the curvature of the free-traffic branch correctly. This agreement with the data allows us to scrutinize the nature of the observed strong scattering of flow-density data points corresponding to congested traffic. First, we compare the *estimated* density using the virtual stationary detector data with the *real* spatial density which, of course, is available in the simulation. The result displayed in Fig. 11.5b reminds us that one has to be very careful when interpreting flow-density data. Moreover, even the scattering of the data points itself is a matter of the way the data is plotted: While the congested traffic data is much more scattered than the free-flow data in the flow-density data in Fig. 11.5a, both branches show similar scattering in the speed-flow diagram 11.5d—in spite of the fact that both diagrams show the *same* data.

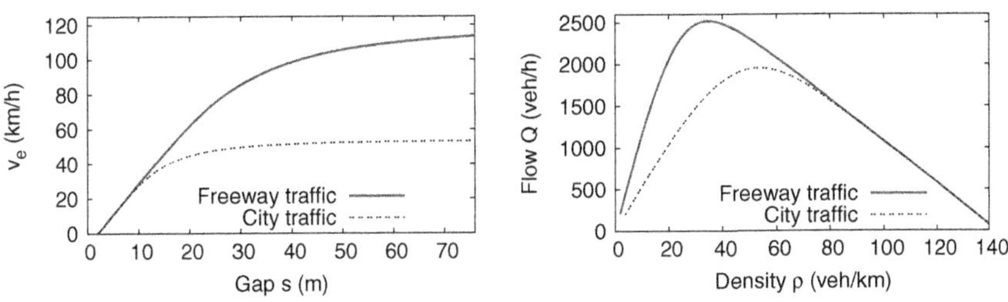

Fig. 11.6 Microscopic (*left*) and macroscopic (*right*) fundamental diagram of the IDM using the parameters shown in Table 11.2

City traffic. In the city traffic simulation (Fig. 11.4c–e) we see a smooth, realistic acceleration/deceleration profile, except in vehicle platoons with speed close to v_0 where followers do not accelerate up to the desired speed and thus the distance between the vehicles does not reach a constant value before the braking maneuver begins. This happens because, when approaching the desired speed, the free acceleration function decreases continuously to zero while the interaction (braking) term s^*/s remains finite (reaching zero only in the limit $s \to \infty$). Thus, for $v \lesssim v_0$, the actual steady-state equilibrium distance (where the free acceleration and the interaction terms cancel each other) is significantly larger than $s^*(v, 0)$. In the next section, we will investigate this more closely and propose a solution in Sect. 11.3.7.

11.3.6 Steady-State Equilibrium

By postulating $\dot{v} = \Delta v = 0$ we obtain the condition for the steady-state equilibrium of the IDM from the acceleration function (11.14):

$$1 - \left(\frac{v}{v_0}\right)^{\delta} - \left(\frac{s_0 + vT}{s}\right)^2 = 0. \tag{11.20}$$

For arbitrary values of δ we can solve this equation in closed form only for s (cf. Fig. 11.6),

$$s = s_e(v) = \frac{s_0 + vT}{\sqrt{1 - \left(\frac{v}{v_0}\right)^{\delta}}}. \tag{11.21}$$

This yields the equilibrium gap $s_e(v)$ with the speed being the independent variable (instead of the equilibrium speed $v_e(s)$ as a function of the gap). Using the micro-macro relation (10.16), $s_e = \frac{1}{\rho_e} - 1$, $v = V$, and $Q_e = \rho_e V$, we obtain the speed-density and the fundamental diagrams shown in Fig. 11.6.

Note that due to the continuous transition between free and congested traffic, the equilibrium gap $s_e(v)$ is *not* given by $s^*(v, 0) = s_0 + vT$. Instead, for $v \lesssim v_0$ it is much larger which can be seen by looking at the denominator of Eq. (11.21). Therefore the fundamental diagram is not a perfect triangle but rounded close to the maximum flow. This causes the curvature in the macroscopic speed-density and flow-density diagrams (Fig. 11.5), but also produces the mentioned unrealistic car-following behavior in platoons with identical driver-vehicle units.

11.3.7 Improved Acceleration Function

Using the IDM as example, this section shows the scientific modeling process, aiming at eliminating some deficiencies of a model while retaining the good and well-tested features and keeping the model parsimonious, i.e., adding as few model parameters as possible.[9] The IDM is unrealistic in following aspects:

- If the actual speed exceeds the desired speed (e.g., after entering a zone with a reduced speed limit), the deceleration is unrealistically large, particularly for large values of the acceleration exponent δ.
- Near the desired speed v_0, the steady-state gap (11.21) becomes much greater than $s^*(v, 0) = s_0 + vT$ so that the model parameter T loses its meaning as the desired time gap. This means that a platoon of identical drivers and vehicles disperses much more than observed. Moreover, not all cars will reach the desired speed (Fig. 11.4c, d).
- If the actual gap is considerably smaller than desired (which may happen if another vehicle cuts too close when changing lanes) the braking reaction to regain the desired gap is exaggerated as illustrated in Problem 11.3.

We will treat the first two aspects here while the third aspect (which is only relevant in multi-lane situations) will be deferred to Sect. 11.3.8.

To improve the behavior for $v > v_0$, we require that the maximum deceleration must not exceed the comfortable deceleration b if there are no interactions with other vehicles or obstacles. The parameter δ should retain its meaning also in the new regime, i.e., leading to smooth decelerations to the new desired speed for low values and decelerating more "robotically" for high values. Furthermore, the free acceleration function $a_{\text{free}}(v)$ should be continuously differentiable, and remain unchanged for $v \le v_0$, i.e., $a_{\text{free}}(v) = \lim_{s\to\infty} a_{\text{IDM}}(s, v, \Delta v)$ for $v \le v_0$. Probably the simplest free-acceleration function meeting these conditions is given by

$$a_{\text{free}}(v) = \begin{cases} a\left[1 - \left(\frac{v}{v_0}\right)^{\delta}\right] & \text{if } v \le v_0, \\ -b\left[1 - \left(\frac{v_0}{v}\right)^{a\delta/b}\right] & \text{if } v > v_0. \end{cases} \tag{11.22}$$

[9] Ideally, no parameters are added as in this example.

To improve the behavior near the desired speed, we tighten the second condition in Sect. 11.3.1 by requiring that the equilibrium gap $s_e(v) = s^*(v, 0)$ should be *strictly equal* to $s_0 + vT$ for $v < v_0$. However, we would like to implement any modification as conservatively as possible in order to preserve all the other meaningful properties of the IDM (especially the "intelligent" braking strategy). Thus, changes should only have an effect

- near the steady-state equilibrium, i.e., if $z(s, v, \Delta v) = s^*(v, \Delta v)/s \approx 1$,[10]
- and when driving with $v \approx v_0$ and $v > v_0$.

We can accomplish this by distinguishing between the cases $z = s^*(v, \Delta v)/s < 1$ (the actual gap is greater than the desired gap) and $z \geq 1$. The new condition requires $\tilde{a}_{\text{mic}} = 0$ for all input values that satisfy $z(s, v, \Delta v) = 1$ and $v < v_0$. The other conditions in Sect. 11.3.1 and the conditions (11.1)–(11.5) are automatically satisfied if $\partial\tilde{a}_{\text{mic}}/\partial z < 0$, and if $\tilde{a}_{\text{mic}}(z)$ is continuously differentiable at the transition point $z = 1$. Probably the simplest acceleration function that fulfills all these conditions for $v \leq v_0$ is given by

$$\left.\frac{\mathrm{d}v}{\mathrm{d}t}\right|_{v \leq v_0} = \begin{cases} a\,(1 - z^2) & z = \frac{s^*(v, \Delta v)}{s} \geq 1, \\ a_{\text{free}}\left(1 - z^{(2a)/a_{\text{free}}}\right) & \text{otherwise.} \end{cases} \tag{11.23}$$

For $v > v_0$, there is no steady-state following distance, and we simply combine the free acceleration a_{free} and the interaction acceleration $a(1 - z^2)$ such that the interaction vanishes for $z \leq 1$ and the resulting acceleration function is continuously differentiable:[11]

$$\left.\frac{\mathrm{d}v}{\mathrm{d}t}\right|_{v > v_0} = \begin{cases} a_{\text{free}} + a(1 - z^2) & z(v, \Delta v) \geq 1, \\ a_{\text{free}} & \text{otherwise.} \end{cases} \tag{11.24}$$

This *Improved Intelligent Driver Model* (IIDM) uses the same set of model parameters as the IDM and produces essentially the same behavior except when vehicles follow each other near the desired speed or when the vehicle is faster than the desired speed. Simulating the standard city traffic scenario with the IIDM shows that all vehicles in the platoon now accelerate up to the desired speed (Fig. 11.7, left) while the self-stabilizing braking strategy and the observance of a comfortable deceleration are still in effect (Fig. 11.7, right).

However, the fundamental diagram is an exact triangle now. Thus, simulating highway traffic will no longer produce curved free-traffic branches in the flow-density, speed-flow, and speed-density diagrams (contrary to the unmodified IDM, cf. Fig. 11.5). Problem 11.6 discusses an alternative cause of this curvature.

---

[10] In fact, this condition is more general since it also includes continuations of the steady state to nonstationary situations.

[11] At $v = v_0$ and $z > 1$, the full acceleration function (11.23), (11.24) is only continuous, but not differentiable with respect to v. It would require a disproportionate amount of complication to resolve this special case of little relevance.

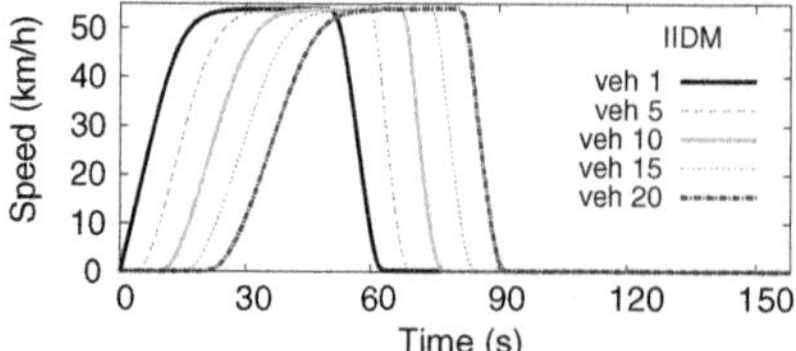

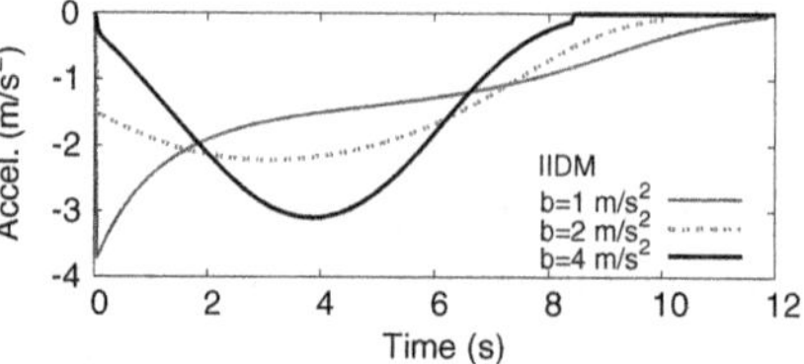

Fig. 11.7 Simulation of the city traffic scenario (*left*) and the situation shown in Fig. 11.3 (*right*) using the Improved Intelligent Driver Model (IIDM) with the parameters listed in Table 11.2

11.3.8 Model for Adaptive Cruise Control

While ACC systems only automate the longitudinal driving task, they must also react reasonably if the sensor input variables—the gap s and approaching rate Δv—change discontinuously as a consequence of "passive" lane changes (another lane-changing vehicle becomes the new leader), and also "active" lane changes (the ACC driver changes lanes manually). This means, the third of the IDM deficiencies mentioned in the previous Sect. 11.3.7—overreactions when the gap decreases discontinuously by external actions—must be taken care of.

The reason for the overreactions is the intention of the IDM (and IIDM) development to be accident-free even in the *worst case*, in which the driver of the leading vehicle suddenly brakes to a complete standstill. However, there are situations characterized by low speed differences and small gaps where human drivers rely on the fact that the drivers of preceding vehicles will *not* suddenly initiate full-stop emergency brakings. In fact, they consider such situations only as mildly critical. As a consequence, a more plausible and realistic driving behavior will result when drivers act according to the *constant-acceleration heuristic* (CAH) rather than considering the worst-case scenario. The CAH is based on the following assumptions:

- The accelerations of the considered and leading vehicle will not change in the near future (generally a few seconds).
- No safe time gap or minimum distance is required at any given moment.
- Drivers (or ACC systems) react without delay, i.e., with zero reaction time.

For actual values of the gap s, speed v, speed v_l of the leading vehicle, and constant accelerations $\dot{v}$ and $\dot{v}_l$ of both vehicles, the maximum acceleration $\max(\dot{v}) = a_{\text{CAH}}$ that does not lead to an accident under the CAH assumption is given by

$$a_{\text{CAH}}(s, v, v_l, \dot{v}_l) = \begin{cases} \frac{v^2 \tilde{a}_l}{v_l^2 - 2s\tilde{a}_l} & \text{if } v_l(v - v_l) \leq -2s\tilde{a}_l, \\ \tilde{a}_l - \frac{(v-v_l)^2 \Theta(v-v_l)}{2s} & \text{otherwise.} \end{cases} \tag{11.25}$$

The effective acceleration $\tilde{a}_l(\dot{v}_l) = \min(\dot{v}_l, a)$ (with the maximum acceleration parameter a) has been used to avoid the situation where leading vehicles with higher acceleration capabilities may cause "drag-along effects" of the form (11.6), or other artefacts violating the general plausibility conditions (11.1)–(11.5). The condition $v_l(v - v_l) = v_l \Delta v \leq -2s\dot{v}_l$ is true if the vehicles have stopped at the time the minimum gap $s = 0$ is reached. The Heaviside step function $\Theta(x)$ (with $\Theta(x) = 1$ if $x \geq 0$, and zero, otherwise) eliminates negative approaching rates Δv for the case that both vehicles are moving at the time t^* of least distance. Otherwise, t^* would lie in the past.

In order to retain all the "good" properties of the IDM, we will use the CAH acceleration (11.25) only as an *indicator* to determine whether the IDM will lead to unrealistically high decelerations, and modify the acceleration function of a model for ACC vehicles only in this case. Specifically, the proposed ACC model is based on following assumptions:

- The ACC acceleration is never lower than that of the IIDM. This is motivated by the circumstance that the IDM and the IIDM are accident-free, i.e., sufficiently defensive.
- If both, the IIDM and the CAH, produce the same acceleration, the ACC acceleration is the same as well.
- If the IIDM produces extreme decelerations, while the CAH yields accelerations in the comfortable range (greater than $-b$), the situation is considered to be mildly critical, and the resulting acceleration should be between $a_{\mathrm{CAH}} - b$ and a_{CAH}. Only for *very* small gaps, the decelerations should be somewhat higher to avoid an overly reckless driving style.
- If both, the IIDM and the CAH, result in accelerations significantly below $-b$, the situation is seriously critical and the ACC acceleration is given by the maximum of the IIDM and CAH accelerations.
- The ACC acceleration should be a continuous and differentiable function of the IIDM and CAH accelerations. Furthermore, it should meet the consistency requirements (11.1)–(11.5).

Probably the most simple functional form satisfying these criteria is given by

$$a_{\mathrm{ACC}} = \begin{cases} a_{\mathrm{IIDM}} & a_{\mathrm{IIDM}} \geq a_{\mathrm{CAH}}, \\ (1-c)\, a_{\mathrm{IIDM}} + c\left[a_{\mathrm{CAH}} + b \tanh\left(\frac{a_{\mathrm{IIDM}} - a_{\mathrm{CAH}}}{b}\right)\right] & \text{otherwise.} \end{cases} \tag{11.26}$$

This *ACC model* has only one additional parameter compared to the IDM/IIDM, the "coolness factor" c. For $c = 0$ one recovers the IIDM while $c = 1$ corresponds to the "pure" ACC model. Since the pure ACC model would produce a reckless driving behavior for very small gaps, a small fraction $1 - c$ of the IIDM is added. It turns out that a contribution of 1 % (corresponding to $c = 0.99$) gives a good compromise between reckless and overly timid behavior in this situation while it is essentially irrelevant, otherwise.

In contrast to the other models of this section, the ACC model has the *acceleration* $\dot{v}_l$ of the leading vehicle as additional exogenous factor (besides the speed, the

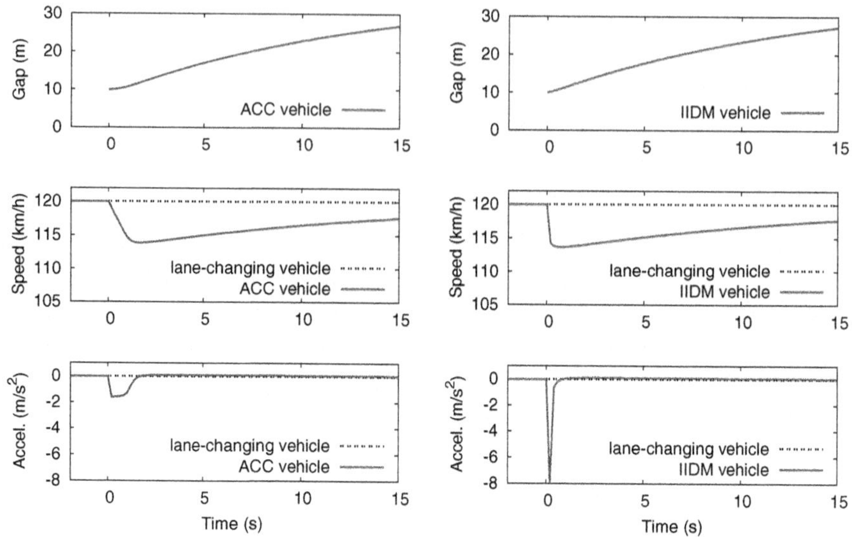

Fig. 11.8 Response of an ACC and an IIDM vehicle (parameters of Table 11.2 and coolness factor $c = 0.99$) to the lane-changing maneuver of another vehicle immediately in front of the considered vehicle. The initial speed of both vehicles is 120 km/h (equal to the desired speed), and the initial gap is 10 m which is about 30 % of the desired gap. This can be considered as a "mildly critical" situation

speed difference, and the gap). This models a behavior similar to the human reactions to brake lights, but in a continuous rather than in an off-on way.[12]

The Figs. 11.8 and 11.9 show the effect of this model improvement: In the mildly critical situation of Fig. 11.8, a lane-changing car driving at the same speed as the considered car cuts in front leaving a gap of only 10 m which is less than one third of the "safe" gap $s_0 + vT = 35.3$ m. While the IIDM (and IDM) will initiate a short emergency braking maneuver in this situation, the ACC reflects a relaxed reaction by braking at about the comfortable deceleration. In contrast, if the situation becomes really critical (Fig. 11.9), both the (I)IDM and the ACC model will initiate an emergency braking.

When implementing all these modifications, one needs to bear in mind that, in most situations, the ACC model should behave very similarly to the IDM so as to retain its well-tested good properties. To verify this, Fig. 11.10 shows the familiar *fact sheet* for the two standard situations. As expected, there is little difference compared to the IDM (and the IIDM) since the modified behavior kicks in only in the highway scenario, and only at the merging region of the on-ramp.

In summary, the ACC model can be considered as a minimal fully operative control model for ACC systems. With minor modifications, it has been implemented in real cars and tested on test tracks as well as on public roads and highways.

---

[12] Since the acceleration $\dot{v}_l$ cannot be measured directly, it is obtained by numerical differentiation of the approaching rate and the speed-changing rate. Care has to be taken to control the resulting discretization errors.

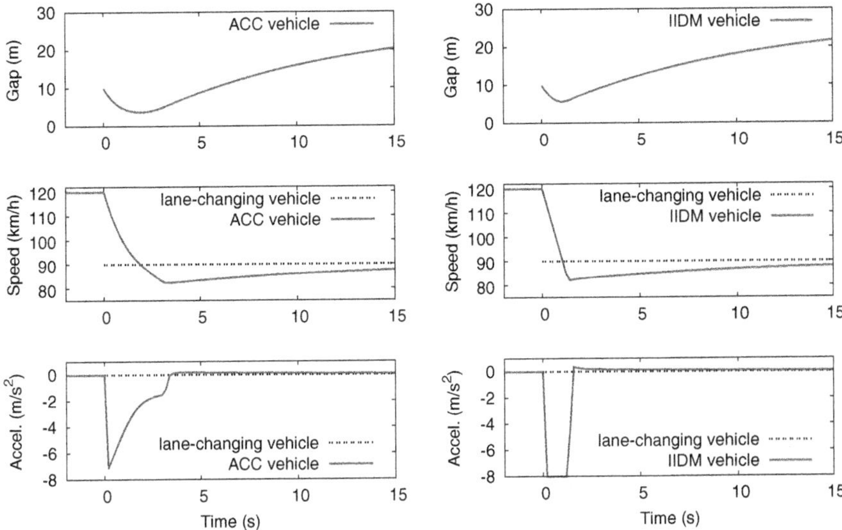

Fig. 11.9 Response of an ACC and an IIDM vehicle to a dangerous lane-changing maneuver of a slower vehicle (speed 90 km/h) immediately in front of the considered vehicle driving initially at $v_0 = 120$ km/h. The decelerations are restricted to $8\,\mathrm{m/s^2}$. The parameters are according to Table 11.2 and a coolness factor of $c = 0.99$ for the ACC vehicles. The desired speed of the lane-changing vehicle is reduced to 90 km/h

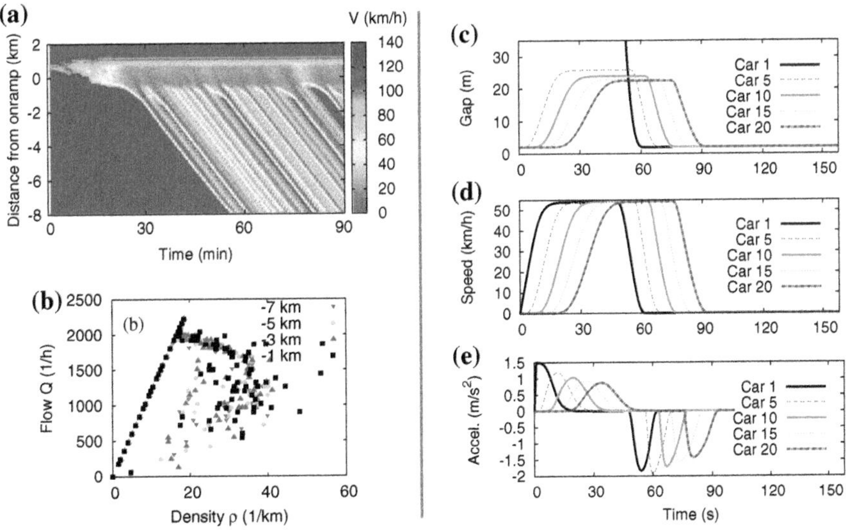

Fig. 11.10 Fact sheet of the ACC-model. The two standard scenarios "highway" (*left*) and "city traffic" (*right*) are simulated with the parameter values listed in Table 11.2 and the "coolness factor" $c = 0.99$. See Sect. 10.5 for a detailed description of the simulation scenarios

Problems

11.1 Conditions for the microscopic fundamental diagram
Use the consistency conditions (11.1)–(11.5) to derive the conditions (11.7) that have to be fulfilled by the steady-state speed-distance relation $v_e(s)$ (microscopic fundamental diagram).

11.2 Rules of thumb for the safe gap and braking distance

1. A common US rule for the safe gap is the following: "Leave one car length for every ten miles per hour of speed". Another rule says "Leave a time gap of two seconds". Compare these two rules assuming a typical car length of 15 ft. For which car length are both rules equivalent?
2. In Continental European countries, one learns in driving schools the following rule: "The safe gap should be at least half the reading of the speedometer". Translate this rule into a safe time gap rule and compare it with the US rule stated above. Take into account that, in Continental Europe, speed is commonly expressed in terms of km per hour.
3. A rule of thumb for the braking distance says "Speed squared and divided by 100". If speed is measured in km/h, what braking deceleration is assumed by this rule?

11.3 Reaction to vehicles merging into the lane
A vehicle enters the lane of the considered car causing the gap s to fall short of the equilibrium gap s_e by 50%. Both vehicles drive at the same speed. Find the resulting (negative) accelerations produced by the simplified Gipps' model and the IDM, assuming the parameter values $\Delta t = 1\,\mathrm{s}$, $b = 2\,\mathrm{m/s^2}$, $a = 1\,\mathrm{m/s^2}$, $\delta = 4$, and $v = v_0/2 = 72\,\mathrm{km/h}$ for all vehicles. (No other parameter values are needed for this problem.)

11.4 The IDM braking strategy
Derive Eq. (11.19) for the explicit description of the self-regulating braking strategy when approaching a standing obstacle ($\Delta v = v$). Assume that the IDM acceleration can be reduced to the braking term $\dot{v} = -b_{\text{kin}}^2/b$ for this case. *Hint*: Keep in mind that $\dot{s} = -\Delta v$.

11.5 Analysis of a microscopic model
Assume a car-following model that is given by the following acceleration equation:

$$\frac{\mathrm{d}v}{\mathrm{d}t} = \begin{cases} a & \text{if } v < \min(v_0, v_{\text{safe}}), \\ 0 & \text{if } v = \min(v_0, v_{\text{safe}}), \\ -a & \text{otherwise}, \end{cases} \qquad v_{\text{safe}} = -aT + \sqrt{a^2T^2 + v_l^2 + 2a(s - s_0)}.$$

As usual, v_l is the speed of the leading vehicle, and s the corresponding bumper-to-bumper gap.

1. Explain the meaning of the parameters a, s_0, v_0, and T by examining (i) the acceleration on a free road segment, (ii) the driving behavior when following another vehicle with constant speed and gap, and (iii) the braking maneuver performed when approaching a standing vehicle.
2. Find the steady-state speed $v_e(s)$ as a function of the distance assuming $v_0 = 20\,\mathrm{m/s}$, $a = 1\,\mathrm{m/s}^2$, $T = 1.6\,\mathrm{s}$, and $s_0 = 3\,\mathrm{m}$. Also, sketch the fundamental diagram for vehicles of length 5 m.
3. Assume that a vehicle standing at position $x = 0$ for $t \leq 0$ accelerates for $t > 0$ and then stops at a red traffic light at $x = 603\,\mathrm{m}$. Derive the speed function $v(t)$ for this scenario, assuming the parameter values $v_0 = 20\,\mathrm{m/s}$, $a = 1\,\mathrm{m/s}^2$, $T = 0$, and $s_0 = 3\,\mathrm{m}$. (Hints: The traffic light is modeled by a standing "virtual" vehicle; the vehicle will reach its desired speed in this scenario.)

11.6 Heterogeneous traffic

For identical vehicles and drivers, the modified IDM with its strictly triangular fundamental diagram (IIDM) does not produce the pre-breakdown speed drop observed in Fig. 11.5. Is it possible to produce the speed drop by introducing a combination of different desired speeds or the possibility of passing maneuvers?

11.7 City traffic in the modified IDM (IIDM)

On a road segment with two traffic lights, a number of vehicles is standing in front of the first traffic light. When the light turns green, the vehicles accelerate but have to stop again at the second traffic light. The upper panel shows trajectories of all 15 vehicles and the red lights (horizontal lines). The lower panel shows the corresponding speeds of the two bold trajectories.

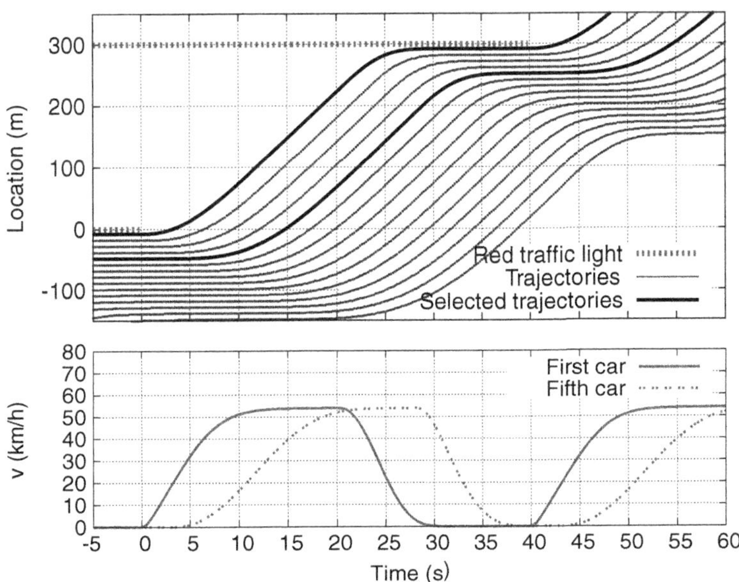

1. Estimate the capacity C of the free road segment (without traffic lights) by finding the maximum possible flow.
2. How many vehicles are able to pass the traffic light at $x = 0$ if the green light is on for (i) 5 s, (ii) 15 s, or (iii) 40 s? Find appropriate τ_0 and β such that $\tau(n) = \tau_0 + \beta n$ is the time the light has to be green to let n vehicles pass.
3. Estimate the velocity c_{cong} of the transition "standing traffic" $\rightarrow$ "starting to move" from the shown trajectories.
4. Estimate the IDM parameters v_0, $l_{\text{eff}} = l + s_0$, $T = 1/(\rho_{\text{max}} c_{\text{cong}})$, a, and b used in the simulation. Add appropriate tangents to the speed diagram to find the accelerations.

Further Reading

- Gipps, P.G.: A behavioural car-following model for computer simulation. Transportation Research Part B: Methodological **15** (1981) 105–111
- Krauss, S.: Microscopic Modeling of Traffic Flow: Investigation of Collision Free Vehicle Dynamics. Ph.D. Thesis, University of Cologne, Cologne, Germany (1997)
- Treiber, M., Hennecke, A., Helbing, D.: Congested traffic states in empirical observations and microscopic simulations. Physical Review E **62** (2000) 1805–1824
- Kesting, A., Treiber, M., Helbing, D.: Enhanced Intelligent Driver Model to access the impact of driving strategies on traffic capacity simulations. Philosophical Transactions of the Royal Society A **368** (2010) 4585–4605

Chapter 12
Modeling Human Aspects of Driving Behavior

It takes 8 460 bolts to assemble an automobile, and one nut to scatter it all over the road.

Author unknown

Abstract The driving characteristics described by the microscopic models of the previous chapters correspond, from a formal perspective, to semi-automated driving as realized by *adaptive cruise control* (ACC). On the one hand, human drivers are less efficient than ACC systems since reaction times, attention time spans, and estimation errors play a significant role. On the other hand, humans can take into account more input stimuli than acceleration controllers, for example: brake lights, turning signals, next-nearest neighbors, and external conditions. Moreover, in contrast to the present-day ACC systems reflected by the previous models, they can *anticipate* the situation for the next few seconds. All these specific human aspects will be formulated in terms of psycho-physiological extensions to the previous car-following models, in particular Gipps' model and the Intelligent Driver Model. Another class of psycho-physiological models explicitly take into account finite perception thresholds leading to sudden changes in accelerations whenever the difference from the ideal acceleration becomes significant. We present the Wiedemann model as a representative of this model class.

12.1 Man Versus Machine

The exogenous (input) variables of the models presented in Chaps. 10 and 11 are the own speed v, the (bumper-to-bumper) gap s to the leading vehicle, and its speed v_l (For the ACC model (11.26), the acceleration $\dot{v}_l$ is an additional input). The model output is the acceleration $a_{\text{mic}}(s, v, v_l)$ or the targeted speed $v_{\text{mic}}(s, v, v_l)$ for time-continuous or discrete models, respectively. Remarkably, both the input and the output are essentially the same as that required by the acceleration controllers for

M. Treiber and A. Kesting, *Traffic Flow Dynamics*,
DOI: 10.1007/978-3-642-32460-4_12,

semi-automated driving. Such driver-assistance systems, often called *adaptive cruise control* (ACC), are already available for many makes of cars. ACC systems obtain their input directly from the rotation rate of the tires (v), and by forward-looking radar or infrared-laser range finders (s and v_l). The acceleration $\dot{v}_l$ can be calculated from the speed v_l by taking the numerical time derivative (and smoothing the result).[1] The acceleration output of the car-following models corresponds to the signals the ACC system sends to the controllers for the engine and the brakes. The car-following model itself corresponds to the core control logic of the ACC system.[2]

In summary, the ACC-like driving styles generated by the models of the previous chapters are characterized by negligible estimation errors and response times, and an unshakeable attention. However, the restriction to the three input variables v, s, and v_l corresponds to a certain shortsightedness (only the immediate leader is considered but no vehicles driving further ahead), a tunnel vision (external circumstances as well as vehicles driving on neighboring lanes or behind are ignored), and ignorance brake lights, turning signals, horns, or headlight flashes are not taken into account).

In the following sections, we describe and model human driving characteristics that have not been captured in the models of the previous chapters:

- *Finite reaction time*. For attentive drivers, it is about $T_r = 1$ s. Depending on the driver and the situation, it can be significantly greater.
- *Estimation errors*. Gaps and speeds can only estimated with limited accuracy.
- *Temporal anticipation*. Experienced drivers can predict the traffic situation for the next few seconds.
- *Multi-vehicle anticipation* or *spatial anticipation*. The driver takes into consideration next-nearest and further vehicles ahead. Sometimes, vehicles on other lanes (particularly when driving through highway work zones) or the vehicle behind (when tailgating) play a role as well.
- *More input signals* like brake lights, direction indicators, horns.
- *Context sensitivity*. The driving style depends on the present and past overall traffic situation (*memory effect*). For example, after driving for some time in congested traffic, the time gap to the leading vehicle becomes greater and the driver's alertness wears down. In contrast, when approaching a lane closure and driving on one of the through lanes, drivers react to vehicles squeezing into their lane in a more relaxed manner, and drivers temporarily accept a smaller time gap than normal.
- *Finite perception threshold*. Humans cannot perceive small changes in exogenous factors but only respond to significant changes.
- *Courtesy and cooperation*. This aspect of human drivers is particularly relevant in situations with mandatory lane changes: Human drivers move out of the inside lane to allow the vehicle from the on-ramp to enter, or they brake (or accelerate) to

[1] In future, it may also be possible to directly measure accelerations with inertial sensors fused with gyroscopic sensors and others.

[2] This can be taken literally: During one of the authors' projects with a car manufacturer, the ACC model in Eq. (11.26) was implemented (with some modifications) as kernel of an ACC system of a mid-range car, and was tested on public city streets and highways.

make space for other drivers. This will be discussed in Sect. 14.3.3 in the context of modeling lane changes.

Because human driver models combine physiological restrictions (reaction times, estimation errors, perception thresholds) and psychological aspects (anticipation heuristic, context sensitivity, driving strategy in general), they are also referred to as *psycho-physiological car-following models*.

Remarkably, it turns out that the destabilizing effects caused by human imperfections are essentially compensated for by the stabilizing effects of anticipation and context sensitivity. In many situations this allows us to describe the traffic flow of human drivers by "machine-like" acceleration models of the type (10.3).

On the other hand, when investigating the strengths and weaknesses of human drivers and their influence on the efficiency and stability of traffic flow, the psycho-physiological models to be described below become necessary tools. Similarly to the lane-changing and decision models to be described in Chap. 14, we will formulate the above aspects (with the exception of finite perception thresholds) in terms of *extensions* to the ACC-like car-following models such as the IDM or the simplified Gipps' model. Another class of psycho-physiological model explicitly takes into account finite perception thresholds. We will briefly describe the Wiedemann model as most prominent representative of this model class.

12.2 Reaction Times

From the traffic flow modeler's point of view the reaction time T_r is composed of the following contributions:

- The *mental processing time* which, in turn, is composed of the sensation time ("there is a moving object on the road"), the perception or recognition time ("the object is a pedestrian"), the "situation awareness time" needed to recognize and interpret the scene ("I am moving towards the pedestrian possibly resulting in a crash"), and the decision time ("I will brake instead of do nothing or steering to the left").
- The *movement* or *action time*, e.g., to lift the foot from the throttle pedal, move it to the brake pedal, and apply pressure.
- And the *technical response time* of the respective vehicle components (about 100 ms for the brakes, several 100 ms for the accelerator).[3]

The reaction time depends on many factors such as the age and experience of the driver, visibility conditions, degree of surprise, and urgency of action. Nevertheless, nearly all models and simulators assume a constant common value of the order of 1 s for all drivers in all situations.

In time-continuous models, the driver's response is directly given by the acceleration function a_{mic}. So it is straightforward to introduce a reaction time: One simply

[3] The technical response time also applies to ACC systems.

calculates the acceleration function with time delayed input stimuli (gap, own speed,[4] and speed difference),

$$\dot{v}(t) = a_{\text{mic}}\big[s(t - T_r), v(t - T_r), v_l(t - T_r)\big]. \tag{12.1}$$

From the mathematical point of view, the differential equations of the time-continuous models become *delay-differential equations*. It is known from control theory that systems become more unstable with increased *dead times*. In the traffic context this means that traffic flow becomes more unstable with increased reaction times; this is intuitively plausible.

Equation (12.1) cannot be integrated analytically, but it is straightforward to approximatively solve it numerically. This is true even if the reaction time T_r is not a multiple of the update time Δt of the numerical scheme (10.9), (10.10). One simply approximates the quantity $u(t - T_r)$ (where u may stand for s, v, Δv, or v_l) in the update step $i = t/\Delta t$ by linear interpolation:

$$u(t - T_r) = r u_{i-j-1} + (1 - r) u_{i-j}, \quad j = \text{int}\left(\frac{T_r}{\Delta t}\right), \quad r = \frac{T_r}{\Delta t} - j. \tag{12.2}$$

Here int(x) and r denote the integer and fractional parts of x, respectively. In order to implement this formula, it is necessary to temporarily save the past $j + 1$ values $u_{i-j-1}, \ldots, u_i$ of the dynamical variables u in a buffer. With this approach, it is possible to model variations of the reaction time over the drivers and circumstances, as discussed above.

When modeling traffic flow by time-discrete iterated coupled maps or cellular automata (Chap. 13), the reaction time is often identified with the update time. However, this corresponds to another concept of the driver's reaction: While Eq. (12.1) represents a permanently attentive driver who (together with his or her vehicle) always needs the same reaction time to put the input stimuli into action, the speed function v_{mic} together with the update rules (10.7) and (10.8) of the iterated maps corresponds to a vehicle with zero response time and an instantly reacting driver who, however, looks at the traffic situation only at fixed time instances which are an *attention span* Δt apart. At all other times, the drivers are inattentive and do nothing, i.e., do not change the acceleration.[5]

Both the proper reaction time T_r and the attention span Δt contribute to the *effective reaction time* T_{eff}. In order to compare the relative effects of these factors, we reinterpret the numerical update time of time-continuous models as a model parameter in its own right (instead of an auxiliary numerical quantity) which has the same meaning as in the discrete-time models.

---

[4] One can argue that the own speed is known a priori without delay, so the associated sensation and recognition times are zero. However, as discussed above, these times make up only a fraction of the reaction time, so the speed argument of the acceleration function should be given by $v(t - T_r')$ with $T_r' < T_r$. Some models nevertheless assume $T_r' = 0$. In Eq. (12.1), we have assumed $T_r' = T_r$.

[5] This interpretation assumes that the acceleration is directly given by the pressure on the throttle or brake pedals. For time intervals within Δt this is a good approximation.

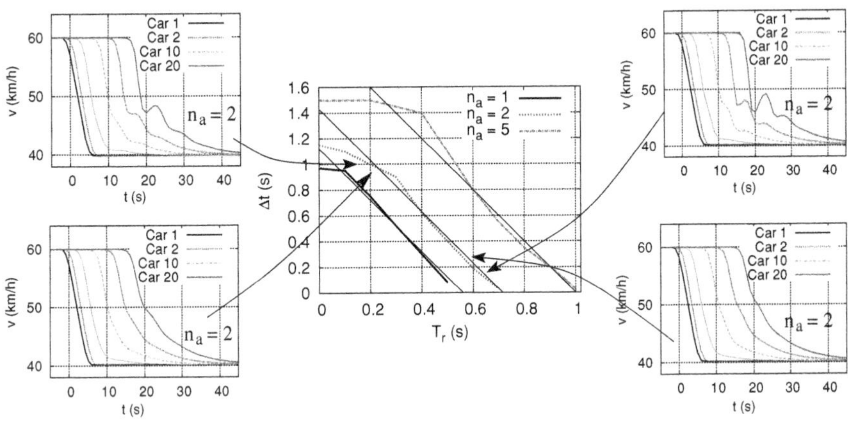

Fig. 12.1 Influence of the reaction time T_r, the update time interval Δt, and the number n_a of anticipated leading vehicles on the string stability (stability against the formation of traffic waves) of a vehicle platoon. The thick *solid lines* of the central diagram show the limits of string stability for $n_a = 1$, 2, and 5 anticipated vehicles. The curve for $n_a = 1$ corresponds to the IDM (see Sect. 11.3) with time delay. Traffic is string stable in the region below the corresponding line, and string unstable, otherwise. The diagrams at the *four corners* show the speed profiles of selected vehicles responding to a braking maneuver of the leading vehicle (Car 1) for $n_a = 2$ anticipated vehicles and values of T_r and Δt indicated by the *arrows* pointing to the central diagram

Figure 12.1 shows the reaction of a platoon of vehicles to a braking maneuver of a leading vehicle (Car 1) that cannot be overtaken. The solid thick lines of the central diagram of this figure shows the limits of string stability as a function of the reaction time T_r and the attention span Δt. In a wide range, this limit is approximatively given by the linear relation (thin solid lines)

$$T_r + \frac{\Delta t}{2} = T_{\text{eff}} = \text{const.} \tag{12.3}$$

We conclude that the destabilizing effect of T_r is *twice* as large as that of Δt. This can be understood by averaging the varying effective delaying effect $T_a(t)$ of the finite attention span over time: Immediately after an update, i.e., at times $t = i\,\Delta t,\ i = 1, 2, \ldots$, the acceleration response is instantaneous ($T_a = 0$). In contrast, the unchanged acceleration just before an update corresponds to a maximum delay $T_a = \Delta t$, so the average delay $\langle T_a \rangle = \Delta t/2$. Adding the reaction time T_r and assuming that the destabilizing effect grows linearly with the average total response time $T_r + \langle T_a \rangle = T_r + \Delta t/2 = T_{\text{eff}}$ leads to the relation (12.3). The central diagram in Fig. 12.1 shows that this relation is not satisfied for very large values of Δt where the assumption of linearly growing destabilization effects breaks down.

We conclude that the update time of iterated maps (or of the numerical integration method of time-continuous models) acts as an effective reaction time when considering its global effects. However, the value $\langle T_a \rangle = \Delta t/2$ of the effective reaction time is only *half* the numerical value of the update time.

12.3 Estimation Errors and Imperfect Driving Capabilities

Apart from the delay by reaction times, human driving behavior deviates from the machine-like driving style modeled by the acceleration function a_{mic} in two further ways: Firstly, humans make errors when estimating the input stimuli s, v, and v_l of the acceleration function. Secondly, the driving behavior is not completely rational, i.e., even if the same input stimuli are estimated for different times, drivers will behave differently.

Here and in the following, we formulate the specific human driving elements in terms of modifications of the acceleration function. This does not restrict the analysis to time-continuous models: The considerations carry over to the speed function $v_{\mathrm{mic}}(s, v, v_l)$ of time-discrete models by means of Eq. (10.11) mapping the acceleration function to the speed function *via* the relation $v_{\mathrm{mic}}(s, v, v_l) = v + \Delta t\, a_{\mathrm{mic}}(s, v, v_l)$.

12.3.1 Modeling Estimation Errors

When exclusively considering estimation errors, the acceleration function a_{mic} itself remains unchanged while the true values of its independent variables s and v_l are replaced by the estimated values s^{est} and v_l^{est}, respectively. Since the driver's own speed v can be estimated with sufficient accuracy by looking at the speedometer, we will neglect the associated errors.[6] The magnitude of the estimating errors $s^{\mathrm{est}} - s$ and $v_l^{\mathrm{est}} - v_l$ depends on the driving situation, on the driver, and on external circumstances such as illumination and visibility. It can be determined by traffic psychological experiments in a driving simulator.[7] In the following, we will consider how the driving situation influences the errors when estimating the gap to and the speed of the leader.

Estimation error of the gap. In most driving situations, the *relative* estimation error for the gap, or, equivalently, the error of the logarithm of the gap, turns out to be essentially constant:

$$\ln s^{\mathrm{est}} - \ln s = V_s w_s(t). \tag{12.4}$$

The model parameter V_s describes the relative standard deviation of s_{est} from the true value s, also known as statistical *variation coefficient*. Typical values are of the order of 10 %. The error is assumed to have no bias. The (0,1)-normally distributed stochastic variable $w_s(t)$ describing the temporal evolution of the error will be discussed below.

---

[6] In fact, the speed indicated by the speedometer can be greater than the true speed by up to 5 %. Due to its systematic nature, this error will be taken care of automatically at model calibration time, so there is no need to consider it explicitly in the model development.

[7] Notice that generating and modeling surrounding traffic for driving simulators is one of the applications of microscopic traffic flow models; the effects of the driver's actions themselves (such as steering or braking) are described by sub-microscopic models, cf. Table 1.1.

Estimation error of the speed of the leading vehicle. The driver estimates the speed v_l of the leader relative to his or her own speed (whose true value is assumed to be known) by the change of the apparent optical angle $\phi \approx w_{\text{veh}}/s$ under which the leading vehicle of width w_{veh} is seen. Based on experiments, we assume the error of the *rate of relative angular change* to be constant,

$$r = \frac{\mathrm{d}\phi/\mathrm{d}t}{\phi} = \frac{\frac{w_{\text{veh}}}{s^2}\Delta v}{w_{\text{veh}}/s} = \frac{\Delta v}{s} = \frac{1}{\tau_{\text{TTC}}}. \qquad (12.5)$$

Interestingly, the rate r of relative change is the inverse of the *time-to-collision* $\tau_{\text{TTC}} = s/\Delta v$ which is an important safety indicator.

The safety indicator *time-to-collision* (TTC) $\tau_{\text{TTC}} = s/\Delta v$ is defined by the hypothetic time interval to a collision if neither vehicle accelerates or brakes. Values in the range $0 < \tau_{\text{TTC}} \leq 4\,\text{s}$ are generally considered as critical. TTC is only meaningful for positive approach rates.

Assuming a constant standard deviation σ_{r} of the relative approach rate (of the order of $r = 0.01\,\text{s}^{-1}$) we obtain

$$v_l^{\text{est}} - v_l = -\left(\Delta v^{\text{est}} - \Delta v\right) = -s\left(\frac{1}{\tau_{\text{TTC}}^{\text{est}}} - \frac{1}{\tau_{\text{TTC}}}\right) = -s\,\sigma_{\text{r}}\,w_l(t). \qquad (12.6)$$

In analogy to the quantity $w_s(t)$, the stochastic quantity $w_l(t)$ describes the distribution of the error and its change in time. In the following, we will look more closely how to model $w_l(t)$ and $w_s(t)$.

A model for the time dependence of estimation errors. The Eqs. (12.4) and (12.6) reduce the estimation errors for the gap and the speed difference to the time-dependent stochastic quantities $w_s(t)$ and $w_l(t)$. Each driver has his or her own set of stochastic variables $\{w_s(t), w_l(t)\}$, which all are independent from each other.

Generally, time-dependent stochastic variables are defined by a *stochastic process*. In the following, we will assume that both $w_s(t)$ and $w_l(t)$ are instances of a stationary process $w(t)$ which is defined by (i) the distribution function at a given time (which is time independent), (ii) the *autocorrelation* function describing the correlation of the stochastic process at different times as a function of the time difference. As distribution function, we assume a standardized Gaussian

$$w(t) \sim N(0, 1), \quad \langle w(t) \rangle = 0, \quad \left\langle w^2(t) \right\rangle = 1. \qquad (12.7)$$

On average, the estimation errors are zero. This is no restriction since systematic errors can be mapped to changes of the model parameters which will be taken care of when calibrating the model.

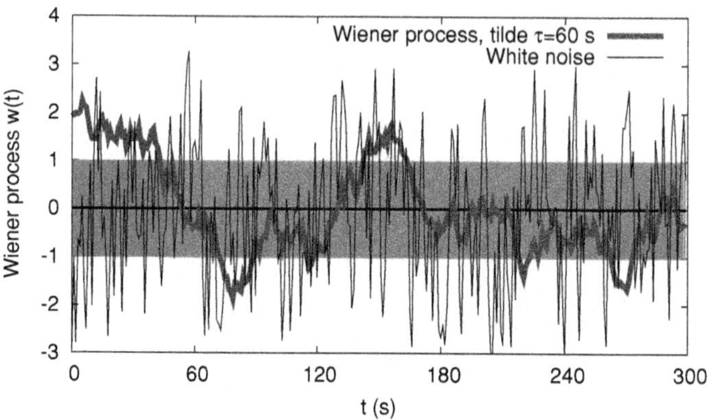

Fig. 12.2 Time series of a Wiener process ($\tilde{\tau} = 60\,\text{s}$) generated by Eq. (12.11). Also shown is the range of one standard deviation around the mean $\langle w \rangle(t) = 0$ and white noise

When describing the autocorrelation function, we take into account that human errors are characterized by a certain *persistence*: If a driver, say, underestimates the gap at a given time, he or she is likely to underestimate it in the next second as well. In mathematical terms, the errors at two times are positively correlated for small time differences of a few seconds up to one minute. This can be described by following autocorrelation function:

$$\langle w(t)w(t') \rangle = \exp\left(-\frac{|t-t'|}{\tilde{\tau}} \right), \tag{12.8}$$

where the persistence time $\tilde{\tau}$ is a model parameter of the order of several seconds (cf. Table 12.1 and Fig. 12.2). In Problem 12.1, we show that a stochastic process $\{w(t)\}$ satisfying the conditions (12.7) and (12.8) can be generated by following stochastic differential equation also known as *Wiener process*:

$$\frac{\mathrm{d}w}{\mathrm{d}t} = -\frac{w}{\tilde{\tau}} + \sqrt{\frac{2}{\tilde{\tau}}}\xi(t). \tag{12.9}$$

The "standardized white noise" $\xi(t)$ is characterized by

$$\langle \xi(t) \rangle = 0, \quad \langle \xi(t)\xi(t') \rangle = \delta(t-t'). \tag{12.10}$$

Here, Dirac's δ-function $\delta(t)$ is equal to zero for all $t \neq 0$. For $t = 0$, the function diverges in a way that any integral $\int_a^b \delta(t)\mathrm{d}t = 1$ as long as the integration range includes zero, $a < 0 < b$.[8]

[8] Strictly speaking, $\delta[\cdot]$ is a *functional*, i.e., a mapping of a function $f(x)$ to a number: $\delta[f(x)] = f(0)$. This functional can be represented by a definite integral: $\delta[f(x)] = \int_{-\infty}^{\infty} \delta(x) f(x)\mathrm{d}x = f(0)$. Consequently, the δ-function does only make sense inside an integral.

The stochastic differential equation (12.9), (12.10) for the Wiener process allows for a simple and efficient numerical integration scheme. The update rule to generate the quantity $w_i = w(i\Delta t)$ of the ith step is given by

$$w_i = \mathrm{e}^{-\Delta t/\tilde{\tau}} w_{i-1} + \sqrt{\frac{2\Delta t}{\tilde{\tau}}}\,\eta_i, \tag{12.11}$$

where η_i are instances of computer-generated i.i.d. pseudo-random numbers with expectation zero and unit variance.[9] Since the prefactors $\exp(-\Delta t/\tilde{\tau})$ and $\sqrt{2\Delta t/\tilde{\tau}}$ can be calculated in advance at the beginning of the simulation, this is a very efficient numerical scheme. There are two independent Wiener processes $w_s(t)$ and $w_l(t)$ for each driver which are initialized using the pseudo-random number generator as well, $w_0 = \eta_0$.

Get yourself the idea that a persistence time going to infinity, $\tilde{\tau} \to \infty$, corresponds to simulating heterogeneous traffic of deterministically behaving drivers instead of homogeneous traffic of stochastically behaving drivers: Each driver has his or her individual acceleration function which is defined by the initialization of the two respective Wiener processes.

12.3.2 Modeling Imperfect Driving

Driving errors and irregularities in driving style result in erratic components of the driver's action, i.e., acceleration. This can be modeled by adding to the acceleration function a_{mic} itself some *acceleration noise* of standard deviation σ_a whose time dependence is modeled by a third Wiener process. Including the estimation errors, the resulting acceleration is given by

$$\dot{v}(t) = a_{\text{mic}}(s^{\text{est}}, v, v_l^{\text{est}}) + \sigma_a w_a(t). \tag{12.12}$$

Besides modeling imperfections of the drivers, acceleration noise can also represent corrections due to factors that are not explicitly considered, or due to deficiencies of the model itself.[10]

In most stochastic models, the persistence of the acceleration noise is ignored and its time dependence is modeled by white noise $\delta(t)$ (or a time-discrete version $\{\eta_i\}$ of it) rather than by a correlated Wiener process $w_a(t)$. Particularly, this ansatz is

---

[9] The numbers may be normally distributed but it is not required. For $\tilde{\tau} \gg \Delta t$, the central limit theorem guarantees that w_i is Gaussian for any distribution of η_i satisfying $\langle \eta_i \rangle = 0$ and $\langle \eta_i^2 \rangle = 1$, for example, a uniform distribution.

[10] In this sense, this random term represents an admission that not everything can be modeled precisely. It has the same meaning as the additive stochastic term of econometric models.

adopted for essentially all cellular automaton models (see Chap. 13) but also for some stochastic variants of Gipps' model such as Krauss' model. However, since driving errors have a certain persistence as well, this is unrealistic. For each application, one has to decide whether the increased simplicity of using white noise compensates for the deficiencies in describing the phenomena.

12.4 Temporal Anticipation

Figure 12.1 shows that simulating time-continuous ACC-like car-following models with reaction-time delays of the order of the normal time gap, or even less, generally leads to crashes. This is true even if none of the human imperfections described above (Sect. 12.3) are applied. From this point of view, it is remarkable that humans drive essentially accident-free in spite of reaction times of this order, and additional estimation errors, and driving imperfections. Moreover, in dense but not yet congested traffic, traffic is free of instabilities (traffic oscillations) even if the average time gap s/v is *smaller* than the reaction time T_r (cf. Fig. 4.8). Simulations and the analytical analysis of Chap. 15 show that this is only possible when considering several leading vehicles (multi-anticipation, see Sect. 12.5) and when anticipating the traffic situation for the next few seconds as described in the following. We can model the anticipation ability of experienced drivers by assuming that they adopt following simple *heuristic*:[11]

1. The own speed and acceleration is known. Furthermore, the acceleration will not change during the anticipation time horizon assumed to be equal the reaction time T_r to bridge the delay caused by the reaction time.[12] This *constant-acceleration heuristic* for the movement of the own vehicle corresponding to a linear forward projection of the speed:

$$v^{\text{prog}}(t) = v^{\text{est}}(t - T_r) + T_r\,\dot{v}(t - T_r). \tag{12.13}$$

   Here, $\dot{v}(t - T_r)$ is the acceleration realized at time $t - T_r$.

2. The acceleration of the leading vehicle is difficult to estimate since only the binary information "brake lights on or off" is available.[13] Most models do not consider brake lights as an exogenous factor at all (see Sect. 12.6 for an exception).

[11] Heuristic refers to experience-based assumptions and strategies for finding a sufficiently good solution to a problem with limited knowledge in a short time. Generally, a heuristic cannot be justified or "proven" in any precise sense.

[12] Regarding the driver's actions, this essentially corresponds to the "do nothing" assumption: The pressure on the throttle or the brake pedal remains unchanged.

[13] In some cases, further information on the acceleration may be available from the situational context ("truck approaching a hill", "vehicle approaching an exit", "roadworks ahead"), or from multi-anticipation ("red brake lights several vehicle ahead", "jam ahead"). These factors are described in the Sects. 12.5 and 12.7.

Therefore, we adopt the *constant-speed heuristic* for the preceding vehicle (and for all further vehicles if multi-anticipation is considered). To first order in the reaction time, this also corresponds to a linear forward projection of the gap:

$$v_l^{\text{prog}}(t) = v_l^{\text{est}}(t - T_r), \quad s^{\text{prog}}(t) = s^{\text{est}} - T_r\, \Delta v^{\text{est}}(t - T_r). \tag{12.14}$$

12.5 Multi-Vehicle Anticipation

Human drivers do not only form hypotheses about the traffic state in the near future (temporal anticipation) but they take into account several vehicles ahead whenever this is possible.[14] This anticipation is also denoted as spatial anticipation or *multi-anticipation*.

To express this in mathematical terms by a generalization of the existing model, we divide the acceleration function $a_{\text{mic}}(s, v, v_l)$ of time-continuous models into a *free-flow acceleration* a_{free} and an *interaction acceleration* a_{int} representing the obstructions caused by other vehicles:

$$a_{\text{mic}}(s, v, v_l) = a_{\text{free}}(v) + a_{\text{int}}(s, v, v_l), \quad a_{\text{free}}(v) = \lim_{s\to\infty} a_{\text{mic}}(s, v, v_l). \tag{12.15}$$

By means of the relation $v_{\text{mic}}(s, v, v_l) = v + \Delta t\, a_{\text{mic}}(s, v, v_l)$, multi-anticipation can be defined for time-discrete models in analogy. The general plausibility conditions (11.1)–(11.5) imply that a_{int} is non-positive. In this sense, the acceleration function (and the speed function of time-discrete models) can be considered as a composition of two *social forces*: The free-flow acceleration reflects the driver's desire to drive at a certain speed. The decelerating interaction a_{int} exerted on the driver of the subject vehicle by the leading vehicle reflects the necessity to avoid crashes and keep a minimum time gap.

When considering $n_a \geq 1$ leading vehicles $\alpha - 1, \ldots, \alpha - n_a$, it is most straightforward to add the decelerating social forces caused by these vehicles as though the vehicles in between do not exist.[15] For vehicle α, this results into the acceleration function[16]

[14] Often, one can recognize vehicles further ahead through the windows of the immediate leader, or by looking past the sides of vehicles when the road is curved. If this is not possible (e.g., when driving behind trucks or SUVs), driving is perceived as less comfortable and the gap will be increased measurably.

[15] In physics, this corresponds to a linear superposition of all forces without shielding effects. Examples include gravitational forces or electrostatic forces for non-polarizable particles.

[16] In this context, the vehicle indices are relevant, so they will no longer be omitted as has been done previously for readability. Notice that, when modeling heterogeneous traffic the acceleration functions of each driver-vehicle unit (not only the function arguments) are different, so each functions should have an index on its own. This index will still be omitted.

$$\dot{v}_\alpha = a_{\text{free}}(v_\alpha) + \sum_{\beta=\alpha-n_a}^{\alpha-1} a_{\text{int}}(s_{\alpha\beta}, v_\alpha, v_\beta), \quad s_{\alpha\beta} = \sum_{j=0}^{\alpha-\beta-1} s_{\alpha-j}. \tag{12.16}$$

Since the vehicle lengths play no role in the dynamics, the total gap $s_{\alpha\beta}$ is calculated by summing only over the individual gaps $s_{\alpha-j}$, i.e., the lengths of all vehicles in between are excluded.

One should realize that, for accelerations satisfying the general plausibility criteria of Sect. 11.1, the immediate leader exerts the largest decelerating social force, in agreement with experience.

Directly implementing this ansatz, however, leads to unwanted consequences: The total social force appearing in Eq. (12.16) becomes more negative when increasing the number of considered vehicles for a given configuration of the vehicle positions and speeds. As a consequence, the fundamental diagram changes and the modeled road capacity decreases. In order to directly compare the behavior with and without multi-anticipation, we require that the fundamental diagram does not change. This can be realized by multiplying all social forces with a common prefactor $c \le 1$ (which possibly depends on the speed) such that following condition for the steady-state equilibrium $v_\alpha = v_\beta = v$, $s_\alpha = s_{\alpha-j} = s_e(v)$ is satisfied:

$$c(v) \sum_{j=1}^{n_a} a_{\text{int}}(j s_e(v), v, v) = a_{\text{int}}(s_e(v), v, v). \tag{12.17}$$

For the IDM, this leads to the speed independent reduction factor (see Problem 12.3)

$$c_{\text{IDM}} = \left(\sum_{j=1}^{n_a} \frac{1}{j^2} \right)^{-1}. \tag{12.18}$$

Even when considering infinitely many leading vehicles, $c_\infty = 6/\pi^2 \approx 0.61$ is nonzero reflecting the fact that the infinite sum of all social forces converges. In this case, the immediate leader is responsible for 61 % of the total interactions, and all other vehicles for the remaining 39 %.

In summary, the multi-anticipative acceleration function is given by

$$a_{\text{multi}}(s_\alpha, v_\alpha, \{s_\beta\}, \{v_\beta\}) = a_{\text{free}}(v_\alpha) + c \sum_{\beta=\alpha-n_a}^{\alpha-1} a_{\text{int}}(s_{\alpha\beta}, v_\alpha, v_\beta). \tag{12.19}$$

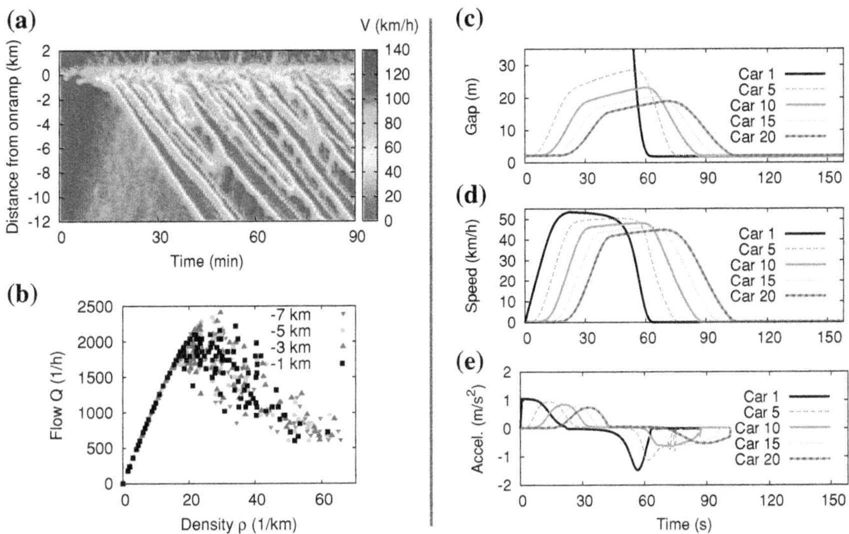

Fig. 12.3 Fact sheet of the Human Driver Model (12.20). The reaction time T_r, the number of considered vehicles, and error standard deviations and persistence times are given in Table 12.1. The IDM parameters are that of Table 11.2. The simulation scenarios are discussed in detail in Sect. 10.5

When simultaneously considering estimation errors, temporal anticipation, and reaction times, one inserts into the acceleration function the estimated arguments $s_\alpha^{\text{prog}}(t)$, $v_\alpha^{\text{prog}}(t)$, and $v_\beta^{\text{prog}}(t)$. The human-driver extensions (12.19), (12.13), and (12.14) can be applied to the acceleration and speed functions of any time-continuous and time-discrete model, respectively.

When applying the extensions to the IDM, we obtain the Human Driver Model (HDM):

$$\boxed{\dot{v} = a_{\text{free}}^{\text{IDM}}(v_\alpha) + c_{\text{IDM}} \sum_{\beta=\alpha-n_a}^{\alpha-1} a_{\text{int}}^{\text{IDM}}(s_{\alpha\beta}^{\text{prog}}, v_\alpha^{\text{prog}}, v_\beta^{\text{prog}}) \quad \text{HDM.}} \tag{12.20}$$

The speed estimate v_α^{prog} is evaluated using Eq. (12.13), and the gap and leading speed estimates $s_{\alpha\beta}^{\text{prog}}$ and v_β^{prog} by Eq. (12.14), respectively. The free-flow and interaction accelerations are given by

$$a_{\text{free}}^{\text{IDM}}(v) = a\left[1 - \left(\frac{v}{v_0}\right)^\delta\right], \quad a_{\text{int}}^{\text{IDM}}(s, v, v_l) = -a\left(\frac{s^*(v, v - v_l)}{s}\right)^2 \tag{12.21}$$

with s^* from Eq. (11.15).

Figure 12.3 shows simulations of the HDM for the two standard scenarios discussed in detail in Sect. 10.5. The most notable change with respect to the original

Table 12.1 Parameters of the human driver extensions to the acceleration function

Parameter	Typical value
Reaction time T_r	0.6 s
Number of anticipated vehicles n_a	5
Variation coefficient of gap estimation error V_s	10 %
Estimation error for the inverse TTC σ_r	0.01 s^{-1}
Magnitude of acceleration noise σ_a	0.1 m/s^2
Persistence time of the estimation errors $\tilde{\tau}$	20 s
Persistence time of the acceleration noise $\tilde{\tau}_a$	1 s

IDM (cf. Fig. 11.4) is the increased wavelength of the traffic waves in the freeway scenario, and the decrease of the maximum braking decelerations for the vehicles further behind in the queue. Both are essentially caused by multi-anticipation. The finite reaction time and the temporal anticipation essentially cancel each other while the estimation errors (10 % for the gap and 0.01 s^{-1} for the relative approaching rate, see Table 12.1) have little influence. Increasing the errors, however, will eventually lead to drastic effects or even accidents.

Figure 12.1 demonstrates that the limit of string stability of a platoon of vehicles as a function of the reaction time T_r and the attention span Δt depends strongly on the number of considered vehicles. When considering $n_a = 5$ leaders, the critical effective reaction time $T_r + \Delta t/2$ at the stability limit is about twice as large as the corresponding critical value without multi-anticipation ($n_a = 1$). Particularly, traffic can be stable even if the reaction time exceeds the average time headway. This agrees with everyday observations but cannot be realized in simulations without multi-anticipation. There are limits, however: Anticipating more than five vehicles ahead will change the dynamics insignificantly.

We emphasize that, in accordance with the design principles presented in Sect. 11.3.7, the model extensions have been formulated with as few additional parameters as possible. Apart from the specification of the estimation errors, there are only two additional parameters which both have an intuitive meaning and plausible values: reaction time and the number of anticipated vehicles.

12.6 Brake Lights and Further Exogenous Factors

Several cellular automata (see Chap. 13) and a few continuous models include *brake lights* as a further binary input variable: Brake lights on or off:

$$z_b = \begin{cases} 1 & \dot{v}_l < a_c, \\ 0 & \text{otherwise.} \end{cases} \tag{12.22}$$

The parameter a_c (typical values are around $-0.2\,\text{m/s}^2$) corresponds to the deceleration if neither the throttle nor the brake pedal are touched. If the lights are on ($z_b = 1$), the drivers adapt their driving style or their anticipation heuristic to a

Table 12.2 Implementing changes of the driving mode for some simple models in response to brake lights of the leader ($z_b = 1$) or flashing headlights of the follower ($z_t = 1$)

Parameter	Reference	Brake lights ($z_b = 1$)	Tailgating ($z_t = 1$)
Desired speed v_0 (all models)	120 km/h	120 km/h	140 km/h
Time gap T (OVM, FVDM, IDM)	1.0 s	1.5 s	1.0 s
Acceleration a (Gipps, IDM)	1.0 m/s^2	1.0 m/s^2	2.0 m/s^2
Comfortable deceleration b (Gipps, IDM)	1.5 m/s^2	1.0 m/s^2	1.5 m/s^2

more defensive mode compared to the situation with "brake lights off" ($z_b = 0$). Table 12.2 shows a possible implementation of this behavioral change for some elementary models.[17] Notice that z_b can be considered as a discrete version of the acceleration $\dot{v}_l$ of the leading vehicle which is already an exogenous variable of the ACC model (cf. Sect. 11.3.8).

Analogously, including direction indicators in the model allows one to simulate cooperative and anticipative lane changing strategies (see Chap. 14). Extending multi-anticipation to include the rear vehicle allows one to simulate forward social forces caused by tailgating drivers, e.g., by parameter changes as given in Table 12.2. Tailgating can be characterized by the binary exogenous variable

$$z_t = \begin{cases} 1 & s_{\alpha+1}/v_{\alpha+1} < T_c \text{ or follower flashes headlamps,} \\ 0 & \text{otherwise.} \end{cases} \tag{12.23}$$

This means a driver is subject to tailgating if the time gap $s_{\alpha+1}/v_{\alpha+1} < T_c$ of the follower is below some critical value T_c or if the follower flashes headlamps. Beside forward social forces, one can also include social forces urging a lane change if the subject vehicle is on the faster lane.[18]

12.7 Local Traffic Context

After driving for some time in congested or jammed traffic, most drivers become less alert, the accelerations decrease, and the time gaps increase. This *resignation effect* implies a less efficient driving style which reduces the flow downstream of the bottleneck compared to the situation before the breakdown. This so-called *capacity drop* leads to a positive feedback of jam formation that is often observed in reality (cf. Sects. 4.3 and 4.4). Figure 12.4 shows the result of modeling the resignation effect with the IDM by gradually reducing its acceleration parameter a and increasing the

[17] In Gipps' model, the time gap is "hardwired" to the update time step and therefore is not available to model a behavioral change of some vehicles.

[18] In some countries and in some situations, it may be useful to include using the horn or displaying rude gestures as further exogenous binary variables.

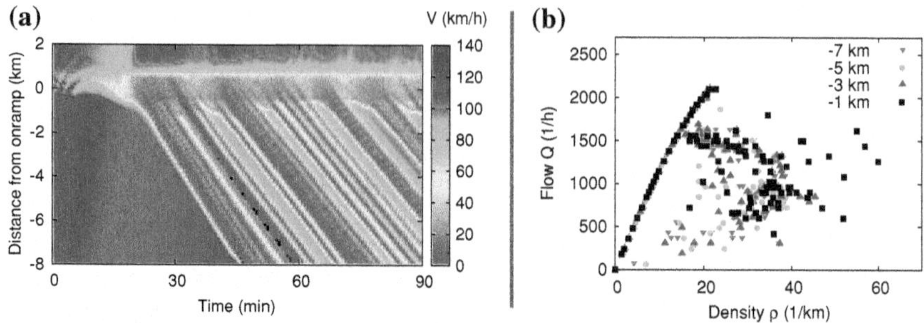

Fig. 12.4 Simulations of the standardized highway scenario using the IDM with *memory effect*. When driving in a jam, the time gap parameter T is increased up to 140 %, and the IDM acceleration parameter a decreased down to 50 % of the corresponding reference values on a time scale of 10 min. The reference parameters are given by Table 11.2. See Sect. 10.5 for a detailed description of the highway scenario

time gap parameter T over time scales of several minutes. When comparing this figure with the IDM simulation of Fig. 11.4a, one observes that the inverse-λ shape of the flow-density data of the virtual detectors indicating the capacity drop is more pronounced. The plot of the spatiotemporal local speed (Fig. 12.4a) shows more congestions (red areas) than that for the IDM. This indicates that resignation effects aggravate the congestion.

On the flip side, if it were possible to locally invert the sign of this effect near bottlenecks, this would open the possibility to dynamically "fill" the local "capacity holes" constituting the bottleneck. In Sect. 21.5, we will present simulations of a future driver assistance system that makes use of this possibility by increasing the agility near bottlenecks or when leaving a congested zone, and adopting a more defensive driving style when approaching roadworks or jams.

More generally, one can interpret the influencing regions of variable speed limits and other control measures as a local traffic context, or simply the distinction between city and freeway traffic. By changing the model parameters, it is straightforward to model or simulate a change of the driving mode caused by a new context. This can be done gradually on time scales of several minutes as in the memory effect discussed above or when lightning or weather conditions change. Other situations require an instantaneous change such as passing a new speed limit sign, entering/leaving a tunnel, approaching a zone of roadworks, or entering/leaving the city limits.

12.8 Action Points

In the previous models, drivers are assumed to react to the exogenous stimuli of the traffic environment in a continuous way, no matter how small their changes. However, it is well known from physio-psychological investigations that humans have finite *perception thresholds* in discriminating different gaps, speeds, or speed differences.

Perception thresholds can be modeled by the concept of *action-point models*: drivers react actively (by changing the pressure on the throttle or brake pedals or switching the pedals) only if the current action deviates significantly from the action considered as ideal for the given situation. For car-following models described by an acceleration function this essentially amounts to a constant acceleration most of the time until the actual acceleration differs significantly from the "ideal" acceleration given by the acceleration function $a_{\rm mic}$. Alternatively, one can describe the event triggering a conscious action in terms of perceptible changes of the exogenous variables s, v, or Δv.[19] Once the threshold is exceeded, the driver reverts his or her acceleration to the value of the acceleration function $a_{\rm mic}$ for the present situation and keeps the new acceleration until the perception threshold is again exceeded. The thresholds and the associated acceleration or behavioral changes at irregular time instances define the *action points*. Trajectory data show that not only the time intervals between two action points but also the thresholds and the actions itself (i.e., the acceleration changes) are stochastic quantities. Generally, acceleration changes caused by action points are hard to distinguish from the effects of correlated acceleration noise. To date, the existence of action points in the data remains controversial.

12.9 The Wiedemann Car-Following Model

A model that considers both the local traffic context and action points is the time continuous *Wiedemann-Model*. It serves as basis for some commercial traffic flow simulators.[20] This model describes the psycho-physiological aspects of the driving behavior in terms of four discrete driving regimes: (i) free flow, (ii) approaching slower vehicles, (iii) car-following near the steady-state equilibrium, and (iv) critical situations requiring stronger braking actions. In each of these regimes k, different acceleration functions $a_{\rm mic}^{(k)}(s, v, \Delta v)$ apply. The boundaries between the regimes are given by nonlinear equations of the form $f_{k'}(s, v, \Delta v) = 0$ defining curved areas in three-dimensional state space $(s, v, \Delta v)$ spanned by the exogenous variables (see Fig. 12.5). Additionally, acceleration noise of the type described in Sect. 12.3.2 is superimposed. In accordance with the philosophy of action points, the acceleration changes abruptly at the boundaries of the regimes to the new acceleration function representing the driving mode in the new regime. Nevertheless, the Wiedemann model does not implement the concept of action points in its pure form: Firstly, the regime boundaries, i.e., the conditions for action points, are deterministically fixed by the conditions $f_{k'}(s, v, \Delta v) = 0$ rather than stochastic. Secondly, the "do nothing" philosophy between the action points is replaced by the acceleration

[19] If the model includes binary exogenous factors, their changes (e.g., brake lights on or off) would trigger a conscious action as well, see Sect. 12.6.

[20] The actual acceleration functions of these simulators strongly deviate from the original formulation of the Wiedemann model.

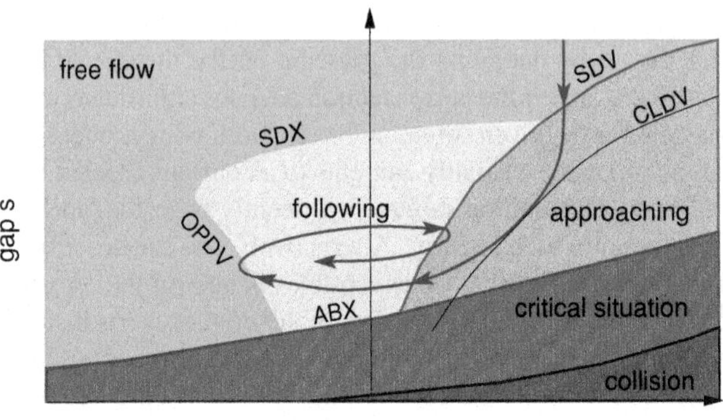

Fig. 12.5 Schematic and simplified representation of the regimes of the Wiedemann model in the three-dimensional state space spanned by s, v, and Δv. Shown are the intersections of the regimes and their boundaries with the plane $v_l = v - \Delta v =$ constant (the leader drives at constant speed v_l). The *blue line* shows the trajectory of a vehicle approaching a slower vehicle in the projected state space. The speed-difference thresholds CLDV ("closing in"), OPDV ("opening"), SDV ("sensitivity threshold"), and the gap-related thresholds ABX and SDX (minimum and maximum gap in car-following regime) are denoted as in the literature

functions $a_{\text{mic}}^{(k)}(s, v, \Delta v)$ of the different driving modes.[21] In spite of its psycho-physiological nature, the Wiedemann model does not contain explicit reaction times. The model is complex since four acceleration functions, several nonlinear equations for defining the boundaries of each regime, and the acceleration noise have to be specified.

Simulating the Wiedemann model typically leads to oscillations in state space with quasi-periodic transitions between different regimes as illustrated by the blue trajectory of Fig. 12.5.[22] This trajectory represents a car approaching a slower vehicle that cannot be overtaken. Initially, the driver of the car is in the free-flow regime (acceleration function $a_{\text{free}}(v)$) cruising at his or her desired speed. This corresponds to a constant approaching rate $\Delta v > 0$. Once the perception threshold of recognizing the leading vehicle is crossed, the driver enters the approaching regime. The driving mode in this regime reflects the strategy to simultaneously reach the speed of the leader and the desired gap by braking accordingly. Here, this strategy is successful and the driver enters the car-following regime. Once in this regime, the trajectory oscillates around the ideal state $\Delta v = 0$ and $s = s_e(v_l)$. Depending on the model variant, the boundaries of this regime may be reached or slightly exceeded (as in the schematic Fig. 12.5).

For comparison, Fig. 12.6 shows the trajectory in state space $(s, \Delta v)$ of a vehicle approaching a slower vehicle as simulated with the original IDM and for an IDM

[21] Notice that this implies a response to infinitesimal changes which is at variance to the pure idea of action points.

[22] In order to reflect basic kinematic constraints, the qualitative shape of the regime boundaries and the vehicle trajectory have been modified with respect to the figure in the original publication.

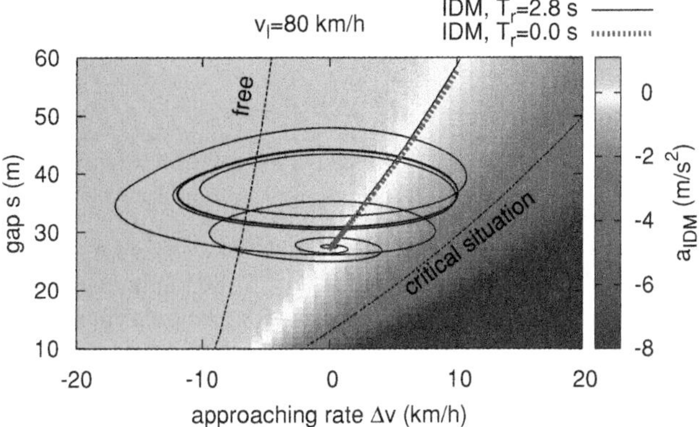

Fig. 12.6 Changes of the state space coordinates s and Δv of a simulated vehicle (desired speed 120 km/h) approaching a slower vehicle driving at constant speed $v_l = 80$ km/h. The *solid curve* is for the IDM with the parameters of Table 11.2 and a reaction time $T_r = 2.8$ s (no other HDM elements added). The *dotted line* is for the reference IDM with no delay. The color-coded surfaces depicts the IDM acceleration function. The critical approaching regime is defined by IDM accelerations below $-2b = -3\,\mathrm{m/s}^2$

variant with finite reaction time (all other aspects of the HDM have been omitted). The IDM trajectory shows two qualitative differences:

- There is no vertical part of the trajectory since the leader exerts a nonzero social force on the follower even if the gap is significantly larger than the safety gap,
- There are no oscillations.[23]

When introducing instabilities by adding a large reaction time to the (otherwise unchanged) IDM, the resulting trajectories in state space are qualitatively similar to that of the Wiedemann model. Similar trajectories can also be observed for any car-following model when simulating string unstable traffic flow (stop-and-go waves). Again, this shows that it is hard to empirically determine whether action points and discrete driving modes play a significant role in describing human-driven traffic flow.

Problems

12.1 Statistical properties of the Wiener process

Verify that the stochastic differential equation (12.9) for the stochastic process $w(t)$ leads to the autocorrelation function (12.8) once stationary conditions are reached. In order to show this, use the formal solution

---

[23] In Chap. 15, we will show that traffic flow is strongly string unstable if even the slightest (and strongly over-damped) oscillations appear in this situation.

$$w(t) = \sqrt{\frac{2}{\tilde{\tau}}} \int_{-\infty}^{t} e^{-(t-t')/\tilde{\tau}} \xi(t')\mathrm{d}t'$$

and evaluate $\langle w(t)w(t')\rangle$ for $t' \le t$ making use of the relation $\langle \xi(t_1)\xi(t_2)\rangle = \delta(t_1 - t_2)$.

12.2 Consequences of estimation errors
Consider a platoon of several identical cars α (desired speed $v_0 = 120$ km/h) driving in steady-state equilibrium behind a truck at $v_\alpha = 80$ km/h. Assume that the cars drive according to the Optimal Velocity Model with triangular fundamental diagram with the parameters $s_0 = 0$, $T = 1.4$ s, $v_0 = 120$ km/h, and $\tau = 0.65$ s. What are the effects on the actual steady-state gap when all drivers constantly overestimate the gap by $V_s\, w_s = 10\,\%$? To illustrate this, calculate the steady-state gap with and without this estimation error. What are the effects on the steady-state gap when there are no estimation errors but a constant additional acceleration of 0.4 m/s^2 (acceleration noise for $\tilde{\tau} \to \infty$)?

12.3 Multi-anticipation for the IDM
Derive Eq. (12.18) for the reduction factor c of the interaction part of the IDM acceleration. Furthermore, show that, instead of multiplying the interaction acceleration by the reduction factor, one could multiply the model parameters s_0 and T by a factor of $\sqrt{c}$.

Further Reading

- Green, M.: "How long does it take to stop?" Methodological analysis of driver perception-brake times. Transportation Human Factors **2** (2000) 195–216
- Treiber, M., Kesting, A., Helbing, D.: Delays, inaccuracies and anticipation in microscopic traffic models. Physica A **360** (2006) 71–88
- Kesting, A., Treiber, M.: How reaction time, update time and adaptation time influence the stability of traffic flow. Computer-Aided Civil and Infrastructure Engineering **23** (2008) 125-137
- Treiber, M., Helbing, D.: Memory effects in microscopic traffic models and wide scattering in flow-density data. Physical Review E **68** (2003) 046119
- Treiber, M., Kesting, A., Helbing, D.: Understanding widely scattered traffic flows, the capacity drop, and platoons as effects of variance-driven time gaps. Physical Review E **74** (2006) 016123
- Barceló, Jaume (Ed.): Fundamentals of Traffic Simulation. Springer, ISBN 978-1-4419-6141-9 (2010)
- Wiedemann, R.: Simulation des Straßenverkehrsflusses. In: Heft 8 der Schriftenreihe des IfV. Institut für Verkehrswesen, Universität Karlsruhe (1974)

Chapter 13
Cellular Automata

As far as the laws of mathematics refer to reality, they are not certain, as far as they are certain, they do not refer to reality.
Albert Einstein

Abstract A cellular automaton (CA) describes traffic dynamics in a completely discrete way: Space is subdivided into cells, time into time steps, and derived quantities such as speed or acceleration are integer multiples of the corresponding basic units. Cellular automata are easy and fast to simulate. However, due to their discrete nature, they reproduce real-life traffic only in a schematic way. This is particularly true for the Nagel-Schreckenberg Model as the simplest and most generic representative of traffic-related CA. Besides this model, two more refined cellular automata are presented.

13.1 General Remarks

Cellular automata (CA) describe all aspects of dynamical systems by using (generally small) integers. Space is subdivided into cells and time into time steps. At any time, each cell is in one of a small number of states. In the simplest case, there are only two states such as zero and one, black and white, or occupied and empty. In general, the interaction is local, i.e., the new state of a cell is determined by only a few cells in the neighborhood of this cell. Mathematically, CAs belong to the class where space, time, and internal states are discrete. The connection to other mathematical model classes is summarized in Table 13.1.[1]

[1] This summary is not complete. For example, traffic flow can also be described by stochastic queuing models which are discrete in the space and state variables and continuous or discrete in time (master equations or Markov chains, respectively). These models are not discussed here.

M. Treiber and A. Kesting, *Traffic Flow Dynamics*,
DOI: 10.1007/978-3-642-32460-4_13, © Springer-Verlag Berlin Heidelberg 2013

Table 13.1 Mathematical model classes

Mathematical class	Time	Space	State variables
Acceleration models	Continuous	Continuous	Continuous
Lane changing rules for acceleration models	Continuous	Discrete	Discrete
Iterated coupled maps	Discrete	Continuous	Continuous
Cell-transmission model	Discrete	Discrete	Continuous
Cellular automata	Discrete	Discrete	Discrete

History. Cellular automata were already proposed in the 1940s by von Neumann. In the 1970s, a two-state, two-dimensional CA named *The Game of Life* became widely known among the early computing community. The basis of traffic-related CAs was laid by Stephen Wolfram in 1983, who systematically investigated the simplest nontrivial class of CAs called "elementary cellular automata". This class is defined by a one-dimensional string of cells where each cell has only two states (since the context was pattern formation, they are denoted as "black and white") and interacts only with itself and its two next neighbors. Since the triplet within interaction range has $2^3 = 8$ possible states or patterns and each pattern can be mapped to one of two possible new states of the center cell, there are $2^8 = 256$ different rules defining the class of elementary cellular automata.

One member of this class called, *Rule 184*, can be regarded as the most generic traffic-related cellular automaton. It is defined as follows:

Current pattern	111	110	101	100	011	010	001	000
New state of the center cell	1	0	1	1	1	0	0	0

In the traffic context, the states 1 and 0 denote cells occupied by a car and empty cells, respectively. In each time step, the cars move one cell to the right if the new cell is empty ("free traffic"), and remain in their old cells, otherwise ("jam"). The name *Rule 184* becomes evident when looking at the bottom row of the table which is just the binary representation of the decimal number 184.

Cellular automata for traffic flow. For traffic-flow related CAs, the relation between the physical and the discrete-valued variables can be described as follows (cf. Figs. 13.1 and 6.2):

- The physical position x_{phys} along the arc length of the road is subdivided into cells of length Δx_{phys}. A given position along the road is denoted by the cell index i (stationary frame of reference used for macroscopic models) or by the position x_α of the front bumper of vehicle α (microscopic models)[2]

[2] Notice that, in the context of cellular automata, $i = x_\alpha$, t, and v are index-like integer quantities. Consequently, equations such as $x(t+1) = x + v$ are formally correct although they may look terribly erroneous to any physics teacher. Since this naming is a de-facto standard in the scientific literature on microscopic traffic CA, we have adopted it although variable names such as i_α for x_α and j for t would make the index-like nature of the variables more explicit. In order to avoid

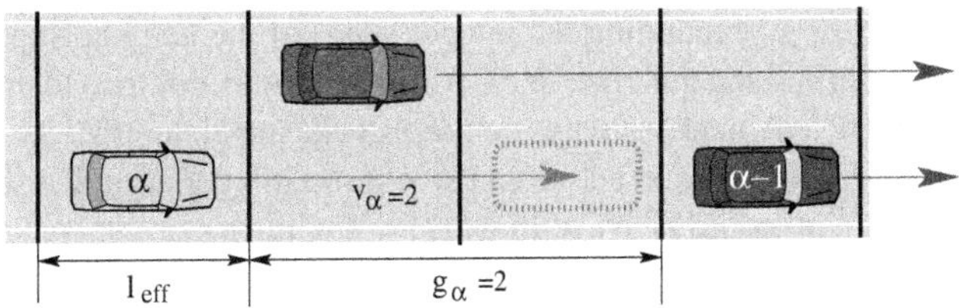

Fig. 13.1 Cellular automata describe the longitudinal and transverse logical coordinates of a road section by dividing it into cells of a fixed size. Shown is the simplest case where one vehicle occupies exactly one cell

$$x_{\text{phys}} = i\,\Delta x_{\text{phys}}, \quad x_\alpha^{\text{phys}} = x_\alpha\,\Delta x_{\text{phys}}.$$

For multi-lane models, the lane index l describes the lateral position.[3] When simulating networks, a road index m is added for each link in each direction. In the most general case, the position of the front bumper of a vehicle in physical space is given by the functions mapping the arc length x_{phys} and the lane index l (the logical coordinates) of road m to geo-referenced coordinates.

- As for iterated maps, time is subdivided into time steps of constant duration. The physical time t_{phys} is related to the time index t by

$$t_{\text{phys}} = t\,\Delta t_{\text{phys}}.$$

- At each time step t, each cell i can be occupied ($\rho_i(t) = 1$), or empty ($\rho_i(t) = 0$). In the traffic context, occupied cells typically have further integer-valued state variables such as the speed v. The relation to the physical speed v_{phys} is given by

$$v_{\text{phys}} = v\,\frac{\Delta x_{\text{phys}}}{\Delta t_{\text{phys}}}.$$

This discretization carries over to other state variables. For example, possible values for the accelerations are multiples of $\Delta x_{\text{phys}}/(\Delta t_{\text{phys}})^2$.

While most traffic-flow related CAs represent microscopic models, macroscopic formulations are possible as well.

Macroscopic CA models. The update rules for macroscopic CAs are cell-based and of the form of the first CA proposed by Von Neumann and Wolfram: The new state $Z_i(t+1)$ of cell i at the new time step $t+1$ depends on the states $Z_j(t)$ of the neighboring cells at the old time step t:

$$Z_i(t+1) = Z_{\text{CA}}\big(\{\rho_j(t)\}, \{Z_j(t)\}, \xi(t)\big). \tag{13.1}$$

---

confusion, in this chapter, dimensional physical quantities are denoted with a sub- or superscript "phys".

[3] Most continuous models describe the lateral position in a discrete way with a lane index, as well.

Here, the function $Z_{CA}(\cdot)$ mapping the integer states of the neighboring cells to the new integer state defines the macroscopic CA in question. In contrast to the originally proposed CA, traffic CA nearly always contain stochastic elements.[4] To reflect this, the mapping $Z_{CA}(\cdot)$ contains an additional stochastic integer argument $\xi(t)$. A way to define this argument could be by setting $\xi(t)=1$ with probability p, and $\xi(t)=0$, otherwise.

The state Z of a cell i reflects the local traffic flow properties: The density (occupation number) ρ_i and, additionally, the speed index V_i for second-order models. This information can be mapped to the integer state Z, e.g., by the coding $Z=\rho$ (first-order models), or $Z=\rho+(\rho_{\max}+1)V$ (second-order models).

When interpreted macroscopically, $V_i\,\Delta x_{\text{phys}}/\Delta t_{\text{phys}}$ denotes the local speed $[V(x,t)]_{\text{phys}}$ at position $x_{\text{phys}}=x\,\Delta x_{\text{phys}}$ and time $t_{\text{phys}}=t\,\Delta t_{\text{phys}}$, and $\rho_i/\Delta x_{\text{phys}}$ denotes the corresponding local density $[\rho(x,t)]_{\text{phys}}$. Cells of macroscopic CAs typically can be multiply occupied ($\rho>1$). For example, when assuming a cell length $\Delta x_{\text{phys}}=100\,\text{m}$ and a maximum dimensional density $\rho_{\max}^{\text{phys}}=150\,\text{km}^{-1}$, the maximum occupation number is $\rho_{\max}=15$.

Microscopic CAs models. When a CA represents a microscopic traffic flow model, a cell can be occupied at most once. An occupied cell represents a vehicle (or parts of it, cf. Sect. 13.3) driving at the physical speed $v_{\text{phys}}=v\,\Delta x_{\text{phys}}/\Delta t_{\text{phys}}$. In contrast to the classical *cell-based* formulation adopted for macroscopic models, microscopic CAs are generally formulated in terms of *particle-based* update rules for the location x_α, speed v_α, and further state variables z_α of the driver-vehicle "particle" α corresponding to occupied cells:

$$\begin{aligned}
v_\alpha(t+1) &= v_{CA}\big(\{x_{\alpha'}(t)\},\{v_{\alpha'}(t)\},\{z_{\alpha'}(t)\},\xi(t)\big),\\
x_\alpha(t+1) &= x_\alpha(t)+v_\alpha(t+1),\\
z_\alpha(t+1) &= z_{CA}\big(\{x_{\alpha'}(t+1)\},\{v_{\alpha'}(t+1)\},\{z_{\alpha'}(t)\}\big).
\end{aligned} \tag{13.2}$$

In this particle formulation, the cell-based *Rule 184* constituting the basis of all microscopic CAs for traffic becomes

$$\begin{aligned}
v_\alpha(t+1) &= \begin{cases} 1 & \text{if } x_{\alpha-1}(t)-x_\alpha(t)>1 \\ 0 & \text{otherwise} \end{cases}\\
x_\alpha(t+1) &= x_\alpha(t)+v_\alpha(t+1).
\end{aligned} \tag{13.3}$$

As in the mathematical description of car-following models, the first vehicle has the smallest index α. In many CAs, the cell length $\Delta x_{\text{phys}}=l_{\text{eff}}$ is equal to the (average) effective vehicle length, i.e., the actual vehicle length plus the minimum gap. There are also CAs where one vehicle may occupy several cells (Sect. 13.3). The interactions are generally local, i.e., the set $\{\alpha'\}$ of interacting vehicles consists of a few vehicles. If no multi-anticipation or lane changing is considered, we have $\{\alpha'\}=\{\alpha,\alpha-1\}$. The integer state indicator z_α codes internal state variables such as a switch for an

[4] Simply because it is hard to obtain meaningful results, otherwise.

active or defensive driving style, brake lights on or off, or direction indicators. The variable z_α and the associated third equation of (13.2) are not present for the simple models discussed below, with the exception of the KKW model (Sect. 13.3.2).

Verify that the cell-based *Rule 184* as given in the Table on page 226 is equivalent to the particle-based formula (13.3).

Realize that, apart from the scaling of the variables, the particle-based formulation of cellular automata is not conceptionally different from discrete-time car-following models formulated as iterated maps (these can contain additional state variables such as brake lights as well).

In each time step, the vehicles of microscopic CAs are moved by the distance $v_\alpha(t+1)$ rather than by the distance $\frac{1}{2}[v_\alpha(t)+v_\alpha(t+1)]$ corresponding to the positional update of continuous car-following models. Why is this latter rule not applied to cellular automata although it is of higher order?

13.2 Nagel-Schreckenberg Model

The first, most popular, and simplest CA which is actually used for traffic flow simulations is the *Nagel-Schreckenberg Model* (NSM). This model generalizes *Rule 184* to a stochastic model with more than two speed levels. In its basic form, it describes single-lane traffic consisting of identical vehicles of effective length 7.5 m. Since each vehicle occupies exactly one cell, this is also the cell length. The update time step is $\Delta t_{\text{phys}} = 1$ s and the update is performed by following equations of motion (cf. Fig. 13.2):

1. Deterministic acceleration as a function of the rear-bumper-to-front-bumper gap g_α (number of empty cells),[5] and the desired speed v_0. The new speed is the minimum of the speed $v+1$ obtained when accelerating with the free-flow acceleration, the desired speed v_0, and the safe speed $v_{\text{safe}} = g$:

$$v_\alpha^*(t+1) = \min\left(v_\alpha(t)+1,\ v_0,\ g_\alpha\right). \tag{13.4}$$

2. *Dawdling* by not accelerating, or braking more than necessary, with a certain dawdling probability p:

[5] To be consistent with the literature, we denote the gap by g instead of s, in this chapter.

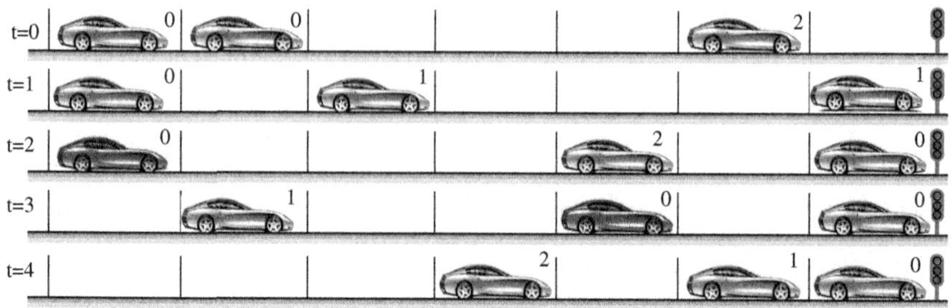

Fig. 13.2 Visualization of the update rules of the Nagel-Schreckenberg model for $v_0 = 2$. The numbers in the *boxes* indicate the speed realized in the past time step. The vehicles painted in *red* have dawdled at this time

Table 13.2 Parameters of the Nagel-Schreckenberg Model and the *slow-to start* extension by Barlovic et al. (see Sect. 13.3.1)

Parameter	Typ. value highway	Typ. value city
Cell length $\Delta x_{\text{phys}} = l_{\text{eff}}$	7.5 m	7.5 m
Time step Δt_{phys}	1 s	1 s
Dawdling probability v_0	5	2
Dawdling probability p	0.2	0.1
Dawdling probability p_0 when stopped (only for the Barlovic model)	0.4	0.2

$$v_\alpha(t+1) = \begin{cases} \max\left(v^*_\alpha(t+1) - 1,\ 0\right) & \text{with probability } p, \\ v^*_\alpha(t+1) & \text{otherwise.} \end{cases} \tag{13.5}$$

3. Driving:

$$x_\alpha(t+1) = x_\alpha(t) + v_\alpha(t+1). \tag{13.6}$$

As indicated by the time indices, the update is performed simultaneously (*parallel update*), i.e., all updates are based on the same *old* situation. While it is possible to formulate consistent upstream or downstream *sequential update* rules for one link, sequential updating becomes ill-defined for road networks, so it is rarely used.

Try to understand the meaning of the NSM equations of motion with the model parameters of Table 13.2 from the viewpoint of a driver: Which values have the desired speed and the maximum acceleration and deceleration, in physical units? How many seconds is the typical time gap in car-following mode? Are there reactions to the speed difference? Is the model accident free? What is the maximum density in physical units?

The discretization of space and time of the NSM is chosen such that one cell corresponds to the effective vehicle length, and one time step to a typical time gap in car-following mode (Table 13.2). These coarse discretizations do not allow for a realistic description of single-vehicle dynamics since speed and acceleration can take on only multiples of 7.5 m/s and 7.5 m/s^2, respectively. Keeping aside the implicit parameters Δx_{phys} and Δt_{phys} which are only relevant for the scaling and the physical interpretation, the NSM has two parameters influencing the dynamics itself: The dawdling probability p, and the desired speed v_0.

Show that the NSM reduces to *Rule 184* for $p=0$ (deterministic limiting case) and $v_0=1$.

The fact sheet of the NSM shown in Fig. 13.3 displays the same two standard highway and city scenarios that have been simulated with the models of the previous chapters. The desired speed $v_0=2$ of the city scenario corresponds to 15 m/s = 54km/h, while the value $v_0=5$ for highways corresponds to 37.5 m/s = 126 km/h. From a macroscopic perspective, the simulation of the highway scenario shows some plausible results: The speed field of Fig. 13.3a displays a congestion caused by the on-ramp near road kilometer zero. Furthermore, the patterns inside the congested region move upstream with a realistic velocity of about −15 km/h. In Problem 13.5, the reader can verify that the propagation velocity is bounded from below by

$$c_{\text{cong}}^{\text{phys, NSM}} = -(1-p)\frac{\Delta x_{\text{phys}}}{\Delta t_{\text{phys}}}. \tag{13.7}$$

Consequently, one can calibrate the propagation velocity by varying the dawdling parameter p (and Δx_{phys} and Δt_{phys}). However, Fig. 13.3a shows significantly less negative propagation velocities as well. As a consequence, the congestions simulated by the NSM do not consist of persistent stop-and-go waves but of short-living statistical structures. Moreover, some of these structures appear in regions of free flow (downstream of the ramp at $x>0$ and upstream of the congested region, e.g., at $x_{\text{phys}} \approx -6$ km and $t_{\text{phys}} \approx 20$ min) where they are never observed in reality, and even cannot exist for purely kinematic reasons.[6] Consequently, these structures are model artifacts.

Although the stochastic component of the model is always active, the fundamental diagram in Fig. 13.3b shows a strong scattering in congested traffic, only—in agreement with observations. This can be explained by looking at the consequences of dawdling: In free traffic, there are few interactions and the speed fluctuations of the vehicles are independent from each other, with a variance $p(1-p)$. The mean speed over n vehicles passing a virtual detector within one aggregation interval has an even smaller variance $p(1-p)/n$ which carries over to small fluctuations of the flow-density data (cf. Sect. 4.4). In congested traffic, however, the dawdling may cause a

[6] The propagation velocity of perturbations of free traffic is always positive.

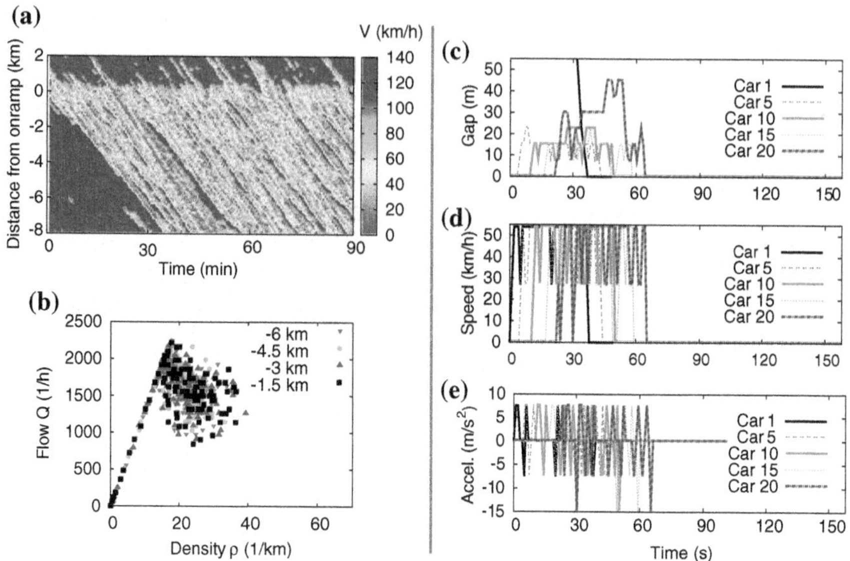

Fig. 13.3 Fact sheet of the Nagel-Schreckenberg Model for the parameters of Table 13.2. See Sect. 10.5 for a detailed description of the simulation scenarios

breakdown implying greater and, moreover, correlated fluctuations. Both increase the fluctuations of the mean speed and hence the fluctuation of the flow-density points.

Because of the stochastic nature of the NSM and other cellular automata, there exist no steady-state equilibrium and, consequently, no fundamental diagram in the strict sense. However, one can define steady states in the stochastic sense by calculating speed expectation values and variances (see Problems 13.4 and 13.5).

The simulation of the city scenario in the Fig. 13.3c–e shows the limits of the NSM in simulating single vehicles. Due to its coarse-grained nature—there are only three possible values for the speed and four values for the acceleration—the time series of gap, speed, and acceleration are not realistic. In spite of being a microscopic model, the NSM can only be used for describing the macroscopic dynamics. There are many extensions and refinements to tackle this deficiency. One simple and one more complex extension are presented in the next section.

13.3 Refined Models

13.3.1 Barlovic Model

A simple extension to make the behavior of the Nagel-Schreckenberg Model somewhat more realistic is to increase the dawdling probability of Eq. (13.5) for slow or

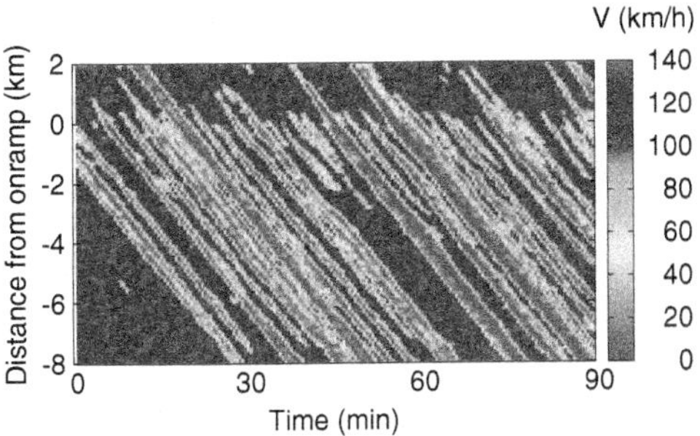

Fig. 13.4 Simulation of the highway scenario with the Barlovic model for the parameters of Table 13.2

standing vehicles corresponding to a *slow-to-start* rule. In the simplest case first proposed by Barlovic et al. the dawdling probability is increased for standing vehicles, only:

$$p(v) = \begin{cases} p & v > 0, \\ p_0 > p & v = 0. \end{cases} \tag{13.8}$$

This models a simple form of context sensitivity: If drivers are stuck in a jam, they become less agile and start more slowly.[7] Thus, this mechanism self-enforces the spurious structures of the NSM such that more realistic long-living stop-and-go waves can be expected for suitable values of p_0 (cf. Table 13.2).

The speed field of the highway scenario in Fig. 13.4 displays the expected persistence of the jam waves. However, since the mechanism for triggering the congested structures is the same as in the NSM, the problem of the unrealistic jam structures in free traffic ($x > 0$ in the figure) is not resolved. Moreover, the positive feedback of the slow-to-start rule makes these structures long-living as well.

13.3.2 KKW Model

As an example of a more complex CA with a very fine cell discretization we present the *KKW model* proposed by Kerner, Klenov, and Wolf.

The cells in this model are just 0.5 m long. Consequently, a single vehicle occupies several cells. Generally, one simulates with an effective vehicle length of 15 cells corresponding to a physical value $l_{\text{eff}} = 7.5$ m as in the NSM. The time step is $\Delta t_{\text{phys}} = 1$ s as well. The resulting fine speed and acceleration steps of 0.5 m/s = 1.8 km/h and 0.5 m/s^2, respectively, allow for a more differentiated and realistic simulation of single vehicles. However, at a desired speed $v_0 = 67$ for the highway scenario ($v_0^{\text{phys}} = 120$ km/h), i.e., 68 speed levels, the KKW model is far away from the original philosophy of cellular automata making sense for a small

[7] Remember that cellular automata only make sense for aggregated phenomena. If $p_0 > p$, the *average* acceleration of stopped vehicles is lower than that of moving vehicles.

number of internal states, only. Instead, it behaves *de facto* like a (coarse-grained) time-discrete car-following model.

The model equations are outside the scope of this book and are therefore only described in general terms.[8] In contrast to the Nagel-Schreckenberg and Barlovic models, the KKW model includes responses to speed differences. Moreover, in contrast to all models presented up to now, there is a range of indifference for the time gaps in congested traffic: When following a vehicle at the same speed and when this speed is less than the desired speed ($v = v_l < v_0$), there are no deterministic accelerations if the (rational-valued) bumper-to-bumper time gap $T = g/v$ to the preceding vehicle is in the range $[1, k]$. The model parameter $k > 1$ indicates the size $k - 1$ of the range of indifference. Notice that this condition implies that, in physical units, the minimum time gap is given by $T_{\rm min} = 1$ (in physical units of 1 s). Consequently, the model does not possess a unique fundamental diagram, even for the deterministic limiting case.

Furthermore, the model is consistent with the so-called "three-phase traffic theory" proposed by Kerner. This theory essentially states that, in addition to free traffic, there are exactly two qualitatively different "phases" of congested traffic flow, "synchronized traffic", and "jams".[9] The properties of the simulated synchronized traffic essentially depend on the range factor k (defining the range of time gaps in this phase) and on the speed threshold v_p (indicating typical speeds in regions of synchronized traffic). Like in other traffic-flow CAs, the deterministic acceleration is superseded by stochastic elements. Here, they depend on v and v_l and contain five probabilities as model parameters, including the dawdling probability p and the slow-to-start probability p_0.

The simulation of the highway scenario shown in the left column of Fig. 13.5 reveals a clearly more realistic behavior than the Nagel-Schreckenberg and Barlovic models. Particularly, the flow-density diagram shows a more realistic scattering of the data points, and the stop-and-go waves emerging for $t > 30\,\mathrm{min}$ have a correct propagation velocity and a sufficient wavelength. However, the propagation velocities $c \approx -40\,\mathrm{km/h}$ inside the synchronized traffic state (the yellow lightly congested region with little structures appearing before the stop-and-go waves) and the velocity of the transition zone between free and synchronized traffic are significantly too negative.[10]

The simulation of the city scenario in the right column of Fig. 13.5 exemplifies the effect of the fine discretization and the similarity of the KKW model to car-following models: The speed and acceleration profiles of single vehicles are meaningful. However, the initial accelerations are too low (they are restricted to $\Delta x_{\rm phys}/(\Delta t_{\rm phys})^2 = 0.5\,\mathrm{m/s^2}$), and the decelerations during the braking phase at the red traffic light are unrealistically high. This behavior is due to the safe-speed rule

[8] See pp. 411 in the book *The Physics of Traffic* by B.S. Kerner (Springer, 2004).

[9] For details, the reader is referred to the two monographes of Kerner on this topic, "The Physics of Traffic" (Springer, 2004), and "The Long Road to Three-Phase Traffic Theory" (Springer, 2009).

[10] This artifact vanishes if the synchronized state is eliminated by reducing the range factor k from its published value 2.55 to $k = 1$ or by setting $v_p = v_0$.

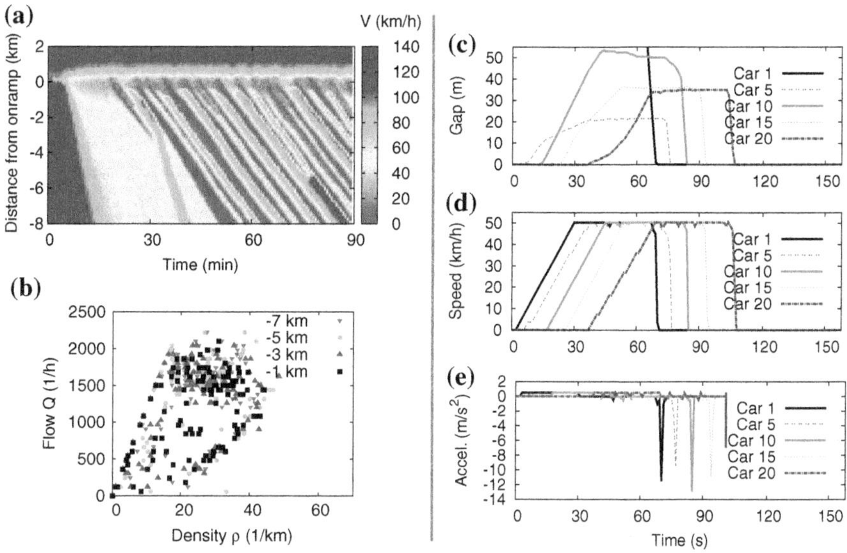

Fig. 13.5 Simulation of the two standard scenarios "highway" (*left*) and "city traffic" (*right*) for the KKW model with standard model parameters taken from Kerner's books. For the highway scenario, the desired speed has been changed to $v_0 = 67$ (corresponding to 120 km/h). For the city scenario, we set $v_0 = 28$ (50.4 km/h) and change the speed threshold v_p to the value 14. See Sect. 10.5 for a detailed description of the simulation scenarios

inherited from the NSM which does not takes into account kinematic constraints. The KKW model has no desired braking deceleration as a model parameter.

Introducing realistic deceleration characteristics to the KKW model. Let us consider the following modification as an example for model development. One of the rules of the KKW model restricts the maximum speed of a vehicle to the safe speed $v_{\text{safe}} = g$ of the Nagel-Schreckenberg Model. In the worst case, this leads to decelerations of $-v_0$ (corresponding to 14 m/s^2 for the city scenario and even more for highway parameters). It is plausible to introduce kinematic restraints by replacing this safe speed rule by a discretized version of the safe speed (11.10) of Gipps' model,

$$v_{\text{safe}}(s, v_l) = \text{floor}\left[\min\left(g, -b + \sqrt{b^2 + v_l^2 + 2bg}\right)\right]. \tag{13.9}$$

This rule gives the maximum speed given a reaction time of 1 (1 s in physical units), and a braking deceleration of b. The minimum condition is necessary since, otherwise, the stochastic terms of the KKW model may led to collisions for some circumstances.

The simulation of the city scenario (right column of Fig. 13.6) shows, in fact, realistic decelerations of the order given by the new model parameter b. The outcome for the highway scenario (left column) has not changed qualitatively.

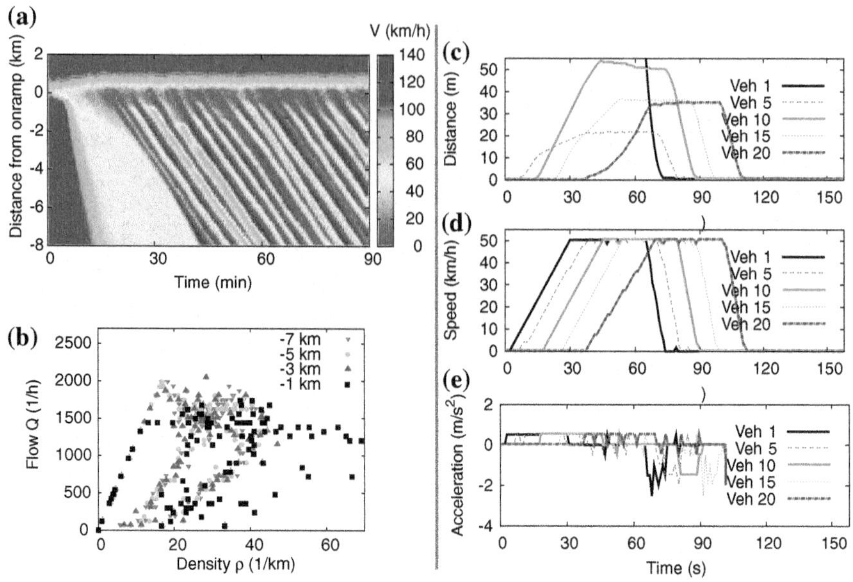

Fig. 13.6 Simulation of the two standard scenarios "highway" (*left*) and "city traffic" (*right*) for the refined KKW model where the original safe speed rule $v_{\text{safe}} = g$ is replaced by Eq. (13.9). The KKW model parameters are the same as in Fig. 13.5. The new parameter $b = 4$ corresponds to a deceleration $\dot{v} = -2\,\text{m/s}^2$

13.4 Comparison of Cellular Automata and Car-Following Models

The advantages and disadvantages of CAs with respect to other mathematical model classes are essentially due to the discrete scaling of time, space, and state variables.

This leads to a few advantages. Firstly, because of their simplicity, cellular automata are easier to implement, especially for complex road networks or if the movements are truly two-dimensional as in pedestrian traffic.[11]

Secondly, simple CAs have a speed advantage over continuous models. However, this advantage becomes less pronounced for the more elaborated models, especially if they contain many rules, or if they are so fine-grained as the KKW model where 15 update steps correspond to a single update step of a continuous model. Moreover,

[11] When implementing continuous social-force pedestrian models, the main problem is the bookkeeping to obtain references to the pedestrians of the local environment. This is typically solved by a virtual grid such that, at any time, each pedestrian is associated with an element of this grid. Since CAs are based on a fixed regular grid with predetermined neighboring relations, no such problems arise for CAs.

with ever-growing processor power and the increasing use of graphics processors for the core calculations, this speed advantage is no longer relevant, for most purposes.[12]

The grid-like nature of cellular automata also has some disadvantages. Since CAs are phenomenological rather than based on intuitive concepts or strategies, most model parameters are not intuitive and their values are not realistic.[13] Furthermore, traffic CAs need ad-hoc stochastic terms (such as the dawdling term of the NSM) to avoid artifacts or extreme sensitivities with respect to model parameters or initial values. This lack of *robustness* makes them unsuitable for many applications. Finally, because of their discrete nature, CAs are not suitable for modeling the movement of individual vehicles which is necessary when investigating different driving styles or assessing the influence of driver-assistance systems or infrastructure-based control measures on traffic flow (see Chap. 21). Even refined models show unrealistic behavior in some situations, especially when approaching slower or standing vehicles, or red traffic lights.

Since many details are averaged out on grander scales, CAs can and are successfully applied as kernel for a model-based short-term traffic state prediction (see Chap. 18). However, macroscopic models such as the Cell-Transmission Model are competitors for this type of applications.

Problems

13.1. Dynamic properties of the Nagel-Schreckenberg Model

The numerical value of v_0 for the highway scenario given in Table 13.2 corresponds to which physical value? How long does it take for an unimpeded vehicle to accelerate from zero to 100 km/h (i) in the deterministic Nagel-Schreckenberg Model ($p=0$) and (ii) on average in the stochastic model with $p=0.4$?

13.2. Approaching a red traffic light

Describe the approaching strategy of a driver modeled with the deterministic Nagel-Schreckenberg Model ($p=0$) when approaching a red traffic light if there are no other influencing vehicles. Which braking decelerations are reached in physical units for city traffic ($v_0=2$)?

13.3. Fundamental diagram of the deterministic NSM

Draw the fundamental diagram of the deterministic NSM for $v_0=2$ (cities) and $v_0=5$ (highways). What are the physical values of (i) the desired speed, (ii) the propagation velocity of perturbations in congested traffic, (iii) the time gap, and (iv) the maximum density? Which qualitative and quantitative changes can be expected for the stochastic model (dawdling probability $p>0$)?

---

[12] Graphics processors are optimized for calculations with real numbers, so the speed advantage of using integers is essentially lost.

[13] When presenting CA simulations, many authors even do not bother to translate the results into physical units.

13.4. Macroscopic desired speed
Imagine completely free traffic flow (no interactions between vehicles) and a stationary situation where expectation values do not change with time. Show that, in physical units, the macroscopic speed (vehicle speed averaged over many vehicles and/or time steps) of the NSM is given by

$$V_{\max}^{\text{phys}} = \frac{(v_0 - p)\Delta x_{\text{phys}}}{\Delta t_{\text{phys}}}.$$

13.5. Propagation velocity of downstream jam fronts
The most negative propagation velocity is realized for the position of vehicles starting from a queue of waiting vehicles ($\rho_{\max} = 1$) when the traffic light is green. Show that the velocity of this "starting wave" is given by Eq. (13.7).

Further Reading

- Von Neumann, J., Burks, A.: Theory of self-reproducing automata. (1966)
- Schadschneider, A., Chowdhury, D., Nishinari, K.: Stochastic Transport in Complex Systems: From Molecules to Vehicles. Elsevier 2010
- Chowdhury, D., Santen, L., Schadschneider, A.: Statistical physics of vehicular traffic and some related systems. Physics Reports **329** (2000) 199–329
- Maerivoet, S., DeMoor, B.: Cellular automata models of road traffic. Physics Reports **419** (2005) 1–64
- Nagel, K., Schreckenberg, M.: A cellular automaton model for freeway traffic. Journal de Physique I France **2** (1992) 2221–2229
- Barlovic, R., Santen, L., Schadschneider, A., Schreckenberg, M.: Metastable states in cellular automata for traffic flow. Euro. Phys. Journal **5** (1998) 793
- Kerner, B.S.: The Physics of Traffic: Empirical Freeway Pattern Features, Engineering Applications, and Theory. Springer (2004)
- Kerner, B.S.: The Long Road to Three-Phase Traffic Theory. Springer (2009)

Chapter 14
Lane-Changing and Other Discrete-Choice Situations

Imagination is more important than knowledge.
Albert Einstein

Abstract Simulating any nontrivial traffic situation requires describing not only acceleration and braking but also lane changes. When modeling traffic flow on entire road networks, additional discrete-choice situations arise such as deciding if it is safe to enter a priority road, or if cruising or stopping is the appropriate driver's reaction when approaching a traffic light which is about to change to red. This chapter presents a unified utility-based modeling framework for such decisions at the most basic operative level.

14.1 Overview

From the driver's point of view, there are three main actions that directly influence traffic flow dynamics: Accelerating, braking, and steering.[1] The dynamics of steering is part of the vehicle dynamics and therefore the domain of *sub-microscopic models* (cf. Table 1.1). Traffic flow dynamics describes the dynamics one level higher by directly modeling lane-changing decisions and the associated actions. At this level, the set of possible actions is discrete, i.e., performing a lane change, or not. Details of the lane-changing maneuver such as duration or lateral accelerations are not resolved, and the lane-changing itself is assumed to take place instantaneously.[2]

[1] Further actions such as using direction indicators, flashing headlights, or applying the horn, are only considered in very detailed models.

[2] In reality, the duration of a lane-changing maneuver is of the order of a few seconds. In many microscopic traffic flow simulators, lane changes are *represented* graphically as a smooth process but *simulated* as an instantaneous jump to the target lane. For the other drivers, the car is already on the target lane when the visualized lane change begins.

M. Treiber and A. Kesting, *Traffic Flow Dynamics*,
DOI: 10.1007/978-3-642-32460-4_14,

Discrete decisions and actions can also pertain to the longitudinal dynamics, in parallel to the continuous actions modeled by the acceleration function $a_{\mathrm{mic}}(s, v, v_l)$: When approaching a yellow traffic light which is about to turn red, the driver has to decide whether it is safe to pass this traffic light without changing speed, or if it is necessary to stop. Furthermore, lane changes generally influence the longitudinal acceleration of the decision maker (e.g., preparing for a lane change) or that of the other affected drivers (e.g., cooperatively making a gap to enable a change, or restoring the safety gap afterwards). Discrete choices in the traffic-flow context involve several levels:

1. The *strategic level* (destination choice, mode choice, and route choice) is modeled within the domain of transportation planning (cf. Table 1.1).
2. The *tactical level* includes anticipatory measures to enable or facilitate operative actions such as changing lanes or entering a priority road. This includes cooperative behavior such as allowing another vehicle to merge at a point of lane closure (*zipper mode* merging). Modeling the tactical level is notoriously difficult and is only attempted in the most elaborate commercial simulators.
3. On the *operative level*, the actual decision is made.
4. Finally, in the *post-decision phase*, the actions pertaining to this decision are simulated, e.g. performing the lane change or keeping to one's lane, waiting or entering a priority road, or cruising versus stopping at the traffic light.

In this chapter, we restrict the description to the operative level and the post-decision phase. We model the different discrete-choice situations consistently in terms of maximizing *utility functions* associated with each alternative. The utility of a given alternative increases with the (hypothetical) longitudinal acceleration that would be possible once this alternative had been adopted.

Using accelerations as utility ensures the compatibility between the acceleration and discrete-choice models. Furthermore, this ansatz is parsimonious since it minimizes the number of parameters and assumptions. For example, when the acceleration model is parameterized to simulate aggressive drivers, the lane-changing style of these drivers becomes aggressive as well, without introducing further parameters. Generally, any aspect considered in the longitudinal model carries over to the decision model. Specifically, the lane-changing considerations take into account speed differences, brake lights, or anticipative elements if, and only if, these exogenous factors are included in the acceleration model.

The approach reaches its limits when tactical and cooperative measures are crucial. One example of such a situation is zipper-like merging, or, more generally, merges to a congested target lane.

14.2 General Decision Model

We assume that, at a given moment, the driver can chose from a discrete set K of alternatives k. In the context of lane changes, the alternatives would be the (active) decisions to change to the left or right, and the (passive) decision not to change.

When about to enter a priority road, the alternatives would be to initiate the merging, or stopping and waiting for a sufficient gap between the main-road vehicles. We assume that the drivers are aware of the consequences of their decisions, i.e., they can anticipate, for each alternative, the speeds and gaps of all involved vehicles. This allows us to calculate all relevant accelerations (i.e., the utilities) using the normal acceleration functions of these vehicles. If the acceleration model is formulated as an iterated map or cellular automaton, the acceleration is calculated using Eq. (10.11).

In the decision process, the driver maximizes his or her utility (incentive criterion) subject to the condition that the action is safe (safety criterion). Both criteria are based on the acceleration function as follows:

Safety criterion. None of the drivers β affected by the consequences of opting for alternative k (including the decision maker α) should be forced to perform a critical maneuver as a consequence of a decision for alternative k. A maneuver is deemed to be critical, if it entails braking decelerations exceeding the *safe deceleration* b_{safe}:

$$a_{\text{mic}}^{(\beta,k)} > -b_{\text{safe}}. \tag{14.1}$$

The value of the model parameter b_{safe} (of the order of $2\ \text{m/s}^2$) is comparable to the comfortable deceleration b of the IDM or Gipps' model. For reason of parsimony, the safe deceleration can be inherited from these models ($b_{\text{safe}} = b$), if applicable.

Incentive criterion. Choosing among all safe alternatives k', the driver α selects the option of maximum utility U:

$$k_{\text{selected}} = \arg\max_{k'} U^{(\alpha,k')}. \tag{14.2}$$

As in most other discrete-choice models, the incentive criterion is based on a rational decision maker (also called *homo oeconomicus*) who maximizes his or her utility. In the simplest case, the utility is directly given by the acceleration function,

$$U^{(\alpha,k)} = a_{\text{mic}}^{(\alpha,k)}. \tag{14.3}$$

In contrast to the standard framework for discrete decisions (multinomial Logit and Probit models and their variants) we do not assume explicit stochastic utilities unless the acceleration model itself contains stochastic terms.[3]

For some discrete-choice situations such as discretionary lane changes, one needs an additional threshold preventing all active decisions (e.g., a decision to change lanes rather than to stay put) when the associated advantage is only marginal. Such a threshold prevents unrealistically frequent withdrawals of an active decision taken in the last time step which could, for example, lead to frantic lane-changing actions.

---

[3] The rationale behind stochastic utilities is to include in a global way all uncertainties of the decision process and contingencies in evaluating the utilities. In microscopic traffic flow simulations, there are so many *directly considered* contingencies in form of the positions and types of the involved vehicles influencing the decision process that further stochastic elements are superfluous.

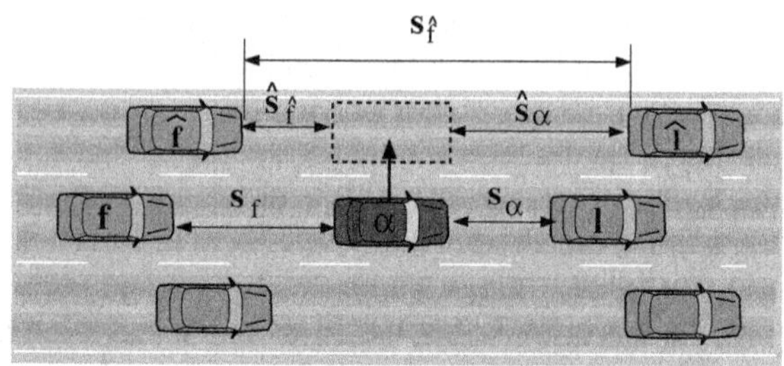

Fig. 14.1 Notation for a lane change of the *center* vehicle α to the *left*. All quantities with a hat pertain to the new situation after the (possibly hypothetical) lane change

Traffic rules (such as a "keep right" directive) may also enter the utility. Finally, one can include all consequences of a decision to other drivers by introducing a *politeness factor* (Sect. 14.3.3).

14.3 Lane Changes

Figure 14.1 depicts the general situation. The vehicle α of the decision maker (speed v_α) is located in the center. There are three alternatives: Change to the right, change to the left, and no change. Without loss of generality, we compare only the last two alternatives. Here, and in the following, we denote the vehicle of the decision maker with α, the leading vehicle with l, and the following vehicle with f. All accelerations, gaps, or vehicle indices with a hat refer to the new situation after the lane change has been completed while quantities without a hat denote the old situation.[4]

14.3.1 Safety Criterion

Assuming that the present situation (i.e., the alternative "no change") is safe, the safety criterion (14.1) refers to the acceleration $\hat{a}_{\hat{f}}$ of the new follower ($\beta = \hat{f}$) after a possible change, and also to the new acceleration $\hat{a}_\alpha$ of the decision maker him or herself ($\beta = \alpha$). For the follower, this criterion becomes

$$\boxed{\hat{a}_{\hat{f}} = a_{\text{mic}}(\hat{s}_{\hat{f}}, v_{\hat{f}}, v_\alpha) > -b_{\text{safe}} \qquad \text{safety criterion.}} \tag{14.4}$$

In order that this condition also prevents lane changes whenever there are following vehicles on the target lane at nearly the same longitudinal position (the gap $\hat{s}_{\hat{f}}$ is

[4] Examples: $a_{\hat{f}}$ denotes the acceleration of the new follower in the old situation, $\hat{a}_f$ the acceleration of the old follower in the new situation, and $\hat{s}_\alpha$ the gap of the decision maker's vehicle in the new situation. Notice that $\hat{\alpha} = \alpha$ (the decision maker is the same before and after the lane change), and $\hat{v} = v$ (lane changes are modeled as instantaneous jumps without changes of the speed).

negative, i.e., a change would result in an immediate accident), the acceleration function $a_{\mathrm{mic}}(s, v, v_l)$ should return prohibitively negative values if $s < 0$. The parameter b_{safe} indicates the maximum deceleration imposed on the new follower which is considered to be safe (cf. Problem 14.2). If one simulates heterogeneous traffic with individual acceleration functions, the acceleration function $\hat{a}_{\hat{f}}$ of the new follower $\hat{f}$ is calculated with the function and parameters of *this* driver-vehicle unit.[5]

Regarding the safety of the decision maker him- or herself ($\beta = \alpha$), condition (14.4) prevents changes if the new gap $\hat{s}_\alpha$ is dangerously low such that $\hat{a}_\alpha = a_{\mathrm{mic}}(\hat{s}_\alpha, v_\alpha, v_{\hat{l}}) < -b_{\mathrm{safe}}$. Since this condition is less restrictive than the incentive criterion to be discussed below, there is no need to explicitly check this condition. In any case, the condition on the acceleration function to return prohibitively negative values for negative gaps guarantees that changes are prohibited if the leader on the target lane is essentially at the same longitudinal position ($\hat{s}_\alpha < 0$) which would result in an immediate crash.

14.3.2 Incentive Criterion for Egoistic Drivers

Most lane-changing models formulate the incentive criterion exclusively from the perspective of the decision maker ignoring the advantages and disadvantages to the other drivers. Furthermore, the lane-changing behavior depends on the legislative regulations of the considered countries. For example, a *right-overtaking ban* is in effect on most European highways.[6] Here, we will restrict to the simpler situations of lane changes on highways in the United States, or more generally to lane changes in city traffic, where lane usage is only mildly asymmetric.[7] Then, the incentive criterion for the egoistic driver reads

$$\hat{a}_\alpha - a_\alpha > \Delta a + a_{\mathrm{bias}}, \tag{14.5}$$

where

$$a_\alpha = a_{\mathrm{mic}}(s_\alpha, v_\alpha, v_l) \quad \hat{a}_\alpha = a_{\mathrm{mic}}(\hat{s}_\alpha, v_\alpha, v_{\hat{l}}). \tag{14.6}$$

The lane-changing threshold Δa prevents lane changes when the associated advantage is only marginal (cf. Table 14.1). Furthermore, the constant weight a_{bias} introduces a simple form of asymmetric behavior. If a keep-right directive is to be modeled,

---

[5] One may object that—lacking mind-reading abilities—drivers do not know the acceleration function of others. However, the evidence allows for a coarse judgement. At the least, one can distinguish between cars and trucks, and between normal and evidently very sluggish or agile drivers.

[6] For asymmetric "European" lane-changing rules we refer to the literature (Kesting, A., Treiber, M., Helbing, D.: General lane-changing model MOBIL for car-following models).

[7] Often, a "keep to the right" directive is in effect in countries with right-hand traffic. Furthermore, in the United States, one should preferably overtake on the left lanes. However, this is not enforced and, *de facto*, overtaking takes place to the left and to the right.

Table 14.1 Parameters of the lane-changing models 14.4–14.7

Parameter	Typical value
Limit for safe deceleration b_{safe}	2 m/s^2
Changing threshold Δa	0.1 m/s^2
Asymmetry term (keep-right directive) a_{bias}	0.3 m/s^2
Politeness factor p (MOBIL lane-changing model)	0.0–1.0

The parameters b_{safe} and Δa apply to any changing model, $a_{\text{bias}} \neq 0$ only if asymmetric driving rules are to be modeled, and $p \neq 0$ if the drivers are not purely egoistic

a_{bias} would be positive for changes to the left, and reverses its sign for changes to the right. This contribution should be relatively small ($|a_{\text{bias}}| \ll b_{\text{safe}}$) but greater than Δa. Otherwise, vehicles would not change to the right lanes if the highway was essentially empty (see Table 14.1).

Jamming paradox: The grass is always greener on the other side. A motivation to change lanes in jammed situations is the observation that the other lanes are faster, most of the time, suggesting that these lanes are "better". In Problem 14.1 we show that this is a fallacy: Even if the travel times on all lanes are the same, the fraction of the time one finds oneself on the slower lane is greater than 50 % *on any lane*. The fallacy is resolved by observing that, when the other lanes are slower, the active overtaking rate (overtaken vehicles per time unit) is greater than the passive overtaking rate in the periods where the other lanes are faster. Since the models presented here do not include tactical components, the simulated drivers also succumb to this fallacy and tend to change lanes unnecessarily often.

14.3.3 Lane Changes with Courtesy: MOBIL Model

The changing conditions (14.4) and (14.5) characterize purely egoistic drivers who consider other drivers only via the safety criterion. If the lane change is mandatory as in lane-closure or merging situations, this behavior is plausible (and, additionally, the changing threshold $\Delta p = 0$). On the other hand, if the lane change is not necessary (also termed a *discretionary* lane change), most drivers refrain from changing lanes if their own advantage is disproportionally small compared to the disadvantage imposed on others, even if the safety criterion is satisfied. This can be modeled by augmenting the balance of the incentive criterion with the utilities of the affected drivers, weighted with a *politeness factor* p,

$$\boxed{\hat{a}_\alpha - a_\alpha + p\left(\hat{a}_{\hat{f}} - a_{\hat{f}} + \hat{a}_f - a_f\right) > \Delta a + a_{\text{bias}} \quad \text{MOBIL incentive.}} \tag{14.7}$$

For the special case when politeness $p = 1$ (corresponding to a rather altruistic driver), no bias ($a_{\text{bias}} = 0$), and negligible threshold ($\Delta a = 0$), a lane change takes place if the sum of the accelerations of all affected vehicles increases by this maneuver.[8] Hence the acronym for this model:

> MOBIL—Minimizing **o**verall **b**raking deceleration **i**nduced by **l**ane changes.

The central component of the MOBIL criterion is the politeness factor indicating the degree of consideration of other drivers if there are no safety restraints. Since a degree of consideration amounting to $p = 1$ is rare (which would correspond to "Love thy neighbor as thyself"), sensible values are of the order 0.2.[9]

> What is your estimate for the politeness factor p of the two drivers sketched in Fig. 11.2? Is it possible to describe by appropriate, possibly event-driven values of the politeness p following situations: (i) purely altruistic drivers (exclusively caring for the well-being of others), (ii) malign drivers (accepting own disadvantages to obstruct others), (iii) self-righteous drivers (obstructing other speeding drivers to "teach" them the traffic rules), (iv) timid drivers quickly making way when tailgated by others?

14.3.4 Application to Car-Following Models

The general lane-changing criteria presented above return explicit rules only when combined with a longitudinal acceleration model. In principle, the safety criterion (14.4) and the incentive criteria (14.5) or (14.7) are compatible with any longitudinal model providing the acceleration function a_{mic} either directly (time-continuous car-following models) or indirectly via Eq. (10.11) (time-discrete iterated coupled maps, see Sect. 10.2, or cellular automata, see Chap. 13).

When applying the safety criterion (14.4) to any acceleration model satisfying the general plausibility conditions discussed in Sect. 11.1, we obtain a minimum condition for the *lag gap* $\hat{s}_{\hat{f}}$ of the new follower behind the changing vehicle on the new lane,

$$\hat{s}_{\hat{f}} > s_{\text{safe}}(v_{\hat{f}}, v_\alpha). \tag{14.8}$$

The *safe gap function* $s_{\text{safe}}(v_f, v)$ is obtained by solving the equation defining marginal safety of the follower,

---

[8] As always, the safety criterion must be satisfied unconditionally. However, safety is nearly always given if the incentive criterion for $p = 1$ is satisfied.

[9] On the interactive simulation website www.traffic-simulation.de, one can simulate traffic flow with variable degree of politeness (altruism) which can be controlled by the user. The underlying acceleration model is the IDM (see Sect. 14.3.4).

$$a_{\mathrm{mic}}\left(s_{\mathrm{safe}}, v_f, v\right) = -b_{\mathrm{safe}}, \tag{14.9}$$

for the gap s_{safe}. Notice that a unique solution s_{safe} exists by virtue of the plausibility condition (11.2) stating that, in the interaction range, the function a_{mic} increases strictly monotonically with respect to s. This means, the safety criterion allows changes if the following gap (lag gap) on the target lane is greater than some minimum value depending on the speeds of the changing vehicle and the new follower $\hat{f}$, i.e., the safety criterion becomes a generalized *gap-acceptance* rule for the lag gap.

Similarly, the general incentive criterion (14.5) of egoistic drivers can be written as a generalized gap-acceptance rule for the *lead gap* of the changing vehicle on the new lane,

$$\hat{s}_{\mathrm{lead}} = \hat{s}_\alpha > s_{\mathrm{adv}}(s_\alpha, v_\alpha, v_l, v_{\hat{l}}). \tag{14.10}$$

The *advantageous gap function* $s_{\mathrm{adv}}(s, v, v_l, v_{\hat{l}})$ is obtained by solving the equation

$$a_{\mathrm{mic}}\left(s_{\mathrm{adv}}, v, v_{\hat{l}}\right) - a_{\mathrm{mic}}(s, v, v_l) = \Delta a + a_{\mathrm{bias}} \tag{14.11}$$

defining a marginal change of utility, for s_{adv}. Again, condition (11.2) ensures that a unique solution s_{adv} exists if $a_{\mathrm{mic}}(s, v, v_l) + \Delta a + a_{\mathrm{bias}} < a_{\mathrm{free}}(v)$ where $a_{\mathrm{free}}(v) = a_{\mathrm{mic}}(\infty, v, v_l)$ is the free-flow acceleration function (11.3). In contrast to the safety condition, however, this is not always satisfied. Then, s_{adv} is not unique or even does not exist. Obviously, this corresponds to an infinite advantageous gap reflecting the fact that there is no need to change lanes because one can either drive freely on the old lane, or there is an obstruction but it is so small that the finite threshold $\Delta a + a_{\mathrm{bias}}$ prevents lane changing for marginal utility improvements, even if the target lane is free. In the following, we discuss the application to three specific longitudinal models.

Rules for the Optimal Velocity Model. Introducing the OVM acceleration $\dot{v} = (v_{\mathrm{opt}}(s) - v)/\tau$ into the safety criterion (14.8) with Eq. (14.9), we obtain the condition

$$\hat{s}_{\hat{f}} > s_{\mathrm{safe}}^{\mathrm{OVM}}(v_{\hat{f}}) = s_e\left(v_{\hat{f}} - \tau b_{\mathrm{safe}}\right) \tag{14.12}$$

for the minimum safe lag gap of the OVM driver. Here,

$$s_e(v) = v_{\mathrm{opt}^{-1}}(s) \tag{14.13}$$

is the inverse function of the optimal-velocity function indicating the steady-state gap for a given speed, i.e., $v_{\mathrm{opt}}(s_e(v)) = v$ (see the paragraph below Eq. (10.13) for details). This means that, after the (yet hypothetical) change, the optimal velocity of the follower on the target lane must not be smaller than the actual speed of this follower minus the safe deceleration multiplied by the speed adaptation time, $v_{\mathrm{opt}}(\hat{s}_{\hat{f}}) > v_{\hat{f}} - \tau b_{\mathrm{safe}}$.

In analogy, the incentive criterion (14.10) for egoistic drivers with Eq. (14.11) leads to a further gap-acceptance rule for the lead gap between the changed vehicle and its new leader:

$$\hat{s}_{\text{lead}} > s_{\text{adv}}^{\text{OVM}}(s_\alpha) = s_e\left[v_{\text{opt}}(s_\alpha) + \tau(\Delta a + a_{\text{bias}})\right]. \tag{14.14}$$

For the special case of the optimal velocity function (10.22) corresponding to a triangular fundamental diagram, and its inverse function $s_e(v) = s_0 + vT$ for $v < v_0$, we obtain the three conditions[10]

$$\begin{aligned} \hat{s}_{\hat{f}} &> s_0 + T\left(v_{\hat{f}} - b_{\text{safe}}\tau\right), \\ \hat{s}_{\text{lead}} &> s_\alpha + T\,\tau\,(\Delta a + a_{\text{bias}}), \\ s_\alpha &< s_0 + T\,[v_0 - \tau\,(\Delta a + a_{\text{bias}})]. \end{aligned} \tag{14.15}$$

Notice that the parameters T and τ are both of the order of 1 s (cf. Table 10.1, and the accelerations b_{safe} and Δa are of the order of 1 m/s^2 or less, respectively (cf. Table 14.1). Consequently, all contributions in the conditions (14.15) containing products of τT are of the order of 1 m or less, i.e., negligible compared to the gaps s_α, $\hat{s}_\alpha$, and $\hat{s}_{\hat{f}}$. Effectively, this results in

$$\hat{s}_{\hat{f}} > s_e(v_{\hat{f}}), \quad \hat{s}_{\text{lead}} > s_\alpha, \quad v_{\text{opt}}(s_\alpha) < v_0. \tag{14.16}$$

This means, there are three conditions for a lane change to be safe and desirable: (i) the new lag gap is greater than the safe gap, (ii) the lead gap on the target lane is larger than the actual lead gap, and (iii) an obstruction exists. We emphasize that, for $b_{\text{safe}} = \Delta a = 0$, rules that are identical to the gap acceptance rules (14.16) can be derived for any longitudinal model with a unique steady-state speed function if the speed difference does not enter as an exogenous factor. This includes Newell's model (Sect. 10.8), and even some cellular automata such as the Nagel-Schreckenberg Model (Sect. 13.2).

However, the gap-acceptance rules (14.16) are unrealistic since, in real situations, the minimum lead and lag gaps depend crucially on speed differences: Given a certain lag gap, it makes a big safety difference whether the new follower drives at about the same speed or approaches quickly. Therefore, we now apply the general safety and incentive gap-acceptance rules (14.8) and (14.10) to acceleration models taking into account the speed difference. We expect that this new exogenous factor carries over to the resulting lane-changing rules in a consistent way.

Rules for the Full Velocity Difference Model. The acceleration of the Full Velocity Difference Model (FVDM) is that of the OVM plus a contribution depending linearly on the speed difference, $a_{\text{FVDM}}(s, v, v_l) = a_{\text{OVM}}(s, v) - \gamma(v - v_l)$ (cf. Sect. 10.7). Applying the safety condition (14.8) and the incentive criterion (14.10) to this acceleration function gives

[10] The third condition ensures that s_{adv} is defined. Otherwise, there is never an incentive for changing, see the text below (14.11) for details.

$$\hat{s}_{\hat{f}} > s_{\text{safe}}^{\text{FVDM}}(v_{\hat{f}}, v_\alpha) = s_e\left[v_{\hat{f}} - \tau b_{\text{safe}} + \tau\gamma(v_{\hat{f}} - v_\alpha)\right], \tag{14.17}$$

$$\hat{s}_{\text{lead}} > s_{\text{adv}}^{\text{FVDM}}(s_\alpha, v_l, v_{\hat{l}}) = s_e\left[v_{\text{opt}}(s_\alpha) + \tau\ \left(\Delta a + a_{\text{bias}} + \gamma(v_l - v_{\hat{l}})\right)\right]. \tag{14.18}$$

As a result, the minimum lead and lag gaps on the target lane allowing a change depend on the speed difference, i.e., they are consistent with the acceleration law.

We emphasize that the speed difference contributions are significant as illustrated by following example: With typical value of the FVDM speed adaptation time $\tau = 5\,s$ and the sensitivity $\gamma = 0.6\,\text{s}^{-1}$ for speed differences (cf. the caption of Fig. 10.6),[11] we have $\gamma\tau = 3$, i.e., the speed differences in the arguments of the equilibrium gap function $s_e(v)$ are weighted thrice with respect to the respective speeds themselves. Assuming furthermore a linear steady-state gap function $s_e(v) = s_0 + vT$ (corresponding to the congested branch of the triangular fundamental diagram) and bound traffic on either lane (no gap is larger than $s_0 + v_0 T$), the above conditions become

$$\hat{s}_{\hat{f}} > s_{\text{safe}}^{\text{FVDM}}(v_{\hat{f}}, v_\alpha) = s_0 + T\left[v_{\hat{f}} - \tau b_{\text{safe}} + \gamma\tau(v_{\hat{f}} - v_\alpha)\right] \quad \text{safety}, \tag{14.19}$$

$$\hat{s}_{\text{lead}} > s_{\text{adv}}^{\text{FVDM}}(s_\alpha, v_l, v_{\hat{l}}) = s_\alpha + T\tau\left[\Delta a + a_{\text{bias}} + \gamma(v_l - v_{\hat{l}})\right] \quad \text{incentive}. \tag{14.20}$$

With the steady-state time gap $T = 1.4$ s, the lane-changing threshold $\Delta a = 0.1$ m/s^2, and asymmetry $a_{\text{bias}} = [0.3] m/s^2$ (cf. Tables 10.1 and 14.1), this leads to following relations:

- The minimum lag gap implied by the safety rule increases by 1.4 m when all affected drivers drive faster by 1 m/s. Furthermore, it increases by 4.2 m if only the new follower drives faster by 1 m/s.
- If the old and new leaders drive at the same speed, the incentive criterion implies that a lane change to the left is desirable if the new lead gap is larger than the old one by at least 2.8 m while the asymmetry term makes changes to the right desirable even if there is a smaller lead gap provided it is smaller by at most 1.4 m. Furthermore, every 1 m/s speed advantage on the new lane compensates for a decrease of the new lead gap by 4.2 m.

Rules for the Intelligent Driver Model. In contrast to most versions of the OVM or FVDM, the IDM safe gap function $s_{\text{safe}}(v_{\hat{f}}, v_\alpha)$ defining the gap-acceptance rule (14.8) for the lag gap can be analytically calculated,

$$\hat{s}_{\hat{f}} > s_{\text{safe}}^{\text{IDM}}(v_{\hat{f}}, v_\alpha) = \frac{s^*(v_{\hat{f}}, v_{\hat{f}} - v_\alpha)}{\sqrt{\frac{1}{a}\left(a_{\text{free}}(v_{\hat{f}}) + b_{\text{safe}}\right)}}. \tag{14.21}$$

[11] Because of the additional anticipation introduced by the speed difference term, larger and more realistic values of the speed adaptation time τ can be assumed for the FVDM compared to the OVM which would lead to crashes if τ exceeds the order of 1 s.

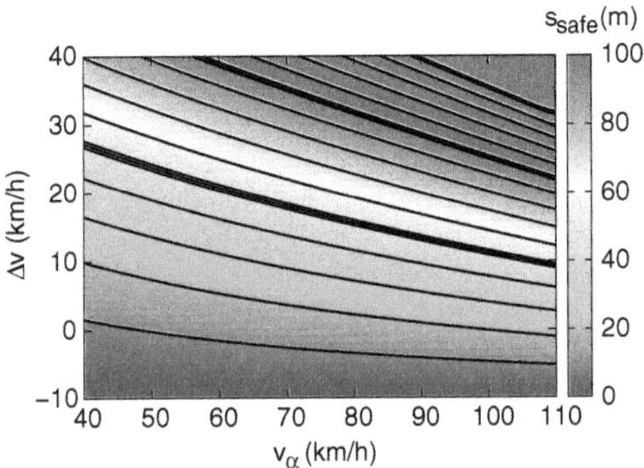

Fig. 14.2 Minimum lag gap (14.21) of the IDM as a function of the speed v_α of the changing vehicle, and the speed difference $\Delta v = v_{\hat{f}} - v_\alpha$ of the new follower with respect to this vehicle. The distance between two *thin* and *thick lines* correspond to a change of the minimum gap by 10 and 50 m, respectively

In the improved IDM (IIDM, Sect. 11.3.7), the condition becomes even simpler and does not depend on the desired speed parameter,

$$\hat{s}_{\hat{f}} > s_{\text{safe}}^{\text{IIDM}}(v_{\hat{f}}, v_\alpha) = \frac{s^*(v_{\hat{f}}, v_{\hat{f}} - v_\alpha)}{\sqrt{1 + \frac{b_{\text{safe}}}{a}}}. \tag{14.22}$$

In any case, the denominators are of the order unity, so the minimum lag gap is essentially given by the dynamical desired IDM gap s^*, Eq. (11.15), evaluated for the new follower.

Figure 14.2 shows the minimum lag gap (14.21) for the highway IDM parameters of Table 11.2 and $b_{\text{safe}} = b = 1.5\,\text{m/s}^2$. If, for example, both the changing vehicle and the following vehicle on the target lane drive at 50 km/h, the minimum lag gap according to the safety criterion is about 10 m. At this gap, the new follower would have to brake with a deceleration b_{safe} in order to regain his or her safe gap which is of the order of 17 m.[12]

If, however, the new follower drives at 70 km/h, i.e., approaches by a rate of $\Delta v = 20$ km/h, the safety criterion displayed in Fig. 14.2 gives a minimum acceptable gap of about 40 m amounting to an increase of about 6 m for every 1 m/s the follower drives faster. As in the FVDM, the IDM minimum gap depends crucially on the speed *difference*. In contrast to the former, the dependence is a nonlinear one and takes into account kinematic facts such as the quadratic dependence of the braking distance on the speed.

[12] In fact, investigations on trajectory data show that drivers temporarily accept shorter gaps after a change and only brake minimally such that the gap increases only gradually to the normal gap. This is part of the tactical behavior which is not considered here.

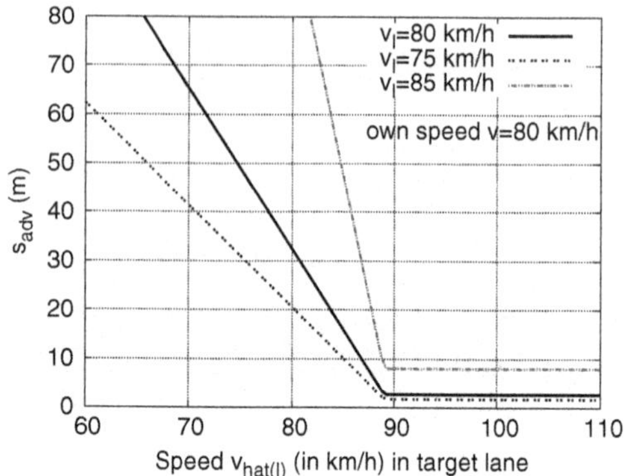

Fig. 14.3 Minimum lead gap on the target lane where a lane change is considered as advantageous as a function of the speeds of the leaders on the old and new lanes. The own speed is fixed to $v_\alpha = 80$ km/h, and the actual gap $s_\alpha = 30$ m

The minimum gap $s_{\text{adv}}(s_\alpha, v_\alpha, v_l, v_{\hat{l}})$ for an advantageous lane change appearing in the incentive criterion (14.10) is also accessible to analytical treatment resulting in

$$\hat{s}_{\text{lead}} = \hat{s}_{\text{adv}}^{\text{IDM}}(s, v, v_l, v_{\hat{l}}) = \frac{s^*(v, v - v_{\hat{l}})}{\sqrt{\left(\frac{s^*(v, v-v_l)}{s}\right)^2 - \frac{\Delta a + a_{\text{bias}}}{a}}}. \tag{14.23}$$

In agreement with intuition, the minimum advantageous gap depends strongly on the speed difference between the actual and the new leader, i.e., essentially on the speed difference driven on the new and old lanes: The higher the speed on the new lane, the smaller the accepted gaps for a change to this lane (cf. Fig. 14.3). Notice that the IDM gap-acceptance criterion accepts very small gaps if the leading vehicles are significantly faster. Nevertheless, the situation remains safe: After all, it is certain that the gaps will increase in the following seconds.[13]

14.4 Approaching a Traffic Light

When approaching a signalized intersection and the traffic light switches from green to yellow, it is necessary to decide whether it is better to cruise over the intersection with unchanged speed, or to stop (Fig. 14.4). This can be modeled within the general discrete-choice framework of Sect. 14.2. Since incentives are not relevant for this

[13] Such "close shaves" may not feel comfortable to most drivers, however. In order to suppress such a behavior, it is most straightforward to modify the minimum condition of Eq. (11.15) for the dynamical desired IDM gap $s^*(v, \Delta v)$, e.g., by not allowing the dynamical IDM gap s^* to be below $s_0 + \frac{1}{2}vT$.

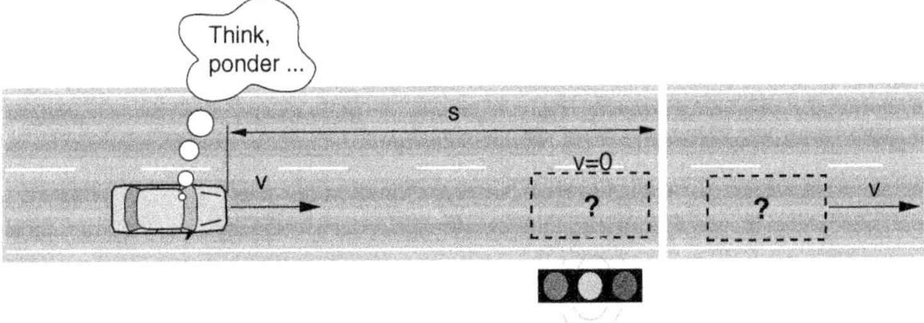

Fig. 14.4 Illustration of the decision to stop or to cruise at a traffic signal about to go *red*

situation,[14] the decisions are determined by the safety criterion alone: "Stop if it is safe to do so". In our general framework, the decision to stop is considered as safe if the anticipated braking deceleration will not exceed the safe deceleration b_{safe} at any time of the braking maneuver. For models with a plausible braking strategy, it is sufficient to consider the braking deceleration for this option *at decision time*.[15]

To calculate this deceleration, we model the traffic light as a standing virtual vehicle ($v_l = 0$, $\Delta v = v$, desired speed $v_0 = 0$) of zero extension such that s denotes the distance of the front bumpers to the stopping line. This results in the simple rule

$$\begin{array}{ll} \text{cruise} & \text{if } \; a_{\text{mic}}(s, v, v) < -b_{\text{safe}} \Leftrightarrow s < s_{\text{crit}}(v), \\ \text{stop} & \text{otherwise.} \end{array} \tag{14.24}$$

Obviously, the critical distance s_{crit} where the decision changes is a special case of the safe gap function (14.8) of the safety criterion,

$$s_{\text{crit}}(v) = s_{\text{safe}}(v, 0). \tag{14.25}$$

It is particularly instructive to apply this rule to the IDM safe gap (14.21) for the common situation when the driver approaches the signalized intersection at his or her desired speed, and the IDM parameters satisfy approximatively $a = b = b_{\text{safe}}$. In this case, $s_{\text{crit}}(v) = s^*(v, v)$ is equal to the dynamic desired IDM gap for $\Delta v = v$, and condition (14.24) becomes

$$\begin{array}{ll} \text{cruise} & \text{if } s < s^* = s_0 + v_0 T + \frac{v_0^2}{2b}, \\ \text{stop} & \text{otherwise.} \end{array} \tag{14.26}$$

[14] Sometimes, one observes that drivers pass yellow traffic lights, or even accelerate, if they could safely stop. Obviously, there *are* incentives at work. We will not attempt to model this behavior.

[15] If larger decelerations are unavoidable for kinematic reasons, the drivers described by such models will attempt to bring the situation under control as soon as possible, i.e., they will brake hardest at the beginning. The IDM belongs to this model class.

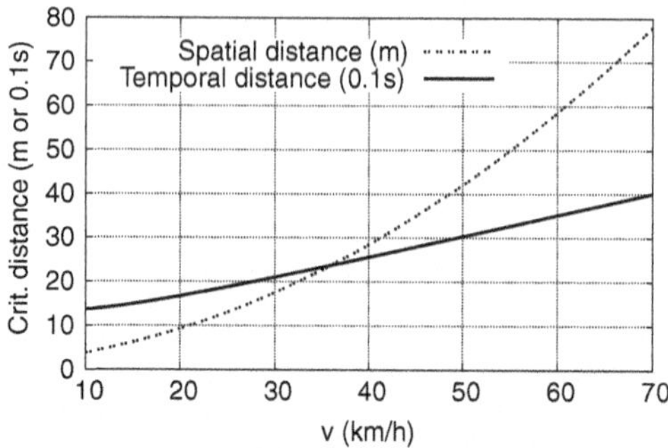

Fig. 14.5 Critical distance to the stopping line of a traffic light at decision time as a function of the speed for the IDM with $b = 2\,\mathrm{m/s^2}$, $b_{\mathrm{safe}} = 3\,\mathrm{m/s^2}$, and further IDM parameters taken from Table 11.2. Also shown is the associated TTC value

When setting the desired time gap T equal to the driver's reaction time, this means that one stops if, at decision time, the distance to the stopping line is greater than the *stopping distance* (11.8). This is perfectly consistent since this distance (which we have already introduced when formulating Gipps' model) is necessary to stop in a controlled way taking into account reaction time.

Figure 14.5 shows that, also for more general parameter settings, the critical distance increases quadratically with the speed, while the critical *time-to-collision* (TTC) (here defined as the time to reach the stopping line at unchanged speed) increases essentially linearly. We emphasize that the critical TTC of 3 s for 50 km/h, and 4 s for 70 km/h is consistent with European legislative regulations for the minimum duration of yellow phases of traffic lights at streets with the respective speed limits (Problem 14.2).

14.5 Entering a Priority Road

This situation can be considered as a special case of mandatory lane-changing decisions:

- The (nearest lane of the) main-road corresponds to the target lane of the lane-changing situation.
- The merging action to enter the road corresponds to the lane-changing maneuver.
- The speed of the merging/lane-changing vehicle is very low or zero (the latter is true if there are stop signs, or the entering vehicle is already waiting).
- And the incentive criterion is always satisfied.[16]

In contrast to normal (discretionary) lane-changing decisions, entering a priority road implies *two* safety criteria, one for the new follower and one for the merging vehicle

[16] After all, a driver wants to enter the main-road as soon as it can be done safely.

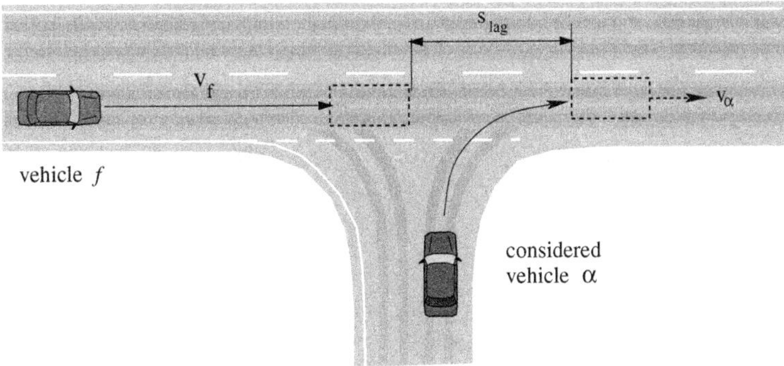

Fig. 14.6 Illustration of the safety criterion for the decision "stopping/waiting or merging" when entering a priority road

itself. The latter was not necessary for discretionary lane changes since, there, a fulfilled incentive criterion automatically implies safety for the decision maker him or herself (cf. the last paragraph of Sect. 14.3). When formulating the criteria, we assume that the driver of the merging vehicle can anticipate his or her speed v_α, the speeds v_f and v_l of the follower and leader, and the corresponding gaps s_f and s_l, respectively, at merging time (Fig. 14.6)[17]

$$\begin{array}{ll} \text{merge} & \text{if } s_f > s_{\text{safe}}(v_f, v_\alpha) \text{ AND } s_{\text{lead}} > s_{\text{safe}}(v_\alpha, v_l), \\ \text{stop or wait} & \text{otherwise.} \end{array} \tag{14.27}$$

Problems

14.1. Why the grass is always greener on the other side?

Give the reason why, when driving in congested conditions, one generally spends more time in the slower lanes and that a lane change does not help. Consider following situation:

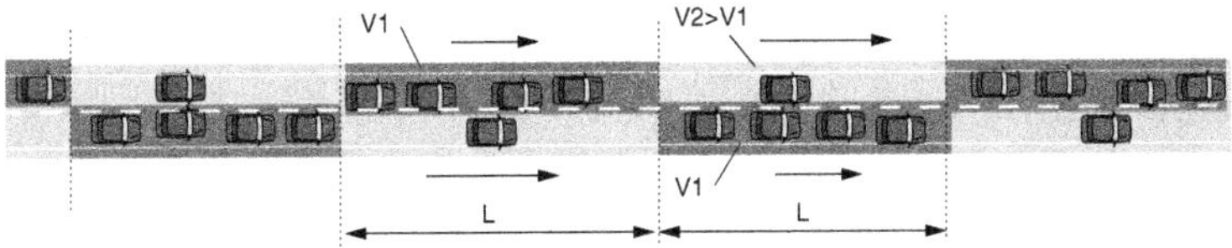

The figure shows two-lane traffic with staggered traffic waves of length L containing jammed traffic creeping at average speed V_1, and the regions in between (of length L as well) where traffic flows more quickly ($V_2 > V_1$) but yet not freely (the

[17] Since there are no relevant cars to consider for the alternative "stop or wait", we have dropped all hats denoting the new situation, for simplicity.

congested branch of the fundamental diagram remains relevant). Assume a triangular fundamental diagram and negligible speed adaptation times (which is a good approximation for the OVM or Newell's model). Calculate the fraction p_{slow} of the time one drives inside a wave, i.e., the other lane is faster, as a function of V_1, V_2, and the model parameters T and l_{eff}. Be astounded by the result!

Hint: You can solve this problem by calculating the relative velocity between the driven speed and the wave velocity. Or simply count the vehicles.

14.2. Stop or cruise?
A decision strategy abiding traffic regulation and restricting braking decelerations to minimal values while taking into account reaction times is the following [cf. the IDM condition (14.26)]: Anticipate if you can pass the traffic light at unchanged speed before it turns red. If so, cruise. Otherwise, brake smoothly with constant deceleration so as to stop just before the stopping line. When the speed limit is 50 km/h, the minimum duration of the yellow phase prescribed by law is $\tau_y = 3$ s. What is the maximum deceleration the legislative authority expects you to use, assuming a complex reaction time of 1 s?

14.3. Entering a highway with roadworks
Assume a highway on-ramp whose merging section, due to roadworks, is nearly nonexistent, i.e., one has to merge at essentially zero speed. The relevant safety criterion is defined by the OVM with a triangular fundamental diagram ($s_0 = 0$ s, $T = 1$ s) assuming a safe deceleration threshold $b_{safe} = 0\,\text{m/s}^2$. The driver of the considered car waits at the merging position while another car approaches on the main-road at 72 km/h. At decision time, the safety condition is just satisfied (the lag gap is only marginally greater than s_{safe}), so the driver begins to merge. Calculate the minimum deceleration at which the driver of the main-road vehicle has to brake in order to avoid a collision. Assume that the entering car accelerates with $2\,\text{m/s}^2$ for the first few seconds. Furthermore, assume that the driver on the main-road reacts instantaneously and is able to anticipate the situation perfectly. Discuss whether the OVM safe gap-acceptance criterion is really "safe", particularly, when assuming finite reaction times.

14.4. An IDM vehicle entering a priority road
Formulate the IDM safety criterion for merging into a priority road with a stop sign (the initial speed of the merging vehicle is zero). Calculate the safe lead and lag gaps for the IDM parameters $a = b = b_{safe} = 2\,\text{m/s}^2$, $T = 1$ s, $v_0 = 50$ km/h, and $s_0 = 0$ m if the speed on the main road is 50 km/h.

Further Reading

- Ben-Akiva, M., Lerman, S.: Discrete choice analysis: Theory and application to travel demand. MIT press (1993)
- Gipps, P.G.: A model for the structure of lane-changing decisions. Transportation Research Part B: Methodological **20** (1986) 403–414

- Kesting, A., Treiber, M., Helbing, D.: General lane-changing model MOBIL for car-following models. Transportation Research Record: Journal of the Transportation Research Board **1999** (2007) 86–94
- Nagel, K., Wolf, D., Wagner, P., Simon, P.: Two-lane traffic rules for cellular automata: A systematic approach. Physical Review E **58** (1998) 1425–1437
- Thiemann, C., Treiber, M., Kesting, A.: Estimating acceleration and lane-changing dynamics from next generation simulation trajectory data. Transportation Research Record: Journal of the Transportation Research Board **2088** (2008) 90–101
- Redelmeier, D., Tibshirani, R.: Why cars in the next lane seem to go faster. Nature **401** (1999) 35.

Chapter 15
Stability Analysis

Mathematics is the key and door to the sciences.

Galileo Galilei

Abstract Second-order macroscopic models and most car-following models are able to reproduce traffic waves or other observed instabilities of traffic flow. After an intuitive introduction, we define the relevant instability concepts: Local instability, convective string and flow instability, and absolute string and flow instability. We give general analytic criteria for the occurrence of these instabilities for microscopic and macroscopic models. The formulation is more comprehensive than the various accounts in the specialized literature and can be evaluated for any traffic flow model with a well-defined acceleration function. The stability criteria allow us to characterize the influencing factors of traffic flow instabilities and answer the question of if, and in which way, the driving behavior (or new driver-assistance systems) influence traffic flow stability.

15.1 Formation of Stop-and-Go Waves

Instabilities of traffic flow resulting in *traffic waves*, also termed *stop-and-go waves*, are caused by the delays in adapting the speed to the actual traffic conditions. These *delays* are the consequence of finite acceleration and braking capabilities, and also result from finite reaction times of the drivers. If traffic density is sufficiently high, this delay leads to a positive feedback on density and speed perturbations.[1] We will now

[1] Generally, delays in a feedback control system favor instabilities. This can be experienced intuitively when taking a shower and controlling the water temperature, particularly, if the response time between the controlling action and the result (a change of the water temperature) is rather long.

M. Treiber and A. Kesting, *Traffic Flow Dynamics*,
DOI: 10.1007/978-3-642-32460-4_15,

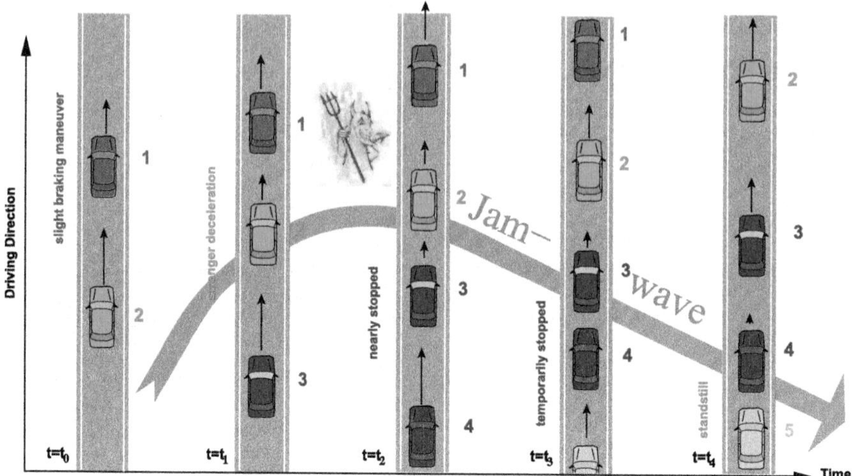

Fig. 15.1 The *vicious circle*: In order to regain the safety gap, the driver of every following vehicle needs to brake harder than his or her predecessor. The *numbers* beside the vehicles denote the vehicle index

intuitively explain this *vicious circle* with the help of Fig. 15.1 (see also Fig. 15.8):

- The scenario starts with a platoon of cars initially in steady-state equilibrium at speed v_e. At time $t = t_0$, the driver of car 1 brakes slightly (for whatever reason) and continues driving at a slightly lower speed $v_1 < v_e$.
- As a result, the new optimal speed for car 2 is given by v_1 as well. So the driver of this car reduces his or her speed from v_e to v_1 in a finite time interval ending at time t_1.
- If traffic is sufficiently dense, or if the speed adaptation time is sufficiently long, the gap of car 2 at time t_1 is smaller than the steady-state gap $s_e(v_1)$ at the speed of the leading car 1. In order to regain his or her desired gap, the driver of car 2 has to brake *more*, i.e., he or she decelerates temporarily to a speed $v_2 < v_1$ in the time interval between t_1 and t_2. The degree of this *overreaction* increases with the sensitivity to changes of the gap which is given by $|v_e'(s)|$ and $V_e'(\rho)$ for microscopic and macroscopic models, respectively.
- Since the driver of the next car 3 also needs some time to adapt the speed, the gap between car 2 and car 3 may become smaller than the steady-state gap $s_e(v_2)$. Therefore, the driver of car 3 decelerates further to a minimum speed $v_3 < v_2$ at time t_2.
- This positive feedback continues when going to the next car 4 which has to stop completely (time t_3).
- The resulting traffic wave dissolves only if the number of new vehicles approaching the wave from behind decreases.

As a result, a stop-and-go wave emerges "out of thin air" giving rise to the name *phantom jam* for this phenomenon (see also Sect. 18.2 and the right diagram of

Fig. 15.10). At sufficiently low traffic density, or when traffic consists predominantly of agile drivers, the vicious circle is broken. In this case, the drivers have already equilibrated their speed to the new situation at the time where a new vehicle comes within interaction distance, so the stop-and-go mechanism is not effective. As a result, all drivers following car 1 decelerate to v_1 but not further (see the left diagram of Fig. 15.10 for an example). From the qualitative consideration, it follows that the stop-and-go mechanism is never effective in models describing instantaneous speed adaptations and zero reaction times as in first-order macroscopic models (LWR models), or in Newell's microscopic model. As a result, density perturbations never grow in such models, so they cannot describe traffic instabilities.[2]

In summary, the qualitative argumentation suggests that the tendency to traffic flow instabilities increases with

- increasing speed adaptation time,
- increasing traffic density,
- and increasing sensitivity $|v_e'(s)|$ or $V_e'(\rho)$ for changes of the gap.

The stability analysis expounded below agrees with this reasoning.[3]

15.2 Mathematical Classification of Traffic Flow Instabilities

We emphasize that all types of instabilities discussed in this chapter describe a tendency to oscillations, traffic waves, stop-and-go traffic and the like. However, they do *not* correspond to accidents (which would be characterized by negative gaps or densities exceeding the maximum density $\rho_{\max}$ in the microscopic and macroscopic descriptions, respectively). Generally, simulated accidents only occur if the instability thresholds are exceeded extremely. However, in some models representing "short-sighted" drivers (such as the OVM), accidents may happen even for parameters corresponding to *perfectly stable* traffic.

Moreover, the *physical* instabilities of real traffic discussed below have to be distinguished from *numerical instabilities*. The latter result from integration steps being too large, or by applying an unsuitable numerical update method (see Sect. 9.5

---

[2] The only way to generate a traffic breakdown in such models is by simulating a bottleneck and assuming upstream boundary conditions corresponding to an inflow exceeding the bottleneck (footnote 2 Continued) capacity. Then, as soon as the flow at the bottleneck exceeds its capacity, the density immediately upstream of the bottleneck jumps to the congested branch of the fundamental diagram at a flow corresponding to the bottleneck capacity (cf. Sect. 8.5).

[3] Notice that, in some models, the speed adaptation time may depend on traffic density getting shorter for increased density. This can more than compensate for the destabilizing effects of traffic density itself, so congested traffic may be unstable for most densities but *restabilize* for high densities near the maximum.

for details). In contrast, real traffic instabilities are the consequence of physical delays due to finite accelerations and reaction times.[4]

In the following, we distinguish categories of traffic instabilities depending on criteria for their existence and the type of resulting congestion pattern.

Number of involved vehicles: local versus string instability. Local instability relates to the car-following dynamics of a single or a few vehicles following a leader with a predetermined trajectory (typically introducing a perturbation by a temporary speed drop while driving at constant speed for the rest of the time). This system is *locally unstable*, if the gap and speed fluctuations of the follower(s) increase with time, or, at least, do not decay (cf. Fig. 15.2). Otherwise, it is locally stable. Since this definition refers to a platoon of a finite number of vehicles, one also speaks of *platoon (in)stability*. Obviously, this stability concept is only applicable for microscopic models. For practical purposes, it is relevant when developing the feedback controllers of ACC systems.[5]

In contrast, the ubiquitous traffic waves are the result of *string instability*.

Traffic flow is string stable if local perturbations decay *everywhere* even in *arbitrarily long* vehicle platoons. Otherwise, it is string unstable. As illustrated by Fig. 15.3, string stability is a much more restrictive concept compared to local stability: Traffic flow may be string unstable even if speed fluctuations within a vehicle platoon of finite size decay quickly, or even if there are *no local oscillations at all*. An example of this latter case is given in the two simulations displayed in the middle of Fig. 15.2, and in Fig. 15.6 (cf. Problem 15.3). This has immediate practical implications for developers of ACC controllers: Even if the ACC is optimized to be perfectly free of oscillations when following a "test hare vehicle" driving a prescribed speed profile, traffic flow mainly consisting of such ACC vehicles may be *absolutely string unstable*.

Since string instability is defined in terms of a collective phenomenon, it can be applied to both microscopic and macroscopic models. To emphasize its macroscopic nature, one also speaks of *collective instability*, or *flow instability*.[6]

---

[4] In particular, both physical and numerical instabilities include so-called *convective instabilities* which are discussed in the Sects. 9.5 and 15.5, respectively. Convective physical and numerical instabilities have no commonalities, whatsoever.

[5] At least, if the penetration rate of ACC equipped vehicles is sufficiently small. Otherwise, the influence of ACC-driven vehicles on the string instability becomes relevant as will be discussed in the main text below.

[6] Some authors stress that there is a conceptual difference between string instability (relevant for microscopic models), and flow instability (macroscopic models). However, observed differences are merely a consequence of an imperfect equivalence between microscopic and macroscopic models with respect to macroscopic phenomena (notice that microscopic models can describe macroscopic phenomena but not vice versa). The unified instability criteria to be developed in the next sections show that the concepts of string and flow instability are identical in a precisely defined sense: For each microscopic model displaying string instabilities in a subset of the space spanned by the model parameters and the steady-state traffic density, there exists a micro-macro relation to a macroscopic model displaying flow instability for exactly the same subset.

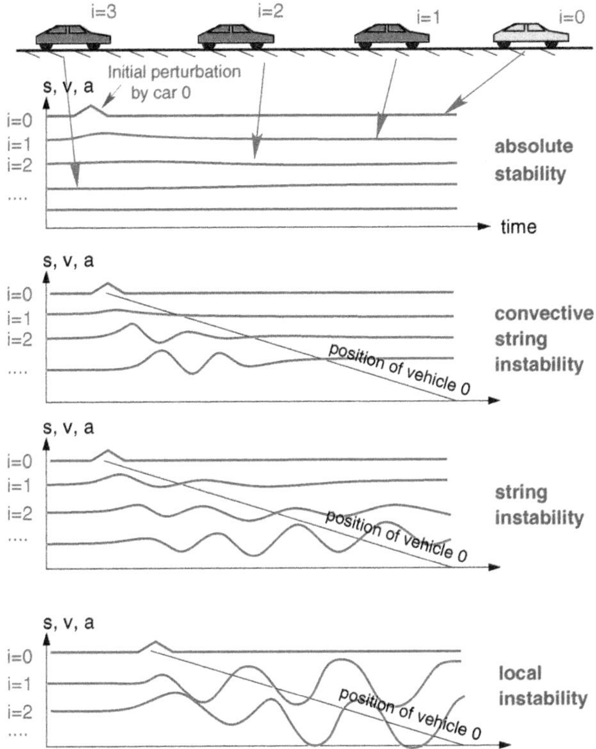

Fig. 15.2 Schematic sketch of the different instability concepts in terms of speed time series of single vehicles: If traffic flow is *convectively string unstable*, perturbations grow but propagate only upstream. Consequently, all vehicles drive smoothly at the time they pass the location of the initial perturbation (indicated by the *thin black line*). If traffic flow is absolutely string unstable, the perturbation eventually spreads everywhere but any given vehicle eventually drives smoothly, i.e., it can follow a vehicle with predetermined trajectory without sustained oscillations (platoon stability). In the presence of *local instabilities* or *platoon instabilities*, even following a single vehicle leads to sustained oscillations

Types of perturbation and asymptotic state: Ljapunov, asymptotic and structural stability. If we require that any sort of sufficiently small initial perturbations remain small forever, we speak of *Ljapunov stability*. If we additionally require that sufficiently small perturbations tend to zero for $t \to \infty$, the system is *asymptotically stable*. If we allow not only initial perturbations but also small persistent fluctuations and all trajectories remain close to the unperturbed trajectories, the system is *structurally stable*. These stability concepts are mainly used by mathematicians[7] and apply to arbitrary dynamical systems. For the traffic flow models with smooth acceleration functions considered here, Ljapunov and asymptotic stability are equivalent

[7] We do not give the precise mathematical definitions.

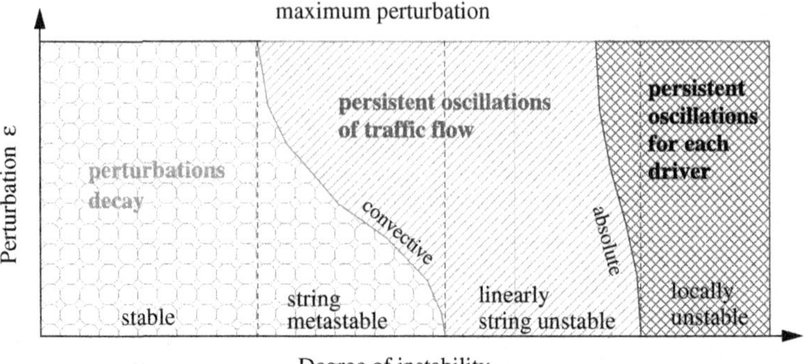

Fig. 15.3 Sequence of stability types as a function of the inherent tendency to instabilities (*horizontal axis*) and the amplitude of the initial perturbation. The string instability is convective in the limit of reaching the boundary to stability, and absolute at the boundary of local instability

concepts.[8] Since Ljapunov and asymptotic stability are defined in terms of sufficiently small but otherwise arbitrary initial and asymptotic perturbations, these concepts refer to linear string (or flow) stability when applied to traffic flow models. When explicitly defining the latter in Eq. (15.2), we refer to asymptotic rather than Ljapunov stability, i.e., we require that all small perturbations have negative growth rates, i.e., they tend to zero.

Amplitude of perturbation: linear versus nonlinear instability. If arbitrarily small perturbations increase in the course of time, one speaks of *linear instability*. If small perturbations decay but sufficiently severe perturbations (caused, e.g., by hard braking maneuvers or inconsiderate lane changes) develop to persistent traffic waves, this corresponds to *nonlinear instability*. As illustrated in Fig. 15.3, car-following models or second-order macroscopic models generally have parameter ranges where, for a certain range of steady-state densities, traffic flow is linearly stable and simultaneously nonlinearly unstable, i.e., small perturbations decay and larger ones develop to stop-and-go waves. This is termed *metastability*. As a consequence of this type of instability, the future dynamics depends not only on the present and future exogenous conditions but also on the past—for arbitrarily long times. For example, given the same traffic demand profile, there may be growing regions of congested traffic (a traffic breakdown occurred in the past), or completely free traffic (no breakdown in the past). This dependence on the past ("path dependence") is also called *hysteresis*.

In order that a perturbation can develop to a persistent jam, the outflow from the congested region must be smaller than the inflow, i.e., the bottleneck capacity under congested conditions (also known as *active* or *activated bottleneck*) must be smaller than the maximum possible flow through the bottleneck under free-flow

[8] The distinction may become relevant for models with non-smooth or even non-continuous acceleration functions. Typically, this is the case when the model formulation involves several distinct traffic regimes (e.g., Gipps' model or the Wiedemann model). Such models may be Ljapunov but not asymptotically stable.

conditions (*static capacity*). Observed values for the difference between the static and dynamic capacities, the so-called *capacity drop*, are of the order of 10 % (cf. Chap. 4). Consequently, the fundamental diagram is not unique for densities in the metastable range. Instead, there are two values for the flow, a higher one for free traffic, and a lower one for congested traffic. For the graph of the fundamental diagram, this leads to the characteristic shape of a mirrored Greek λ, also referred to as the *inverse lambda* shape,[9] cf. Figs. 4.11 and 4.12.

Formally, we define linear and nonlinear string instability in macroscopic terms by considering an infinite system initially in steady state at density ρ_e and looking at the spatiotemporal development of the response $U(x,t)$ of a temporary and localized perturbation $U_\varepsilon(x,0)$ of amplitude ε denoting, e.g., the difference between the actual and steady-state local speed fields. If the initial perturbation corresponds to a sudden change ε of speed of a single vehicle located at $x = 0$, the macroscopic initial perturbation $U_\varepsilon(x,0)$ of the speed field is

$$U_\varepsilon(x,0) = U_\varepsilon(x) = \begin{cases} \varepsilon & \text{if } |x| < \frac{1}{2\rho_e}, \quad \varepsilon > 0, \quad x \in IR, \\ 0 & \text{otherwise.} \end{cases} \tag{15.1}$$

This means, the speed field is changed by ε in a region whose width $\Delta x = 1/\rho_e$ corresponds to the distance between two vehicles, i.e., to the effective space attributed to one vehicle. Traffic flow is linearly unstable if

$$\lim_{t\to\infty} \max_x U(x,t) > 0 \quad \text{for all } \varepsilon > 0. \tag{15.2}$$

It is nonlinearly unstable or *metastable*, if there exists a minimum perturbation amplitude $\varepsilon_{\text{nl}} > 0$ such that

$$\lim_{t\to\infty} \max_x U(x,t) = \begin{cases} U_0 > 0 & \text{if } \varepsilon > \varepsilon_{\text{nl}}, \\ 0 & \text{if } \varepsilon \in [0, \varepsilon_{\text{nl}}]. \end{cases} \tag{15.3}$$

As illustrated in Fig. 15.3, the limit between linear instability and metastability is defined by $\varepsilon_{\text{nl}} \to 0$, while the limit between metastability and absolute stability is given by Eq. (15.3) for the limit of a maximum perturbation, e.g., $|\varepsilon| = V$ (braking to a complete stop).[10]

Propagation of the perturbation: absolute versus convective instability. If traffic flow is (linearly or nonlinearly) string unstable, the region of perturbations as considered from a stationary observer can propagate in both directions (*absolute string instability*), or exclusively upstream or downstream which is termed *upstream* and *downstream convective instability*, respectively (see Fig. 15.4).

Convective instabilities were originally observed in open systems of fluid flows such as water in pipes. In this case, the convective instability is of the downstream

[9] Although this is not correct: The Greek λ is mirrored and not upside down.

[10] To make the perturbation more massive, the duration of the perturbation must be increased such that it results in a fully-formed initial jam.

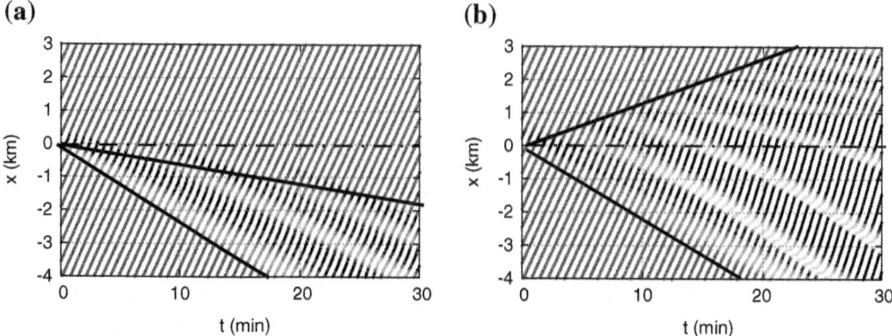

Fig. 15.4 Visualization of the spatiotemporal evolution of **(a)** convective upstream string instability, and **(b)** absolute string instability by vehicle trajectories in a space–time plot. Shown are IDM simulations with $l = 5$ m, $v_0 = 120$ km/h, $s_0 = 2$ m, and $b = 1.5\,\mathrm{m/s^2}$, $T = 1.5$ s, and acceleration parameters $a = 1.1\,\mathrm{m/s^2}$ and $a = 0.9\,\mathrm{m/s^2}$ for the plots **(a)** and **(b)**, respectively. Shown are the trajectories of every 20th vehicle

type: perturbations leave the system together with the fluid after some time, i.e., they are *convected out* of the system.[11]

In traffic flow, however, one observes that perturbations generally grow *against* the driving direction and leave the system, i.e., the road section under consideration, by the upstream boundary. Of course, this is particularly true for stop-and-go traffic waves moving backwards at a constant velocity (cf. Sect. 18.3). We emphasize that this propagation direction is not obvious: While the asymmetric interactions of drivers (reacting essentially to the leading and hardly to the following vehicle) ensure that, when considering a system *comoving with the drivers*, string instability is always of the upstream convective type[12] (cf. Figs. 15.4a and 15.10) both types of convective instability are theoretically plausible in the fixed system. In fact, both types can be reproduced in simulations. However, downstream convective instability is not robust against nonlinear effects (cf. Fig. 15.5), so only upstream convective instability is actually observed.

The distinction between convective and absolute instability is relevant since traffic flow relates to an *open system* where absolute and convective instability leads to qualitatively different congestion patterns:

- If traffic flow is absolutely string unstable, the perturbed region will sooner or later cover the whole road section under consideration.
- If traffic flow is convectively string unstable, the perturbations eventually will leave the system. Thus, in a given section, oscillations resulting from temporary perturbations are not persistent *even in the presence of linear instability*. If there are persistent local perturbations (e.g., lane changes near ramps or lane closures), the oscillations are, of course, persistent as well. However, they are small near the

[11] This technical term originates from the Latin *convehi*: to move together.

[12] At least, if traffic flow is locally stable which is safe to assume.

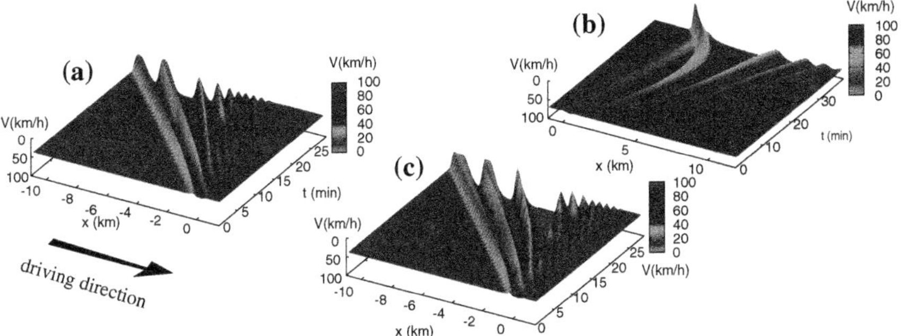

Fig. 15.5 Speed response functions for the initial localized perturbation (15.1). Shown are IDM simulations of **(a)** convective instability propagating upstream, **(b)** linear convective instability propagating downstream which is destroyed by nonlinearities, and of **(c)** the limit between convective and absolute instability

location of the perturbations (generally at a bottleneck), and increase in amplitude further upstream. All this is markedly different in closed systems (ring roads) where there is no qualitative long-term difference between these stability types.

Convective instability is a widespread phenomenon. For example, *all* oscillations and traffic waves on the German highway A5 (Fig. 5.1) are the consequence of convective instability driven by persistent perturbations near the bottlenecks. In contrast, city traffic flow generally is stable and stop-and-go conditions are the trivial consequence of the operations of traffic lights. In Sect. 15.5 we show that string instability *always* starts as a convective instability (cf. Fig. 15.3). From a multitude of observations, we conclude the following:

> The vast majority of all instabilities of highway traffic flow is of the convective type.

Formally, one can define convective instability in terms of the dynamics of the perturbation field $U(x,t)$ for a given localized and temporary initial perturbation $U_\varepsilon(x)$ according to Eq. (15.1): Homogeneous flow is convectively unstable with respect to this perturbation, if

$$\lim_{t\to\infty} \max_x U(x,t) > 0 \quad \text{and} \quad \lim_{t\to\infty} U(0,t) = 0. \tag{15.4}$$

The first condition is true if traffic flow is (linearly or nonlinearly) string unstable, i.e., the initial perturbation does not decay to zero, at least somewhere in the system. The second condition states that the perturbations eventually vanish at the location of the initial triggering point.[13]

[13] More generally, the perturbations eventually vanish at *any* fixed location x for $t \to \infty$.

By analogy one can define absolute string instability by requiring the first condition to be true, and the second to be false. By suitably combining these conditions with Eq. (15.3), one can define in a straightforward way convective and absolute linear instability and metastability (nonlinear instability).

Stability in stochastic models. We can determine the behavior in the presence of acceleration noise or other fluctuations described by stochastic models from the above definitions of convective and nonlinear instability:

- If the noise is of sufficiently low amplitude to allow a linear analysis, we obtain persistent fluctuations for any type of linear string or flow instability. In contrast, if a deterministic model describes convectively unstable traffic flow in open systems, all initial perturbations are eventually convected out of the system. This means, small fluctuations change the qualitative behavior in such systems while they have not much influence, otherwise.
- If the noise is of sufficient amplitude to warrant a nonlinear description, it can trigger nonlinear instabilities. This means, larger-amplitude noise can change the qualitative system behavior with respect to the deterministic description if the system is convectively or absolutely metastable.

Wavelength of the perturbations. Since traffic flow represents an extended system which can be abstracted to an infinitely long homogeneous road, there is, in principle, an infinite multitude of perturbations leading to instabilities. In Sect. 15.4, we show, that the perturbations can be arranged in two branches or "modes" of periodic perturbations with arbitrary real-valued wavelengths. However, we can only observe the perturbations becoming first unstable when increasing the traffic density (or making the model more unstable): Once nonlinearities become effective (saturation, capacity drop, reversal of the propagation velocity), all other perturbations are suppressed.

Depending on the nature of the onset of the "first" instability, we distinguish two categories: In the presence of *short-wavelength instabilities*, the first instability has a finite and typically short wavelength of only a few vehicle distances, i.e., each wave consists of only a few vehicles. In contrast, if there is a *long-wavelength instability*, the wavelength of the "first" unstable perturbation tends to infinity. Since vehicle conservation implies that the growth rate tends to zero when the wavelength tends to infinity *regardless of the degree of (in-)stability*, the practically observed waves originating from long-wavelength instabilities, i.e., the perturbations with maximum growth rate, are large but finite (of the order of 1 km or more). Mathematicians have shown that instabilities are always of the long-wavelength type

- for continuous-in-time car-following models containing no explicit reaction time (such as the OVM or the IDM)
- for second-order local macroscopic models such as the Payne Model or the Kerner–Konhäuser Model.

In contrast, the instability may be (but need not to be) of the short-wavelength type if

- time is discrete (iterated maps, e.g., Gipps' model),

- explicit reaction times are modeled (HDM, cf. Sect. 12.2),
- or nonlocal macroscopic models (such as the GKT model) are considered.

This means, the first instability may be of the short-wavelength type if the model contains some nonlocalities in space or time. Since observed instabilities are always of the long-wavelength type, one can restrict the further analysis to this category. Conversely, if one observes short-wavelength instabilities in the simulations,[14] this must be considered as an artifact of the model, or the consequence of an erroneous (or erroneously parameterized) numerical integration method.

15.3 Local Instability

We consider a situation where a leading vehicle drives at constant speed and investigate small changes $y(t)$ and $u(t)$ of the gap and speed of a single follower with respect to the steady-state equilibrium:

$$s(t) = s_e + y(t), \tag{15.5}$$

$$v(t) = v_e + u(t). \tag{15.6}$$

When analyzing local stability, it is essential that the leading vehicle does not exhibit persistent perturbations since the question whether persistent perturbations are amplified when transferred to the following vehicles refers to string instability. Furthermore, instead of considering an initial perturbation of the leader, we can investigate an unperturbed leader and an initial perturbation of the follower as specified above. Inserting this ansatz into the general formulation (10.3), (10.6) of time-continuous models, we obtain, in zeroth order of the perturbations ($y = u = 0$), the steady-state conditions

$$a_{\text{mic}}(s_e, v_e, v_e) = 0 \quad \text{and} \quad \tilde{a}_{\text{mic}}(s_e, v_e, 0) = 0 \tag{15.7}$$

for the two forms a_{mic} and $\tilde{a}_{\text{mic}}$ of the acceleration function, respectively (cf. Sect. 10.3). These conditions define the *microscopic fundamental diagram* in terms of the steady-state gap s_e for a certain constant speed v_e which can be written as $v_e(s_e)$ (steady-state speed for a given gap), or $s_e(v_e)$ (steady-state gap for a given speed).

In first order of the perturbations y and u of the follower, we obtain, for models defined by the acceleration function a_{mic}, the following system of ordinary linear differential equations:

$$\frac{\mathrm{d}y}{\mathrm{d}t} = u_l - u = -u, \tag{15.8}$$

$$\frac{\mathrm{d}u}{\mathrm{d}t} = a_s y + a_v u + a_{v_l} u_l = a_s y + a_v u. \tag{15.9}$$

---

[14] For example, Gipps' model in its original formulation exhibits a short-wavelength instability with the smallest possible wavelength of two car distances.

Notice that the assumed constant speed of the leading vehicle implies $u_l = 0$. The coefficients a_s, a_v, and a_{v_l} of the linearization originate from a first-order Taylor expansion of the acceleration function a_{mic} with respect to its three independent variables around the steady-state equilibrium,

$$a_{\text{mic}}(s, v, v_l) = a_{\text{mic}}(s_e, v_e, v_e) + a_s y + a_v u + a_{v_l} u_l + \text{ higher orders} \tag{15.10}$$

with the expansion coefficients

$$\boxed{a_s = \left.\frac{\partial a_{\text{mic}}}{\partial s}\right|_e, \quad a_v = \left.\frac{\partial a_{\text{mic}}}{\partial v}\right|_e, \quad a_{v_l} = \left.\frac{\partial a_{\text{mic}}}{\partial v_l}\right|_e.} \tag{15.11}$$

The subscript e denotes that the derivatives are evaluated at the steady-state point $s = s_e$ and $v = v_l = v_e(s_e)$.

By virtue of condition (15.7) describing a one-dimensional manifold of steady-state solutions $v_e(s)$, the three Taylor coefficients are not independent of each other. Moving along the space of steady-state solutions by simultaneously changing s and $v = v_l$ must not change the acceleration (which is always zero), i.e.,

$$a_s \mathrm{d}s_e + \left(a_v + a_{v_l}\right) \mathrm{d}v_e = a_s \mathrm{d}s_e + \left(a_v + a_{v_l}\right) v_e'(s_e) \mathrm{d}s_e = 0 \tag{15.12}$$

resulting in

$$a_s = -v_e'(s_e)(a_v + a_{v_l}). \tag{15.13}$$

Expanding the general acceleration equation (10.3) for the alternative acceleration function $\tilde{a}_{\text{mic}}(s, v, \Delta v)$ to first order leads to the linear system

$$\frac{\mathrm{d}y}{\mathrm{d}t} = u_l - u = -u, \tag{15.14}$$

$$\frac{\mathrm{d}u}{\mathrm{d}t} = \tilde{a}_s y + (\tilde{a}_v + \tilde{a}_{\Delta v})\, u - \tilde{a}_{\Delta v} u_l = \tilde{a}_s y + (\tilde{a}_v + \tilde{a}_{\Delta v})\, u. \tag{15.15}$$

with the Taylor expansion coefficients

$$\boxed{\tilde{a}_s = \left.\frac{\partial \tilde{a}_{\text{mic}}}{\partial s}\right|_e, \quad \tilde{a}_v = \left.\frac{\partial \tilde{a}_{\text{mic}}}{\partial v}\right|_e, \quad \tilde{a}_{\Delta v} = \left.\frac{\partial \tilde{a}_{\text{mic}}}{\partial \Delta v}\right|_e.} \tag{15.16}$$

Comparing Eq. (15.9) with Eq. (15.15), it is evident that one needs to consider only one formulation of the acceleration function which we chose to be $a_{\text{mic}}(s, v, v_l)$. Formulations for the alternative acceleration function $\tilde{a}_{\text{mic}}(s, v, \Delta v)$ can be obtained from that for a_{mic} by the following set of replacements:

$$a_s = \tilde{a}_s, \quad a_v = \tilde{a}_v + \tilde{a}_{\Delta v}, \quad a_{v_l} = -\tilde{a}_{\Delta v}. \tag{15.17}$$

This is valid for all expressions in this chapter, including these for string instability. As an example, when applying the replacement rules to the steady-state condition (15.13), we obtain

$$\tilde{a}_s = -v_e'(s)\tilde{a}_v \tag{15.18}$$

for the interdependence of the expansion coefficients of $\tilde{a}_{\text{mic}}$.

Equations (15.8) and (15.9) describe a harmonic damped oscillator. To see this explicitly, we write them as a single second-order differential equation by taking the time derivative of Eq. (15.8) and inserting Eq. (15.9),

$$\frac{\mathrm{d}^2 y}{\mathrm{d}t^2} + 2\eta \frac{\mathrm{d}y(t)}{\mathrm{d}t} + \omega_0^2 y(t) = 0. \tag{15.19}$$

The damping constant η and the angular oscillation frequency ω_0 are given by

$$\eta = -\frac{a_v}{2} = -\frac{(\tilde{a}_v + \tilde{a}_{\Delta v})}{2}, \quad \omega_0^2 = a_s = \tilde{a}_s. \tag{15.20}$$

Assuming the exponential ansatz

$$y = y_0 \mathrm{e}^{\lambda t} \tag{15.21}$$

we arrive at the quadratic equation

$$\lambda^2 + 2\eta\lambda + \omega_0^2 = 0 \tag{15.22}$$

for the (generally complex) growth rate $\lambda = \sigma + \mathrm{i}\omega$ ($\mathrm{i} = \sqrt{-1}$ is the imaginary unit) with the solutions

$$\lambda_{1/2} = -\eta \pm \sqrt{\eta^2 - \omega_0^2}. \tag{15.23}$$

The dynamics of the follower is locally stable if both solutions decay, i.e., the real parts are negative, $\sigma_{1/2} = \mathrm{Re}(\lambda_{1/2}) \leq 0$. This is satisfied if $\eta > 0$, or, with the definitions (15.20)

$$\boxed{a_v < 0 \quad \text{or} \quad \tilde{a}_v + \tilde{a}_{\Delta v} < 0 \quad \text{Local stability.}} \tag{15.24}$$

Since, by virtue of condition (11.1), $a_v < 0$ for all plausible models, we conclude that time-continuous car-following models without additional delay by explicit reaction times are unconditionally locally stable.

As a more restrictive condition on the local behavior, we can require that all deviations from the steady-state decay are without oscillations, not even damped ones. This is the case if the imaginary parts of the growth rates are zero leading to

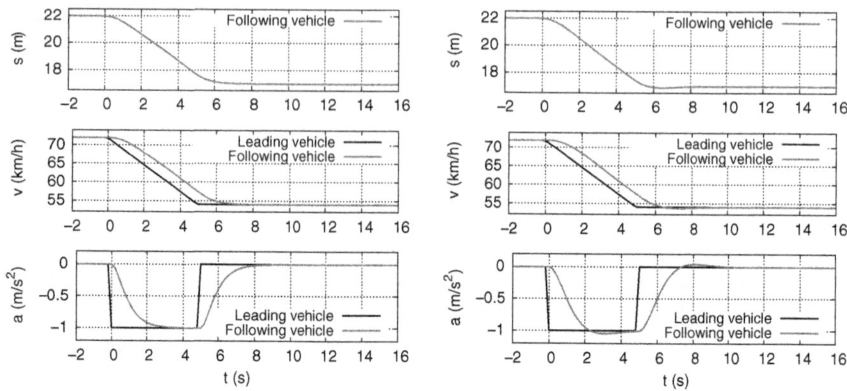

Fig. 15.6 Response of an OVM vehicle to a speed reduction of the leading vehicle (driving a fixed speed profile) from 72 to 54 km/h. *Left* at the limit of an oscillation-free response ($v_e' = 1\,\mathrm{s}^{-1}$, $\tau = 0.25\,\mathrm{s}$). *Right* limit of string instability ($v_e' = 1\,\mathrm{s}^{-1}$, $\tau = 0.5\,\mathrm{s}$)

$\omega_0^2 < \eta^2$, or $a_s \leq a_v^2/4$. Expressing a_s by the sensitivity $v_e'(s)$ to changes of the gap, we obtain the following no-oscillation conditions (cf. the left column of Fig. 15.6):

$$a_s \leq \frac{a_v^2}{4} \quad \text{or} \quad v_e'(s) \leq \frac{-\tilde{a}_v}{4}\left(1 + \frac{\tilde{a}_{\Delta v}}{\tilde{a}_v}\right)^2 \quad \text{No local oscillations.} \tag{15.25}$$

Here, the transformation rules (15.17) and (15.18) have been applied to arrive at the second condition for models given in terms of the acceleration function $\tilde{a}_{\text{mic}}(s, v, \Delta v)$. In summary, we can make the following statements on local instability:

- Since $a_v < 0$ for all sensible models and the above considerations are valid for time-continuous car-following models without explicit reaction-time delay, such models are *always* locally stable. However, this need not to be the case for iterated maps (Gipps' Model), or when considering explicit reaction times by delay-differential equation as in the HDM.
- The more restrictive no-oscillation condition (15.25) is not always satisfied. For example, we obtain for the Optimal Velocity Model[15] the condition

$$v_e'(s)_{\text{OVM}} < \frac{1}{4\tau}. \tag{15.26}$$

This condition is more restrictive as the condition $v_e'(s)_{\text{OVM}} < 1/(2\tau)$ for string stability to be derived in the following section. As can be seen by the derivation, this relation between the thresholds of over-damped local stability and string stability

[15] The partial derivatives of the acceleration function are $a_v = -1/\tau$, $a_{v_l} = 0$, $a_s = -v_e'(s)a_v$.

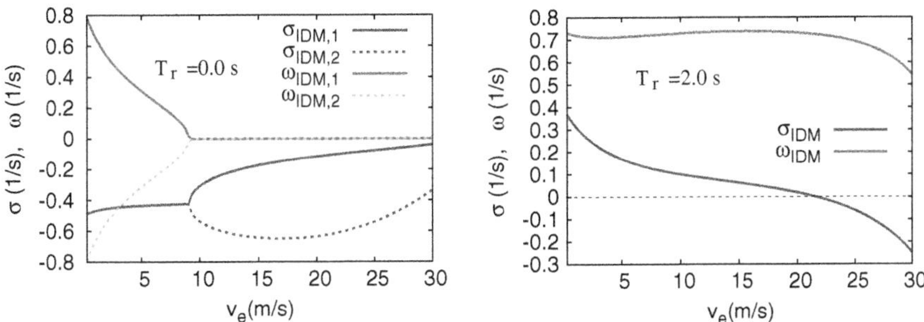

Fig. 15.7 *Left* the two branches of the linear growth rate $\lambda = \sigma + \mathrm{i}\,\omega$ according to Eq. (15.23) for the conventional IDM with the standard highway parameters of Table 11.2 as a function of the steady-state speed. *Right* most unstable branch of the solutions to Eq. (15.27) for the IDM with an additional delay by the reaction time $T_r = 2.0$ s (no other human driver properties added)

is valid for any car-following model without sensitivity to speed differences that is formulated by ordinary differential equations.

- Near the threshold to string instability, the oscillations of a single vehicle when approaching the local steady state are hardly recognizable (cf. right column of Fig. 15.6). Reasoning in the converse direction, we conclude that when a vehicle driving with adaptive cruise control shows recognizable oscillations, it is nearly certain that traffic flow consisting of such vehicles is string unstable, even if the oscillations of the single vehicle are strongly damped. Considering models with the speed difference as exogenous factor, the model may even be *completely* free of oscillations in the local context, and simultaneously string unstable when considering traffic flow with many vehicles (cf. Problem 15.3).

In contrast to time-continuous models of the form (10.3) as discussed above, time-continuous models with delay, i.e., delay-differential equations of the form (12.1) modeling a *finite reaction time*, or time-discrete models (iterated maps) of the form (10.7) may become locally unstable. Performing the same stability analysis as above for models of the form (12.1), i.e., models whose acceleration equation is of the form $\frac{\mathrm{d}}{\mathrm{d}t} v(t + T_r) = \tilde{a}_{\text{mic}}(s(t), v(t), \Delta v(t))$, we obtain

$$\lambda^2 + \mathrm{e}^{-\lambda T_r} \left(2\eta\lambda + \omega_0^2\right) = 0. \tag{15.27}$$

In spite of its simple appearance, solving this equation for the growth rate $\lambda = \sigma + \mathrm{i}\,\omega$ is nontrivial and can be done only numerically. For sufficiently high delay times (more than 2.0 s for the IDM with the highway parameters of Table 11.2), the real part σ of the most unstable solution becomes positive for some steady-state situations, i.e., the model becomes locally unstable (Fig. 15.7).

15.4 String Instability

Even if a system consisting of a single or a few vehicles following a leader with a fixed speed profile is well within the stable range, the oscillations may increase with each following vehicle, i.e., traffic flow is *string unstable* (cf. Figs. 15.1, 15.2 and 15.8).

Generally, the resulting oscillations or waves have a wavelength of 1 km or more, i.e., a single wave contains many vehicles corresponding to a *long-wavelength instability* This is fortunate since it allows compact analytical expressions for the stability thresholds of time-continuous car-following models and macroscopic models.

15.4.1 String Instability Conditions for Car-Following Models

We start with the general formulation (10.3), (10.6) of time-continuous car-following models without delay and without multi-anticipation.[16] Furthermore, we consider identical driver-vehicles on a homogeneous infinite road, i.e., the same acceleration functions and identical parameter sets for all vehicles. The set of coupled equations for the gap s_α and the speed v_α reads

$$\frac{\mathrm{d}s_\alpha}{\mathrm{d}t} = v_{\alpha-1} - v_\alpha, \tag{15.28}$$

$$\frac{\mathrm{d}v_\alpha}{\mathrm{d}t} = a_{\text{mic}}\left(s_\alpha(t), v_\alpha(t), v_{\alpha-1}(t)\right). \tag{15.29}$$

As in the analysis for local instability, we assume, for all vehicles α, small deviations y_α and u_α from the steady-state gap s_e and speed v_e, respectively,

$$s_\alpha = s_e + y_\alpha(t), \tag{15.30}$$

$$v_\alpha = v_e + u_\alpha(t). \tag{15.31}$$

In zeroth order with respect to y_α and u_α, we obtain the same result as for the local analysis: The microscopic fundamental diagram $v_e(s)$ and the relations (15.13), (15.17) and (15.18) remain valid.

In first order, we obtain the following system of coupled linear differential equations with constant coefficients:

$$\frac{\mathrm{d}y_\alpha}{\mathrm{d}t} = u_{\alpha-1} - u_\alpha, \tag{15.32}$$

[16] Including multi-anticipation, i.e., considering several leading (or trailing) vehicles, does not pose any technical problem: Starting the investigation from Eq. (12.19) with Eq. (12.15) and proceeding (footnote 16 Continued) along the following lines is straightforward. We restrict ourselves to a single leader so as to not clutter the presentation.

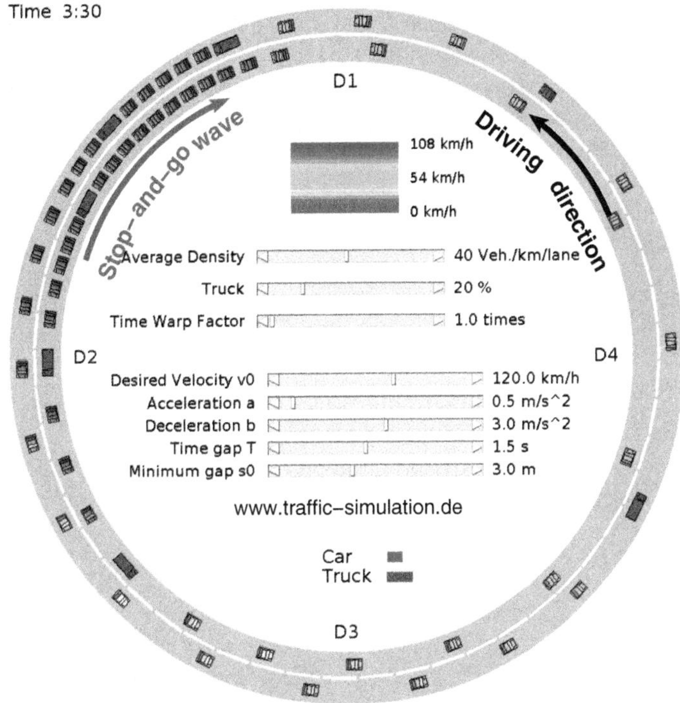

Fig. 15.8 Interactive simulation of stop-and-go waves with the Intelligent Driver Model (IDM) on the authors' website

$$\frac{\mathrm{d}u_\alpha}{\mathrm{d}t} = a_s y_\alpha + a_v u_\alpha + a_{v_l} u_{\alpha-1}, \tag{15.33}$$

where the partial derivatives a_s, a_v, and a_{v_l} are given by (15.11). We emphasize that, in contrast to the local stability analysis, the perturbation $u_{\alpha-1}$ constituting the coupling between the vehicles cannot be set to zero. The set of Eqs. (15.32) and (15.33) can be solved using the *Fourier-Ansatz*

$$\begin{pmatrix} y_\alpha(t) \\ u_\alpha(t) \end{pmatrix} = \begin{pmatrix} \hat{y} \\ \hat{u} \end{pmatrix} \mathrm{e}^{\lambda t + \mathrm{i}\alpha k}. \tag{15.34}$$

This ansatz corresponds to linear waves of strict periodicity and contains the following elements:

- $\mathrm{i} = \sqrt{-1}$ is the imaginary unit.
- $\lambda = \sigma + \mathrm{i}\omega$ is the complex growth rate. The real part σ denotes the growth rate of the oscillation amplitude while the imaginary part ω indicates the angular frequency *from the perspective of the driver*. The driver passes a complete wave in the time $2\pi/\omega$.

- The dimensionless *wave number* $k \in [-\pi, \pi]$ indicates the phase shift of the traffic waves from one vehicle to the next at a given time instant. Consequently, the number of vehicles per wave is given by $2\pi/k$. Since the steady-state distance between the front bumpers of two vehicles is equal to $s_e + l = 1/\rho_e$, the physical wavelength is given by $(s_e + l)2\pi/k$.
- The phase velocity is defined by the movement of points of constant phase, i.e., by a constant imaginary part $\omega t + \alpha k$ of the exponent of Eq. (15.34). This gives rise to following quantities:
  - The *passing rate*
  $$\dot{\alpha} = -\frac{\omega}{k} \tag{15.35}$$
  denotes the vehicle flux through the waves in a coordinate system moving with the waves,[17] i.e., with points of constant phase $\omega t + \alpha k$.[18]
  - In physical space, the relative propagation velocity *in the system comoving with the vehicles* is given by[19]
  $$\tilde{c}_{\text{rel}}(k) = (s_e + l)\frac{\omega}{k} = \frac{\omega}{\rho_e k}. \tag{15.36}$$
  - In the fixed system of a stationary observer at the road side, the steady-state speed of the vehicles has to be added to the relative velocity,
  $$\tilde{c}(k) = v_e(s_e) + \tilde{c}_{\text{rel}}(k). \tag{15.37}$$
  This road-based propagation velocity is the one that can be derived from traffic data. In order to be consistent with observations, the long-wavelength limit $\tilde{c} = \lim_{k\to 0} \tilde{c}(k)$ should be of the order of -15 km/h, in congested situations.
- The traffic waves include periodic changes of both gap and speed. The fraction $\hat{u}/\hat{y}$ of the prefactors indicates the relation between the respective amplitudes. For example, a traffic wave described by $\hat{y} = 0$ would consist of speed changes, only.

Inserting the *traffic wave ansatz* (15.34) in the linear system (15.32), (15.33) results in

$$\begin{pmatrix} \lambda & 1 - \mathrm{e}^{-ik} \\ -a_s & \lambda - \left(a_v + a_{v_l}\mathrm{e}^{-ik}\right) \end{pmatrix} \cdot \begin{pmatrix} \hat{y} \\ \hat{u} \end{pmatrix} = 0. \tag{15.38}$$

---

[17] This technical term has to be distinguished from passing in the sense of overtaking which is completely unrelated.

[18] Beware of the signs: Assuming positive wave numbers k and our convention to assign to the first vehicle the lowest vehicle index, the angular frequency will generally be negative. This means, the passing rate is positive: The waves propagate in the direction of increasing vehicle indices, i.e., opposite to the movement of the vehicles.

[19] The sign is reversed with respect to Eq. (15.35), i.e., negative. The waves propagate upstream, i.e., in negative x direction.

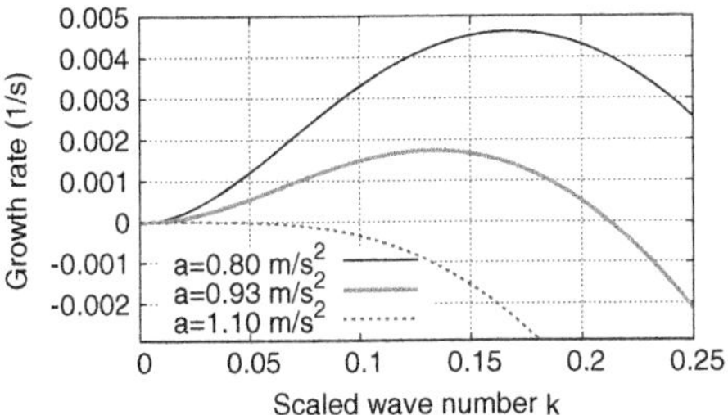

Fig. 15.9 Linear growth rate $\lambda(k)$ of the more unstable branch of perturbations (15.40) of a steady state corresponding to congested traffic ($v_e = 48$ km/h) for the Intelligent Driver Model (IDM) as a function of the wave number (phase shift) k for three values of the IDM acceleration parameter a. The remaining IDM parameters are $v_0 = 120$ km/h, $T = 1.5$ s, $s_0 = 2$ m, $b = 1.3\,\mathrm{m/s^2}$, and the vehicle length is $l = 5$ m

This linear-homogeneous 2×2 system for the amplitudes has only nontrivial solutions if the determinant of the matrix of coefficients is equal to zero. The resulting solvability condition assumes the form of a quadratic equation

$$\lambda^2 + p(k)\lambda + q(k) = 0 \tag{15.39}$$

for the complex growth rate λ with solutions given by (cf. Fig. 15.9)

$$\lambda_{1/2}(k) = -\frac{p(k)}{2}\left(1 \pm \sqrt{1 - \frac{4q(k)}{p^2(k)}}\right), \tag{15.40}$$

where

$$p(k) = -a_v - a_{v_l}\mathrm{e}^{-ik}, \tag{15.41}$$
$$q(k) = a_s\left(1 - \mathrm{e}^{-ik}\right).$$

For a given phase shift k between two consecutive vehicles, only two growth rates are possible. For each of the growth rates, the actual solution $(\hat{y}, \hat{u})$ (the *eigenvector*) gives the amplitudes and the phases of the gap and speed oscillations. Since the eigenvector is only defined up to a (complex) common factor, it essentially gives the relation of the amplitudes and the relative phase between the speed and gap oscillations. Typically, one solution is fast decaying (Re λ is strongly negative) while the other decays more slowly, or even grows. The latter solution branch which is also called the "slow mode" will be of interest. The model is string stable if the real parts $\sigma = \mathrm{Re}\,\lambda$ are negative

- for both solution branches $\lambda_1(k)$ and $\lambda_2(k)$ and
- for all relative phase shifts (wave numbers) in the range $k \in [-\pi, \pi]$.

Mathematically, one can prove that the first instability of time-continuous models without delay times always occurs for phase shifts $k \to 0$, i.e., for the limit of the wavelengths and periods going to infinity. Since only waves of a finite wavelength can have finite growth rates, the resulting wavelengths are finite but consist of many vehicles. We illustrate this by Fig. 15.9. The middle curve corresponds to a maximum of the growth rate at $k_0 \approx 0.13$ corresponding to $2\pi/k_0 \approx 50$ vehicles per wave This means, the instability is of the *long-wavelength type*, in agreement with observations of real traffic waves.

Restrict the further investigations to wave numbers $|k| \ll \pi$, we expand the coefficients of the quadratic equation for $\lambda(k)$ in a Taylor series around $k = 0$[20]:

$$\begin{aligned} p(k) &= p_0 + p_1 k + \mathcal{O}(k^2), \\ q(k) &= q_1 k + q_2 k^2 + \mathcal{O}(k^3), \end{aligned} \tag{15.42}$$

with

$$\begin{aligned} p_0 &= -(a_v + a_{v_l}) = -\tilde{a}_v, \\ p_1 &= \mathrm{i} a_{v_l} = -\mathrm{i} \tilde{a}_{\Delta v}, \\ q_1 &= \mathrm{i} a_s = \mathrm{i} \tilde{a}_s = \mathrm{i} v_e'(s_e) p_0, \\ q_2 &= \frac{a_s}{2} = \frac{\tilde{a}_s}{2} = \frac{v_e'(s_e)}{2} p_0. \end{aligned} \tag{15.43}$$

The prefactors p_0 and q_2 are real-valued while p_1 and q_1 are purely imaginary. Notice that the expressions for q_1 and q_2 on the right-hand sides of the last equal sign follow from Eqs. (15.13) and (15.18). Since there are no zero-order terms of $q(k)$, and the general criteria for sensible microscopic models imply that $p_0 = -\tilde{a}_v$ is strictly positive, the real part of λ can (in lowest order) become positive only for the solution with the negative sign of the square root of Eq. (15.40). Expanding this solution around $k = 0$ to quadratic order making use of the expansion

$$\sqrt{1-\varepsilon} = 1 - \frac{1}{2}\varepsilon - \frac{1}{8}\varepsilon^2 + \mathcal{O}(\varepsilon^3) \tag{15.44}$$

for complex-valued ε, we arrive at the general expression

$$\lambda = -\frac{q_1}{p_0} k + \left(\frac{q_1 p_1}{p_0^2} - \frac{q_2}{p_0} - \frac{q_1^2}{p_0^3} \right) k^2 + \mathcal{O}(k^3), \tag{15.45}$$

---

[20] The "order" symbol $\mathcal{O}(\cdot)$ defines how fast the symbolized contributions converge to zero. Specifically, if a contribution $f(k)$ is of the order $\mathcal{O}(k^\gamma)$, then $\lim_{k\to 0} k^{-\gamma} f(k)$ is finite, and $\lim_{k\to 0} k^{-\gamma+\varepsilon} f(k) = 0$ for any positive real-valued ε.

which is also valid for the second-order macroscopic models to be discussed below. Inserting into this expression the coefficients (15.43) for time-continuous microscopic models finally gives

$$\lambda = -\mathrm{i}v_e'(s_e)\,k + \frac{v_e'(s_e)}{a_v + a_{v_l}}\left[\frac{1}{2}(a_{v_l} - a_v) - v_e'(s_e)\right]k^2 + \mathcal{O}(k^3). \tag{15.46}$$

The growth rate λ tends to zero for $k = 0$. This is a direct consequence of the continuity equation. By virtue of the conservation of the number of vehicles, traffic waves of infinite wavelength (or wave number $k = 0$) cannot dissolve since there is simply no way for the vehicles to leave the wave.

The contribution linear in k is purely imaginary and therefore describes the *propagation properties* of the waves for small phase shifts between consecutive vehicles. Setting $\lambda = \sigma + \mathrm{i}\omega$ and writing $\omega = -v_e'(s_e)k + \mathcal{O}(k^3)$, we arrive at the following simple expression for the relative propagation velocity (15.36) of the traffic waves:

$$\tilde{c}_{\mathrm{rel}}(k) = (s_e + l)\frac{\omega}{k} = -(s_e + l)v_e'(s_e) + \mathcal{O}(k^2). \tag{15.47}$$

Notice that, in this equation, the acceleration function of the model enters only indirectly via the gradient $v_e'(s) = -\tilde{a}_s/\tilde{a}_v$ of the microscopic fundamental diagram $v_e(s)$. Since $v_e'(s) \ge 0$ and, consequently, $\tilde{c}_{\mathrm{rel}} \le 0$, the waves propagate against the direction of the flow, at least in the coordinate system comoving with the drivers. This is plausible since the considered class of car-following models represents drivers reacting only to the leading but not to the following vehicle. In the limit of completely interaction-free traffic corresponding to $v_e = v_0, v_e'(s) = 0$, we have $\tilde{c}_{\mathrm{rel}} = 0$, i.e., the waves move with the vehicles. In fact, the waves can be interpreted as independently moving vehicle clusters, in this limiting case.

The second-order contribution of the growth rate (15.46) is purely real, and therefore describes the *growth properties* of the waves. In particular, traffic flow is string stable if this term is negative. Since $v'(s_e) \ge 0$ and $a_v + a_{v_l} = \tilde{a}_v < 0$, string stability implies that the bracket of Eq. (15.46) is positive. This results in the following criterion for string stability:

$$\boxed{v_e'(s_e) \le \frac{1}{2}\left(a_{v_l} - a_v\right) \quad \text{String stability for } \dot{v}_\alpha = a_{\mathrm{mic}}(s_\alpha, v_\alpha, v_{\alpha-1}).} \tag{15.48}$$

For models whose acceleration function is of the form $\tilde{a}(s, v, \Delta v)$, we apply the replacements (15.17) and obtain the alternative condition

$$\boxed{v_e'(s_e) \le -\frac{\tilde{a}_v}{2} - \tilde{a}_{\Delta v} \quad \text{String stability for } \dot{v}_\alpha = \tilde{a}_{\mathrm{mic}}(s_\alpha, v_\alpha, \Delta v_\alpha).} \tag{15.49}$$

Discussion. The above criteria for string stability directly point to the three main factors determining the stability of traffic flow with respect to collective perturbations. It is most convenient to extract these factors from the formulation (15.49).

Firstly, a necessary condition for string *instability* is a sufficient sensitivity $v_e'(s) \geq 0$ to changes of the gap (left-hand side of Eq. 15.49): Without this sensitivity, there is no feedback, and the instability mechanism discussed qualitatively in Sect. 15.1 would break down already in the first step. The drivers simply ignore the vehicles in front of them. Since this is a plausible behavior for low traffic densities, only, it explains why a minimum traffic flux and density is necessary for generating traffic flow instabilities.

Secondly, string instability implies that the sensitivity $-\tilde{a}_v/2 > 0$ to speed changes, i.e., the first term of the right-hand side of Eq. (15.49) remains below a certain threshold $v_e'(s_e) + \tilde{a}_{\Delta v}$. In terms of the driver's behavior this means that *responsive* or *agile* drivers corresponding to high values of $-\tilde{a}_v$ tend to suppress string instabilities.

Thirdly, string instabilities are only possible if the sensitivity $-\tilde{a}_{\Delta v}$ to speed *differences* remains below a certain threshold. In agreement with common sense, drivers without any sensitivity to speed differences drive very short-sightedly and tend to make traffic flow more unstable. Since future gaps can be estimated by speed differences, one can conclude that $-\tilde{a}_{\Delta v}$ describes a simple form of anticipation.

In summary, the stability analysis shows that the factors favoring string instability are (i) sufficiently dense or congested traffic, (ii) drivers with little agility, and (iii) a driving style characterized by little anticipation . This observation can be used as a starting point for increasing traffic flow stability by driver-assistance systems (cf. Sect. 21.5), or for formulating rules for effective driving to be taught in driving schools. Even if stop-and-go conditions prevail, anticipative drivers (or suitable ACC systems) react earlier to braking maneuvers of the preceding vehicles than their more short-sighted peers thereby reducing the inflow to the traffic waves. Moreover, responsive and anticipative drivers (or ACC-driven vehicles) leave traffic waves faster than their more sluggish and short-sighted contemporaries. With less inflow and more outflow, even existing traffic waves eventually will dissolve.

Interactive simulations. All three factors of string instability can be interactively simulated at the authors' website[21] using the ring-road scenario depicted in Fig. 15.8. In the default setting, traffic flow is unstable and traffic waves emerge after some time. These waves can be suppressed by each of the following actions:

- Reducing the "average density" via the top scrollbar. This reduces the overall interactions and thus the positive destabilizing feedback characterized by $v_e'(s)$.
- Increasing the "acceleration a" by controlling the corresponding scrollbar. This makes the drivers more agile and corresponds to increasing of the sensitivity $-\tilde{a}_v$ (see also Fig. 15.9).

[21] see: www.traffic-simulation.de.

- Decreasing the (comfortable) "deceleration b". Since one needs to react earlier in order to reduce decelerations, this corresponds to increasing the level of anticipation $-\tilde{a}_{\Delta v}$.

The latter two actions can also be applied to the other simulation scenarios.

15.4.2 Flow Stability of Macroscopic Models

For investigating macroscopic flow instability, i.e., the equivalent of the microscopic string instability, we start from the general acceleration equation (9.1) of second-order macroscopic models combined with the continuity equation (9.10) for homogeneous road sections,[22] including possible diffusion terms. We rewrite the acceleration equation such that all partial derivatives and nonlocalities contributing to actual accelerations appear explicitly as independent variables of the acceleration function. Together with the continuity equation, this gives

$$\frac{\partial \rho}{\partial t} + \frac{\partial(\rho V)}{\partial x} = D\frac{\partial^2 \rho}{\partial x^2}, \tag{15.50}$$

$$\frac{\partial V}{\partial t} + V\frac{\partial V}{\partial x} = A\left(\rho, V, \rho_a, V_a, \rho_x, V_x, \rho_{xx}, V_{xx}\right). \tag{15.51}$$

The partial derivatives and nonlocalities of the density field are given by

$$\rho_x = \frac{\partial \rho(x,t)}{\partial x}, \quad \rho_{xx} = \frac{\partial^2 \rho(x,t)}{\partial x^2}, \quad \rho_a(x,t) = \rho(x_a,t) \text{ with } x_a > x. \tag{15.52}$$

The derivatives V_x, V_{xx} and nonlocalities V_a of the speed field are defined in analogy.

As for the microscopic models, we expand Eqs. (15.50) and (15.51) around the steady-state solution (ρ_e, V_e). The steady-state condition itself defines the fundamental speed-density relation $V_e = V_e(\rho)$ by

$$A\left(\rho, V_e(\rho), \rho, V_e(\rho), 0, 0, 0, 0\right) = 0. \tag{15.53}$$

Moving along the one-dimensional space of steady-states,

$$\mathrm{d}A = \left(A_\rho + A_{\rho_a}\right)\mathrm{d}\rho + \left(A_V + A_{V_a}\right)\frac{\mathrm{d}V_e}{\mathrm{d}\rho}\mathrm{d}\rho = 0, \tag{15.54}$$

we obtain the following relation between the partial derivatives of the acceleration function:

$$\frac{\mathrm{d}V_e(\rho)}{\mathrm{d}\rho} = V_e' = -\frac{A_\rho + A_{\rho_a}}{A_v + A_{V_a}}. \tag{15.55}$$

[22] Otherwise, Fourier modes cannot be used and the analysis becomes more complicated.

Here, the partial derivatives of the acceleration function (including these appearing in Eq. 15.60 below) are given by

$$A_\rho = \left.\frac{\partial A}{\partial \rho}\right|_e, \quad A_{\rho_a} = \left.\frac{\partial A}{\partial \rho_a}\right|_e, \quad A_{\rho_x} = \left.\frac{\partial A}{\partial \rho_x}\right|_e, \quad A_{\rho_{xx}} = \left.\frac{\partial A}{\partial \rho_{xx}}\right|_e. \tag{15.56}$$

The derivatives A_V, A_{V_a}, A_{V_x}, and $A_{V_{xx}}$ are defined in analogy. The subscript "e" denotes that the functions are evaluated at the steady-state point $(\rho_e, V_e(\rho_e))$. Linearizing Eqs. (15.50) and (15.51) using the ansatz

$$\rho(x,t) = \rho_e + \tilde{\rho}(x,t), \tag{15.57}$$

$$V(x,t) = V_e + \tilde{V}(x,t), \tag{15.58}$$

leads to the linear partial (and possibly nonlocal) differential equations

$$\frac{\partial \tilde{\rho}}{\partial t} = -\rho_e \frac{\partial \tilde{V}}{\partial x} - V_e \frac{\partial \tilde{\rho}}{\partial x} + D\frac{\partial^2 \tilde{\rho}}{\partial x^2}, \tag{15.59}$$

$$\begin{aligned}\frac{\partial \tilde{V}}{\partial t} &= -V_e \frac{\partial \tilde{V}}{\partial x} + A_\rho \tilde{\rho} + A_V \tilde{V} + A_{\rho_a}\tilde{\rho}_a + A_{V_a}\tilde{V}_a \\ &\quad + A_{\rho_x}\frac{\partial \tilde{\rho}}{\partial x} + A_{V_x}\frac{\partial \tilde{V}}{\partial x} + A_{\rho_{xx}}\frac{\partial^2 \tilde{\rho}}{\partial x^2} + A_{V_{xx}}\frac{\partial^2 \tilde{V}}{\partial x^2}\end{aligned} \tag{15.60}$$

which $\tilde{\rho}_a(x,t) = \tilde{\rho}(x_a,t)$ and $\tilde{V}_a(x,t) = \tilde{V}(x_a,t)$.

The general ansatz to solve this system of equations consists of linear waves (*Fourier modes*) of wave number k and a growth rate $\lambda(k)$,

$$\begin{pmatrix}\tilde{\rho}_k(x,t)\\ \tilde{V}_k(x,t)\end{pmatrix} \propto \begin{pmatrix}\hat{\rho}\\ \hat{V}\end{pmatrix} \mathrm{e}^{\lambda t - \mathrm{i}kx} = \begin{pmatrix}\hat{\rho}\\ \hat{V}\end{pmatrix} \mathrm{e}^{(\sigma + \mathrm{i}\omega)t - \mathrm{i}kx}. \tag{15.61}$$

In contrast to the microscopic ansatz (15.34), the macroscopic Fourier modes are defined in the stationary (road) system. Furthermore, the quantity k is dimensional with the unit m^{-1}. Specifically:

- The wavelength is given by $2\pi/k$, i.e., k is consistent with the physical definition of a *wave number*. This has to be contrasted with the physical wavelength $2\pi(s_e+l)/k$ of microscopic models.
- With I lanes, a wave contains $I\rho_e 2\pi/k$ vehicles.
- The points of constant phase $\phi = \omega t - kx$ (the waves), move with the velocity $\tilde{c}(k) = \omega/k$ in the stationary system. This has to be contrasted with the physical propagation velocity $\tilde{c}_{\text{mic}}(k) = v_e(s_e) + (s_e + l)\frac{\omega}{k}$ of microscopic waves in the stationary system.

Similarly to the analysis of car-following models, inserting the ansatz (15.61) into Eqs. (15.59) and (15.60) results in an algebraic linear system of equations for the amplitudes $\hat{\rho}$ and $\hat{V}$ of the density and speed oscillations, respectively:

$$\begin{aligned}\lambda\hat{\rho} &= (\mathrm{i}kV_e - Dk^2)\hat{\rho} + \mathrm{i}k\rho_e\hat{V},\\ \lambda\hat{V} &= \left(A_\rho + A_{\rho_a}\mathrm{e}^{-\mathrm{i}ks_a} - \mathrm{i}kA_{\rho_x} - k^2A_{\rho_{xx}}\right)\hat{\rho}\\ &\quad + \left(\mathrm{i}kV_e + A_V + A_{V_a}\mathrm{e}^{-\mathrm{i}ks_a} - \mathrm{i}kA_{V_x} - k^2A_{V_{xx}}\right)\hat{V}.\end{aligned}$$

Here, $s_a = x_a - x$ is the anticipation distance of nonlocal models.

As in the analysis of the car-following models, the solvability condition for this homogeneous linear system leads to a quadratic equation of the form (15.39) for λ. The long-wavelength expansion of the more unstable branch (given by Eq. 15.40 with the negative sign of the square root) around $k = 0$ proceeds in exact analogy to the analysis of car-following models in Sect. 15.4.1. Again, the result takes on the general form (15.45) but now with the macroscopic expansion coefficients

$$\begin{aligned}p_0 &= -(A_V + A_{V_a}),\\ p_1 &= \mathrm{i}(A_{V_x} + s_aA_{V_a} - 2V_e),\\ q_1 &= \mathrm{i}V_e(A_V + A_{V_a}) - \mathrm{i}\rho_e(A_\rho + A_{\rho_a}) = -\mathrm{i}Q'_e p_0,\\ q_2 &= V_e(A_{V_x} + s_aA_{V_a}) - \rho_e(A_{\rho_x} + s_aA_{\rho_a}) - V_\mathrm{e}^2 - DA_V.\end{aligned} \tag{15.62}$$

To arrive at the second equality sign of the expression for q_1, we have applied Eq. (15.55):

$$q_1 = -\mathrm{i}V_e p_0 + \mathrm{i}\rho_e p_0\frac{A_\rho + A_{\rho_a}}{A_V + A_{V_a}} = -\mathrm{i}p_0(V_e + \rho_e V'_e) = -\mathrm{i}Q'_e p_0.$$

In first order of the wave number k, the general long-wavelength expansion (15.45) yields a purely imaginary contribution and results in the phase velocity

$$\tilde{c}(k) = \frac{\omega}{k} = -\frac{q_1}{p_0} + \mathcal{O}(k^2) = Q'_e + \mathcal{O}(k^2). \tag{15.63}$$

As in the LWR models, the propagation speed of waves of low wave number ($k \ll \pi$) is given by the gradient Q'_e of the fundamental diagram. In contrast to these models where $\tilde{c} = Q'_e$ is valid for any perturbation, the wave velocity of second-order macroscopic models changes with the wavelength and $\tilde{c} = \lim_{k\to 0}\tilde{c}(k) = Q'_e$ is only a linear and long-wavelength approximation. In Problem 15.2 we will show that expression (15.63) for the physical phase velocity in the road-based system also applies to the considered car-following models.

The second-order term of Eq. (15.45) is purely real and provides the stability properties. By demanding that this term is negative,

$$\frac{q_1p_1}{p_0^2} - \frac{q_2}{p_0} - \frac{q_1^2}{p_0^3} < 0,$$

and inserting the macroscopic expansion coefficient $q_1 = -iQ'_e p_0$, we obtain a simple yet general macroscopic condition for flow stability (string stability) in terms of the gradient Q'_e of the fundamental diagram and the remaining coefficients p_1 and q_2 (p_0 cancels out):

$$(Q'_e)^2 - \mathrm{i}p_1 Q'_e - q_2 \le 0. \tag{15.64}$$

Local models. Local macroscopic models are defined by $A_{\rho_a} = A_{V_a} = 0$. Inserting the macroscopic expansion coefficients (15.62) for p_1 and q_2 and replacing $Q'_e = V_e + \rho_e V'_e$, we obtain the stability condition

$(\rho_e V'_e)^2 \le -\rho_e \left(V'_e A_{V_x} + A_{\rho_x}\right) - D A_V$	Flow stability for local macroscopic models.

(15.65)

We emphasize that this criterion does not depend on $A_{V_{xx}}$ or $A_{\rho_{xx}}$. So, contrary to intuition, flow stability is *not* enhanced by diffusion terms in the equation for the speed field. It is enhanced, however, by the diffusion term proportional to D in the density equation. If the macroscopic model can be written in the form (9.11), i.e., the acceleration function does not contain speed gradients, and all other gradients can be written in terms of a complete differential $-\frac{1}{\rho}\partial P/\partial x$ of a *traffic pressure* $P(\rho(x,t))$ depending on density, only, Eq. (15.65) assumes the form

$$(\rho_e V'_e)^2 \le P'_e - D A_V \quad \text{where} \quad P'_e = P'(\rho_e). \tag{15.66}$$

Nonlocal models. Since the nonlocal terms containing $\rho_a(x,t) = \rho(x + s_a, t)$ or $V_a(x,t) = V(x+s_a, t)$ constitute anticipative elements ($s_a = x_a - x > 0$), they play the role of the gradient terms of the local models and it does not make sense to include the latter in nonlocal models: After all, nonlocal models have been proposed to overcome some conceptual and numerical problems that are inherent to the gradients of local models.[23] Therefore, we can set $A_{\rho_x} = 0$, $A_{V_x} = 0$, and $D = 0$. However, this applies to gradients related to accelerations of single vehicles, only. Gradients arising from kinematic reasons (the advective term $V\frac{\partial V}{\partial x}$), or representing purely statistical effects (pressure term $-1/\rho \frac{\partial P}{\partial x}$, cf. Sect. 9.3.4) are retained. Consequently, the nonlocal models considered in the following (including the GKT model) have acceleration equations of the form

$$\frac{\partial V}{\partial t} + V\frac{\partial V}{\partial x} + \frac{1}{\rho}\frac{\partial P(\rho)}{\partial x} = A\left(\rho, V, \rho_a, V_a\right). \tag{15.67}$$

[23] Diffusion terms imply infinite speeds. Furthermore, in the presence of speed gradients, negative speeds cannot be excluded. Moreover, local models are numerically more unstable than gradient-free nonlocal models.

Evaluating the general stability condition (15.64) for this model class leads to

$$(\rho_e V_e')^2 \le P_e' - \rho_e s_a \left(V_e' A_{V_a} + A_{\rho_a} \right) \quad \text{Stability condition for nonlocal macro-models.} \tag{15.68}$$

Discussion. As for the microscopic models, macroscopic models tend to become more instable with increasing gap sensitivity $|V_e'(\rho)|$ representing the degree of interaction between drivers, i.e., completely free traffic is never unstable. Furthermore, like in car-following models, anticipation in the form of gradients ($A_{V_x} > 0$, $A_{\rho_x} < 0$) or nonlocalities ($A_{V_a} > 0$, $A_{\rho_a} < 0$) enhance stability. By comparing the stability conditions (15.68) and (15.65) it becomes evident that the nonlocalities A_{ρ_a}, A_{V_a} of the nonlocal models directly correspond to the gradients A_{ρ_x}, A_{V_x} of the local models.

In contrast to microscopic models, the speed sensitivity A_V alone does not influence stability since it appears only in combination with the density diffusion D which is zero, in most macroscopic models. We conclude:

Without gradients or nonlocalities, macroscopic models are unconditionally unstable.

Furthermore, linear stability does not depend on diffusion terms characterized by $A_{\rho_{xx}}$ and $A_{V_{xx}}$ but only on density diffusion characterized by the coefficient D. Nevertheless, diffusion terms in the speed equation tend to stabilize perturbations of higher amplitude and/or frequency that are outside the limits of this linear long-wavelength analysis. Therefore, such terms are included into some local macroscopic models, e.g., the Kerner–Konhäuser Model (9.21).

15.4.3 Application to Specific Models

In the following, we apply the general stability criteria to some of the car-following and macroscopic models presented in the Chaps. 9 and 10, respectively.

Optimal velocity model and extensions. We analyze the Full Velocity Difference Model (FVDM) presented in Sect. 10.7 which is a generalization of the Optimal Velocity Model (OVM). Its acceleration function $\tilde{a}_{\text{mic}}(s, v, \Delta v) = (v_{\text{opt}}(s) - v)/\tau - \gamma \Delta v$ is of the form $\tilde{a}_{\text{mic}}(s, v, \Delta v)$, so Eq. (15.49) is the suitable criterion for string stability. With $\tilde{a}_v = -1/\tau$ and $\tilde{a}_{\Delta v} = -\gamma$, we obtain

$$v_e'(s) \le \frac{1}{2\tau} + \gamma. \tag{15.69}$$

For bound and congested traffic, the left-hand side $v_e'(s)$ is of the order of the inverse of the time gap. Specifically, for the optimal-velocity relation (10.22), it is directly given by the inverse $1/T$ of the desired time gap T.

Traffic flow modeled with the OVM (γ=0) is only string stable if $\tau < \frac{1}{2}v_e'(s)$, i.e., the speed adaptation time τ must be smaller than half the time gap of the order of 1–2 s. Since this implies unrealistically agile drivers and unphysically high accelerations, the OVM cannot describe realistic driving behavior. The speed difference sensitivity γ of the FVDM partially resolves this problem since sensitivities γ of the order of 1 s^{-1} are realistic in car-following mode if speed differences are not too large. However, as discussed in Sect. 10.7, the FVDM is not complete since the sensitivity to speed differences does not tend do zero when gaps tend to infinity.

Newell's model. Newell's model (10.25) is formulated in terms of an iterated coupled map, so the results of Sect. 15.4.1 cannot be applied directly. Proceeding as in this section, the resulting solvability condition for the growth rate λ contains algebraic terms but also exponentials $e^{\lambda \Delta t}$, and therefore cannot be solved analytically.[24] As a consequence, no compact analytic stability criterion can be derived. Moreover, in contrast to models formulated as differential equations but similarly to time-delay differential equations, the first instability may be of a short-wavelength type.

If one assumes a priori that short-wavelength instabilities are not relevant, it is a good approximation to replace difference quotients by time derivatives using Eq. (10.11). Thus, Newell's speed update rule $v_\alpha(t+T) = v_e(s_\alpha(t))$ can approximatively be formulated by the time-continuous acceleration equation

$$a_{\text{mic}}^{\text{Newell}}(s, v) = \frac{v_e(s) - v}{T}. \tag{15.70}$$

It is identical to the OVM if one identifies the speed adaptation time τ with the update (reaction) time T. We conclude that Newell's model is stable with respect to long-wavelength string instabilities if

$$v_e'(s) \leq \frac{1}{2T}. \tag{15.71}$$

Gipps' model. As Newell's model, Gipps' model is formulated as an iterated map and the same restrictions and *caveats* apply. Particularly, the first instability may be of the short-wavelength type which is, in fact, true for the original formulation of Gipps' model. In the simplified version (11.11), (11.10) presented here, however, long-wavelength instabilities appear first, so the stability limits can approximatively be investigated by the time-continuous version of Gipps' model obtained using Eq. (10.11). For congested traffic,[25] its acceleration function is given by

---

[24] The equation for λ is of a similar form as the condition (15.27) for local instability of time-delay differential equations.

[25] No instabilities are possible for free traffic since $v_e'(s_e) = 0$ in this case.

$$a_{\text{mic}}^{\text{Gipps}}(s, v, v_l) = \frac{v_{\text{safe}}(s, v_l) - v}{T} = \frac{\sqrt{b^2T^2 + v_l^2 + 2bs} - bT - v}{T}. \tag{15.72}$$

The gap sensitivity $v_e'(s)$ pertaining to this acceleration function and the partial derivatives are given by are

$$v_e'(s) = \frac{1}{T}, \quad a_s = \frac{b}{T(bT + v_e)}, \quad a_v = -\frac{1}{T}, \quad a_{v_l} = \frac{v_e}{T(bT + v_e)}. \tag{15.73}$$

The acceleration function is of the type $a_{\text{mic}}(s, v, v_l)$ and the suitable stability criterion (15.48) results to the string stability criterion

$$\frac{1}{T} \leq \frac{1}{2T}\left(1 + \frac{v_e}{v_e + bT}\right) \tag{15.74}$$

for congested traffic while free traffic is unconditionally stable. Since condition (15.74) is never satisfied, congested traffic represented by this model is *always* unstable. However, the instabilities are always of the convective type (cf. Sect. 15.5). Moreover, for reasonable values of the deceleration parameter b, the growth rates are so small ($1/\sigma$ is of the order of one hour) that the perturbations need several kilometers of propagation to grow significantly (cf. Fig. 11.1). In many cases, the critical road sections are shorter, so the perturbations leave these sections before growing into fully developed traffic waves. As a result, the model is de facto *marginally stable* if the deceleration parameter b is of the order of 1 m/s^2 or less.

Intelligent driver model. The IDM acceleration function is of the type $\tilde{a}(s, v, \Delta v)$. Since the partial derivative $\tilde{a}_v$ with respect to the vehicle speed would result in a markedly longer analytic expression than the derivative with respect to the gap s, we make use of relation (15.18) and set $\tilde{a}_v = -\tilde{a}_s/v_e'(s_e)$. Then, Eq. (15.49) reads

$$v_e'(s_e) \leq \frac{\tilde{a}_s}{2v_e'(s_e)} - \tilde{a}_{\Delta v}. \tag{15.75}$$

With the partial derivatives

$$\tilde{a}_s^{\text{IDM}} = \frac{2a}{s_e}\left(\frac{s_0 + v_eT}{s_e}\right)^2, \quad \tilde{a}_{\Delta v}^{\text{IDM}} = -\frac{v_e}{s_e}\sqrt{\frac{a}{b}}\left(\frac{s_0 + v_eT}{s_e}\right), \tag{15.76}$$

we obtain the string stability criterion (Fig. 15.10)

$$(v_e'(s_e))^2 \leq \frac{a(s_0 + v_eT)}{s_e^2}\left[\frac{s_0 + v_eT}{s_e} + \frac{v_e v_e'(s_e)}{\sqrt{ab}}\right]. \tag{15.77}$$

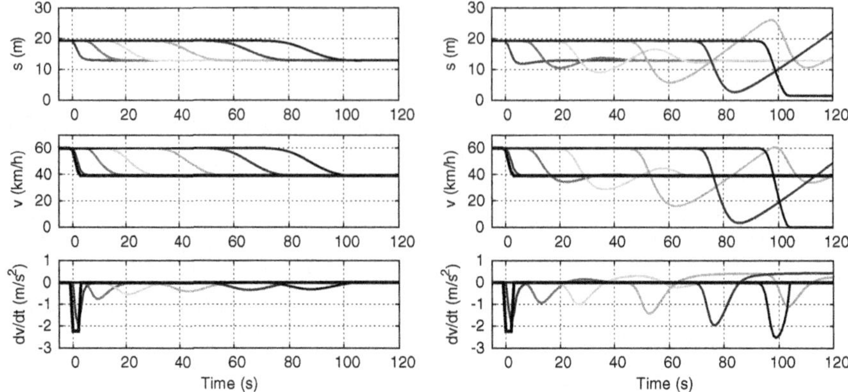

Fig. 15.10 String instability of the IDM visualized by the reaction of a sequence of vehicles driving in steady-state equilibrium far below the desired speed ($v_0 = 120$ km/h) behind a leading vehicle whose driver reduces his or her speed from 60 to 40 km/h. Shown is the first, 10th, 20th, 40th, 60th and 80th follower. *Left* traffic flow is string stable ($T = 1$ s, $s_0 = 2$ m, $a = b = 2$ m/s^2); *right* traffic flow becomes unstable by reducing the acceleration parameter from traffic flow $a = 2$–0.6 m/s^2

This condition reflects the three influencing factors for string stability discussed in Sect. 15.4.1 on page 272:

- The tendency to instability increases with the sensitivity $v_e'(s)$ to changes of the gap fueling the feedback mechanism.
- The tendency to instability decreases with the driver's agility characterized by the acceleration parameter a.
- And it decreases with decreasing comfortable deceleration b, i.e., with increasing level of anticipation.

Notice that $v_e'(s_e) \approx 1/T$ for $v \ll v_0$, so the desired time gap T is the main influencing factor to the gap sensitivity (besides the actual traffic state): Lower values of T lead to higher sensitivities $v_e'(s_e)$ and to a higher tendency to instabilities: In agreement with common sense, traffic flow becomes more unstable if the time gaps in car-following mode are comparatively short.[26]

For the limiting case $v_e \to 0$, or equivalently, $s_e \to s_0$ and $v_e'(s_0) = 1/T$, we obtain the simple explicit stability condition

$$a \geq \frac{s_0}{T^2}. \tag{15.78}$$

If the stability condition (15.78) is satisfied but traffic flow is string unstable for congested traffic of finite steady-state speed v_e, one speaks of *restabilization*. In this case, mildly congested traffic resulting from comparatively small bottlenecks is

[26] On the other hand, short gaps lead to a higher dynamic capacity, see Sect. 11.3.6.

unstable while nearly standing traffic behind severe bottlenecks is stable, creeping slowly. This will be discussed in Sect. 17.2.

Payne's model. The acceleration function of Payne's model (9.18) is given by

$$A(x,t) = \frac{V_e(\rho) - V}{\tau} + \frac{V_e'(\rho)}{2\rho\tau}\frac{\partial \rho}{\partial x}.$$

With the partial derivatives $A_{\rho_x} = V_e'/(2\rho\tau)$, $A_{V_x} = 0$ and $D = 0$, the macroscopic stability condition (15.65) for local models gives

$$\rho V_e'^2 \le -A_{\rho x} = -\frac{V_e'}{2\rho\tau}$$

and (watch the signs keeping in mind that $V_e' = V_e'(\rho) < 0$)

$$-V_e'(\rho) = |V_e'(\rho)| \le \frac{1}{2\rho^2\tau}. \tag{15.79}$$

Again, stability of traffic flow increases with increasing agility of the drivers (decreasing speed adaptation time τ), and decreasing sensitivity $|V_e'(\rho)|$ to density changes which is the macroscopic equivalent to the microscopic gap sensitivity $v_e'(s)$.

LWR models. In the limiting case $\tau \to 0$, Payne's model tends to the LWR model with diffusion (cf. Sect. 9.4.1). According to Eq. (15.79), this model is *unconditionally stable*. In contrast, the stability properties of the classical LWR model without diffusion terms are undefined. However, since even the smallest finite diffusion makes the model unconditionally stable and integration schemes typically introduce a finite amount of *numerical diffusion* (cf. Sect. 9.5 on page 145), the LWR models can be considered as unconditionally stable, for practical purposes.

Kerner–Konhäuser model. Since this model explicitly contains the "traffic pressure" $P(\rho) = \rho\theta_0$, it is convenient to apply the form (15.66) of the stability criterion. With $P'(\rho) = \theta_0$ and $D = 0$, we obtain

$$\left(\rho_e V_e'\right)^2 \le \theta_0. \tag{15.80}$$

As in Payne's model, flow stability is enhanced by decreasing the sensitivity $|V_e'(\rho)|$ to density changes. Furthermore, stability grows with the drivers' level of anticipation which is characterized by the prefactor θ_0 of the traffic pressure.[27] We emphasize that, at variance with expectations, the model parameter τ representing the driver's agility drops out of the stability condition. This makes the model somewhat counterintuitive.

---

[27] In a statistical interpretation, θ_0 formally denotes the speed variance in analogy to the corresponding term $\theta = \alpha(\rho)V^2$ of the GKT model. However, in the Kerner–Konhäuser Model, θ_0 is usually interpreted as a purely phenomenological anticipation term.

GKT model. In spite of the more complex GKT acceleration function given by the right-hand side of Eq. (9.24), the partial derivatives necessary for the nonlocal stability criterion (15.68) can be expressed in a compact form,

$$A_{\rho_a} = \frac{\partial A}{\partial \rho_a} = -\frac{2(V_0 - V_e)\rho_{\max}}{\tau \rho_e(\rho_{\max} - \rho_e)}, \tag{15.81}$$

$$A_{V_a} = \frac{\partial A}{\partial V_a} = \frac{2(V_0 - V_e)}{\tau \sigma_V(\rho_e)\sqrt{\pi}}. \tag{15.82}$$

Here, the speed variance $\sigma_V^2(\rho) = \alpha(\rho)V_e^2(\rho)$ is given by Eq. (9.22). Inserting the partial derivatives into the stability criterion results in the following condition for GKT flow stability

$$\left(\rho_e V_e'\right)^2 \le P_e' + \frac{2s_a(V_0 - V_e)}{\tau}\left[\frac{\rho_{\max}}{\rho_{\max} - \rho_e} - \frac{\rho_e V_e'}{\sigma_V \sqrt{\pi}}\right] \tag{15.83}$$

with $s_a = \gamma V_e T$, and $P_e' = \sigma_V^2 + \rho\alpha'(\rho)V_e^2$ is taken at steady-state conditions, $\rho = \rho_e$. Notice that, in the limit of zero anticipation ($\gamma \to 0$), the GKT flow stability criterion reverts to that for the Kerner–Konhäuser Model but the stability increases for increasing anticipation distance $s_a = \gamma v_e T$ and increasing driver agility $1/\tau$, in agreement with the general qualitative discussion on the influencing factors of string stability in Sect. 15.4.1 on page 272.

Near the maximum density, we can approximate this GKT stability criterion and express it in terms of a simple condition for the anticipation factor γ (cf. Problem 15.6),

$$\gamma > \frac{\tau}{2T^2 \rho_{\max} V_0 \left(1 + (\alpha_{\max}\pi)^{-1/2}\right)}. \tag{15.84}$$

This condition makes explicit that, in the GKT model, stability

- increases with γ characterizing the level of anticipation,
- decreases with increasing τ, i.e., reducing the driver's agility,
- increases with increasing desired time gap T, i.e., reducing the aggressiveness,
- increases with the desired speed V_0,
- and increases with the sensitivity to speed differences which is characterized by $\alpha^{-1/2}$.

Notice that all influencing factors are plausible, i.e., change the stability in the expected direction.

15.5 Convective Instability and Signal Velocities

In order to arrive at an approximate analytical criterion between convective and absolute instability, we start directly with definition (15.4) and investigate whether an initial transient and localized perturbation propagates in both directions (absolute instability), or only in one direction (upstream or downstream convective instability). Since all considerations are based on Eq. (15.39) and this quadratic equation applies equally to car-following and macroscopic models (cf. Eq. 15.41 and the solvability condition derived from Eq. 15.62, respectively), the analysis to be developed below applies to macroscopic flow stability as well as to microscopic string stability. The macroscopic approach allows for a more compact analytical representation, so we will use it in the following.

Equation (15.39) has two solution branches (linear complex dispersion relations) $\lambda_{1/2}(k)$ of which one is always decaying. Since we are interested in growing perturbations, we will consider the more unstable branch, only, by setting

$$\lambda(k) = \begin{cases} \lambda_1(k) & \text{if } \mathrm{Re}(\lambda_1(k)) > \mathrm{Re}(\lambda_2(k)), \\ \lambda_2(k) & \text{otherwise.} \end{cases} \tag{15.85}$$

Generally, the more unstable branch is given by Eq. (15.40) with the negative sign of the square root.

In contrast to the investigations on the instability threshold, the growth rates will no longer be expanded around the wave number $k = 0$ of the *firstly unstable* perturbation but around the wave number

$$k_0 = \arg\max_k (\mathrm{Re}\ \lambda(k)) \tag{15.86}$$

of the *fastest growing* perturbation. Since this investigation only makes sense if there is a linear instability at all, the associated maximum growth rate

$$\sigma_0 = \sigma(k_0) = \mathrm{Re}\ \lambda(k_0) \tag{15.87}$$

is positive. Due to vehicle conservation, waves of infinite wavelength corresponding to $k = 0$ always have a growth rate of zero, so the wave number k_0 of the fastest growing mode is nonzero as well. The qualitative picture is exemplified by Fig. 15.9 displaying the growth rate $\sigma(k) = \mathrm{Re}\,\lambda(k)$ for the IDM as a function of the wave number k and the distance from the linear instability threshold (corresponding to an IDM acceleration parameter $a = 1.10\,\mathrm{m/s^2}$):

- For reasons of symmetry, not only $\sigma(0)$ is 0 but also the tangent slope $\sigma'(0) = 0$.
- At the instability threshold, the first unstable mode has a wave number $k \to 0$, so $k_0 \to 0$. Above the linear threshold, k_0 grows with increasing distance.
- For reasonable parameter settings, the instability retains its long-wavelength nature also above the threshold. In the example of Fig. 15.9, the wave number k_0 of the fastest growing mode at the limit between convective and absolute instability (corresponding to the middle curve) represents traffic waves of wavelength $(l_{\mathrm{veh}} + s_e)$ $2\pi/k_0 \approx 1.3$ km. In other words, each wave contains $2\pi/k_0 \approx 47$ vehicles. Furthermore, although significantly above the threshold, the associated growth rate

$\sigma_0 = 0.0017\,\mathrm{s}^{-1}$ corresponds to a remarkably slow growth by a factor of e^1 every ten minutes.[28]

In order to determine the limits of convective instability, we determine the spatiotemporal evolution $U(x,t)$ of the perturbation amplitude, and check whether it spreads only upstream, only downstream, or in both directions. The amplitude $U(x,t)$ is defined by the system (15.59), (15.60) of linear partial differential equations to be solved in the infinitely extended system with the localized initial perturbation (15.1), or the corresponding microscopic linear equations. This initial-value problem is approximatively solved in the following steps:

- The initial perturbation $U(x,0)$ is partitioned into linear waves by Fourier transforming the initial condition with respect to space. Since the initial perturbation is localized within the space available for one vehicle and the interesting Fourier modes have much greater wavelengths, the integral over x determining the complex amplitude of the modes (Fourier transform) is the same for all relevant modes, and can be set to unity.
- The Fourier modes are evolved in time by the Eqs. (15.34) or (15.61) for microscopic and macroscopic models, respectively
- In the case of microscopic models, the Fourier modes are transformed in a fixed system with dimensional space coordinates. In any case, the development of the complex speed components of the Fourier modes is now given by $\tilde{V}_k(x,t) = \mathrm{e}^{\lambda t - ikx}$ (cf. Eq. 15.61).
- Summing over the speed components $\tilde{V}_k(x,t)$ of the Fourier modes, i.e., performing an inverse Fourier transformation, gives the complex perturbation amplitude $\tilde{U}(x,t) = \int \tilde{V}_k(x,t)\mathrm{d}k$. Taking the real part finally gives the spatiotemporal evolution $U(x,t) = \mathrm{Re}\,\tilde{U}(x,t)$.

While the first three steps are straightforward, the last step can only be evaluated analytically if one expands the complex growth rate to second order around $k = k_0$ and solves the resulting complex Gaussian integral. This rather lengthy calculation results in (cf. Fig. 15.11)

$$U(x,t) = \mathrm{Re}(\tilde{U}(x,t)), \tag{15.88}$$

$$\tilde{U}(x,t) \propto \exp\left[\mathrm{i}(k_0^{\mathrm{phys}} x - \omega_0 t)\right] \exp\left[\left(\sigma_0 - \frac{\left(v_g - \frac{x}{t}\right)^2}{2(\mathrm{i}\omega_{kk} - \sigma_{kk})}\right) t\right]. \tag{15.89}$$

The expansion coefficients are summarized in the following table:

[28] Notice that this is another hint that it may take some time until an initial perturbation develops to high-amplitude traffic waves, or a traffic breakdown.

Quantity in Eq. (15.89)	Microscopic models	Macroscopic models
k_0^{phys}	$\rho_e k_0 = \rho_e \arg\max_k \mathrm{Re}\,\lambda(k)$	$k_0 = \arg\max_k \mathrm{Re}\,\lambda(k)$
σ_0	$\mathrm{Re}\,\lambda(k_0)$	$\mathrm{Re}\,\lambda(k_0)$
ω_0	$v_e \rho_e k_0 + \mathrm{Im}\ \lambda(k_0)$	$\mathrm{Im}\,\lambda(k_0)$
v_g	$v_e + \mathrm{Im}\,\lambda'(k_0)/\rho_e$	$\mathrm{Im}\,\lambda'(k_0)$
σ_{kk}	$\mathrm{Re}\,\lambda''(k_0)/\rho_e^2$	$\mathrm{Re}\,\lambda''(k_0)$,
ω_{kk}	$\mathrm{Im}\,\lambda''(k_0)/\rho_e^2$	$\mathrm{Im}\,\lambda''(k_0)$.

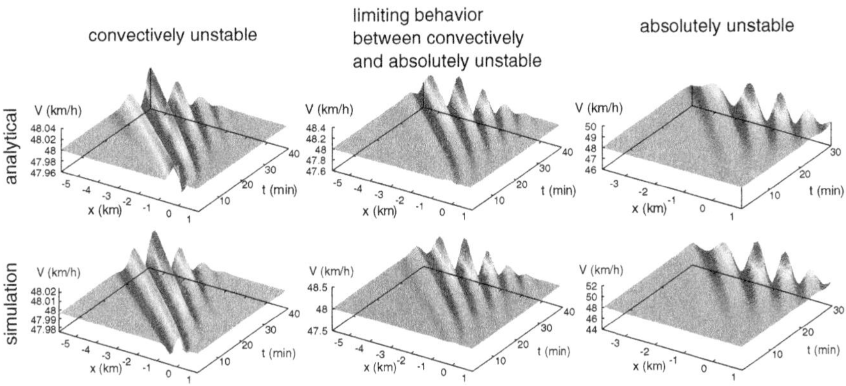

Fig. 15.11 Spatiotemporal propagation $U(x,t)$ of a localized perturbation of the steady-state traffic flow (speed $v_e = 48$ km/h) as simulated with the IDM. The parameter settings of the left column (acceleration parameter $a = 1\ \text{m/s}^2$, further IDM parameters as in Fig. 15.9) correspond to convectively unstable traffic, the right column ($a = 0.85\ \text{m/s}^2$) to absolutely unstable traffic, and the middle column ($a = 0.93\ \text{m/s}^2$) to the limit between convective and absolute instability. For each parameter settings, the analytical result Eq. (15.89) (*top row*) is compared with an IDM simulation (*bottom row*)

As visualized by Fig. 15.11, expression (15.89) represents a localized group of waves with the following properties:

- Single waves propagate with the *phase velocity* $v_\phi = \tilde{c}(k_0) = \omega_0/k_0^{\text{phys}}$ (first factor of $\tilde{U}(x,t)$).
- The center of the perturbation propagates with the *group velocity* v_g (second factor).
- The amplitude at the center of the perturbation grows with the rate σ_0.

Figures 15.12 and 15.13 show that phase and group velocity are different from each other (and also different from the LWR propagation speed $\tilde{c} = \lim_{k\to 0} \tilde{c}(k)$): Since v_g is larger (less negative) than v_ϕ, the waves emerge at the downstream boundary of the perturbation, propagate through the perturbed region, and vanish at the upstream boundary.[29] In spite of the many approximations made in deriving Eq. (15.89), this analytical expression agrees with the simulation result *in fine detail*.

[29] This is similar to a group of water waves triggered by a localized perturbation, e.g., by a stone thrown at the water surface.

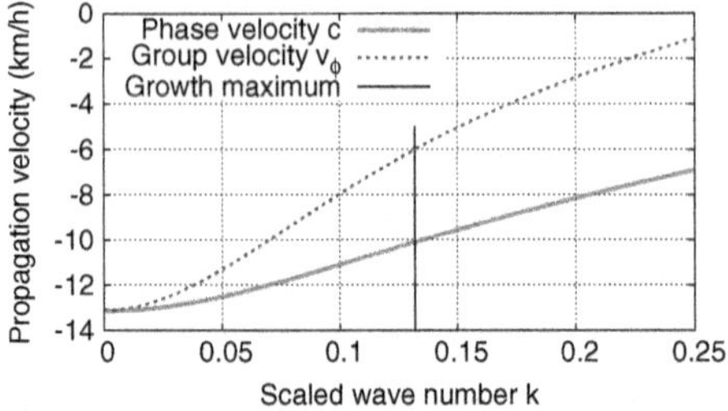

Fig. 15.12 Propagation velocity $\tilde{c}(k)$ and group velocity $v_g(k)$ for the IDM with acceleration parameter $a = 0.93\,\mathrm{m/s^2}$. The steady-state speed v_e and the other IDM parameters are given in the caption of Fig. 15.9

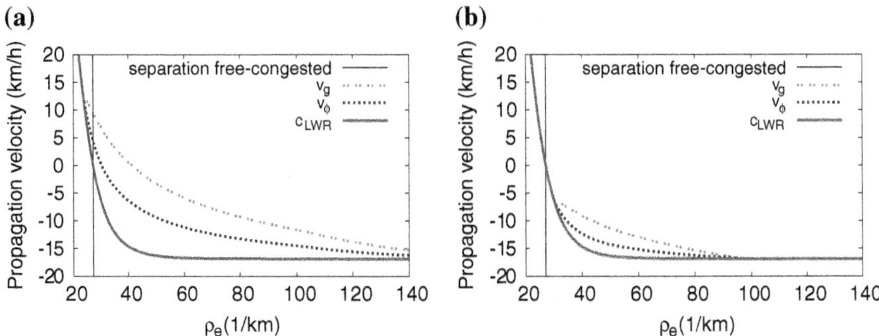

Fig. 15.13 Propagation velocities of linear perturbations as a function of the steady-state density ρ_e. Shown are the group velocity v_g, the phase velocity v_ϕ and, for comparison, the LWR propagation velocity $Q_e'(\rho)$ for the IDM. In Plot **(a)**, traffic flow at capacity is unstable (stability class 1, $a = 0.8\,\mathrm{m/s^2}$, the other parameters are as in Fig. 15.9) while, in diagram **(b)**, traffic flow at capacity is linearly stable (stability class 2, $a = 1.1\,\mathrm{m/s^2}$)

By applying the definition (15.4) of convective instability to the solution (15.89), we finally arrive at the following analytic criteria for convective instability:

$$\boxed{0 < \sigma_0 \leq \frac{v_g^2}{2D_2}, \quad D_2 = -\sigma_{kk}\left(1 + \frac{\omega_{kk}^2}{\sigma_{kk}^2}\right) \quad \text{Convective instability.}} \tag{15.90}$$

The first inequality sign states that traffic flow must be linearly (string or flow) unstable while the second inequality ensures that the perturbations propagate in only one direction. Notice that Eq. (15.90) depends only on the square of the group velocity, so it does not distinguish between upstream and downstream convective instability. The latter information is directly contained in the analytical solution (15.89): A steady-state flow satisfying Eq. (15.90) is convectively upstream unstable if $v_g < 0$, and convectively downstream unstable, otherwise.

Remarkably, the range of growth rates corresponding to convective instability increases with the *square* of the group velocity v_g and with the inverse of the second-order effective dispersion coefficient D_2[30]: If $v_g \approx 0$ (corresponding to the transition between free and congested traffic or to congested traffic of comparatively low density, cf. Fig. 15.13a, the instability is always absolute. For congested traffic sufficiently far away from the transition point, $v_g < 0$ and the instabilities are *nearly always* of an upstream convective nature. Finally, if the model parameter settings imply linear instabilities on the left-hand side of the fundamental diagram ("dense" but technically free traffic flow, $v_g > 0$), Eq. (15.90) allows for convective downstream instabilities, similarly to the original hydrodynamic systems where the concept of convective instability comes from. However, unlike the upstream type, downstream convective instabilities are not robust with respect to nonlinear effects: Downstream propagating growing waves reverse their propagation direction once nonlinearities kick in so the system effectively becomes absolutely unstable (cf. Fig. 15.5). This reversal, also called the *boomerang effect* can also be observed in traffic data (cf. Fig. 18.3). We conclude that, unlike upstream convective instabilities, downstream convective instabilities are not relevant for traffic flow dynamics.

Signal velocities. The signal velocities are defined as the slopes of rays $x = c_s t$ in space–time along which the linear amplitude of instabilities triggered by a localized and instantaneous perturbation at $x = t = 0$ neither grows nor shrinks. Generally, there are two such velocities representing the motion of the two boundaries of the instability region. In Fig. 15.4, these boundaries are indicated by solid black lines.

In order to extract the signal velocities from the perturbation field $U(x,t)$, we consider the amplitude of $U(x,t)$ along rays $x = c_s t$ and determine c_s such that the growth of the amplitude along this ray is equal to zero. This means, we replace $x = c_s t$ in the expression (15.88) for $U(x,t)$ and set the real part of its exponent equal to zero:

$$\sigma_0 - \mathrm{Re}\left(\frac{(v_g - c_s)^2}{2(\mathrm{i}\omega_{kk} - \sigma_{kk})}\right) = \sigma_0 - \left(\frac{(v_g - c_s)^2}{2D_2}\right) = 0.$$

For $\sigma_0 > 0$ (i.e., traffic flow is string unstable which we require anyway), this leads to two signal velocities,

$$c_\mathrm{s}^{\pm} = v_\mathrm{g} \pm \sqrt{2D_2\sigma_0}. \tag{15.91}$$

From this relation, we learn the following:

- The center of the region of significantly perturbed traffic flow propagates with the group velocity.
- The perturbed region grows spatially at a constant rate $2\sqrt{2D_2\sigma_0}$.
- As expected, the spatial growth rate increases with the overall level of instability σ_0 and with the effective dispersion coefficient D_2.

---

[30] This dispersion has the same unit, order of magnitude ($100\,\mathrm{m}^2/\mathrm{s}$), and effect, as the diffusion terms of some macroscopic models.

- The special case $c_s = 0$ leads us to the threshold condition $\sigma_0 = v_g^2/(2D_2)$ between absolute and convective instability which agrees with (15.90).

The latter point indicates that signal velocities are related to convective instability. Moreover, they provide an intuitive, and yet mathematically stringent, approach to distinguish between the upstream and downstream types of convective instability:

$$\text{traffic flow is} \begin{cases} \text{absolutely string unstable} & c_\mathrm{s}^- < 0 < c_\mathrm{s}^+ \\ \text{upstream convectively unstable} & c_\mathrm{s}^- < 0,\, c_\mathrm{s}^+ < 0 \\ \text{downstream convectively unstable} & c_\mathrm{s}^- > 0,\, c_\mathrm{s}^+ > 0. \end{cases} \tag{15.92}$$

The upstream type of convective instability (Fig. 15.4a) is often observed in the traffic flow context while the downstream type is related to the hyrodynamical context where the very concept of convective instability originates.

15.6 Nonlinear Instability and the Stability Diagram

The analytical investigation of the previous sections refer to small perturbations, i.e., to *linear instability*. Few analytical results are available for large-amplitude perturbations or fully developed traffic waves.[31] Instead, one investigates nonlinear effects directly by simulations of well-defined systems that are as simple as possible. The most popular of such *toy systems* is a closed single-lane ring road populated which identical drivers and vehicles.[32] In order to avoid *finite-size effects*, the system should contain more than 500 vehicles. As a further abstraction, one can also consider a ring road with a circumference tending to infinity or, equivalently, an infinitely extended homogeneous road. The only *control parameter* is the global (average) density ρ_e. By simulating the qualitative system dynamics in the full range $[0, \rho_{\max}]$ of possible values for the control parameter, one obtains a *stability diagram*.

We emphasize that a ring road does not represent a realistic abstraction of real road networks: Real road networks are open, so the inflow (traffic demand) rather than the density acts as control parameter. Furthermore, bottlenecks are missing on the idealized ring road. Nevertheless, their investigation allows us to draw far-reaching conclusions on more realistic open systems with bottlenecks. A big advantage of stability diagrams derived from ring roads is that they reflect the dynamical properties of a given model-parameter combination independently of the properties of the road network, or the traffic state.

[31] There is a large body of literature proposing and investigating solitary nonlinear waves which can be investigated analytically. However, the conditions to derive equations for such waves (e.g., a *modified Korteweg–de-Vries equation*) are extremely restrictive and nearly never satisfied in real traffic situations.

[32] A ring road must not be confused with a roundabout which, in contrast to the former, represents a comparatively complex network node.

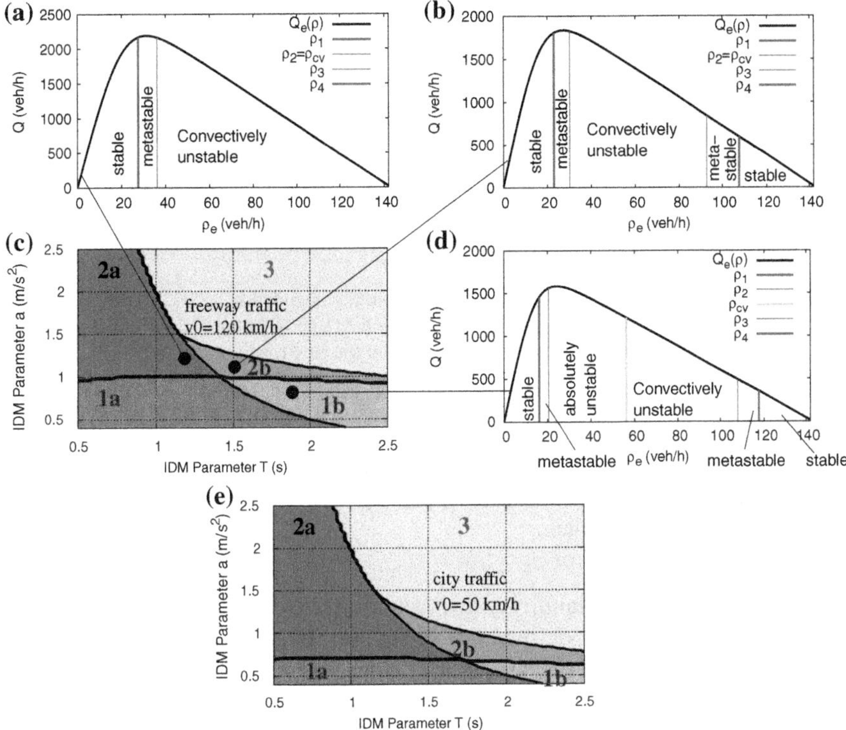

Fig. 15.14 (**c**) Class diagram of the IDM as a function of the time gap T and acceleration a. The other IDM parameters and the vehicle length 5 m are the same as in Fig. 15.9. (**a**), (**b**), (**d**) stability diagrams for three points of the class diagram corresponding to the classes 2a, 2b, and 1b, respectively. (**e**) class diagram for city traffic (v_0 reduced to 50 km/h, everything else unchanged)

To obtain stability diagrams as that of Fig. 15.14a, b, or d, we scan the whole range of global densities $\rho_e \in [0, \rho_{\max}]$. For a given global density, we simulate two scenarios: One is initialized with a very small perturbation, and one with the maximum possible perturbation. Instead of the "linear" scenario initialized with the small perturbation, one could also use the analytical results. However, simulating them represents a good combined test of the simulator code, and of the approximations and assumptions made during the analytical derivations. For each scenario, we check whether the initial perturbation dissolves, or evolves into persistent traffic waves. Generally, the resulting stability diagram is subdivided into the following regions:

- Absolute stability for global densities ρ_e below the lower nonlinear threshold ρ_1.
- *Metastability* in a range $\rho_1 \le \rho_e < \rho_2$ between the lower nonlinear and linear thresholds. In this range, sufficiently small initial perturbations eventually dissolve while higher-amplitude perturbations develop to persistent traffic waves (Fig. 15.15, see also Fig. 15.8).
- Absolute linear instability in a range $\rho_2 \le \rho_e < \rho_{\text{cv}}$.

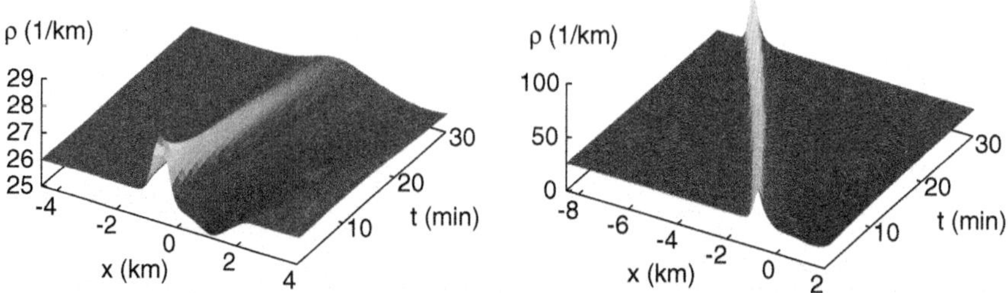

Fig. 15.15 Metastable traffic flow on a ring road with the global density $\rho = 26$ veh/km for IDM parameters as in Fig. 15.14d. Small perturbations dissolve (*left*) while a larger initial perturbation develops to a persistent traffic wave propagating around the ring road (*right*). Notice the different scales of the z-axes

- Convective linear instability in the range $\rho_{\text{cv}} \le \rho_e < \rho_3$.[33]
- Convective metastability in the range $\rho_3 \le \rho_e < \rho_4$ between the upper linear and nonlinear density thresholds.
- And absolute instability for $\rho_e \ge \rho_4$.

Which subset of the above stability types is actually realized when scanning the global density depends on the model-parameter combination. Since this determines the qualitative behavior of congested states in real open road networks (and in particular whether this behavior is realistic or not), the most relevant subsets are attributed to *stability classes* that will be discussed in the next section.

15.7 Stability Classes

While the density regions for the different instability types appear (with few exceptions) always in the order $\rho_1 \le \rho_2 \le \rho_{\text{cv}} \le \rho_3 \le \rho_4$,[34] not all density regions are realized, in general. Particularly, there may be no restabilization for high densities (Fig. 15.14a), no absolute instability ($\rho_2 = \rho_{\text{cv}}$, Fig. 15.14a, b), or no instability at all ($\rho_1 = \rho_4 = \rho_{\text{max}}$). In principle, all ranges apart from the first one ($\rho < \rho_1$) may vanish independently from each other. It is hard, however, to find model-parameter combinations showing metastable regimes but no linear instability at any density.

Analyzing real open systems with bottlenecks, it turns out that the qualitative spatiotemporal behavior, i.e., the set of possible congestion patterns, depends on only a few combinations of existing regimes. Additionally, the relative position of the thresholds with respect to the density ρ_K at capacity (the density where the

[33] Strictly speaking, convective instability is only well-defined in an infinite or open system. However, for practical purposes, the circumference of the ring must be sufficiently large such that no vehicle drives around the complete ring during the simulation time.

[34] For rare combinations of models and parameters, we obtain a region of absolute instability embedded on both sides by regions of convective downstream and upstream instabilities, respectively.

maximum flow is observed) plays an essential role. This leads to the definition of the following *stability classes*:

Class 1a: When increasing the density, traffic flow becomes linearly unstable for densities corresponding to dense but technically yet free traffic. Furthermore, it remains unstable for all higher densities: $\rho_1 \le \rho_2 < \rho_K$, $\rho_3 = \rho_4 = \rho_{\text{max}}$. Since the propagation velocity v_g is 0 for a steady-state density $\rho_e \approx \rho_K$, Eq. (15.90) implies that this class includes density ranges of absolute instability. Typically, the instability remains absolute up to moderately congested traffic and becomes convective for severe congestions near the maximum density.

Class 1b: Traffic flow *restabilizes* for high densities, i.e., traffic flow becomes smoothly creeping rather than oscillatory if severely congested.[35]

Class 2a: Only congested traffic flow (on the "right-hand side" of the fundamental diagram) can become unstable, and there is no restabilization: $\rho_2 > \rho_K$, $\rho_3 = \rho_4 = \rho_{\text{max}}$. Typically, the instability is always of a convective nature. However, a small range of absolutely unstable traffic is possible for congested traffic of comparatively low density.

Class 2b: As Class 2a, but with restabilization, $\rho_3 < \rho_{\text{max}}$.

Class 3: Absolute stability everywhere, $\rho_1 = \rho_{\text{max}}$.

Comparing the patterns simulated in realistic open systems with observations (cf. Chap. 18), we conclude the following:

> Realistic model-parameter combinations for highway traffic flow correspond to stability classes 2a or 2b.

Depending on the parameter set, one and the same model can belong to different stability classes. Figure 15.14 shows that the IDM can assume *all* classes. With he help of this model, we will now discuss the influencing factors leading to the different classes.

Agility. Agility or responsiveness corresponds to the acceleration parameter a. Starting with low agility and increasing the agility by increasing the parameter a, the stability class changes from Class 1 (instabilities are possible even for dense but uncongested traffic, to Class 2 (only congested traffic can become unstable to Class 3 (no instability anywhere). Notice that in some other microscopic or macroscopic models, the agility corresponds to the inverse of the speed adaptation time τ.

Time gap. The capacity of traffic flow (maximum flow) increases with decreasing time gap T in car-following mode. Simultaneously, reducing T also reduces the time

[35] We are aware that, in vehicles with manual transmission, it is hard to drive smoothly at very low speeds where the clutch must be operated even when driving in first gear. While this is considered in sub-microscopic models, it is ignored for the models considered here. In effect, the difficulty to drive very slowly leads to persistent noise at a sub-microscopic level. However, if traffic flow is stable at a microscopic or macroscopic level, these perturbations are not collectively amplified, i.e., traffic data show strong fluctuations but no deterministic signal.

margin of the drivers to react to changing situations, so traffic flow generally becomes more unstable. Remarkably, this does not influence the transition between Classes 1 and 2 which essentially is determined by the acceleration a.

Anticipation. By scaling the IDM appropriately (cf. Problem 15.7), one can show that the dynamics, and particularly the stability class, remains unchanged when simultaneously

- increasing the anticipation by decreasing the comfortable deceleration b by a factor $f_b < 1$,
- decreasing the agility by reducing a by a factor of f_b,
- increasing the time gap T by a factor $1/\sqrt{f_b}$,
- decreasing the desired speed v_0 by a factor of $\sqrt{f_b}$, and
- leaving s_0 unchanged.

As expected, this means that a decrease of agility is compensated for by increasing the responsiveness. Moreover, exact compensation is reached if the ratio a/b remains unchanged. Remarkably, the restabilization properties (subclasses 1a, 2a vs. 1b, 2b, 3) do not depend on the anticipation at all. To see this, we notice that the IDM corresponds to stability subclass a (1a, or 2a) if and only if

$$a < \frac{s_0}{T^2} \tag{15.93}$$

(and to one of the classes 1b, 2b, or 3, otherwise), and that this distinction criterion does not contain b as influencing factor.

15.8 Short-Wavelength Collective Instabilities

When discussing the collective instabilities discussed in the Sects. 15.4–15.7, we have assumed long-wavelength instabilities, i.e., the first instability is always one with respect to waves whose wave number tends to zero and the associated wavelength tends to infinity. Mathematically, it can be shown that this is true for all time-continuous car-following models without explicit reaction time formulated by coupled ordinary differential equations, and for all macroscopic local second-order models, i.e., formulated by partial differential equations for the density and speed fields.

However, many popular models do not belong to one of these mathematical classes. Examples include iterated coupled maps, time-continuous car-following models with reaction times, or nonlocal macroscopic models. Figure 15.16 shows the simultaneous occurrence of long-wavelength and short-wavelength collective instabilities for the IDM with an explicit delay by a reaction time $T_r = 1.2$ s (but no other human driving aspect of Chap. 12 added). We observe that the short-wavelength instabilities propagate faster than the long-wavelength instabilities, so that they "collide" into each other. However, neither these collisions nor the propagation velocity of the short-wavelength modes (about -30 km/h) are realistic. We conclude that

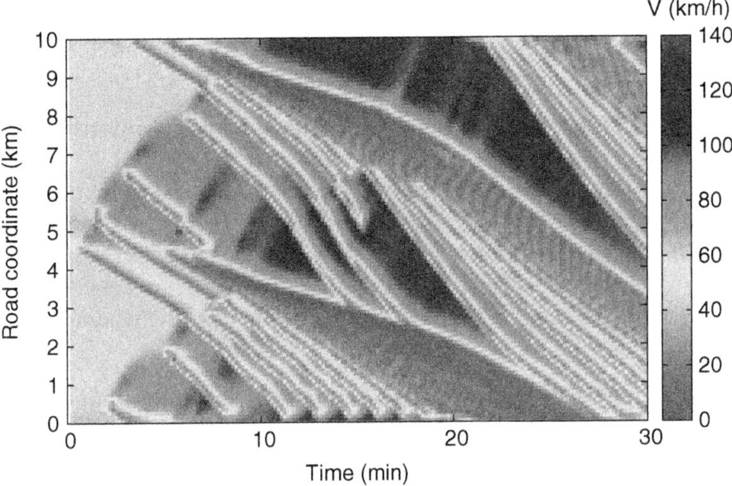

Fig. 15.16 Simultaneous appearance of long-wave and short-wave instabilities in the IDM with reaction time $T_r = 1.2$ s for initially steady-state traffic at $\rho_e = 30\,\text{km}^{-1}$ with a small perturbation. The IDM parameters are $v_0 = 120$ km/h, $T = 1.5$ s, $a = 1\,\text{m/s}^2$, $b = 1.3\,\text{m/s}^2$, $s_0 = 2$ m, and the vehicle length $l = 5$ m

short-wavelength instabilities should not occur for realistic model-parameter combinations.

Finally, we emphasize that, for realistic parameters, the first instability of models including potential short-wavelength instabilities is generally of the long-wavelength type. Since it is often not feasible to test or prove this mathematically, simulations are necessary to check this property.

Problems

15.1. Characterizing the type of instability

Consider the dynamics schematically shown in Fig. 15.1. Is it a local or string instability? If the latter is true: Is the instability absolute or convective, linear or nonlinear?

15.2. Propagation velocity of traffic waves in microscopic models

Show that the long-wavelength limit (15.47) of the microscopic propagation velocity corresponds macroscopically to the gradient $\tilde{c} = Q_e'(\rho)$ of the fundamental diagram. To this purpose, scale the microscopic propagation velocity to dimensional physical units, and transform it from the system comoving with the vehicles to a road-based fixed system. Finally, express the microscopic quantities in terms of macroscopic variables.

15.3. Instability limits for the full velocity difference Model

Consider the acceleration equation (10.23) of the FVDM with a speed adaptation time $\tau = 5$ s and a triangular fundamental diagram given by the microscopic relation

$v_e(s) = \min(s/T, v_0)$, $T = 1$ s. What is the minimum value of the sensitivity γ to speed differences to ensure (i) local stability, (ii) no (damped) oscillations when a single vehicle follows a leader with at a given speed profile, (iii) string stability?

15.4. Stability properties of the optimal velocity model compared to Payne's model
Consider the OVM and Payne's model for general, but equivalent, optimal-velocity (steady-state) relations and show that the conditions for collective (string or flow) stability of both models are equivalent. *Hint*: Find the macroscopic equivalent $V_e(\rho)$ of the microscopic steady-state relation $v_e(s)$, derive a relation between the derivatives $v_e'(s)$ and $V_e'(\rho)$, express the OVM stability condition in macroscopic terms, and compare it with the condition (15.79) for Payne's model.

15.5. Flow instability in Payne's model and in the Kerner–Konhäuser Model
Consider Payne's model and the Kerner–Konhäuser Model with a triangular fundamental diagram $Q_e(\rho) = \min(V_0\rho, 1/T(1 - l_{\text{eff}}\rho))$ and the parameters $l_{\text{eff}} = 6$ m, $V_0 = 144$ km/h and $T = 1.1$ s. (i) Show that Payne's model is unconditionally linearly stable if $\tau < T/2$, and flow unstable in the congested regions, otherwise. (ii) For the Kerner–Konhäuser Model, determine the parameter θ_0 such that this model is string unstable in the density range $\rho \in [20$ vehicles/km, 50 vehicles/km$]$.

15.6. Flow instability of the GKT Model
Consider sufficiently congested traffic such that the speed variation coefficient $\sigma_V/v = \sqrt{\alpha(\rho_{\max})}$ can be considered as constant. Show that, in the local limit of zero anticipation distance ($\gamma = s_a = 0$), the GKT model is unconditionally unstable in this situation for all reasonable parameter values. Furthermore, show that anticipation stabilizes traffic flow by deriving the approximate Condition (15.84) for densities near the maximum density.

15.7. IDM stability class diagram for other parameter values
Calculate the stability class diagram as in Fig. 15.14c but assume a comfortable deceleration $b^* = 2\,\text{m/s}^2$ instead of $b = 1.5\,\text{m/s}^2$, and $v_0^* = 139$ km/h instead of 120 km/h. Is it possible to use this diagram without recalculating anything, just by scaling the axes appropriately?

Hint: Formulate the IDM model equations in scaled units by scaling time in multiples of the unit time $\sqrt{s_0/b}$ (of the order of 1 s), and space in multiples of the minimum gap s_0. Show that the scaled model depends on only three dimensionless parameters

$$\tilde{v}_0 = \frac{v_0}{\sqrt{b s_0}}, \quad \tilde{a} = \frac{a}{b}, \quad \tilde{T} = T\sqrt{\frac{b}{s_0}}, \tag{15.94}$$

and on the scaled vehicle length $\tilde{l}_{\text{veh}} = l_{\text{veh}}/s_0$. Now use the fact that all dynamic properties (and, in particular, the stability class) depend on the scaled parameters and the scaled vehicle length, only. Find appropriate scalings for the two axes of the class diagram.

15.8. Fundamental diagram with hysteresis
Given are the following characteristics of highway traffic flow: Average vehicle length $l = 4.67\,\mathrm{m}$, average gap in car-following situations $s = s_0 + vT$ where $s_0 = 2\,\mathrm{m}$ and $T = 1.6\,\mathrm{s}$, average free-flow speed 120 km/h, and critical density at traffic breakdown (free → congested) $\rho_c = 20$ veh/km per lane. From these data it follows that two values of traffic flow are possible in a certain density range.

1. At which traffic flow does a breakdown occur, i.e., where does the free branch of the fundamental diagram end?
2. Determine the "congested branch" of the fundamental diagram and the density at which it intersects with the free branch. For which density range can free and congested traffic exist simultaneously?
3. The outflowing region of congestions is characterized by the intersection $(\rho_{\text{out}}, Q_{\text{out}})$ of the free and congested branches of the fundamental diagram. Indicate ρ_{out} and calculate Q_{out}. Also calculate the *capacity drop* as the difference between the maximum flow of free traffic and Q_{out}
4. Make a graph of the fundamental diagram showing its mirrored λ-shape.

Further Reading

- Huerre, P., Monkewitz, P.: Local and global instabilities in spatially developing flows. Annual Review of Fluid Mechanics **22** (1990) 473–537
- Treiber, M., Kesting, A.: Evidence of convective instability in congested traffic flow: A systematic empirical and theoretical investigation. Transportation Research Part B: Methodological **45** (2011) 1362–1377
- Wilson, R.: Mechanisms for spatio-temporal pattern formation in highway traffic models. Philosophical Transactions of the Royal Society A **366** (2008) 2017–2032
- Treiber, M., Kesting, A.: Validation of traffic flow models with respect to the spatiotemporal evolution of congested traffic **21** (2012) 31–41

Chapter 16
Calibration and Validation

With four parameters I can fit an elephant, and with five I can make him wiggle his trunk.

Attributed to von Neumann

Abstract Drivers in different countries have different driving styles, drive different types of vehicles, and are subject to different traffic regulations. This means, models need to be adapted to the situations they are to describe by varying their parameters (calibration). Furthermore, it must be verified that this procedure is successful (validation). After introducing the mathematical principles behind calibration, we discuss nonlinear optimization and give hints of how to run a calibration task. We explain the various calibration methods by means of example and also discuss the necessary data preparation. Finally, we introduce validation techniques and point to interpretation pitfalls and the limits of the predictive power of models.

Microscopic traffic flow models describe the driving behavior, local traffic rules, and possible restrictions of the vehicle. Macroscopic models additionally include the driver-vehicle composition and how it changes the collective traffic flow dynamics. This occurs on two levels:

- Selecting a specific model or model combination (e.g., a longitudinal and a lane-changing model) determines the above aspects at a structural or *qualitative* level. For example, mixed traffic of motorized and non-motorized vehicles without distinct lanes (cf. Fig. 6.4) requires a different model class than regular vehicular traffic on lanes. On the other hand, there are situations where lane changes do not play a role at all.
- Changing the model parameters "tunes" the above characteristics *quantitatively*, e.g., by changing the free-flow speed or making the drivers more or less aggressive.

All these aspects vary with the country and with time. For example, drivers in the United States and in Germany have different driving styles, drive different types of vehicles and are subject to different traffic regulations. For drivers in China, the differences are even more pronounced. Moreover, even on a given road in a given

M. Treiber and A. Kesting, *Traffic Flow Dynamics*,
DOI: 10.1007/978-3-642-32460-4_16,

country, the traffic flow characteristics changes with time: In morning rush hours, drivers are generally more alert and drive more effectively than in the evening. At night, the driving styles are different again. The vehicle composition changes with time as well, particularly if there are periods when trucks are not allowed (e.g., in Germany on Sundays or public holidays).

Consequently, not even the best model reproducing the structural aspects can be applied to a specific task with the default parameter set (e.g., that of Tables 8.1, 9.1, 11.1, 11.2, or 12.1). Instead, the traffic flow analyst must change the default values to give a best fit to *training data sets* obtained from observations of situations that are comparable to the problem at hand. This is called *calibration*:

> Calibration is the estimation of parameters to maximize the model's descriptive power to reproduce local driver behavior and/or collective traffic-flow characteristics. The descriptive power is specified by an *objective function* to be applied to the test data.

This task is substantially more difficult than the problem cited in the introductory quote of von Neumann: Calibrating a traffic flow model is tantamount to fitting a herd of running elephants rather than a single standing elephant wiggling its trunk.

16.1 General Aspects

In this section, we discuss aspects that are relevant to all calibration problems: mathematical principles, nonlinear optimization, assessing models, and application hints.

16.1.1 Mathematical Principles

There are two main mathematical approaches to formulate the calibration problem: Least squared errors and maximum likelihood. For online applications, a variant of maximum likelihood is commonly applied: the Kalman filter.

Least squared errors (LSE). In this more intuitive method, also called *regression*, one defines the objective function directly in terms of a *sum of squared errors* (SSE), or, equivalently, the *mean squared error* (MSE) between the test data and the model prediction. The MSE is treated as a function of the parameters while the data are considered to be (and, in fact, are) fixed. In the following, we denote the parameters in shorthand notation by a *parameter vector* $\boldsymbol{\beta}$. For example, the complete IDM parameter vector is $\boldsymbol{\beta}^{\mathrm{T}} = (l, v_0, T, s_0, \delta, a, b)$ where $\delta = 4$ is kept constant, in most cases, and the vehicle length l is only relevant for macroscopic data.[1] If there are n

[1] The superscript T denotes transposition, i.e., a row rather than a standard column vector.

data points y_i^{data}, $i = 1, \ldots, n$, to which to fit the simulation predictions y_i^{sim} (often, i denotes the time step), the simple SSE reads

$$S(\boldsymbol{\beta}) = \sum_{i=1}^{n} \left(y_i^{\text{sim}}(\boldsymbol{\beta}) - y_i^{\text{data}} \right)^2 . \tag{16.1}$$

This reduces the calibration problem to a multi-variate nonlinear optimization problem:

$$\hat{\boldsymbol{\beta}} = \arg \min_{\boldsymbol{\beta}} S(\boldsymbol{\beta}). \tag{16.2}$$

The sought-after parameter estimates contained in $\hat{\boldsymbol{\beta}}$ are the arguments (arg) of $S(\boldsymbol{\beta})$ minimizing this function. Besides the simple (absolute) SSE, there are other SSE objective functions such as relative, mixed, and hybrid SSE which will be discussed in Sect. 16.2.2.

Maximum likelihood (ML) method. In contrast to LSE calibration, the ML method is explicitly based on probabilities making it the more fundamental procedure. To this end, we necessarily need stochastic elements with specified statistical properties, either in the model, or in both model and data (as assumed in the Kalman filter approach). If the model to be calibrated is deterministic, we may add a stochastic term[2], e.g., iid Gaussian acceleration noise[3] for car-following models.

Because of the well-specified statistical properties of the stochastic terms, we can define the probability (or probability density in case of continuous models) $p(\mathbf{y}_i^{\text{sim}}|\boldsymbol{\beta})$ that, at time t_i, the model makes the predictions $\mathbf{y}_i^{\text{sim}}$ subject to a given parameter vector $\boldsymbol{\beta}$ and suitable data-driven conditions for the prior step t_{i-1} or the initial state. The state $\mathbf{y}_i$ to be predicted may include the speed and the gap of one or more vehicles, but also detector counts, travel times, positions of jam fronts, or propagation velocities of traffic waves. This enables us to define the *likelihood function* as the joint probability that the model predicts *all* data points:

$$L(\boldsymbol{\beta}) = \text{prob} \left(\mathbf{y_1}^{\text{sim}}(\boldsymbol{\beta}) = \mathbf{y_1}^{\text{data}}, \ \ldots, \ \mathbf{y_n}^{\text{sim}}(\boldsymbol{\beta}) = \mathbf{y_n}^{\text{data}} \right) . \tag{16.3}$$

In case of continuous models, $L(\boldsymbol{\beta})$ is defined analogously by the multivariate probability density. In order to make the procedure mathematically tractable, one usually assumes that there are no serial correlations, i.e., the deviations $\mathbf{e}_i = \mathbf{y}_i^{\text{data}} - \mathbf{y}_i^{\text{sim}}$ caused by the stochastic terms at time t_i do not depend on deviations at prior time steps $t_j < t_i$.[4] We allow, however, correlations of the components of the deviation

[2] By this step, the ML method looses its first-principles nature and assumes the same ad-hoc nature as the LSE calibration.

[3] iid is an abbreviation for *independently and identically distributed*.

[4] Obviously, this assumption is *not* fulfilled as explicitly described for the Human Driver Model by serially-correlated estimation errors obeying Eq. (12.9). However, it can be shown that violation of this assumption does not change the estimates $\hat{\boldsymbol{\beta}}$ but only their statistical properties.

vector $\mathbf{e}_i$ at any given time. This simplifies Eq. (16.3) to

$$L(\boldsymbol{\beta}) = \prod_{i=1}^{n} p(\mathbf{y}_i^{\text{data}}|\boldsymbol{\beta}). \tag{16.4}$$

The likelihood function is the product of the probabilities that the model reproduces the data (i.e., $\mathbf{e}_i = \mathbf{0}$) at time step i, or, for continuous models, the corresponding product of the probability densities at the data points. As the name implies, maximum-likelihood calibration involves maximizing this function. Since the location of the maximization remains unchanged when applying a strictly monotonously increasing function to L, we can apply any such function with the intention of simplifying L. Particularly suitable is the logarithm function, resulting in the *log-likelihood*

$$\tilde{L}(\boldsymbol{\beta}) = \sum_{i=1}^{n} \ln p(\mathbf{y}_i^{\text{data}}|\boldsymbol{\beta}). \tag{16.5}$$

Now, we can mathematically state the problem of maximum likelihood calibration:

$$\hat{\boldsymbol{\beta}} = \arg\max_{\boldsymbol{\beta}} \tilde{L}(\boldsymbol{\beta}). \tag{16.6}$$

As in the LSE method, this reduces the calibration problem to a multi-variate nonlinear optimization problem. Formally, the LSE and ML methods are equivalent when defining the ML objective function by $S_{\text{ML}}(\boldsymbol{\beta}) = -\tilde{L}(\boldsymbol{\beta})$. If the model contains additive iid Gaussian noise and the predictions are done from time step to time step (local calibration, Sect. 16.2.3), then $-\tilde{L}(\boldsymbol{\beta})$ is even *identical* to a suitably defined SSE.

Online calibration: Extended Kalman filter. The above methods are suitable for offline use, i.e., one estimates the model based on historical data of comparable situations. However, when using models for real-time traffic-state estimation and short-term prediction, we need to tackle the problem that the driving behavior may change unexpectedly, i.e., the underlying model must be calibrated *on the fly*. A suitable method is a variation of the ML method based on the extended (or nonlinear) *Kalman filter*. While the mathematical exposition of this method is beyond the scope of this book, the principle can be described as follows.

A Kalman filter is a statistical procedure to find the most probable value $\hat{\mathbf{y}}_i$ of the true state vector $\mathbf{y}_i$ (e.g., positions and speeds) at the present time t_i given noisy measurements $\hat{\mathbf{y}}_i^{\text{data}}$ and noisy model predictions $\hat{\mathbf{y}}_i^{\text{sim}}$ with specified statistical properties (generally obeying multi-variate Gaussian distributions). Originally, Kalman filters are used for parameter-free estimates, e.g., when estimating the new vehicle position x_α of vehicle α based on noisy GPS measurements (the data) and noisy kinematics (the model). Analogously to Eq. (10.8), the model may have the form

$$\begin{aligned} v_\alpha(t_i) &= v_\alpha(t_i - \Delta t) + \dot{v}_\alpha \Delta t, \\ x_\alpha(t_i) &= x_\alpha(t_i - \Delta t) + v_\alpha(t - \Delta t)\Delta t + \tfrac{1}{2}\dot{v}_\alpha(\Delta t)^2 \end{aligned} \tag{16.7}$$

with a stochastic, Gaussian-distributed acceleration term $\dot{v}_\alpha$. If available, gyroscopic data, steering wheel angles, positions of throttle and braking pedals, or other information available on the vehicle CAN-bus can be included to refine this "ballistic" model.

To use Kalman filters for online calibration, we augment the state space with the parameter vector.[5] Then, the Kalman filter yields a best estimate for the new state and the new parameters simultaneously.

16.1.2 Nonlinear Optimization

Both mathematical approaches to calibration, regression and maximum likelihood, lead to a nonlinear objective function $S(\boldsymbol{\beta})$ to be minimized with respect to the parameter vector $\boldsymbol{\beta}$.[6] Minimizing a nonlinear function of several variables, also known as *nonlinear optimization*, is generally a difficult task and there is no unique "one-size-fits-all" solution method. The suitability of a given method depends essentially on the complexity of the objective function which we will now discuss.

Objective functions. The mathematical properties of the objective function depend on the kind of data used for calibration: We distinguish following categories:

Type I: Smooth and unimodal. When calibrating single vehicle trajectories with LSE or ML techniques and the data are complete in the sense that they contain all relevant traffic situations (cf. Sects. 16.1.3 and 16.2.3), we generally obtain smooth and *unimodal* objective functions, i.e., they are differentiable and have a single global minimum (Fig. 16.1a). The same applies when calibrating parameters of desired speed or time gap distributions to single-vehicle data of free and interacting traffic, respectively.

Type II: Smooth but no unique minimum. When calibrating car-following models to single vehicle trajectories with incomplete data, i.e., the data do not contain all traffic situations the model can describe, the objective function remains smooth but it often does not have a unique minimum. Instead, it is flat along some directions, or may even contain secondary minima and saddle points (Fig. 16.1b).

Type III: Fluctuating and multimodal. When calibrating traffic flow (particularly stop-and-go traffic) to stationary detector data by regression techniques and the SSE is defined in terms of local speed or density differences, the "fitting landscape" generated by the objective function in parameter space contains many secondary minima

[5] Of course, there are no measurements for the parameter vector. However, the Kalman filter does not need empirical data for all state variables.

[6] Since maximizing a function means minimizing the negative function, we will only speak of minimization, henceforth.

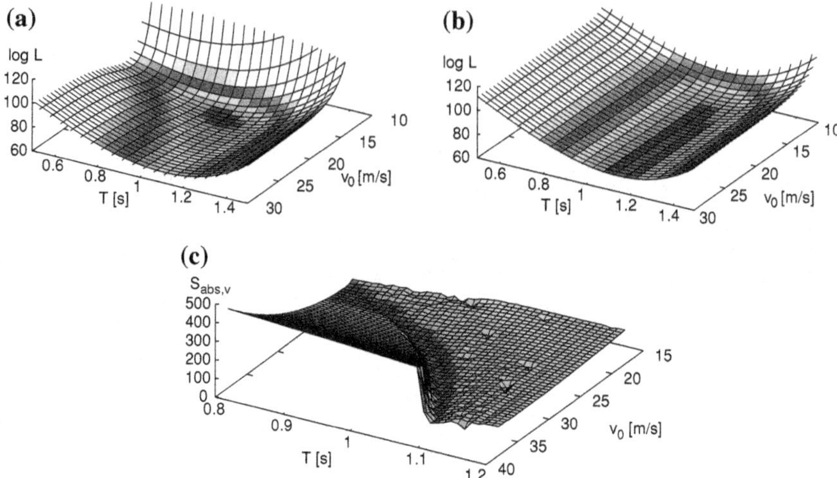

Fig. 16.1 Examples of **a** a smooth objective function with a single global minimum, **b** a smooth function with no definite minimum, **c** an objective function with a rugged fitting landscape. Shown are IDM objective functions in the parameter dimensions v_0 and T for (**a**), (**b**) local trajectory ML calibrations, and (**c**) a global LSE calibration to stationary detector data

and fluctuations (Fig. 16.1c). There are principal reasons for such "rugged" fitting landscapes which are further discussed in Sect. 16.3. Furthermore, any explicitly stochastic model will naturally lead to a stochastic objective function if evaluating the objective function implies simulation.[7] The calibration problem becomes even more difficult if, additionally, the data are incomplete and the smoothed objective function is locally flat as in Type II functions.

Methods. The "best" numerical scheme to find the global minimum in terms of speed and robustness depends on the complexity of the objective function. In any case, the success of the optimization depends crucially on the initial guess which should be as near as possible to the expected optimum. In the following, we briefly discuss six useful schemes for the calibration task.

The simpler deterministic Methods 1–4 require objective functions of Type I, i.e., differentiable and unimodal. If the function is not differentiable, or if it is not practical to calculate the derivatives, the methods may still work with numerical instead of analytical derivatives provided the objective function is unimodal. There are also dedicated methods for non-differentiable functions such as the *amoeba method* which we will not discuss here. Objective functions of the Types II and III are harder to tackle since the first four methods will not converge (Type II) or often converge to a false minimum (Type III). While we can avoid functions of Type II by restricting the calibration to parameters that are relevant for the data set in question,

[7] In contrast, the objective function (negative log-likelihood) obtained by a local maximum-likelihood calibration remains smooth.

Type III may be unavoidable for certain approaches to macroscopic calibration (see Sect. 16.3). Since any deterministic method is bound to get stuck in a local secondary minimum, we need stochastic methods to escape such minima, e.g., Method 5 or 6 of following list.

1. Newton's method. This deterministic method assumes that the objective function is twice differentiable and not too far away from a quadratic form,

$$S(\boldsymbol{\beta}) \approx S_{\min} + (\Delta\boldsymbol{\beta})^{\mathrm{T}} \mathbf{H}\, \Delta\boldsymbol{\beta} \tag{16.8}$$

where $S_{\min} = S(\hat{\boldsymbol{\beta}})$ is the global minimum sought after, $\Delta\boldsymbol{\beta} = \boldsymbol{\beta} - \hat{\boldsymbol{\beta}}$, and $\mathbf{H}$ is the Hessian (matrix of second derivatives) of $S(\boldsymbol{\beta})$ calculated at $\hat{\boldsymbol{\beta}}$. Starting the iteration with $\boldsymbol{\beta}^{(0)} \neq \hat{\boldsymbol{\beta}}$, we have

$$S(\boldsymbol{\beta}) \approx S(\boldsymbol{\beta}^{(0)}) + \mathbf{g}\, \Delta\boldsymbol{\beta} + \frac{1}{2}(\Delta\boldsymbol{\beta})^{\mathrm{T}} \mathbf{H}\, \Delta\boldsymbol{\beta} \tag{16.9}$$

where $\Delta\boldsymbol{\beta} = \boldsymbol{\beta} - \boldsymbol{\beta}^{(0)}$. Both the gradient $\mathbf{g} = \frac{\partial S}{\partial \boldsymbol{\beta}}$ and the Hessian $\mathbf{H}$ are calculated at $\boldsymbol{\beta}^{(0)}$. We calculate the next iteration, $\boldsymbol{\beta} = \boldsymbol{\beta}^{(1)}$, such that the gradient at $\boldsymbol{\beta}^{(1)}$ is equal to zero which is a necessary condition for an extremum:

$$\mathbf{g} + \mathbf{H}\left(\boldsymbol{\beta}^{(1)} - \boldsymbol{\beta}^{(0)}\right) = \mathbf{0}. \tag{16.10}$$

This gives the Newton iteration step $\boldsymbol{\beta}^{(1)} = \boldsymbol{\beta}^{(0)} - \mathbf{H}^{-1}\mathbf{g}$, or generally,

$$\boldsymbol{\beta}^{(k+1)} = \boldsymbol{\beta}^{(k)} - \left(\mathbf{H}^{-1}\mathbf{g}\right)^{(k)}, \tag{16.11}$$

where $\mathbf{H}$ and $\mathbf{g}$ are calculated at $\boldsymbol{\beta}^{(k)}$. Sufficiently near the optimum, this method converges quadratically, i.e. *very* fast.[8] Further away, however, it may not converge at all, not even for the harmless objective function shown in Fig. 16.1a.

2. Gauss-Newton algorithm. This method is a modification of Newton's method specifically developed to minimize sums of squares. Advantageously, second derivatives are no longer required. However, the robustness is similarly poor as that of Newton's method.

3. Method of gradient descent. In this scheme, also known as *method of steepest descent*, the search path proceeds always along the gradient $\mathbf{g}$ at the last iteration point. The minimum along this direction (which can be easily determined by a one-dimensional line search) constitutes the new iteration point. This method is slower but more robust than the Newton and Gauss-Newton methods.

4. Levenberg-Marquardt algorithm. This method tries to combine the advantages of gradient descent (robustness) and Gauss-Newton (fast convergence) by making

[8] It iterates to the exact minimum in one step if the function is purely quadratic, i.e., the Hessian $\mathbf{H}$ does not depend on $\boldsymbol{\beta}$.

a smooth transition between these methods during optimization. The transition is governed by an adaptive "trust region" preventing the Gauss-Newton method from stepping "too far". This is the most popular method for objective functions in form of a differentiable sum of squares, i.e., for standard problems of calibration.

5. *Genetic algorithms (GA).* This method is inspired by evolutionary biology and simulates the main mechanisms of evolution, inheritance, mutation, selection, and recombination. A model with a certain parameter set represents an "individual" (it has as many "genes" as there are model parameters) while the objective function exerts "evolutionary pressure" on the "population" of N such parameter sets. There are many GA schemes. A variant suitable for calibration proceeds as follows:

- *Initialization.* Start with a population of individuals whose parameter sets are randomly selected from the space of plausible values.
- *Evaluation of fitness.* Calculate the objective function for all individuals.
- *Selection and mating.* Select $(N - 1)$ pairs from the population such that fitter individuals are chosen with a higher probability (obviously, an individual can be part of more than one pair). Add to this selection the fittest individual.
- *Offspring.* Create one new individual per pair by randomly recombining the genes (select each parameter individually from either "father" or "mother" parameter set). Allow the fittest individual to reproduce itself on its own.
- *Mutations.* Randomly vary the parameters of all new individuals.
- *Termination.* Calculate at least a fixed minimum number of generations by repeating the four previous steps. Then stop if no further improvement can be found for at least another fixed number of generations.

From this procedure it follows that even "inter-species breeding" is allowed provided the different "species" (models) have identical parameter sets such as the IDM/IIDM/ACC model family.

6. *Kernel-based cross-entropy method (CEM).* This method belongs to the class of *Monte Carlo* approaches where realizations of stochastic variables of a given distribution are drawn during the procedure. Here, the intention is that the resulting fluctuations drive the estimate out of secondary minima. The CEM is also a variant of *simulated annealing*, a class of methods where the magnitude of the fluctuations is reduced during the process to allow for fine-tuning in the last steps. This method is effective when calculating the objective function is costly (as in the macroscopic calibrations of Sect. 16.3.1) and the desire is to calculate it as few times as possible. The method proceeds as follows:

- *Initialization.* Define a uniform sampling distribution of density $f_0(\boldsymbol{\beta})$ over the "bounding box" including all combinations of reasonable parameter values.
- *Selection.* In each step $k > 0$, draw N parameter sets (e.g., $N = 500$) from the distribution function $f_{k-1}(\boldsymbol{\beta})$ calculated in the previous step. Calculate their fitness and select the best n sets $\boldsymbol{\beta}_i$. A good size of this "elite group" is $n = 20$.
- *Determining the kernel density.* For each member of the elite group, define an orthogonal Gaussian kernel in the J-dimensional space spanned by the parameters and add the kernels of all elite members to obtain the kernel density g_k

$$g_k(\boldsymbol{\beta}) = \frac{1}{n} \sum_{i=1}^{n} \prod_{j=1}^{J} g_{jik}(\beta_j) \tag{16.12}$$

where $g_{jik}(\beta_j)$ denotes the univariate Gaussian density for parameter j of set i in iteration step k. Its expectation value β_{jik} equals the realized value for this parameter in set i and step k, i.e., the kernel distribution functions are centered at the values realized in the last elite sample. The heuristic expression for the standard deviation,

$$\sigma_{jk} = 1.06\, n^{-0.2} \hat{\sigma}_j^{\,\text{(sample k)}} \tag{16.13}$$

contains essentially the estimates of the standard deviations within the elite sample.

- *New sampling distribution function.* The distribution used to draw the elite sample of the next step is a weighted average of the old distribution and the new kernel density:

$$f_k(\boldsymbol{\beta}) = (1-\alpha) f_{k-1}(\boldsymbol{\beta}) + \alpha g_k(\boldsymbol{\beta}). \tag{16.14}$$

- *Termination.* Repeat until the distributions and, by virtue of Eq. (16.13), the samples, are sufficiently localized. As halting condition, we may require that the percentaged sample standard deviations drop below 1 % for all parameters.

This method has been applied to calibrate the five GKT model parameters to data of oscillatory congested traffic comparable to that of Fig. 16.5. The method took 20 steps to converge corresponding to $20N = 10{,}000$ evaluations of the objective function. In view of the complexity of the task, this is a small number.

16.1.3 Assessing Models

Naturally, when calibrating different models to the same data, the value of the objective function can be used to rank models according to quality. However, this is subtle and it is easy to draw misleading conclusions. The result depends strongly on the objective function. Moreover, the fitting quality described by the value of the objective function is not the only criterion to assess models. After all, a "good" model is one with a good *predictive* power rather than a good *fitting* power (see Sect. 16.4 below). In the following, we briefly discuss some points that are relevant when assessing models.

Robustness and sensitivity analysis. In spite of a good fit to test data, a model is not suitable to analyze or predict other situations, if the outcome changes rapidly as a result of minor changes of simulation details. This leads to the criterion of robustness.

> A model is *robust* if the simulation outcome does not depend sensitively on small changes of the parameters, the system, or the initial data.

Of course, the outcome in a real traffic network (e.g., the degree of congestion or travel time) depends on the parameters and system data. In order to understand the influencing factors quantitatively and to assess solutions for improving the traffic situation, it therefore makes sense to undertake a *sensitivity analysis* by investigating how traffic flow properties change when changing the driving behavior (i.e., the parameters) or the system (i.e., investigating the effect of speed limits). Only for robust models, does this lead to consistent results. Otherwise, the sensitivity coefficients, i.e., first derivatives of the property to be investigated such as the total travel time, vary wildly and may even assume unplausible signs (e.g., a higher total travel time when demand is reduced).

Choosing the right data. There are a few situations, where the system *itself* depends sensitively on the initial data or on minor changes in infrastructure details. Such a "tipping point" is shown in the "fitting landscape" of Fig. 16.1c at the desired speed $v_0 = 40\,\mathrm{m/s}$ when increasing the desired time gap T. For $T < 1.1\,\mathrm{s}$, no traffic breakdown occurs resulting in a bad fit to the congested data (Fig. 16.5). At $T \approx 1.1\,\mathrm{s}$, traffic breaks down. Due to the nonlinearities of the traffic flow dynamics (capacity drop), a significant congestion results whenever there is a traffic breakdown at all, so there is a discontinuity in the fitting landscape with respect to the parameter T. Since even the best model shows this kind of discontinuity (as real traffic flow does), the data to be used for calibration should not contain such situations. This means, data should either contain significant congestion, or no congestion at all. Furthermore, since it is hard to predict the onset of congestions, it is better to focus on the propagation of existing jams.

Parsimony. If two models have the same fitting quality and robustness, the more parsimonious model is the one with fewer parameters. Generally, the predictive power of a model increases with the level of its parsimony. This is plausible when looking at the extreme case where a model has n parameters and there are n data points to fit (cf. the quote introducing this chapter). Such a model fits exactly and, nevertheless, has nil predictive power. After all, a simple polynomial of order $n-1$ would do the same trick. There are statistical tests for parsimony which quantify the balance "number of parameters versus fit quality" such as likelihood-ratio and F-tests for ML and LSE calibration, respectively. However, both tests assume iid distributed error contributions. Since this condition is rarely fulfilled because of serial correlations, they grossly overestimate the contribution of a new parameter to the fit quality which is tantamount to underestimating the negative effect of each additional parameter on the predictive power. Such correlations are particularly pronounced in high-frequency floating-car data as that considered in Sect. 16.2 (cf. Problem 16.1).

Parameter orthogonality. Ideally, each aspect of the driver's behavior (such as desired speed, time gap and accelerations, degree of experience/anticipation, aggressiveness, and agility) is associated with one model parameter, and changes of one parameter affect other behavioral aspects as little as possible. For example, when characterizing the typical free-flow accelerations indirectly by the speed relaxation time τ via an acceleration term of the form $(v_0 - v)/\tau$, the maximum acceleration

v_0/τ depends on the parameter v_0 for the speed. In contrast, when directly describing typical or maximum accelerations by a parameter a, it is decoupled from the desired speed (which is approximatively true in real traffic). Notice that parameter orthogonality does *not* mean that the errors of the parameter *estimates*, Eq. (16.37), are uncorrelated.[9]

Intra-driver and inter-driver variations. Human drivers are not deterministic automata, so even the best model cannot capture all aspects of real driving. Variations in driving behavior come in two forms:

> *Inter-driver variations* describe differences between the driving styles of different drivers (e.g., relaxed versus aggressive) or driver-vehicle units (car and truck drivers). *Intra-driver variations* reflect that a single individual can change his or her behavior over time or as a result of the traffic environment.

It is possible to capture some of these variations by augmenting the model. By introducing distributed parameter values, the traffic analyst can describe inter-driver variations. By adding time-dependent or event-oriented parameter changes, he or she can describe at least some of the intra-driver variations. For example, some models describe *frustration effects* after being stuck in a jam for a while (see Sect. 12.7), or increase a driver's aggressiveness when a merging maneuver has been unsuccessful for a time by making the drivers accept shorter and only marginally safe gaps.

The limits of calibration. Calibration studies show consistently that there is a residual error of the order of 20 % that not even the best model can beat. Moreover, in some of the studies, even apparently unrealistic models such as the OVM show only marginally worse results. While this is not yet completely understood, two factors evidently play a role. Firstly, intra- and inter-driver variations constitute a *baseline* that no model that does not include these variation can beat. Augmenting the model does not necessarily help since this increases the number of parameters and possibly worsens its predictive power. Secondly, the stalemate between the models may be a result of choosing inappropriate fitting functions that lay too much weight on irrelevant differences such as the phase shift of traffic waves.

16.1.4 Implementing and Running a Calibration

We conclude this general section and summarize the previous discussion on assessing models and calibration methods with some hints for performing a specific calibration task:

---

9 This is also a desirable property. However, orthogonality in the errors of the parameter estimates depends more on the fitting data and on the objective function than on the model.

- Verify that the data do not describe traffic flow near a "tipping point" (e.g., at the verge of congestion or containing the onset of a traffic breakdown) which is unsuitable for calibration.
- Select a model with intuitive parameters and plausible (published) values.
- Identify which parameters of a model are relevant for the traffic situations to be found in the data. Keep the other parameters fixed at published values.
- Restrict the remaining parameter space by a bounding box containing all plausible parameter combinations. At any stage of estimation, the space outside is off-limits.
- Avoid fitting criteria with a sensitive dependence on initial data or parameters since this leads to objective functions of Type III.
- Choose the optimization method according to the objective function. Often, the Levenberg-Marquardt algorithm (possibly with numerical differentiation) is a good choice for unimodal functions, and evolutionary algorithms for functions of Type III.
- Take care to find a good initial guess. If in doubt, start with the published values.
- Check the resulting estimate for plausibility. Plot the fitness landscape around the estimate to verify that there is a global minimum inside the bounding box.

16.2 Calibration to Microscopic Observations

Microscopic traffic flow observations include trajectory and extended floating-car data (Sect. 2.1) and single-vehicle data (Sect. 3.1). Macroscopic observations include aggregated detector data (e.g., one-minute values for flow and average speed) or aggregated trajectory information such as travel times. While macroscopic data allow us to calibrate both microscopic and macroscopic models, microscopic observations are suitable for calibrating microscopic models only.

Single-vehicle stationary detector data allow us to simultaneously calibrate car-following models and mathematical descriptions of inter-vehicle variations by estimating whole *distribution functions* of desired speed v_0 or desired time gaps T. Notice that the corresponding observed distributions of speeds (Fig. 4.6) or time gaps (Fig. 4.8) do not directly reflect these distributions: The data typically result from a mix of freely driving (slower) and interacting (faster) drivers while the parameters for the desired speed and desired time gap pertain to purely free-flowing and interacting traffic, respectively. Therefore, the distributions for v_0 and T cannot be estimated directly from the data but only as part of calibrating a complete model.

Trajectory or extended floating-car data allow us to calibrate car-following models to a single driver thereby eliminating the effect of inter-driver variations. When several trajectories are available, we can

- calibrate several trajectories independently to obtain parameter distributions reflecting inter-driver variations,
- calibrate all trajectories simultaneously resulting in more robust estimates for the aggregated driving style of the drivers represented by these trajectories.

Since calibration to trajectory/floating-car data is more direct and more instructive than calibration to single-vehicle data, we will it now discuss in more detail.

There are two basic approaches: In the *global approach*, we only use the trajectory of the leader while the measured gaps and speeds only serve to initialize the simulation, i.e., the model predicts the driver's behavior for the whole simulated time interval. In the *local approach*, the model only predicts the next time step (or short time intervals) which is tantamount to directly estimating the acceleration and speed functions $a_{\rm mic}(s, v, v_l)$ and $v_{\rm mic}(s, v, v_l)$ of time-continuous and discrete car-following models, respectively (cf. Sect. 10.2).

16.2.1 Data Preparation

Extended floating car data (xFCD) usually come in the form of time series for the directly measurable quantities, i.e., the (arc-length) positions $x_j^{\rm data} = x^{\rm data}(t_j)$, speeds $v_j^{\rm data} = v^{\rm data}(t_j)$, and gaps $s_j^{\rm data}$ at times $t_j = t_0 + j\,\Delta t^{\rm data}$. Here, $\Delta t^{\rm data}$ is the sampling interval (often, $\Delta t^{\rm data} = 1$ s or 0.1 s) corresponding to a sampling rate $1/\Delta t^{\rm data}$.[10] Trajectory data provide the locations (and lanes) of all vehicles in a given spatiotemporal region, so it is straightforward to extract the time series of the considered vehicle $x_j^{\rm data}$ and that of its leader, $x_{lj}^{\rm data} = x_l^{\rm data}(t_j)$. When using xFCD, we calculate the trajectory of the leader assuming that the length l_l of the leading vehicle is equal to zero,[11]

$$x_l^{\rm data}(t_j) = x^{\rm data}(t_j) + s^{\rm data}(t_j). \tag{16.15}$$

Conversely, when calibrating to complete trajectory data, we calculate the gap by

$$s^{\rm data}(t_j) = x_l^{\rm data}(t_j) - x^{\rm data}(t_j) - l_l \tag{16.16}$$

Notice that we require that the data contain the vehicle length of the leader and a definition whether x denotes the position of the front bumper or another position (the above relation is valid for the front bumper position).

Generally, the data are noisy and possibly have a sampling interval $\Delta t^{\rm data}$ that is incompatible with possible simulation update time steps Δt of the model to be calibrated. Then, it is necessary to de-noise and re-sample the data. Furthermore, the data often show all sorts of inconsistencies. Since car-following model calibration poses high quality demands on the data (as opposed to analyses of lane changes, for example), a number of preparatory steps are necessary before using the data.

Check for inconsistencies. This includes negative speeds, negative gaps, unreasonable values for accelerations, unreasonable frequency of sign changes for

---

[10] For notational simplicity, we drop the vehicle index α.

[11] The length drops out: Following the leading vehicle is equivalent to following its rear bumper.

accelerations[12], and sudden "jumps" of vehicles forwards, backwards, or to the side. Since we will calculate speeds and accelerations from the positions by ourselves, only negative gaps and sudden jumps are relevant, at this stage. If negative gaps only appear for very short periods they are likely to vanish after smoothing, so no action is necessary at this stage.

Sudden jumps are more serious. They can be the result of active or passive lane changes: During *passive lane changes*, the leader changes to another lane (cut-out) or another vehicle changes to the considered lane becoming the new leader (cut-in). During *active lane changes*, the driver of the considered vehicle changes lanes him- or herself. In any case, the leader of the considered vehicle changes resulting in a discontinuity in the gap and the leader's position and speed. This can be easily tested by checking for corresponding jumps in the data on other lanes. If this test result is positive, the issue is resolved by identifying individual vehicles and reorganizing the data into data for individual vehicles containing the lane index as additional attribute.[13] However, the jumps can also be real artifacts of the tracking method when automatically generating trajectories from photographic material. Sometimes, this can be traced back to errors in "stitching" images taken from several camera positions into a single image, but there are other possible causes as well.[14] A signature of jumps caused by processing artifacts is that they generally occur at fixed locations. Once such an artifact is identified (i.e., active or passive lane changes are excluded), it can be removed by a "ballistic approach" such as Eq. (16.7), i.e., estimating the new position after a jump from the old position and the old speed rather than using the data.

Smoothing and re-sampling. The standard procedure to suppress noise is kernel-based smoothing which we have already used in formulating the *adaptive smoothing method* for traffic-state reconstruction (Sect. 5.2). Here, we use it to simultaneously suppress the noise and adapt the sampling interval to a feasible simulation time step.[15]

When the task is to smooth the raw data time series y_j^{raw} sampled at times $t_j^{\text{raw}} = t_0 + j\Delta t^{\text{data}}$ over a smoothing region $k\Delta t^{\text{data}}$ (with k a fixed integer) and simultaneously re-sample it to the times $t_i = t_0 + i\Delta t$ of the simulation steps, we apply the transform

$$y_i^{\text{data}} = \frac{1}{N_i} \sum_{j=-k}^{k} \phi_0\left(t_i - t_{j_0+j}^{\text{raw}}\right) y_{j_0+j}^{\text{raw}}, \tag{16.17}$$

[12] In a popular test data set, the sign of the acceleration changes in 80 % of the 0.1-s time intervals.

[13] Sometimes, the data are already organized in this form, as is the case of the data of the well-known NGSIM initiative.

[14] We could trace back such a jump in a common data set to the boundary of a shadow cast from a tall building onto the road which obviously "bamboozled" the tracking software.

[15] If the model allows for a simulation with the data time step, no re-sampling should be undertaken. Otherwise, the simulation time step should be as near as possible to the time step of the data.

where

$$N_i = \sum_{j=-k}^{k} \phi_0\left(t_i - t^{\mathrm{raw}}_{j_0+j}\right), \quad j_0 = \mathrm{round}\left(\frac{i\Delta t}{\Delta t^{\mathrm{raw}}}\right). \tag{16.18}$$

The kernel function $\phi_0(t)$ can be any localized function of range $2k\Delta t^{\mathrm{raw}}$, e.g. the (truncated) Gaussian 5.2

Calculating derived quantities. For calibration, we need the speed time series v_i^{data} and v_{li}^{data} for the considered vehicle and its leader, and sometimes the accelerations $\dot{v}_i^{\mathrm{data}}$ as well. Since we have suppressed high-frequency noise already, we can use direct numerical differentiation,[16] e.g., the symmetrical expressions

$$v^{\mathrm{data}}(t_i) = v_i^{\mathrm{data}} = \frac{x_{i+1}^{\mathrm{data}} - x_{i-1}^{\mathrm{data}}}{2\Delta t}, \quad \dot{v}_i^{\mathrm{data}} = \frac{x_{i+1}^{\mathrm{data}} - 2x_i^{\mathrm{data}} + x_{i-1}^{\mathrm{data}}}{\Delta t^2}, \tag{16.19}$$

and a similar expression for v_{li}^{data}. Some data sets may contain pre-calculated speed and acceleration time series. However, they are rarely documented and often lead to inconsistencies as discussed below. Furthermore, one has no control over the smoothing or other data manipulation methods to be applied. Therefore, we strongly recommend calculating speeds and accelerations directly from the positions by Eq. (16.19), or similar expressions.

Internal and platoon consistency. Generally, the original or processed data contain redundant information, e.g., positions, speeds, accelerations *and* gaps which may contradict each other. On the level of a single trajectory, the elementary definitions of kinematics must be satisfied to exclude such contradictions. This results in following criteria for *internal consistency* (for notational simplicity, we drop the superscript "data")

$$x(t) = x(0) + \int_0^t v(t')\mathrm{d}t', \quad v(t) = v(0) + \int_0^t \dot{v}(t')\mathrm{d}t'. \tag{16.20}$$

When the data set contains several trajectories of vehicles following each other, the gaps need to obey their respective definitions for all times, i.e., the conditions for *platoon consistency* must be additionally satisfied:

$$s(0) = x_l(0) - x(0) - l_l, \quad s(t) = s(0) + \int_0^t \left[v_l(t') - v(t')\right]\mathrm{d}t'. \tag{16.21}$$

Since the data preparation operations (16.17) are based on non-redundant data and Eqs. (16.19), (16.15), and (16.16) obey the kinematic principles, both consistency

[16] Smoothing and differentiation are linear operators, so their order can be exchanged.

conditions are automatically satisfied for x_i^{data}, v_i^{data}, $\dot{v}_i^{\text{data}}$, and s_i^{data} as defined above.

16.2.2 Global Approach

When globally calibrating car-following models to trajectories, a single evaluation of the objective function implies a complete simulation run. Since, due to integration from the accelerations, serial correlations are always present in the position and speed data. So, maximum-likelihood calibration is impractical (though technically possible), and LSE calibration is the method of choice. A complete calibration task consists of data preparation, simulation setup, calculating the objective function, minimizing it, and checking the result for consistency. In the following, we show the principle for estimating a model to a single driver following his or her leader. We denote data values by the superscript "data" but omit the superscript for simulated values, for notational simplicity.

Simulation setup. We initialize the location (the gap) and the speed of the simulated vehicle by the data and let the vehicle follow the fixed trajectory $x_l^{\text{data}}(t)$ of its leader:

$$\frac{\mathrm{d}v}{\mathrm{d}t} = a_{\text{mic}}(s, v, v_l^{\text{data}}; \boldsymbol{\beta}), \quad s(t_0) = s^{\text{data}}(t_0), \; v(t_0) = v^{\text{data}}(t_0). \tag{16.22}$$

As already mentioned, the trajectory data may contain active or passive cut-in or cut-out lane changes involving a change of the relevant leader and discontinuities of the gap $s(t)$ and the leader's speed $v_l(t)$. As an example, the right column of Fig. 16.2 shows a passive cut-out lane change at $t \approx 145$ s. We treat such discontinuities in the same way as a range sensor of an adaptive cruise control (ACC) system detecting a new target would do, i.e., by a discontinuity of the gap described by[17]

$$s(t^+) = s(t^-) + s^{\text{data}}(t^+) - s^{\text{data}}(t^-). \tag{16.23}$$

Objective functions. The LSE calibration aims at minimizing the sum of squared differences between the measured and simulated dynamic variables. In principle, any dynamical variable representing aspects of the driving behavior can serve as an objective function. Candidates are the gap s, speed v, speed difference Δv, or acceleration a (cf. Fig. 16.2). Furthermore, the differences can be formulated as absolute (simple), relative (percent), or mixed differences. Denoting by y the dynamical quantity to be investigated, we define these differences by

[17] In fact, the central ACC control logic must provide consistent acceleration responses to cut-ins, cut-outs, and active lane changes.

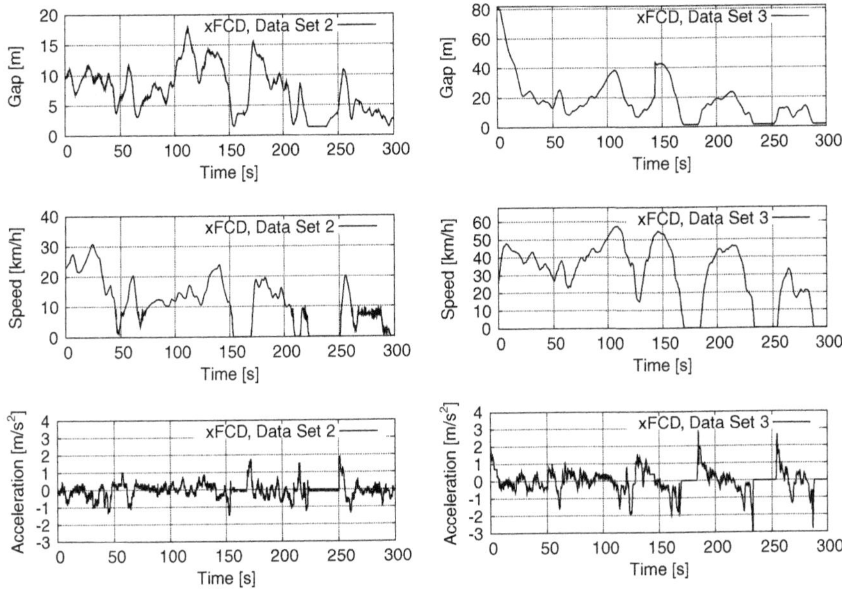

Fig. 16.2 Two sets of extended floating-car data of a car driving in a German city during rush-hour conditions. In the set labelled 3, the leader leaves the lane at $t \approx 145$ s resulting in a discontinuity in the time series for the gap

$$S_y^{\mathrm{abs}}(\boldsymbol{\beta}) = \frac{\sum_{i=1}^n (y_i(\boldsymbol{\beta}) - y_i^{\mathrm{data}})^2}{\sum_{i=1}^n \left(y_i^{\mathrm{data}}\right)^2}, \tag{16.24}$$

$$S_y^{\mathrm{rel}}(\boldsymbol{\beta}) = \frac{1}{n}\sum_{i=1}^n \left(\frac{y_i(\boldsymbol{\beta}) - y_i^{\mathrm{data}}}{y_i^{\mathrm{data}}}\right)^2, \tag{16.25}$$

$$S_y^{\mathrm{mix}}(\boldsymbol{\beta}) = \frac{\sum_{i=1}^n (y_i(\boldsymbol{\beta}) - y_i^{\mathrm{data}})^2/|y_i^{\mathrm{data}}|}{\sum_{i=1}^n |y_i^{\mathrm{data}}|}. \tag{16.26}$$

The dynamical variable itself and the formulation as absolute, relative, or mixed SSE determine the aspects of the data on which to focus. For example, the objective function $S_s^{\mathrm{rel}}(\boldsymbol{\beta})$ focusses on small gaps (slow and standing traffic), $S_s^{\mathrm{abs}}(\boldsymbol{\beta})$ on the larger gaps (periods of cruising), while both traffic situations are relevant for $S_s^{\mathrm{mix}}(\boldsymbol{\beta})$. Similar considerations apply for speed and speed differences. Generally, the gap is preferable since it contains the most degrees of freedom and therefore constitutes the most challenging calibration task: While internal and platoon consistency imply that the cumulated differences between v_i and v_i^{data} or Δv_i and $\Delta v_i^{\mathrm{data}}$ are zero in the long run, on average, no such restrictions apply to the gap.

In principle, we can also devise hybrid multi-criteria objective functions containing more elements of the state vector $\mathbf{y}_i$, e.g.,

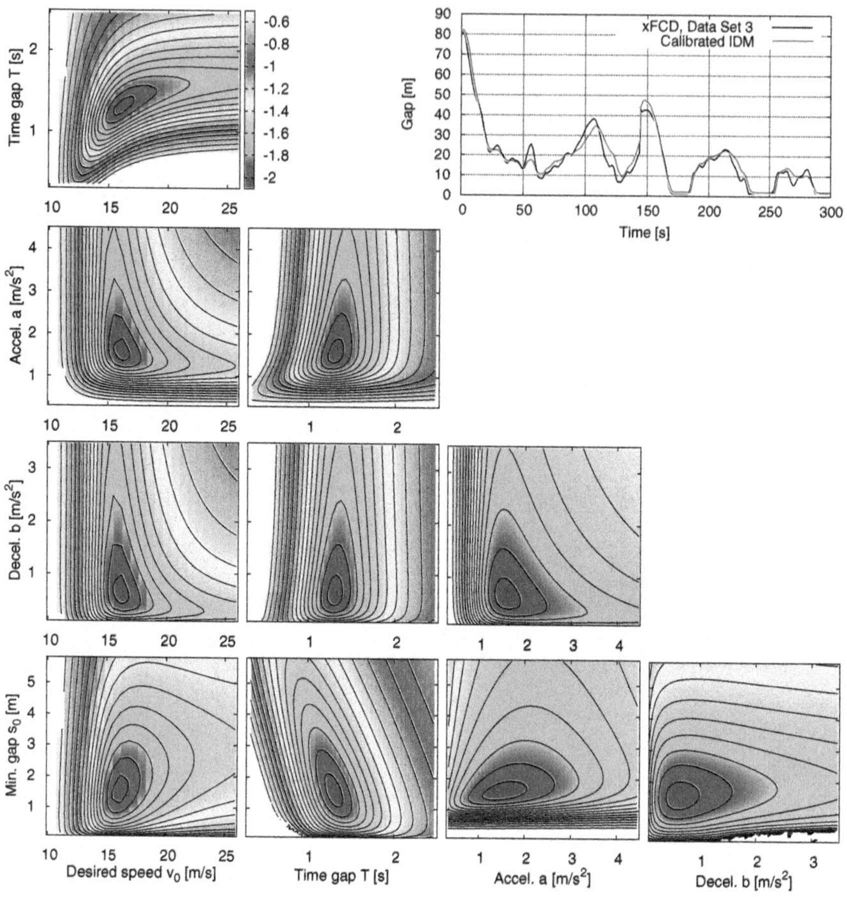

Fig. 16.3 Calibrating the IDM globally to extended floating-car data (see Sect. 2) of a passage on a city street (cf. Fig. 16.3) using the LSE method with the mixed objective function (16.26) for gap differences. Shown is the "fitting landscape" in form of two-dimensional sections of the objective function $S_s^{\text{mix}}(\boldsymbol{\beta})$ in five-dimensional parameter space around the estimate $\hat{\boldsymbol{\beta}}$ (see the main text). The *right upper graphics* compares the observed versus the simulated gaps

$$S^{\text{hybr}}(\boldsymbol{\beta}) = \gamma_1 S_s^{\text{mix}}(\boldsymbol{\beta}) + \gamma_2 S_v^{\text{mix}}(\boldsymbol{\beta}) + (1 - \gamma_1 - \gamma_2) S_{\Delta v}^{\text{mix}}(\boldsymbol{\beta}) \qquad (16.27)$$

with the weights $\gamma_1 \geq 0$ and $\gamma_2 \geq 0$ satisfying $\gamma_1 + \gamma_2 \leq 1$. Notice that we have formulated all objective functions (including the absolute ones) in a dimensionless form allowing a consistent formulation of multi-criteria objective functions such as (16.27).

Calibrating the IDM to extended floating-car data of city traffic. In this example, we calibrate the IDM to xFCD of a car driving through an inner-city street (see Fig. 16.3). Since we want to calibrate the model to both stationary and cruising traffic, we use the mixed objective function $S_s^{\text{mix}}(\boldsymbol{\beta})$. We base it on the gaps since other mixed objective functions such as $S_v^{\text{mix}}(\boldsymbol{\beta})$ or $S_{\Delta v}^{\text{mix}}(\boldsymbol{\beta})$ are not suitable for calibrating the IDM gap parameters s_0 and T. We now proceed to calibrate the IDM to Set 3 (right column of Fig. 16.2). Since this xFCD set is complete in the sense that it contains all the main dynamic situations (cruising, accelerating, braking, and standing), we use the full five-dimensional parameter set

$$\boldsymbol{\beta}_{\text{IDM}} = (v_0, T, a, b, s_0)^{\text{T}}$$

for calibration.[18] Minimizing $S_v^{\text{mix}}(\boldsymbol{\beta})$, e.g., by the Levenberg-Marquardt algorithm with numerical derivatives, yields the parameter estimate $\hat{\boldsymbol{\beta}}$ with the values $\hat{v}_0 = 16\,\text{m/s}$, $\hat{T} = 1.3\,\text{s}$, $\hat{a} = 1.5\,\text{m/s}^2$, $\hat{b} = 0.6]\,\text{m/s}^2$, and $\hat{s}_0 = 1.5\,\text{m}$.

Figure 16.3 shows that the parameter estimate corresponds to a unique global minimum of $S_s^{\text{mix}}(\boldsymbol{\beta})$ at plausible parameter values (only the parameter b is somewhat lower than expected). Furthermore, the fitting to the data is *robust* in the sense that small deviations from the optimal parameters do not significantly deteriorate the fit.

In contrast, calibrating the full IDM parameter set to data Set 2 (left column of Fig. 16.2) would gives an indefinite estimate for v_0 (not shown). This will be discussed in the next subsection on local calibration.

16.2.3 Local Approach

There are three approaches that can be termed local calibration: (i) local linearization, (ii) direct estimate of the acceleration function, and (iii) local maximum-likelihood (ML) calibration.

Local linearization. This method is feasible for sections of trajectory or extended floating-car data corresponding to car-following situations near a steady state. We define deviations from the steady state (s_e, v_e) as in Sect. 15.3 by $y(t) = s(t) - s_e$ and $u(t) = v(t) - v_e$, respectively and consider car-following situations where a single vehicle follows a leader with fixed trajectory $x_l(t)$, i.e., the speed $v_l(t)$ and acceleration profiles $\dot{v}_l(t)$ of the leading vehicle are externally fixed. Adapting the linearized equations (15.8) and (15.9) for time-continuous car-following models to this situation, we arrive at following driven damped harmonic oscillator for the gap deviations $y(t)$:

$$\ddot{y} + 2\eta\dot{y} + \omega_0^2 y = F(t), \tag{16.28}$$

where

$$2\eta = -a_v = -(\tilde{a}_v + \tilde{a}_{\Delta v}), \tag{16.29}$$

$$\omega_0^2 = a_s = \tilde{a}_s, \tag{16.30}$$

[18] The vehicle length l drops out in calibrations to trajectory or extended floating-car data.

$$F(t) = \dot{v}_l(t) - (a_v + a_{v_l})(v_l(t) - v_e) = \dot{v}_l(t) - \tilde{a}_v(v_l(t) - v_e). \tag{16.31}$$

The partial derivatives (sensitivity coefficients) a_s, a_v, a_{v_l} and $\tilde{a}_s$, $\tilde{a}_v$, $\tilde{a}_{\Delta v}$ of the model acceleration functions $a_{\text{mic}}(s, v, v_l)$ and $\tilde{a}_{\text{mic}}(s, v, \Delta v)$ are defined as in the Eqs. (15.11) and (15.16), respectively. In the actual calibration procedure, the sensitivity coefficients are calibrated to the data and associated with (combinations of) the model parameters. Because of the assumption of small deviations from the steady state, only small sections of the data can be calibrated at a time with this method.

Direct data-driven estimate of the acceleration function. When carefully preparing the data according to Sect. 16.2.1 it is feasible to represent the data by a four-dimensional scatter plot $(s_i^{\text{data}}, v_i^{\text{data}}, v_{li}^{\text{data}}, \dot{v}_i^{\text{data}})$ and directly estimate the implied acceleration function $a_{\text{data}}(s, v, v_l)$ by setting $\dot{v}_i^{\text{data}} = a_{\text{data}}(s_i^{\text{data}}, v_i^{\text{data}}, v_{li}^{\text{data}})$ and minimizing an objective function based on the SSE of modeled and measured accelerations.

Maximum-likelihood calibration. At any time step t_i, the model predicts a state $\mathbf{y}_i(\boldsymbol{\beta})$ which may contain gaps, speeds, and accelerations of one or several vehicles. In the following, we restrict ourselves to one vehicle. In the simplest (and nearly exclusively used) case, we assume zero-mean Gaussian multivariate noise with covariance matrix $\boldsymbol{\Sigma}$ which is iid with respect to different time steps (no serial correlation). Then, the log-likelihood (16.5) becomes

$$\tilde{L}(\boldsymbol{\beta}, \boldsymbol{\Sigma}) = \text{const.} - \frac{n}{2}\ln(\det \boldsymbol{\Sigma}) - \frac{1}{2}\sum_{i=1}^{n} \mathbf{e}_i^{\mathrm{T}}(\boldsymbol{\beta})\, \boldsymbol{\Sigma}^{-1}\mathbf{e}_i(\boldsymbol{\beta}) \tag{16.32}$$

where

$$\mathbf{e}_i(\boldsymbol{\beta}) = \mathbf{y}_i^{\text{data}} - \mathbf{y}_i^{\text{sim}}(\boldsymbol{\beta}) \tag{16.33}$$

denotes the vector of deviations. Since we have prescribed the statistical properties of the deviations only qualitatively (multivariate Gaussian) but not quantitatively (values of the covariance matrix), we estimate the matrix and the parameter vector simultaneously by minimizing $\tilde{L}$ with respect to $\boldsymbol{\beta}$ and $\boldsymbol{\Sigma}$. As an intermediate result, this gives the estimate

$$\hat{\boldsymbol{\Sigma}}(\boldsymbol{\beta}) = \frac{1}{n}\sum_{i=1}^{n} \mathbf{e}_i(\boldsymbol{\beta})\, \mathbf{e}_i^{\mathrm{T}}(\boldsymbol{\beta}) \tag{16.34}$$

for the covariance matrix. This allows us to formulate the log-likelihood

$$L^*(\boldsymbol{\beta}) = \tilde{L}\left(\boldsymbol{\beta}, \hat{\boldsymbol{\Sigma}}(\boldsymbol{\beta})\right) \tag{16.35}$$

as a function of the parameter vector $\boldsymbol{\beta}$ alone. The resulting explicit log-likelihood is the basis for parameter estimation:

$$\hat{\boldsymbol{\beta}} = \arg\max L^*(\boldsymbol{\beta}). \tag{16.36}$$

Since the ML method is solidly based on statistics, we can also derive approximate errors for the parameter estimates in form of a covariance matrix,

$$\mathrm{Cov}(\hat{\boldsymbol{\beta}}) \approx -\mathbf{H}_{L^*}^{-1}(\hat{\boldsymbol{\beta}}), \tag{16.37}$$

where $\mathbf{H}_{L^*}$ is the Hessian (matrix of second derivatives) of $L^*(\boldsymbol{\beta})$ at the point $\hat{\boldsymbol{\beta}}$. However, this derivation is only valid if the deviations $\mathbf{e}_i$ are iid which is generally *not* the case. If serial correlations are present, Eq. (16.37) underestimates the true estimation errors. However, at least the estimates themselves as well as their error correlations remain unaffected, in many cases.

Special case: Accelerations. In the simplest case, we let the model predict only one simulation step $i \to i+1$ at a time initializing it with the data for the gap and the speed of the previous step i. Because of the numerical update rules (10.9) and (10.10), the predicted speed v_{i+1}^{sim} and gap s_{i+1}^{sim} depend linearly on the model acceleration function $a_{\mathrm{mic}}(s, v, v_l)$ at time t_i, i.e., this function completely determines the predicted state $\mathbf{y}_{i+1}^{\mathrm{sim}}$. This means, we can reduce the simulated prediction vector $\mathbf{y}_{i+1}^{\mathrm{sim}}$ to a single number, namely the value $a_i^{\mathrm{mic}} = a_{\mathrm{mic}}(s_i, v_i, v_{l,i})$. Augmenting models of the form (10.3) or (10.5) by iid acceleration noise,

$$\frac{\mathrm{d}v}{\mathrm{d}t} = a_{\mathrm{mic}}(s, v, v_l; \boldsymbol{\beta}) + \varepsilon, \quad \varepsilon \sim iid\ N(0, \sigma^2), \tag{16.38}$$

and setting $\mathbf{y}_i^{\mathrm{sim}} = a_{i-1}^{\mathrm{mic}}$ and $\mathbf{y}_i^{\mathrm{data}} = a_{i-1}^{\mathrm{data}}$ for the predicted and measured states, respectively, reduces the log-likelihood (16.32) to

$$\tilde{L}(\boldsymbol{\beta}) = \mathrm{const} - \frac{n}{2}\ln\sigma^2 - \sum_{i=1}^{n} \frac{(a_i^{\mathrm{data}} - a_i^{\mathrm{mic}})^2}{2\sigma^2}. \tag{16.39}$$

Setting $\sigma^2 \approx \hat{\sigma}^2 = 1/n \sum_i (a_i^{\mathrm{data}} - a_i^{\mathrm{mic}})^2$ gives the explicit log-likelihood (16.35)

$$L^*(\boldsymbol{\beta}) = \mathrm{const}^* - \frac{n}{2}\ln\sum_{i=1}^{n}(a_i^{\mathrm{data}} - a_i^{\mathrm{mic}}(\boldsymbol{\beta}))^2 \tag{16.40}$$

where const* does not depend on $\boldsymbol{\beta}$. Maximizing Eq. (16.40) is equivalent to minimizing

$$S_a^{\mathrm{abs}}(\boldsymbol{\beta}) = \sum_{i=1}^{n}\left(a_i^{\mathrm{data}} - a_i^{\mathrm{mic}}(\boldsymbol{\beta})\right)^2, \tag{16.41}$$

so the local ML estimate reduces to the direct data-driven estimate of the acceleration function, and also to the LSE calibration with respect to the sum of absolute acceleration deviations. We emphasize that, in contrast to the global method discussed in Sect. 16.2.2, no simulation is necessary to evaluate $S_a^{\mathrm{abs}}(\boldsymbol{\beta})$. Furthermore,

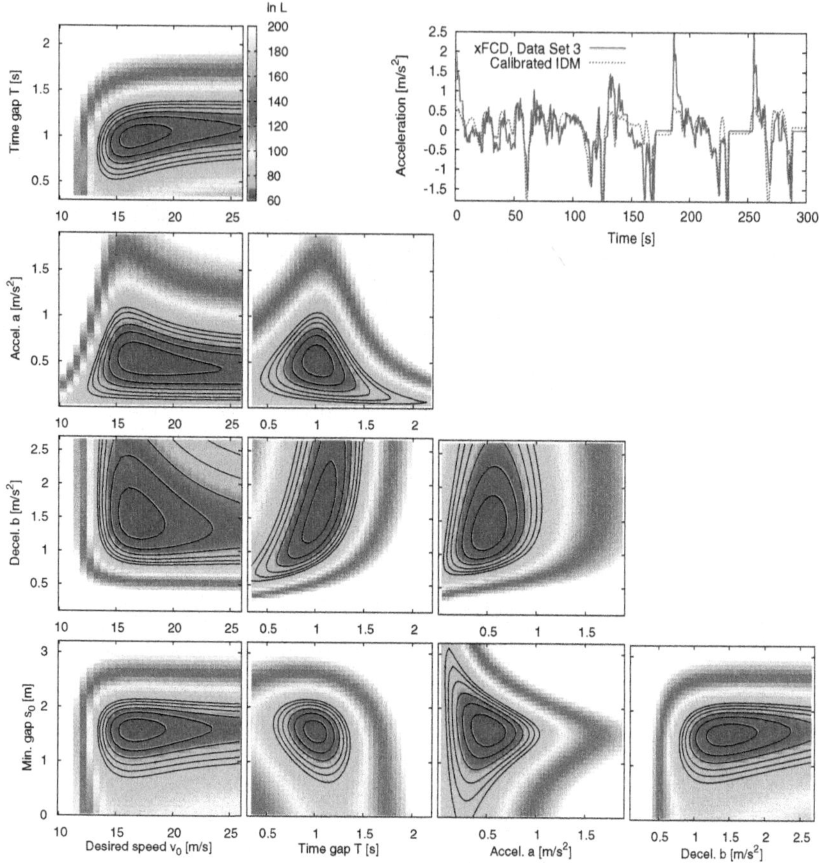

Fig. 16.4 Calibrating the IDM locally to extended floating-car data (see Sect. 16.2.2) of a passage on a city street using the maximum-likelihood method. Shown are two-dimensional sections of the log-likelihood function $\tilde{L}(\boldsymbol{\beta})$ in five-dimensional parameter space around the estimate $\hat{\boldsymbol{\beta}}$ (see the main text). The right upper graphics compares the acceleration obtained from the extended floating-car data with the IDM acceleration for the same exogenous variable values as in the data

it is easy to calculate analytical derivatives, so that minimizing methods such as the Levenberg-Marquardt algorithm can be applied directly.

Calibrating the IDM to extended floating-car data of city traffic. In this example, we use the same data as in Sect. 16.2.2 but calibrate the IDM by minimizing $S_a^{\text{abs}}(\boldsymbol{\beta})$ instead of $S_s^{\text{mix}}(\boldsymbol{\beta})$. Again, we estimate the full five-dimensional IDM parameter vector. As in the global estimate, the minimization reveals a unique global minimum (Fig. 16.4). However, the parameter estimates $\hat{v}_0 = 17\,\text{m/s}$, $\hat{T} = 0.97\,\text{s}$, $\hat{a} = 0.6\,\text{m/s}^2$, $\hat{b} = 1.4\,\text{m/s}^2$, $\hat{s}_0 = 1.5\,\text{m}$ show significantly differences with respect to the global ML calibration. Particularly, the acceleration parameters a and b have swapped their magnitude with a now being lower than expected. This is caused by a few data points producing high penalties when increasing a while the predictions would become better for most of the other points. It also shows the limits of local calibration: Even an acceleration parameter $a = 0$ (the vehicle neither accelerates nor decelerates and the leader is completely ignored) will lead to a reasonable fit while, in the global simulation, such a vehicle could crash or stand all the time, depending on the initialization.

It is instructive to calibrate the full IDM parameter set to the incomplete data Set 2 (Fig. 16.2, left column) lacking free-flow data. Figure 16.1b shows the resulting fitting landscape for $S_a^{\text{abs}}(\boldsymbol{\beta})$ applied to Set 2 with respect to v_0 and T. We observe that the objective function does not depend on the desired speed provided it is greater than $\approx 16\,\text{m/s}$. This can be understood intuitively: In congested city traffic, the speed profile is essentially determined by the leader, so any desired speed significantly above the maximum speed of the leader will produce the same simulated trajectory. In contrast, $S_a^{\text{abs}}(\boldsymbol{\beta})$, and also other objective functions such as $S_s^{\text{mix}}(\boldsymbol{\beta})$, show a global minimum for data Set 2 with respect to the other parameters. This is plausible as well since this set contains all traffic situations except for free traffic. A correct approach for data Set 2 would be setting v_0 to a plausible fixed value and minimizing with respect to the other parameters. This also exemplifies why it is important to choose a model whose parameters can be associated with certain traffic situations: Only in this case, it is possible to exclude some parameters from estimation a priori, if the corresponding situations are not contained in the data.

16.3 Calibration to Macroscopic Observations

Macroscopic traffic flow data for calibrating microscopic or macroscopic models include aggregated stationary detector speed and flow data (Sect. 3.2), travel-time information obtained from floating-car or floating-phone data (Sect. 16.2.2), and trajectory data covering all vehicles on a road segment during a certain time interval. While trajectory data are microscopic in nature, they allow all sorts of macroscopic aggregation by calculating local densities, generating aggregated virtual detector data, or determining travel times.

Calibration to macroscopic data is significantly more difficult than calibration to single trajectories. As a minor complication, we have an additional parameter to calibrate which is irrelevant when calibrating car-following models to trajectories: The vehicle length or, macroscopically, the maximum density. More importantly, however, when naively transferring the successful schemes for microscopic calibration to the more complex macroscopic situations, we generally obtain objective functions of Type III containing many secondary minima and fluctuations (Sect. 16.3.1). This requires more sophisticated schemes for nonlinear optimization and makes the methods less robust. Therefore, it is often better to calibrate to global properties of traffic flow that do not depend sensitively on minor details of the system (Sect. 16.3.2). As well as travel times, this includes the propagation of jam fronts, or global attributes of traffic waves.

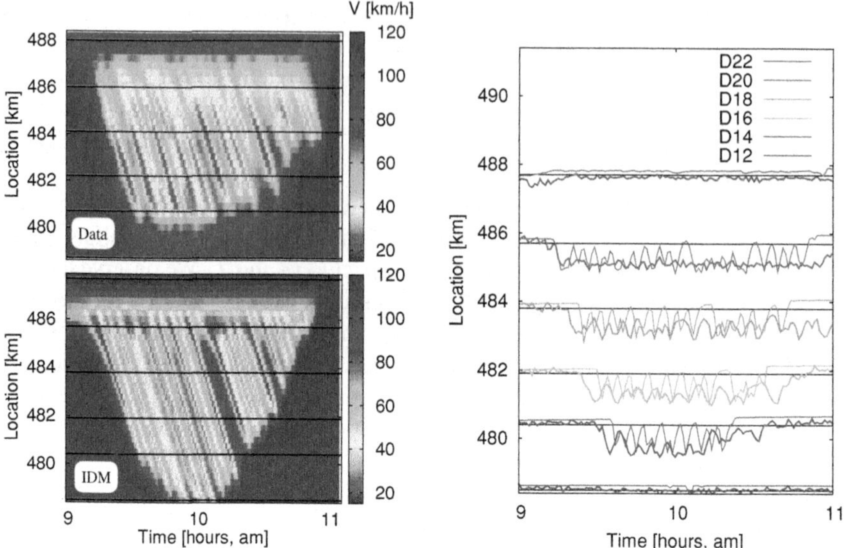

Fig. 16.5 Calibrating the IDM to aggregated stationary detector data of a traffic jam on a German Autobahn. Shown are the local speed reconstructed from the data (*top left*), the simulated local speed (*bottom left*), and one-minute speed averages of some of the real and simulated virtual detectors (*right*). Also shown are the positions of the detectors (*thin black lines*)

16.3.1 Fitting Local Properties of Traffic Flow

In this subsection, we proceed analogously to the calibration to microscopic data and formulate objective functions in terms of deviations of simulated and locally measured quantities such as time series of speed and flow. Vehicle number conservation implies that the integrated flow is fixed while no such restrictions apply for the speed. Therefore, speed differences are generally more suitable.[19]

Calibrating car-following models to stationary detector time series. In the following, we present the procedure in form of an example and calibrate the Intelligent Driver Model (IDM) to lane-averaged one-minute stationary detector data recording a breakdown in oscillatory congested traffic on the German Autobahn A5-South near Frankfurt (Fig. 16.5). The simulation is driven by the data of the most upstream detector serving as in-flowing boundary condition and, near the downstream end of the simulated region, by the flow data of an on-ramp. All other detectors (some of which are indicated by the black lines in Fig. 16.5a) serve for calibration. Generalizing the microscopic definitions, we formulate the objective function in terms of the speed readings of the real and simulated virtual detectors inside the simulated region

[19] Notice that we gave the opposite recommendation when calibrating to extended floating-car data or single trajectory data: There, the integrated speed is externally fixed by the leader while no constraints apply to the gap, i.e., the microscopic density.

by following sum of squared speed differences:

$$S_V^{\text{abs}}(\boldsymbol{\beta}) = \frac{1}{nK} \sum_{k=1}^{K} \sum_{i=1}^{n} \left(V_{ik}(\boldsymbol{\beta}) - V_{ik}^{\text{data}} \right)^2. \tag{16.42}$$

Alternatively, we can formulate the relative objective function

$$S_V^{\text{rel}}(\boldsymbol{\beta}) = \frac{1}{nK} \sum_{k=1}^{K} \sum_{i=1}^{n} \left(\frac{V_{ik}(\boldsymbol{\beta}) - V_{ik}^{\text{data}}}{V_{ik}^{\text{data}}} \right)^2 \tag{16.43}$$

which is also amenable to a hybrid generalization containing both differences of speed and flow, e.g.,

$$S_{\text{hybr}}(\boldsymbol{\beta}) = \gamma S_V^{\text{rel}}(\boldsymbol{\beta}) + (1-\gamma) S_Q^{\text{rel}}(\boldsymbol{\beta}). \tag{16.44}$$

In our example, $K = 11$ denotes the number of detectors that are available for calibration. We consider $n = 120$ one-minute intervals from 9:01 to 11:00 h covering the time interval where congestion occurs. Figure 16.5c shows the differences for six of the eleven detectors. Because of the unavoidable systematic errors when determining the speed by stationary detectors (cf. Sect. 3.2) it is crucial to simulate not only traffic flow but the data aggregation procedure as well. To this end, we place *virtual detectors* at the locations of the real detectors and record one-minute arithmetic speed averages, i.e., exactly what the data provide. The plot of the measured and simulated time series shows that the objective functions defined above sensitively depend on the relative phase shift between the measured and simulated traffic waves. This means they are contingent on minor details of the real and simulated systems shifting the onset of the breakdown by few minutes.[20] Even if a perfect model were able to produce the observed wave period exactly, these shifts would make a difference whether the simulated waves are in or out of sync with the observations.

Figure 16.5c displaying a plot of the "fitting landscape" of Eq. (16.42) in the parameter subspace spanned by V_0 and T shows that, in spite of being derived from an integrated quantity, the objective function is fluctuating, i.e., of Type III. Moreover, it turns out that some parameter settings producing homogeneous congested traffic (i.e., no oscillations at all) lead to better fitting results than any setting corresponding to traffic flow instabilities: While, visually, the spatiotemporal dynamics of the simulation of Fig. 16.5 agrees well with the observations, the parameterization is far away from the values that minimize Eq. (16.42). When lowering the desired speed to unreasonable values, we expect even comparatively good fitting values for parameter settings where the simulation does not produce *any* congestion.

---

[20] This includes fluctuations of the inflow from one minute to the next, whether a driver entering the freeway via the ramp is able to merge at once, or whether there are trucks overtaking each other, or not.

The fitting landscape Fig. 16.1c verifies this: While an indistinct and fluctuating optimum is located at reasonable values ($v_0 \approx 30$ m/s and $T \approx 1.05$ s corresponding to Fig. 16.5), the objective function is nearly as good for desired speeds below 20 m/s (72 km/h) and time gaps of $\approx$0.8 s corresponding to a simulation that does not produce any breakdown.

While we cannot get rid of the fluctuations, we can, at least, eliminate the unwelcome result that a qualitatively wrong dynamics leads to a good objective function: When adding more free-flow situations to the calibration (e.g., the data before 9:00 a.m or after 11:00 a.m), low values for the desired speed v_0 will lead to high penalties "elevating" the corresponding parts of the fitting landscape of Fig. 16.5c resulting in a more distinct optimum.

Calibrating macroscopic models to detector time series. The calibration proceeds similarly to the calibration of car following models: (i) formulate absolute, relative, or hybrid objective function such as Eqs. (16.42)–(16.44) for speed, flow, or density, (ii) run the simulation with the boundary conditions (and possible ramp source terms) driven by detector data, (iii) estimate the parameters by minimizing the objective function.

One difficulty arises because it is not possible to simulate the data aggregation process (virtual detectors) macroscopically. The macroscopic model provides the local density and space mean speed while the detectors provide the time mean speed which is systematically biased towards lower values (cf. Sect. 3.2). The bias is amplified when calibrating the simulated density to the density estimate derived from the speed and flow data (which we do not recommend). In this respect, macroscopic models are harder to calibrate and may even lead to more strongly fluctuating objective landscapes than the car-following models. Therefore, it is essential to choose an efficient nonlinear optimization method. It turns out that the *kernel-based cross-entropy method* (see Sect. 16.1.2) is a good choice.

16.3.2 Calibration to Global Properties

As discussed in the previous section, calibrating microscopic or macroscopic models directly to stationary detector data generally leads to fluctuating objective functions. Furthermore, the calibration results are often neither plausible nor robust. Fortunately, there are better approaches. As a common property, they compare some integrated quantity rather than the original time series. This includes the travel time, the aggregated number of vehicles, or the propagation of congestion fronts. In all these cases, the integration eliminates most of the high-frequency components of the traffic flow dynamics that are mainly responsible for the artifacts discussed above. However, by construction, single traffic waves are eliminated so these methods cannot be used to test the ability of models to describing stop-and-go traffic. We propose a dedicated method based on wave attributes, for this purpose.

Jam front propagation. Consider the situation on the German Autobahn A5-North depicted in Fig. 16.6: Traffic breaks down between road kilometer 481 and 482 at

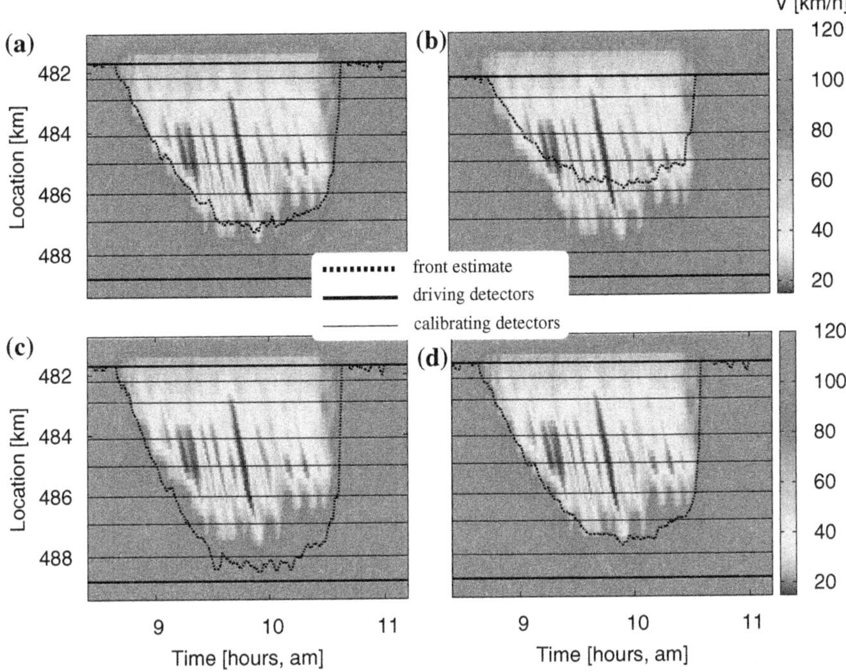

Fig. 16.6 Calibrating the macroscopic LWR model with a triangular fundamental diagram (section-based model) to aggregated detector data on the German Autobahn A5 with respect to the propagation of jam fronts (see the main text for details). **a** Result for the estimated parameters $V_0 = 100\,\text{km/h}$, $c = -17.5\,\text{km/h}$, and $Q_{\text{max}} = 1\,510\,\text{vehicles/h}$; **b** the jam propagation velocity is increased to $c = -13\,\text{km/h}$; **c** the capacity per lane is reduced to $Q_{\text{max}} = 1\,400\,\text{vehicles/h}$; **d** the desired speed is increased to $V_0 = 180\,\text{km/h}$. In the diagrams (**b**)–(**d**), only one parameter is changed with respect to the reference

8:45 a.m, because of the ramps at an interchange. The resulting congestion, or more precisely its upstream front where vehicles enter the jam, propagates upstream for about one hour before the front reverses its propagation direction and crosses the on-ramp bottleneck again at about 10:30 a.m signifying the end of the jam. Taking the location of the jam front as the basis to formulate objective functions is promising for several reasons:

- Models calibrated in this respect make useful predictions: In contrast to details of traffic waves, every driver wants to know the location of a jam front in advance, either to circumnavigate the jam or to carefully approach it.
- The situation involves both free traffic (upstream) and congested traffic (downstream), so parameters pertaining to both regimes can be calibrated.
- Since the position of the front is an integrated quantity (namely the time integral of the propagation velocity), high-frequency fluctuations are damped, and we expect the objective function to be well-defined.

In essence, this approach allows us to calibrate all *kinematic* parameters determining the fundamental diagram. However, it is not sensitive to *dynamic* parameters determining accelerations. Since first-order macroscopic models (LWR models) do not contain such parameters, this approach is particularly suitable for calibrating models of this class.

For the LWR model with a triangular fundamental diagram (section-based model, see Sect. 8.5), the calibration is particularly effective since no explicit simulation is necessary to determine the model predictions. The shock-wave formula (8.19) is all we need. Expressing it in terms of the local flows Q_1 and Q_2 immediately upstream and downstream of the front, respectively, and for the parameters desired speed V_0, congested propagation velocity c, and capacity $Q_{\max}$ per lane, we obtain

$$\frac{\mathrm{d}x_{12}}{\mathrm{d}t} = c_{12} = \frac{(Q_2 - Q_1)V_0 c}{Q_2 V_0 - Q_1 c + Q_{\max}(c - V_0)}. \tag{16.45}$$

We drive the LWR model by the flows $Q_D^{\mathrm{up}}(t)$ and $Q_D^{\mathrm{down}}(t)$ measured by a detector pair at the boundaries of the modeled region at the locations $x = 0$ and $x = L$, respectively. Using relations (8.17) and (8.18) for the propagation velocities, we arrive at following relation between the local and measured lane-averaged flows:

$$Q_1(t) = Q_D^{\mathrm{up}}\left(t - \frac{x_{12}}{V_0}\right), \quad Q_2(t) = Q_D^{\mathrm{down}}\left(t - \frac{x_{12} - L}{c}\right). \tag{16.46}$$

Equations (16.45) and (16.46) constitute a one-dimensional delay-differential equation which is numerically easy to solve. Notice that both delays x_{12}/V_0 and $(x_{12} - L)/c$ are positive, so only past data are needed. Consequently, besides calibration, this approach is also suitable for real-time traffic state estimation.[21]

In our case, we drive the section-based model with data from the upstream detector at road kilometer $x_u = 488.8$ km and the downstream detector at $x_d = 422.2$ km and initialize the jam front to the downstream detector location, $x_{12}(0) = x_d$. Prior to the breakdown, $x_{12}(t)$ remains essentially stationary.[22] As soon as a congestion front crosses x_d, the front $x_{12}(t)$ estimated by Eq. (16.45) starts to propagate upstream along with the real congestion front. Obviously, the fit quality depends on the estimation errors for the location of the front at times where this information can be deduced from the calibrating detectors. A suitable objective function is the root mean square (rms) deviation

$$S_{\delta x}^{\mathrm{abs}}(\boldsymbol{\beta}) = \sqrt{\frac{1}{M}\sum_{m=1}^{M}\left[x_{12}(t_m) - x_{12}^{\mathrm{data}}(t_m)\right]^2}. \tag{16.47}$$

[21] Even for state prediction rather than calibration, we do *not* recommend using speed detector data to estimate densities and use them directly by calculating $c_{12} = (Q_2 - Q_1)/(\rho_2 - \rho_1)$. The bias in estimating the densities makes this approach impractical.

[22] There may be a small drift as a consequence of systematic counting errors at the two detectors. This can be taken care of by applying Eq. (16.45) only if the downstream detector indicates congested traffic. However, this measure was not necessary in our example.

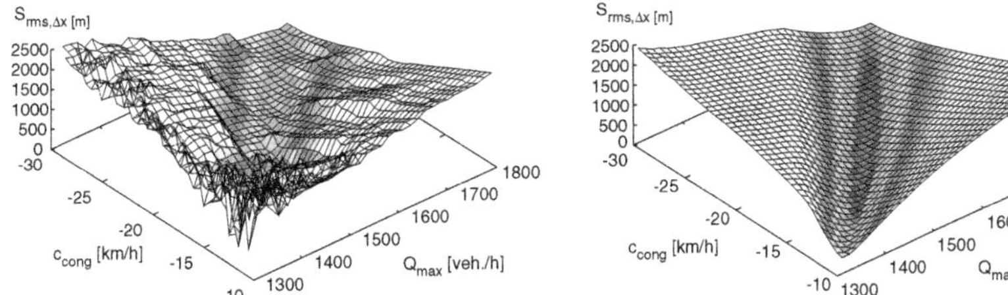

Fig. 16.7 Fitting landscape of the objective function (16.47) for the section-based model (fixed $V_0 = 100$ km/h) applied to the original (*left*) and smoothed detector data (*right*) shown in Fig. 16.6. For the smoothing, we applied a Gaussian filter of width 2 min. The minimum corresponds to Fig. 16.6a

Here, t_m denotes the times where the front crosses one of the calibrating detectors (thin black lines in Fig. 16.6) in either direction. Notice that the number M of such events is not directly related to the number K of calibrating detectors (six, in our example) but generally is limited to $2K$.

A visual inspection of the results for different parameter combinations $\boldsymbol{\beta}^{\mathrm{T}} = (V_0, c, Q_{\max})$ shows that this approach is robust: Varying the parameters by 20 % or more from their respective estimates results in moderate quantitative (but no qualitative) changes. However, the procedure is insensitive to variations of V_0 which, therefore, cannot be estimated (see Fig. 16.6d).

Figure 16.7a displaying the fitting landscape in the dimensions c and $Q_{\max}$ shows, however, that the objective function is fluctuating (Type III). This can be traced back to the upstream free-flow flow data fluctuating wildly by up to 30 % from one minute to the next. Since this is caused by microscopic dynamics (e.g., formation of vehicle platoons following slower vehicles) which is not relevant here, smoothing the data of the driving and calibrating detectors is unproblematic.

Figure 16.7b shows the result after applying a Gaussian smoothing of width 2 min: We obtain a smooth fitting landscape with a unique global minimum at $\hat{c} = -17.5$ km/h and $\hat{Q}_{\max} = 1\,510$ vehicles/h (we kept V_0 at a fixed value of 100 km/h). Inspecting the jam front dynamics for this estimate shows that the smoothing does not change the result in any significant way.

The robustness of this method carries over to the question of which detector pair to choose for driving the model. Generally, any detector pair is suitable as long as the downstream detector is at or upstream of the bottleneck causing the congestion, and the upstream detector is upstream of the maximum extension of the congestion. Furthermore, no ramps are allowed at or between these detectors (Fig. 16.8).[23]

Attributes of traffic flow oscillations.

The above method only calibrates parameters that are related to the fundamental diagram. This does not include parameters related to accelerations such as the speed relaxation time τ of the Optimal Velocity Model and many second-order models,

[23] Generalizing the method to include ramps is straightforward but requires ramp detector data which are rarely available.

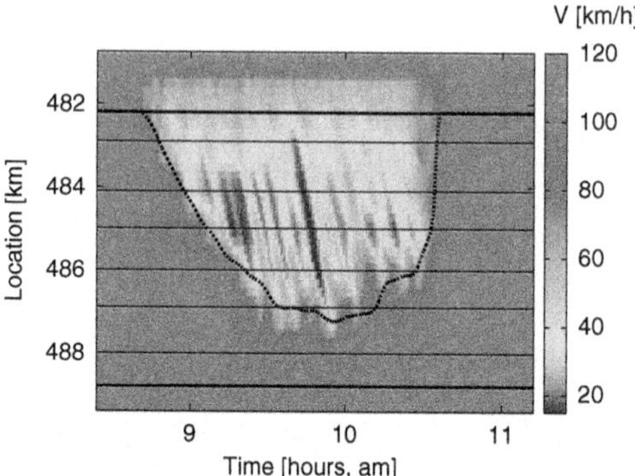

Fig. 16.8 The upstream jam front as predicted by the section-based model driven by the smoothed data (Gaussian filter, width 2 min) of the detectors at road kilometer 489 (*upstream*) and 482 (*downstream*). The parameters are the same as in Fig. 16.6a

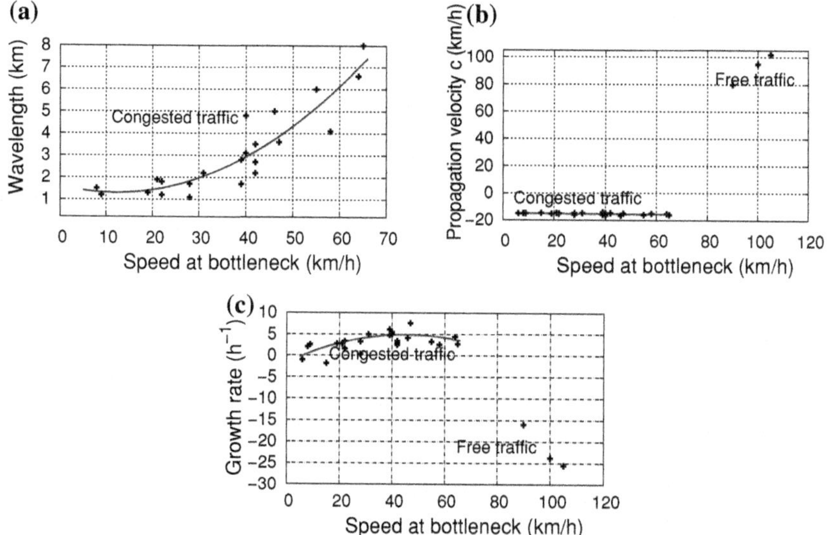

Fig. 16.9 Systematic properties of traveling jam waves that can be used for calibrating microscopic and second-order macroscopic models to traffic flow oscillations

the speed difference sensitivity γ of the Full Velocity Difference Model, or the acceleration parameters a and b of the IDM and Gipps' model.

To calibrate these "dynamic" parameters by macroscopic measurements, we use the fact that these parameters essentially determine the traffic flow instabilities. However, we saw in Sect. 16.3.1 that directly minimizing squared differences to detector time series recording the oscillations is impractical. As an alternative, we define quantitative attributes describing both the observed and simulated traffic waves and construct objective functions out of them. The main attributes are (cf. Fig. 16.9):

- The growth rate σ of developing waves as a function of the average speed $\bar{V}$ at the bottleneck.
- The wavelength L_w of not yet saturated oscillations as a function of $\bar{V}$.
- The wave propagation velocity c.

While this procedure is robust, it has the disadvantage that one instance of a congestion only produces one data point $(\bar{V}, \sigma, L_w, c)$. Furthermore, not all instances of breakdowns are suitable. Finally, already the "measurements" (i.e., the data points of Fig. 16.9) are noisy.

16.4 Validation

In the previous sections, we have discussed *calibration*, i.e., fitting a model to "training data" by estimating its parameters. While this is useful on its own (for example, it allows for a sensitivity analysis), the main intention when using traffic flow models is to reliably predict or assess new situations rather than reconstruct past events. Therefore, we attribute the quality of a model to its reliability or *predictive power* rather than its *fitting power*. Unfortunately, a good fitting power (which is characterized by a low absolute minimum of the objective function) does not automatically carry over to a good predictive power. In the extreme case of estimating m model parameters to fit m data points, the fitting power is 100 % while the predictive power is zero. Therefore, besides calibration, we need additional procedures to assess the reliability and predictive power. This is called *validation*:

> Validation is the process of determining the reliability of a model, i.e., the degree to which it is an accurate representation of the real world from the perspective of the intended uses.

In fact, the reliability of a model (or a certain simulation technique) is more connected to its *robustness* than to its fitting power: If the outcome of a simulation does not change very much when changing the model parameters, it is likely that this model performs well in a new situation where we expect changes of the inherent model parameters (which only a new calibration could reveal, after the act).

Any validation technique must "simulate" the model predictions and compare them with data that are already available. To this end, we split the available data into *training data* on which to calibrate the model, and *test data* or *validation data* on which to simulate predictions of the calibrated model. Ideally, the different data subsets should represent identical situations. Otherwise, the validation results are difficult to interpret since we have no means to separate the effects of model imperfections from differences in the real dynamics (see the second example below).

To assess the prediction quality, we use the same objective function as that for calibration. Depending on how the split into training and validation data is organized, we distinguish following validation techniques:

- *Holdout validation:* One part of the available data is selected for calibration, the other for validation.
- *Cross validation:* The available data are split into K subsets and the holdout validation is repeated K times. In each round, a different subset becomes the test data set while the other $K - 1$ subsets constitute the training data.
- *Inverse cross-validation:* As in cross validation, there are K rounds. In each round, a different subset becomes the training set, and the other $K - 1$ subsets are used for $K - 1$ validations. For $K = 2$, this is equivalent to cross validation. Otherwise, inverse cross validation is more demanding since the model has less training and must predict more than in cross validation.

Cross validation makes maximal use of the available data while inverse cross-validation allows to assess the variability of the parameter estimates when calibrating to independent test data sets. In this way, we can approximate the parameter estimation errors. For the maximum-likelihood method, we can test whether the analytical expression (16.37) for these errors gives plausible values.

In the following, we describe the validation techniques by three examples: Holdout validation of the LWR model on detector data, inverse cross-validation of the IDM on extended FC data, and "synthetic cross-validation" to determine a similarity index of a model pair.

Example 1: Holdout validation of the section-based model. In this example, we test whether the good descriptive power of the section-based model (LWR model with a triangular fundamental diagram) in describing jam fronts (Fig. 16.6) carries over to new situations to which it has not been calibrated.

For this purpose, we interpret the data of the oscillatory congestion on the Autobahn A5-North shown in Fig. 16.6 as training set and let the calibrated model predict the jam front of homogeneous congested traffic on the Autobahn A5-South (Fig. 16.10). While the average deviation $S_{\Delta x}^{\text{rms}} \approx 800\,\text{m}$ is higher than in the training set (about 300 m) and certainly could be calibrated to better results, the predictive power is high considering that the model is driven by just two detectors (which are 12 km apart), and the model does not even "know" about the onset of congestion (we started the integration of Eq. (16.45) at 0:00 h).

Using the two data sets of the A5-South and A-North, we can formulate further validation procedures, e.g.,

- a $K = 2$ cross validation by calibrating the LWR model by the A5-South data and using the A5-North data as test set,
- validation by using different upstream and downstream detector pairs on the same or on the other freeway data set.

Example 2: Inverse cross-validation of the IDM. Table 16.1 shows how one could proceed when validating the IDM to $K = 3$ sets of extended FCD. In each of the three

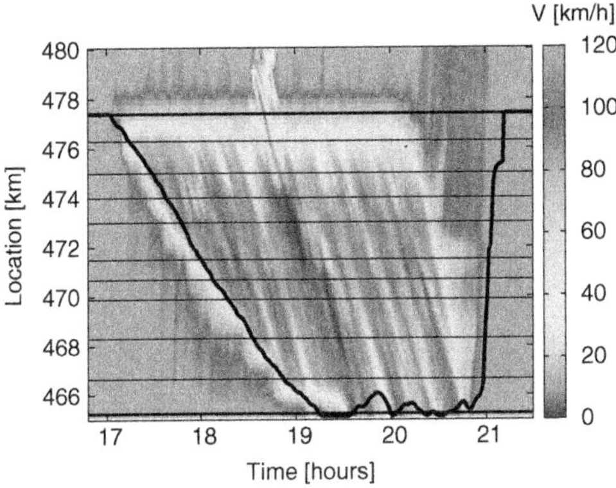

Fig. 16.10 Validating the calibrated LWR-based jam propagation method on data of homogeneous congested traffic (HCT) of the German Autobahn A5-South. We used the parameter values $V_0 = 100\,\text{km/h}$, $c = -17.5\,\text{km/h}$, $Q_{\text{max}} = 1\,510\,\text{vehicles/h}$ resulting from the calibration to data of oscillatory congested traffic (OCT) of the German A5-North (Fig. 16.6a)

Table 16.1 Inverse cross validation of the IDM with respect to the mixed gap objective function $S_s^{\text{mix}}(\boldsymbol{\beta})$ applied to three sets of extended FCD (two of which are shown in Fig. 16.2; the not displayed Set 1 is similar to Set 2)

	Calibration FCD Set 1	Calibration FCD Set 2	Calibration FCD Set 3
v_0 [m/s]	70.0	70.0	16.1
T [s]	1.12	1.44	1.31
s_0 [m]	2.35	2.79	1.52
a [m/s^2]	1.23	0.973	1.56
b [m/s^2]	3.10	0.993	0.626
Validation S_s^{mix}, FCD Set 1	<u>20.8 %</u>	28.7 %	28.6 %
Validation S_s^{mix}, FCD Set 2	35.4 %	<u>26.2 %</u>	40.6 %
Validation S_s^{mix}, FCD Set 3	41.2 %	26.9 %	<u>13.0 %</u>

The upper five data lines and the diagonal elements of the lower 3×3 part show the calibration results while the off-diagonal elements show the validation results

calibration-validation rounds corresponding to the three data columns of this table, we calibrate the IDM to one data set, and apply the calibrated IDM to the two remaining sets. We observe that all parameter estimates except that of the desired speed for Sets 1 and 2 are reasonable (the unrealistic values result from incomplete data, cf. Sect. 16.1.2). Nevertheless, the parameter vectors differ considerably and most of the validation results are significantly worse than the corresponding calibration fitness. However, the three data sets represent different drivers and different situations, so the validation results may reflect data heterogeneities (inter-driver variations) rather than a low predictive power. For example, the low fit quality of the validations to Set 3 are mainly caused by the high value of the desired speed when calibrating to the Sets 1 or 2. This, again, underlines our general recommendation to exclude

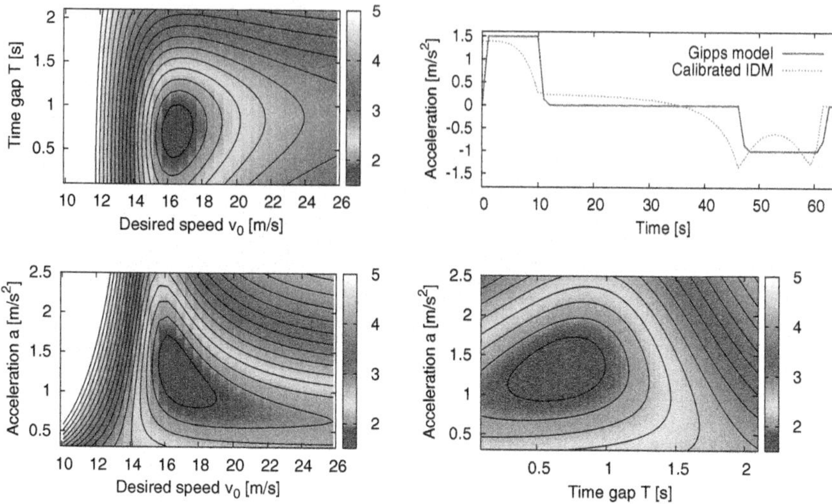

Fig. 16.11 Cross-comparing Gipps' model with the Intelligent Driver Model (IDM). The IDM is locally calibrated to a trajectory generated by Gipps' model for the first vehicle of the city simulation of Fig. 11.1 using the maximum-likelihood method for accelerations. Shown are the acceleration profiles (*right upper graphics*) and three two-dimensional sections of the negative log-likelihood function (16.40) in five-dimensional parameter space around the estimate $\hat{\boldsymbol{\beta}}$ ($\hat{v}_0 = 15.9$m/s, $\hat{T} = 0.7$s, $\hat{a} = 1.4$m/s^2, $\hat{b} = 0.7$m/s^2, $\hat{s}_0 = 1.3$m)

parameters from calibration if the calibrating data do not include the corresponding situations, and to use standard values instead.

If we succeed in validating the model independently on other more homogeneous data subsets (i.e., by splitting a long extended FCD time series of a single driver into training and test sets), we can reverse the line of thought: Then, a poor validation result does no longer mean a substandard model but differences of the driving style of the two drivers associated with the calibration-validation pair. In the last example, we adopt this reasoning to assess the limits of calibration and validation in a controlled environment.

Example 3: Synthetic cross-calibration between models. In this example, we generate synthetic data by simulating a given car-following model and calibrate other models to these data. Strictly speaking, this is no validation since we do not use real data, nor test calibrated models on new data. However, this exercise shows the theoretical limits of model validation due to inter-driver variations in a completely controlled environment.

In all simulations, we assume an identical scenario, namely that depicted in Fig. 11.1: A single vehicle starts at a traffic light, accelerating freely to 15 m/s, and stops at the next traffic light 730 m ahead. Figure 16.11 shows the result when generating the synthetic trajectory by Gipps' model ($v_0 = 15$ m/s, $T = 1.1$ s, $a = 1.5$ m, and $b = 1$ m/s^2), and calibrate the IDM to this trajectory by maximizing the log-likelihood (16.40). Since the situations are identical and there are no intra-driver variations (we do not change parameters during a simulation), all residual deviations are exclusively caused by inter-driver variations. The acceleration time series clearly shows the incompatible driving styles represented by Gipps's model and the IDM. The "IDM driver" is forced to emulate "Gipps driving style" which is only partially successful, even for the best fit. We conclude that each model represents an individual driving style which other models are only partially able to reproduce. We can use this to define a non-symmetric "commonality matrix" whose elements describe the ability of model i to fit the driving style of model j.

Finally let us note that there is evidence that, generally, the driving styles of different human individuals are described best by different models, i.e., on the road, we encounter "Gipps' model drivers", "IDM drivers", and so on. This also means that there are fundamental limits of the predictive power of traffic flow models, at least with respect to microscopic aspects.

Problems

16.1 Influence of serial correlations on measures of parsimony

Consider floating-car data captured at a sampling rate of ten data points per seconds. Each data point is composed of the gap s_i and the speed v_i sampled at time t_i. To "simulate" serial correlation, we assume that the data points are identical for all time steps t_i within each one-second interval. Now consider a LSE estimation of the "models" (i) $\hat{s}(v) = \beta_0$, (ii) $\hat{s}(x) = \beta_0 + \beta_1 v$ to 20 such data points and discuss why the new parameter β_1 explains many more additional data points than it would do if the data were uncorrelated. Also explain why this means that parsimony measures relying on iid errors underestimate the negative effects of the new parameter β_1 on the predictive power for a given fitting power.

Further Reading

- Brockfeld, E., Kühne, R.D., Wagner, P.: Calibration and validation of microscopic traffic flow models. Transportation Research Record: Journal of the Transportation Research Board **1876** (2004) 62–70
- Hoogendoorn, S., Hoogendoorn, R.: Calibration of microscopic traffic-flow models using multiple data sources. Philosophical Transactions of the Royal Society A: Mathematical, Physical, Engineering Sciences **368** (2010) 4497–4517
- Kesting, A., Treiber, M.: Calibrating car-following models by using trajectory data: Methodological study. Transportation Research Record: Journal of the Transportation Research Board **2088** (2008) 148–156
- Ngoduy, D., Maher, M.J.: Calibration of second order traffic models using continuous cross entropy method. Transportation Research Part C: Emerging Technologies **24** (2012) 102–121

- Ossen, S., Hoogendoorn, S.P.: Heterogeneity in car-following behavior: Theory and empirics. Transportation Research Part C: Emerging Technologies **19** (2011) 182–195
- Punzo, V., Borzacchiello, M.T., Ciuffo, B.: On the assessment of vehicle trajectory data accuracy and application to the Next Generation SIMulation (NGSIM) program data. Transportation Research Part C: Emerging Technologies **19** (2011) 1243–1262
- Thiemann, C., Treiber, M., Kesting, A.: Estimating acceleration and lane-changing dynamics from next generation simulation trajectory data. Transportation Research Record: Journal of the Transportation Research Board **2088** (2008) 90–101
- Treiber, M., Kesting, A.: Validation of traffic flow models with respect to the spatiotemporal evolution of congested traffic patterns. Transportation Research Part C: Emerging Technologies **21** (2012) 31–41

Chapter 17
The Phase Diagram of Congested Traffic States

A person with a new idea is a crank until the idea succeeds.
Mark Twain

Abstract In the previous chapter, we characterized the stability properties of traffic flow models by stability diagrams of steady-state traffic as a function of density, and by stability classes distinguishing the most relevant qualitative features of the stability diagrams. Now, we show how this determines the spatiotemporal dynamics of congestion patterns in realistic open systems with bottlenecks. Besides the stability class, the influencing factors are the traffic demand, and the bottleneck strength. We discriminate qualitatively different patterns by regions in a *dynamic phase diagram* spanned by traffic demand and bottleneck strength and compare the theoretical result with simulated and real congestions. It turns out that the observations are described by stability classes displaying instabilities for congested traffic, only.

17.1 From Ring Roads to Open Systems

The *stability diagram* and the stability classes explored in Sect. 15.6 are valid for ring roads and homogeneous infinite systems. For understanding real-world traffic flow, we need to transfer these concepts to open inhomogeneous road networks with bottlenecks. This is possible by combining the knowledge of how to model bottlenecks with LWR models (Sect. 8.5.6) with the stability properties summarized in the stability diagram. The starting point is the observation that, if traffic is congested, the bottleneck determines the congested traffic flow by its capacity. However, unlike the situation in LWR models, the relevant capacity is not given by the *static bottleneck capacity* C_{B} defined by the maximum steady-state flow (reduced by the on-ramp flow

M. Treiber and A. Kesting, *Traffic Flow Dynamics*,
DOI: 10.1007/978-3-642-32460-4_17,

in case of on-ramp bottlenecks) but by the *dynamic capacity* C_B^{dyn} of the *activated* bottleneck,

$$C_B^{\text{dyn}} = C_B(1 - \varepsilon). \tag{17.1}$$

A bottleneck is said to be activated once it has caused a traffic breakdown. As a consequence, the maximum throughput is reduced by the so-called *capacity drop* after the breakdown ("activation" of the bottleneck). Its observed relative value ε is reported to vary wildly but it is of the order of 10 %. In empirical or simulated flow-density data, this is reflected by the mirrored-λ shape, see Sect. 4.4 and Fig. 11.5. In models, the relative capacity drop ε depends on the bottleneck strength and some other properties of the bottleneck such as the length of the merging region in case of on-ramp bottlenecks. It can only be determined numerically by simulation.

In any case, the steady-state of the congested traffic flow upstream of the bottleneck is described by the *congested* branch $\rho_{\text{cong}}(Q)$ at a flow characterized by the dynamic bottleneck capacity per lane. For a congested road section with I lanes, we have

$$Q_e = \frac{C_B^{\text{dyn}}}{I}, \quad \rho_e = \rho_{\text{cong}}(Q_e) \text{ Steady-state of congested traffic in open systems.} \tag{17.2}$$

The steady-state density $\rho_{\text{cong}}(Q)$ of the congested branch of the fundamental diagram plays a crucial role in the transition from closed to realistic open systems:

The dynamics of ring roads or homogeneous infinite roads, i.e., *closed* systems, is controlled by traffic density. The dynamics of realistic roads, i.e., *open* systems with a bottleneck, is controlled by two flow-like quantities: The uncongested road sections are controlled by the inflow per lane, and the congested sections by the (dynamic) bottleneck capacity per lane.

The inversions of the free-flow and congested branches of the fundamental diagram, $\rho_{\text{free}}(Q)$ and $\rho_{\text{cong}}(Q)$, respectively, "translate" the control parameters by expressing the inflow and the bottleneck capacity in terms of the density. They are the central elements for applying the concepts developed for homogeneous roads (stability diagram, stability classes) to realistic infrastructure.

17.2 Analysis of Traffic Patterns: Dynamic Phase Diagram

Relation (17.2) allows us to derive the set of theoretically expected spatiotemporal patterns of congested traffic, i.e., the *dynamic phase diagram*, as a function of the following two control parameters.

Traffic demand. The inflow $Q_{\text{in}} = Q_{\text{demand}}$ indicates the *potential* average traffic flow on the main-road. It is only *realized* if the upstream boundaries of the network are free. Otherwise, the realized inflow is lower.[1]

Bottleneck strength . The severity of the bottleneck is expressed by the difference between the static capacity $C = I\,Q_{\text{max}}$ on the homogeneous road sections upstream of the bottleneck, and the dynamic bottleneck capacity, divided by the number of lanes:

$$\boxed{\Delta Q = \frac{C - C_{\text{B}}^{\text{dyn}}}{I} \quad \text{Bottleneck strength}} \tag{17.3}$$

where $C_{\text{B}}^{\text{dyn}} = C_{\text{B}}(1-\varepsilon)$ indicates the maximum flow through the activated bottleneck that is available for the main-road traffic.

On-ramp bottlenecks. By means of the above definition for $C_{\text{B}}^{\text{dyn}}$, the dynamic capacity decreases and the bottleneck strength increases with increasing ramp flow Q_{rmp}. With the general definition (17.3) and the relations $C = I\,Q_{\text{max}}$ and $C_{\text{B}} = I\,Q_{\text{max}} - Q_{\text{rmp}}$, we obtain for the strength of an on-ramp bottleneck

$$\Delta Q = \frac{C - C_{\text{B}}^{\text{dyn}}}{I} = \frac{C - C_{\text{B}}(1-\varepsilon)}{I} = \frac{Q_{\text{rmp}}(1-\varepsilon)}{I} + \varepsilon Q_{\text{max}}. \tag{17.4}$$

For $I = 1$ lane, and negligible capacity drop $\varepsilon \approx 0$, we obtain the simple relation

$$\Delta Q \approx Q_{\text{rmp}}. \tag{17.5}$$

However, since Q_{rmp} is the influencing exogenous factor for the more general relation (17.4) as well, the phase diagram of on-ramp bottlenecks is often given in terms of the control parameters Q_{in} and Q_{rmp}.

Off-ramp bottlenecks. Off-ramp bottlenecks derive their obstructing power from the higher speed variations and less efficient lane usage due to the lane-changing activity and the decelerations of the vehicles about to leave the main-road. Therefore, the actual bottleneck is located *upstream* of the exit lane of the off-ramp. Regarding the congestion patterns, an off-ramp bottleneck is equivalent to a flow-conserving bottleneck upstream of the actual off-ramp. Since a higher outflow leads to higher perturbations due to lane changes and decelerations, the bottleneck strength of off-ramps increases with the outflow, similarly to that of on-ramps.

Traffic flow upstream of an activated bottleneck. Using Eqs. (17.3) and (17.2), we can write the steady-state density of the congested traffic upstream of the bottleneck by

---

[1] If the accumulated difference between demand and actual inflow is stored in a virtual buffer and this buffer is emptied once demand becomes lower than the maximum possible inflow, the realized inflow may also become higher than the demand.

$$Q_e = Q_{\max} - \Delta Q, \quad \rho_e = \rho_{\text{cong}}(Q_{\max} - \Delta Q). \tag{17.6}$$

This relation shows that the *flow*, and not the density is the controlling factor. Now, we derive the dynamical *phase diagram* by partitioning the two-dimensional space spanned by the control variables Q_{in} and ΔQ (or: Q_{in} and Q_{rmp}) into regions where qualitatively equivalent patterns are observed.[2]

In the following, we will derive the phase diagrams for model-parameter combinations corresponding to one of the stability classes discussed in Chap. 15.

17.2.1 Stability Class 1

This stability class is characterized by linear instability at static capacity, i.e., at a density corresponding to the maximum of the fundamental diagram, $\rho_2 < \rho_C$ (cf. Sect. 15.6). This property allows us to derive the following qualitative spatiotemporal patterns and the associated dynamic traffic phases as a function of the main inflow Q_{in} and the bottleneck strength ΔQ, and possibly history (cf. Fig. 17.1).

Bottlenecks of little obstructing power. Because of the low bottleneck strength ΔQ, the traffic density $\rho_e = \rho_{\text{cong}}(Q_{\max} - \Delta Q)$ [cf. Eq. (17.6)] after breakdown is only a little bit higher than the density ρ_C at capacity. Furthermore, the instability for this stability class and density range is generally absolute (rather than convective), so oscillations propagate in both directions eventually covering the whole region at and upstream of the bottleneck (left plot of Fig. 17.2).

Since the outflow Q_{out} is only insignificantly higher than the bottleneck capacity, the buildup of a new traffic wave upstream of the bottleneck takes some time until it detaches and propagates upstream as a traffic wave ("stop-and-go wave") and the process repeats, so there is free traffic between the waves.[3]

Because of the role of the bottleneck, this type of traffic pattern is called *triggered stop-and-go waves* (TSG). Typically, TSG patterns are triggered by significant perturbations, e.g., passing traffic waves generated elsewhere. In the special case of very weak bottlenecks, the bottleneck capacity can exceed the outflow Q_{out} of a moving traffic wave, and traffic waves generated elsewhere can pass the bottleneck without triggering new waves, i.e., *activating* the bottleneck. We call this state (which includes the limiting case of no bottlenecks) *moving localized clusters* (MLC). Strictly speaking, MLC is not a separate pattern since it cannot be locally distinguished from the isolated traffic waves of the TSG pattern generated elsewhere.

---

[2] We emphasize that the phase diagram as discussed here is not related to physical phase diagrams in the sense of equilibrium thermodynamics. It is just a way of representing qualitatively equivalent patterns in the space of control variables. In a sense, the spatiotemporal patterns can be considered as nonequilibrium dynamical phases, hence the name "dynamical phase diagram".

[3] Strictly speaking, stop-and-go waves violate the validity limits of Eq. (17.6) which assumes a nearly stationary situation.

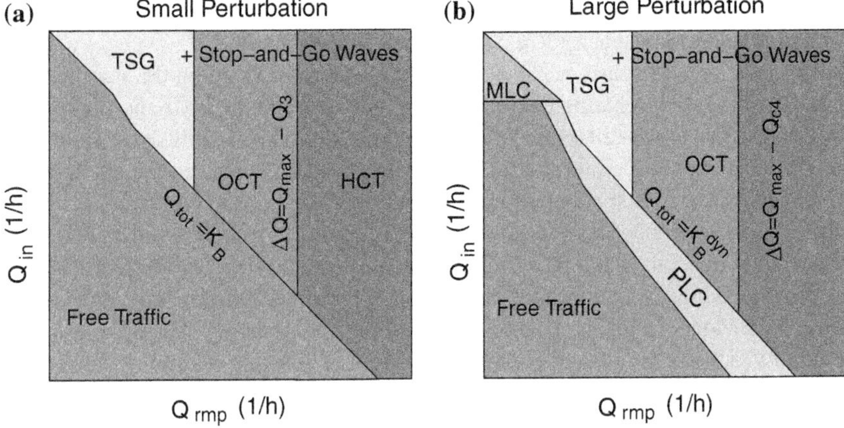

Fig. 17.1 Qualitative phase diagram for model-parameter combinations corresponding to stability class 1b as a function of the inflow Q_{in} and the ramp flow Q_{rmp}, or, equivalently, the bottleneck strength ΔQ. The *left* diagram is valid for a slowly increasing traffic demand without significant perturbations while the *right* diagram depicts the situation when a major perturbation occurred in the past (e.g., by a full-grown traffic wave originating from elsewhere). For stability class 1a, the phase of homogeneous congested traffic (HCT) drops out. For very high demands Q_{in}, traffic flow upstream of the congestion may be metastable or unstable. Therefore, stop-and-go waves may emerge upstream of the main congestion pattern

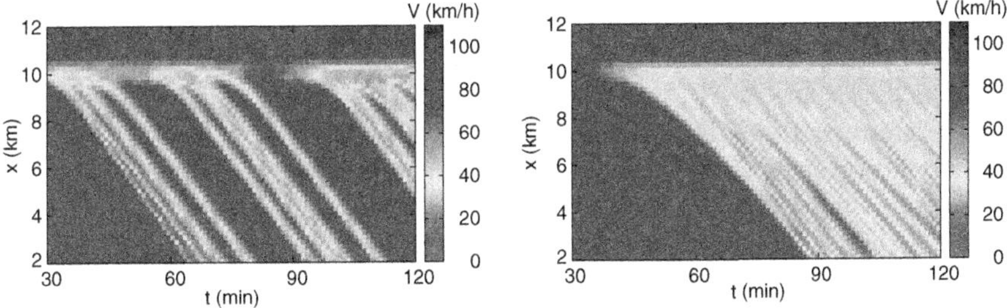

Fig. 17.2 IDM simulation of an on-ramp of inflow $Q_{\text{rmp}}/I = \Delta Q = 100$ veh/h for a parameterization corresponding to absolute instability (class 1, *left*) and convective instability (class 2, *right*). In the phase diagram of parameter space (Fig. 15.15), this corresponds to the points ($T = 1.5$ s, $a = 0.75$ m/s^2) for class 1 and ($T = 1.5$ s, $a = 1.0$ m/s^2) for class 2

Significant bottlenecks. The greater the obstructing effect of the bottleneck (the greater the bottleneck strength ΔQ or the lower its dynamic capacity), the lower is the average traffic flow $Q_{\text{cong}} = C_{\text{B}}^{\text{dyn}}/I = Q_{\text{max}} - \Delta Q$ and the higher is the associated average traffic density $\rho_e = \rho_{\text{cong}}(Q_{\text{cong}})$ as calculated with the congested branch of the inverted fundamental diagram. Consequently, the difference between the outflow Q_{out} of the traffic waves and the effective bottleneck capacity C_{B}/I (per lane) and thus the frequency of the triggered waves increases until there is no longer free traffic between the waves. This defines the traffic pattern *oscillating congested traffic* (OCT).

Severe bottlenecks. When further increasing the bottleneck strength, i.e., reducing the bottleneck capacity $C_{\rm B}$ and the average congested flow $Q_{\rm cong} = C_{\rm B}/I$ (in extreme cases to nearly zero), the resulting patterns depend on the stability subclass. If traffic flow remains unstable for congested traffic of any density provided there is nonzero flow at all (class 1a), the pattern remains OCT, only the frequency of the waves increases. Below a certain threshold flow $Q_{\rm cv}$, however, the instability is no longer absolute but it becomes upstream-convective, i.e., the region of oscillating traffic can only propagate upstream (cf. Fig. 15.15d). As a consequence, a range of nearly homogeneous congested traffic forms near the bottleneck, and traffic waves of growing amplitude emerge further upstream. The mechanism for generating the waves is no longer the triggering mechanism at the bottleneck but the convective instability. This is the typical pattern of traffic flow obeying class 2 dynamics and will be discussed in more detail in Sect. 17.2.2.

If traffic flow obeys the dynamics of stability class 1b, traffic flow reverts to metastable behavior for $\rho > \rho_3$, or, equivalently, for bottleneck strengths satisfying

$$\Delta Q > Q_{\max} - Q_3,$$

and to absolutely stable behavior for $\rho > \rho_4$ or $\Delta Q > Q_{\max} - Q_4$. This *restabilization* leads to slowly creeping non-oscillatory flow, also called the pattern of *homogeneous congested traffic* (HCT).

Traffic demand slightly below the bottleneck capacity. A common feature of the TSG, OCT, and HCT patterns is their extended nature: Theoretically, they can cover arbitrarily long road sections which is a consequence of the traffic demand exceeding the bottleneck capacity, $Q_{\rm in}/I > C_{\rm B}^{\rm dyn}/I = Q_{\max} - \Delta Q$, so the excess demand accumulates over time. If the demand $Q_{\rm in}$ is slightly below the dynamic bottleneck capacity, however, persistent extended patterns cannot occur. Nevertheless, nonlinear effects allow the self-organized formation of *localized* and *standing traffic waves*. Such waves are triggered by significant perturbations, e.g., moving traffic waves (MLC) generated elsewhere that pass the bottleneck. Since these waves are "pinned" at the location of the bottleneck, they are referred to as *pinned localized clusters* (PLC).

Of course, conservation of the vehicles implies that the outflow of this mildest type of traffic congestion is not characterized by the dynamic bottleneck capacity but by the traffic demand $Q_{\rm in}$ (in case of on-ramp bottlenecks by the sum $Q_{\rm in} + Q_{\rm rmp}$) which is below the dynamic bottleneck capacity. Since the PLC pattern follows from truly nonlinear effects, we cannot give analytic expressions for the minimum demand $Q_{\rm in}$ for sustained PLCs. Nevertheless, we can give following plausibility argument for a variable dynamic capacity which is a prerequisite for persistent PLCs: Consider an on-ramp bottleneck with a long merging region and let us assume that the initially free traffic flow downstream of the merging region is metastable, so a significant perturbation can trigger a breakdown with traffic waves propagating upstream. When such a wave moves along the merging region, it passes regions with less and less traffic flow (since an increasing fraction of on-ramp vehicles

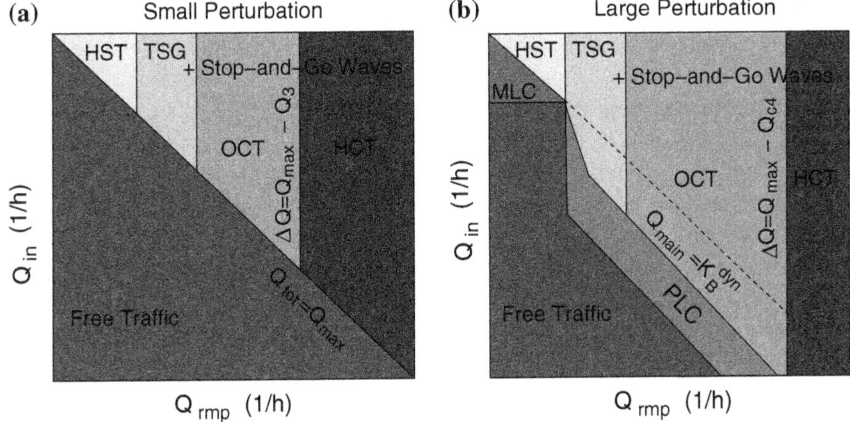

Fig. 17.3 Qualitative phase diagram of congestion patterns for models of stability class 2b as a function of the main-road inflow Q_{in} and on-ramp flow Q_{rmp} (corresponding to the bottleneck strength ΔQ for other types of bottlenecks). The patterns are homogeneous synchronized traffic (HST), triggered stop-and-go waves (TSG), oscillating congested traffic (OCT), homogeneous congested traffic (HCT), moving localized clusters (MLC) and pinned localized clusters (PLC). Traffic flow of stability class 2a exhibits the same patterns except for HCT which is missing

enters downstream) until the local traffic flow becomes stable and, consequently, the wave decreases its amplitude. However, in class 1 dynamics, the wave reverts its direction before dissolving completely, so it can enter the metastable region again. There, the wave grows again reverting its propagation once more. This process continues until the wave eventually settles as a standing wave at some position parallel to the merging lane. Similar considerations apply for other bottleneck types: In any case, the wave gets pinned at the bottleneck. This is the defining feature of a PLC.

17.2.2 Stability Class 2

In contrast to class 1, traffic flow of stability class 2 is stable or metastable at capacity ($\rho_2 > \rho_C$). So, moderately congested traffic caused by small bottlenecks can be homogeneous in the density range $\rho_e \in [\rho_C, \rho_2]$ corresponding to $C_B^{\text{dyn}} > Q_2$, cf. Eq. (17.6). Such a pattern is called *homogeneous synchronized traffic* (HST) , cf. Fig. 17.3.[4]

The qualitative properties are the same as that of the HCT pattern. However, HCT is caused by severe bottlenecks while HST forms behind weak bottlenecks.

[4] Originally, the term *synchronized traffic* was coined to describe the situation where the speed at a given longitudinal position is essentially the same on all lanes, i.e., it is "synchronized" across lanes. However, since this is true for all types of congested traffic, this naming is not very descriptive.

Moreover, in the phase diagram, these states are separated by the oscillatory states TSG and OCT. If HST is metastable, significant perturbations can trigger a transition to stop-and-go waves (MLC or TSG).[5]

For higher traffic densities $\rho_e > \rho_2$, traffic flow is unstable and the resulting patterns TSG, OCT, and HCT (the latter only for class 2b) are the same as that for class 1. One qualitative difference remains: Since string and flow instabilities of class 2 dynamics are nearly always of the convective type, spatial regions containing traffic waves can only propagate upstream (cf. Sect. 15.2). This means we observe essentially stationary traffic flow near the bottleneck (which may extend over several kilometers), and significant traffic waves (OCT or TSG) further upstream (Fig. 17.2 right). This also means that stop-and-go waves can only form after a nearly stationary congested state upstream of the bottleneck has become sufficiently extended (see Footnote 5). In contrast, if traffic flow is absolutely unstable as in class 1 for many situations, oscillatory regions can propagate in both directions resulting in significant oscillations over the complete congested region (Fig. 17.2 left). As in stability class 1, there are two subclasses: In subclass 2b, traffic flow restabilizes for very high densities resulting in HCT while in class 2a such a restabilization and the HCT pattern are missing.

The majority of real congestions exhibit nearly stationary traffic near the bottleneck. For example, in the complex state of Fig. 5.1 showing two instances of OCT patterns and two TSG waves, congested traffic flow is essentially homogeneous in a 3 km long region, for all three instances. We conclude that real traffic flow is best described by model-parameter combination exhibiting a class 2 dynamics.

Finally we notice that, in class 2, patterns with isolated moving or standing waves (TSG, MLC, PLC) are only possible if flow at capacity (and in a small density range corresponding to free traffic) is metastable rather than stable, $\rho_1 < \rho_C$. Otherwise, these patterns are replaced by HST or by free traffic in the phase diagram.

17.2.3 Stability Class 3

This class is characterized by stable traffic flow over the whole density range and by a minimal or zero capacity drop ε [cf. Eq. (17.1)]. This means, the dynamics is, in essential, the same as that of the LWR models, i.e., oscillatory or localized congestion patterns (OCT, TSG, PLC, MLC) are nonexistent. Only free traffic and the HCT/HST pattern (these two states can no longer be distinguished) survive.

---

[5] This is consistent with a postulate of the so-called *three-phase theory*. According to this theory, there are three phases: Free traffic, synchronized traffic, and stop-and-go waves. Furthermore, stop-and-go waves only form via synchronized traffic as intermediate state. In reality, this is often but not always the case: Counterexamples include the Figs. 18.2c and 18.3, and the OCT pattern of Fig. 17.6.

17.3 Simulating Congested Traffic Patterns and the Phase Diagram

In the previous sections, we developed the phase diagram starting from the stability diagram, the stability classes, and the observation of a capacity drop. Since these concepts apply to both microscopic and macroscopic models, we should also be able to simulate the associated patterns with both model classes. To verify this, we simulate open systems with a bottleneck for one representative of each model class: The microscopic Intelligent Driver Model (IDM, Sect. 11.3) and the macroscopic GKT model (Sect. 9.4.3). Both models can be parameterized to represent any of the stability classes 1a, 1b, 2a, 2b, or 3 (cf. the class diagram of Fig. 15.15). Since stability class 2 represents best the observed patterns, we set the models to this class.[6] For the IDM, we use the parameters of Table 11.2 corresponding to class 2a. We parameterize the GKT model by $V_0 = 120$ km/h, $T = 1.7$ s, $\tau = 30$ s, $\rho_{\text{max}} = 140$ vehicles/km, $\gamma = 1$, and the variance-density relation (9.23) by $\alpha_{\text{free}} = 0.012$, $\alpha_{\text{cong}} = 0.036$, $\rho_{\text{cr}} = 0.4\rho_{\text{max}}$, and $\Delta\rho = 0.05\rho_{\text{max}}$ which brings the GKT model to stability class 2b.

Inflow boundary conditions. The *traffic demand* Q_{in}, i.e., the y-axis of the phase diagram, is introduced into the simulations by following upstream boundary conditions:

- Microscopic models (IDM): In case of free traffic at the boundary, introduce a vehicle at a speed $v_{\text{free}}(Q_{\text{in}})$ corresponding to the free branch of the speed-flow diagram whenever the accumulated demand $\sum_t Q_{\text{in}}(t)\Delta t$ exceeds 1. In case of congested traffic, introduce a vehicle at a speed corresponding to the equilibrium speed of the available gap whenever the accumulated demand since the last insertion is greater than unity *and* the available gap is equal to or greater than the steady-state gap associated with the congested branch $s_e(Q_{\text{in}})$ of the microscopic fundamental diagram. In any case, decrement the accumulated demand $\sum_t Q_{\text{in}}(t)\Delta t$ by one after each successful insertion.
- Macroscopic models (GKT): If there is free traffic at the boundary, *or* if traffic demand is less than the flow at the boundary, impose fixed *Dirichlet* boundary conditions $Q = Q_{\text{in}}$ and $\rho = \rho_{\text{free}}(Q_{\text{in}})$. Otherwise, implement homogeneous *Von-Neumann* boundary conditions $\frac{\partial \rho}{\partial x} = 0$, $\frac{\partial Q}{\partial x} = 0$.

A remark on these boundary conditions is in order: Programming and simulation experience teaches us that upstream boundary conditions that reliably work in all situations are notoriously difficult to formulate. Particularly, the boundary conditions describe here may introduce artificial boundary-induced bottlenecks (cf. Problem 7.2). Depending on the situation and the numerical integration scheme, slightly modified rules (e.g., allowing new vehicles if the available gap is greater than 0.8 times the steady-state gap) bring better results.

---

6 For simulated patterns of class 1a, we refer to the first two references in the "Further Reading" list at the end of this chapter.

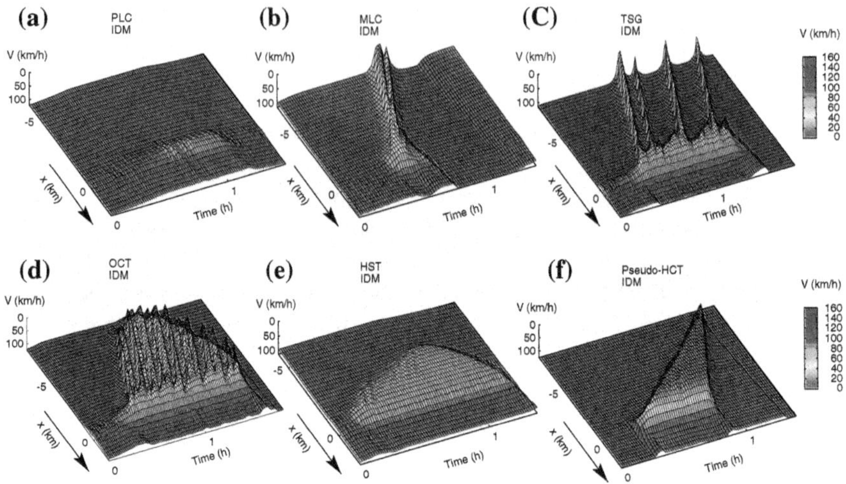

Fig. 17.4 Traffic patterns of stability class 2a as simulated with the microscopic IDM using the parameters of Table 11.2

Modeling the merging process. According to Eq. (17.4), the strength of the on-ramp bottleneck is essentially given by Q_{rmp}, so it can be controlled by changing the ramp inflow. We implement the on-ramp as follows:

- In the microscopic IDM simulation, we could have modeled the merging by the decision model MOBIL (cf. Sect. 14.3.3). Instead, we opt here for the simplest possible mechanism: Whenever the accumulated demand $\sum_t Q_{\text{rmp}}(t)\Delta t$ exceeds 1 and the maximum of the gaps along the merging region exceeds some minimum value, decrement the sum by one and introduce a new vehicle at the center of the largest gap along the merging region at the speed of the new predecessor.
- In the macroscopic GKT model, the on-ramp is realized by the corresponding source term (7.12) of the continuity equation, and by a term A_{rmp} in the acceleration equation according to Eq. (9.17) with $V_{\text{rmp}} = 0.6\,V$ (the ramp vehicles enter at an average speed of 60 % of the speed of the main-road vehicles).

Generating the phase diagrams. We obtain the phase diagrams by a series of simulations of open single-lane systems with on-ramp bottlenecks simulated with unchanged parameters but different values of the inflow Q_{in} and on-ramp flow Q_{rmp}. Since each point $(Q_{\text{in}}, Q_{\text{rmp}})$ of the phase diagram corresponds to a complete simulation, the simulations to generate a phase diagram are quite extensive. Each dynamical phase is defined by a region in phase space $\{(Q_{\text{in}}, Q_{\text{rmp}})\}$ producing qualitatively the same pattern in the simulation.

Results. The simulated phase diagrams for the IDM (Fig. 17.4) and the GKT model (Fig. 17.5) confirm that this concept can be applied to both microscopic and macroscopic models and that the boundaries of the dynamic phases are consistent with the

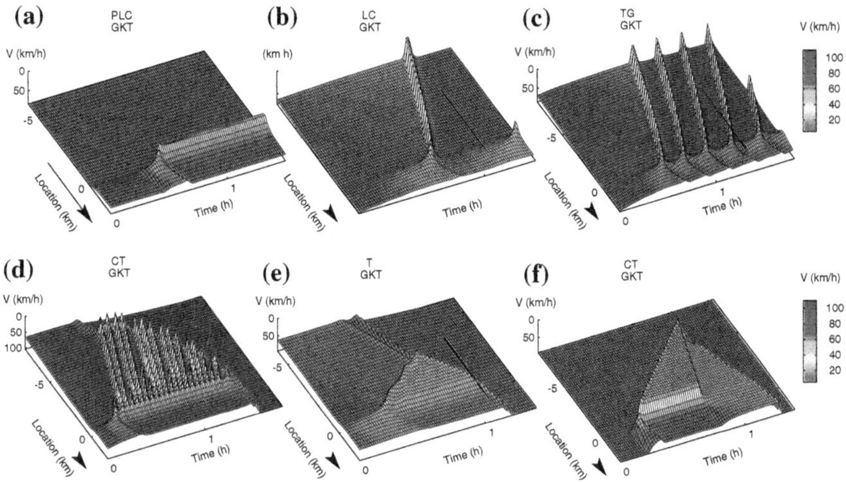

Fig. 17.5 Traffic patterns of stability class 2b as simulated with the macroscopic GKT model. The parameters are given in the main text

analytical considerations of Sect. 17.2. However, the reader might notice an apparent discrepancy in Fig. 17.4: For severe bottlenecks, the IDM produces essentially stationary congested traffic, although its parameter settings correspond to class 2a where no HCT pattern exists. In reality, the IDM produces oscillatory traffic in this situation. It appears to be nearly homogeneous in the plot (hence the name "pseudo-HCT") since the frequency of the oscillations is so high that it is of the order of the temporal resolution of this figure (2 min).

In summary, we can characterize the traffic patterns shown in the plots (a) to (f) of the Figs. 17.4 and 17.5 as follows:

Pinned localized cluster (PLC): A single localized standing traffic wave "pinned" at a bottleneck. This pattern can emerge for a wide range of bottleneck strengths if the inflow is below the critical value that would lead to extended patterns (TSG, OCT, or HCT). It is triggered by a significant perturbation of the traffic flow.

Moving localized cluster (MLC): Isolated single moving traffic waves propagating at a constant velocity against the direction of traffic. Here, the bottleneck is so insignificant that the dynamics is essentially that of a homogeneous road. The wave itself is triggered by a significant perturbation occurring elsewhere, e.g., as part of the TSG pattern of a bottleneck further downstream.

Triggered stop-and-go (TSG): Isolated moving waves detach regularly from a locally congested zone near the bottleneck.

Oscillating congested traffic (OCT): Extended congested zone with moving traffic waves. The wavelength (about 1–3 km) depends strongly on the bottleneck strength ΔQ (the larger ΔQ, the shorter the wavelength) while the propagation velocity is essentially a "traffic flow constant". This is the most frequent pattern in real traffic. OCT and TSG states are commonly referred to as "stop-and-go traffic".

Homogeneous synchronized traffic (HST): As HCT, HST is an extended homogeneous congested traffic pattern. While HCT is caused by severe bottlenecks (e.g., accidents or roadworks with a lane closure), HST implies less significant bottlenecks such as junctions. This state requires a traffic dynamics of class 2. Since HST patterns are observed in real traffic, this is a strong indication that real traffic flow is consistent with class 2 dynamics.

Homogeneous congested traffic (HCT): HCT is an extended homogeneous congested traffic state. In contrast to HST, its density is significantly higher, and the associated speed V lower. Notice that "homogeneous" only means "free of correlated macroscopic oscillations" but not "free of erratic fluctuations". However, if the frequency of the oscillations of an OCT pattern becomes so high that the period of the oscillations is not significantly longer than the typical detector aggregation interval (1 min), the oscillations can no longer be distinguished from noise resulting in "Pseudo-HCT" (Fig. 17.4f).

17.4 Reality Check: Observed Patterns of Traffic Jams

The ultimate check of the concept of the phase diagram can only be a comparison of the predicted (or modeled) patterns with the observed ones (cf. Chap. 18). As can be seen in the Figs. 18.1 and 18.2, real congested traffic patterns are often more complex than the generic "textbook" patterns HCT, OCT, TSG etc. as described above. However, they usually can be decomposed into the generic patterns. Consistent with this interpretation is the observation that congestion patterns generated by different bottlenecks often interact with each other.

Moreover, there are also instances where real traffic congestion corresponds directly to a single generic pattern. Figure 17.6 shows this with a collection of observed jams on the German Autobahn A5. In this figure, the HCT pattern was caused by an accident-induced partial closure while the other patterns are due to intersection, on-ramp, and uphill bottlenecks.

Problems

17.1. Phase diagram for stability class 3
Sketch the phase diagram for LWR models and for model-parameter combinations displaying stable steady-state traffic over the whole density range. Are there differences between small and large perturbations?

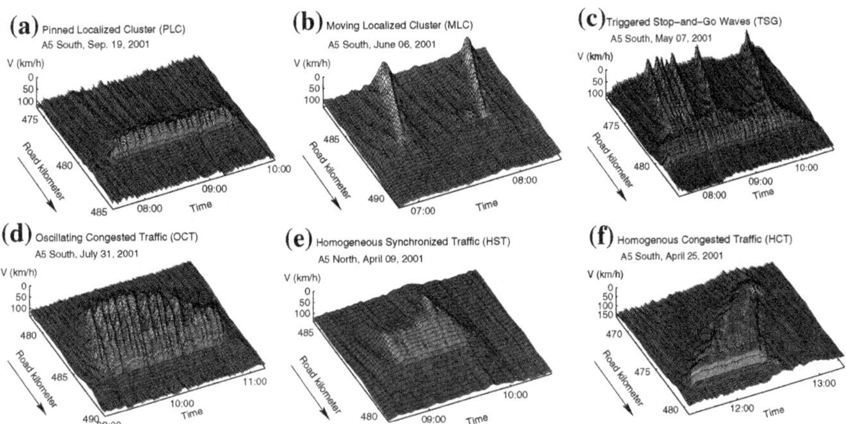

Fig. 17.6 Textbook cases of the different congested traffic patterns as observed on the German highway A5 near Frankfurt/Main

17.2. Boundary-induced phase diagram
We can also define *boundary-induced* phase diagrams where the downstream boundary takes over the supply-limiting function of the bottleneck. In such diagrams, phase space is spanned by the traffic demand Q_{in} (as in the normal bottleneck-induced phase diagram), and by the supply Q_{out} which we define as the maximum flow allowed at the downstream boundary. Obviously, at least partially congested traffic flow arises if $Q_{in} > Q_{out}$. Besides congestion in the bulk, a so-called *maximum-flow state* with a standing wave at the inflow boundary and free traffic, otherwise, emerges for $Q_{in} > C_{dyn}$ and $Q_{out} > C_{dyn}$ if the upstream boundary conditions are defined as in the main text.

Sketch the boundary-induced phase diagram for a single lane and traffic flow dynamics corresponding to class 1b in the range $(Q_{in}, Q_{out}) \in [0, Q_{max}] \times [0, Q_{max}]$. Assume a nonzero capacity drop $C - C_{dyn} = Q_{max} - C_{dyn}$ and C_{dyn} sufficiently large such that the TSG pattern exists in a certain region. Give a qualitative explanation why standing waves may occur at the upstream boundary.

Further Reading

- Lee, H., Lee, H., Kim, D.: Origin of synchronized traffic flow on highways and its dynamic phase transition. Physical Review Letters **81** (1998) 1130
- Lee, H.Y, Lee, H.W., and Kim, D.: Dynamic states of a continuum traffic equation with on-ramp. Physical Review E **59** (1999) 5101
- Helbing, D., Hennecke, A., Treiber, M.: Phase diagram of traffic states in the presence of inhomogeneities. Physical Review Letters **82** (1999) 4360–4363

- Lee, H.Y, Lee, H.W., and Kim, D.: Phase diagram of congested traffic flow: An empirical study. Physical Review E **62** (2000) 4737
- Jiang, R., Wu, Q.S., and Wang, B.H.: Cellular automata model simulating traffic interactions between on-ramp and main road. Physical Review E **66** (2002) 036104
- Kerner, B.S.: The Physics of Traffic: Empirical Freeway Pattern Features, Engineering Applications, and Theory. Springer (2004)
- Kerner, B.S.: Introduction to modern traffic flow theory and control: the long road to three-phase traffic theory. Springer (2009)
- Helbing, D., Treiber, M., Kesting, A., Schönhof, M.: Theoretical versus empirical classification and prediction of congested traffic states. The European Physical Journal B **69** (2009) 583–598
- Treiber, M., Kesting, A., and Helbing, D.: Three-phase traffic theory and two-phase models with a fundamental diagram in the light of empirical stylized facts. Transportation Research Part B: Methodological **44** (2010), 983–1000
- Treiber, M., and Kesting, A.: Evidence of Convective Instability in Congested Traffic Flow: A Systematic Empirical and Theoretical Investigation. Transportation Research Part B: Methodological **45** (2011), 1362–1377

Part III
Applications of Traffic Flow Theory

Chapter 18
Traffic Flow Breakdown and Traffic-State Recognition

Real knowledge is to know the extent of one's ignorance.
Confucius

Abstract Flow and aggregated speed data from stationary detectors and trajectories of floating cars allow us to investigate many aspects of traffic breakdown and jam propagation. In the first two sections, we discuss the three main factors of traffic breakdowns: High traffic flow, bottlenecks, and a disturbances in the traffic flow itself. In the next two sections, we summarize the *stylized facts* of the spatiotemporal evolution of congested traffic patterns, i.e., typical empirical findings that are repeatedly observed on various highways all over the world. In the last section, we apply this knowledge to real-time traffic-state estimation and short-term prediction. While the traffic breakdown as such is a stochastic process and therefore, in principle, only predictable in terms of probabilities, the *stylized facts* allow a quasi-deterministic forecast of the evolution of already congested traffic. The focus is on highways and major roadways but the contents of this chapter is also applicable to other types of roads.

18.1 Traffic Flow Breakdown: Three Ingredients to Make a Traffic Jam

What are the reasons for traffic breakdown on highways and major roadways? Can they be reproduced with traffic flow models? To discover systematic and reproducible mechanisms, traffic flow researchers investigate stationary detector data and other data sources recording instances of traffic breakdowns on many highways and in many countries and reconstruct the spatiotemporal dynamics of congested regions with dedicated methods such as the *adaptive smoothing method*. In order to apply such methods, we need detectors at several consecutive road cross sections which (i) are not farther than 3 km apart, and (ii) are positioned strategically upstream and

M. Treiber and A. Kesting, *Traffic Flow Dynamics*,
DOI: 10.1007/978-3-642-32460-4_18, © Springer-Verlag Berlin Heidelberg 2013

downstream of permanent bottlenecks (cf. Sect. 5.2). Alternatively, the data can come from floating cars, providing they are sufficiently frequent.

Analyzing these data, we observe that nearly all real-world traffic breakdowns are caused by the simultaneous action of three factors: High traffic load, a bottleneck, and disturbances of traffic flow caused by individual drivers. In the following, we describe each of these three "ingredients to make a traffic jam":

High traffic load. This factor is the most obvious one: If the traffic load on the network is low enough then disturbances caused by bottlenecks or abrupt driving maneuvers cannot grow and propagate since traffic is *unconditionally stable* for sufficiently low densities (cf. Chap. 15). Nevertheless, even absolutely stable traffic will break down if the inflow exceeds the static capacity of bottlenecks in the considered road section (cf. Sect. 8.5.6). This, however, sets again a lower limit on the traffic load required for a breakdown (at least when excluding the special case of a complete road closure).

Bottlenecks. While *phantom traffic jams* are theoretically possible and occur regularly when simulating string unstable traffic flow on a homogeneous road, they do not occur on real road networks which are necessarily inhomogeneous (cf. Sect. 18.2). On a real network, there is always a "weakest link" in form of a *bottleneck* leading us to the following definition:

> We define a *bottleneck* as a local reduction of the road capacity (cf. Fig. 21.6). Bottlenecks can be *permanent* attributes of the infrastructure (e.g., on-ramps, off-ramps, roadworks, etc.) or *temporary*, e.g., when caused by accidents.

In a figurative analogy, we can represent free and congested traffic by the gaseous and liquid phases of fluids, respectively. In this analogy, the bottlenecks act as "condensation seeds" triggering the breakdown if the gas is supercooled (corresponding to a sufficient demand) and if disturbances, e.g., in form of vibrations (corresponding to abrupt driving maneuvers) are present.

Bottlenecks come in many forms. The most common instances on highways and major roadways are the following[1]:

- On-ramps and off-ramps.
- Lane closures, road narrowings and curves (particularly at roadworks and road construction sites).
- Uphill and downhill gradients.
- Temporary obstructions caused by accidents (including complete road closures).
- spectacular accidents or jammed traffic on the opposite side of the road.

[1] In cities, traffic lights are the most common form of (temporary) bottlenecks.

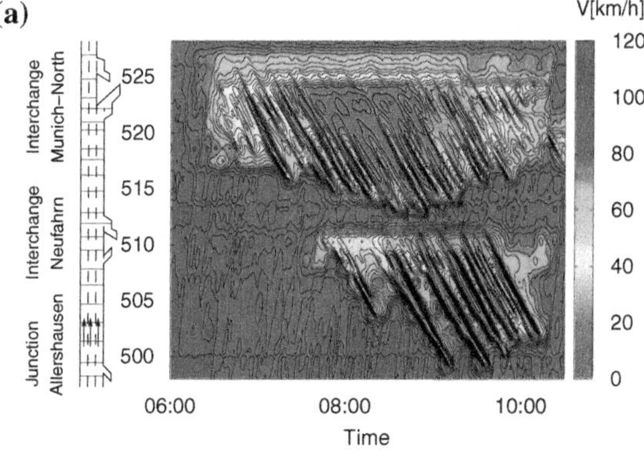

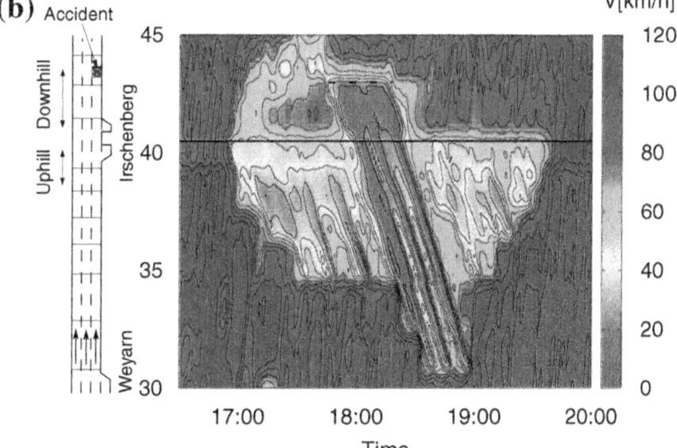

Fig. 18.1 Spatiotemporal local speed on the German highways A9-South (during the *morning rush hour*) and A8-East (in the *evening hours*)

The first three types are *permanent bottlenecks* while the bottlenecks caused by accidents are *temporary*.[2] All but the first bottleneck above are *flow-conserving bottlenecks*. The last type is remarkable since this bottleneck consists of neither permanent or temporary physical obstructions. Instead, this *behaviorally induced bottleneck* is caused by a *locally changed* (and less effective) driving style due to rubbernecking.

When traffic breaks down at the bottleneck and congested traffic has formed upstream of it, one says that the bottleneck is *activated*. Typically, bottleneck activation is accompanied by a drop in the traffic flow through the bottleneck, i.e., the bottleneck capacity, of the order of 10–20 %. This *capacity drop* from the static to the dynamic bottleneck capacity is also observed in simulations of microscopic and second-order macroscopic models (see, e.g., Fig. 11.4).

The Figs. 18.1a and 18.2 show instances of activated on-ramp bottlenecks. In Fig. 18.1a, the on-ramps of two interchanges at road 511 and 525 km of the German Autobahn A9-South produce oscillatory congested traffic (OCT). The interchange

[2] Remember the time scales of traffic-flow investigations displayed in Table 1.1. On the scales of months or years, roadworks bottlenecks are temporary as well.

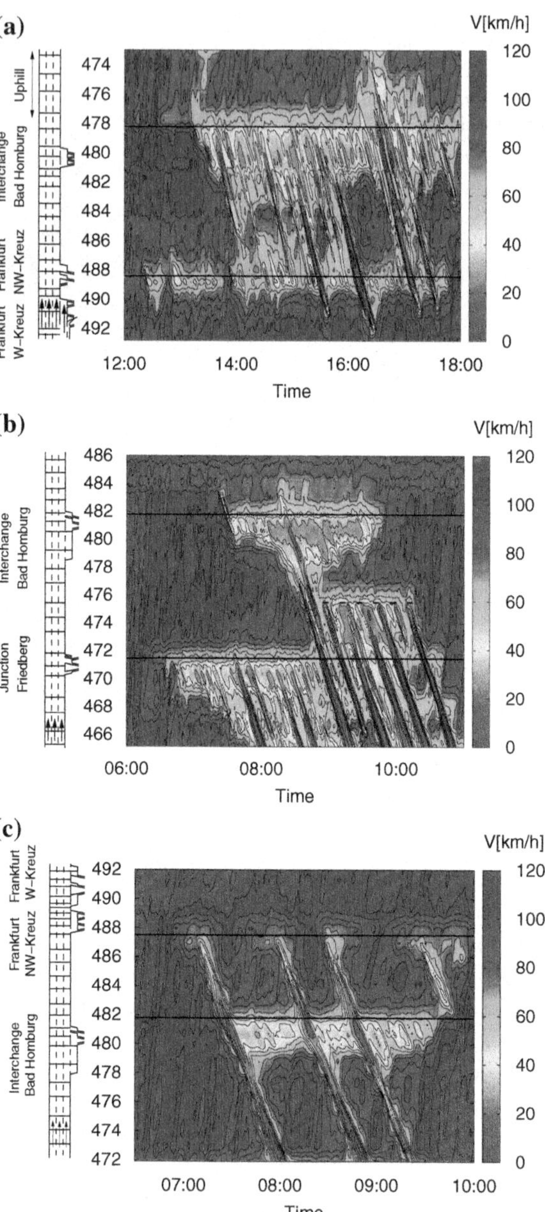

Fig. 18.2 Spatiotemporal local speed on the German highways A5-South and A5-North to the north of Frankfurt/Main

at road 482 km on the German A5-South produces pinned local clusters (PLC) and one moving localized cluster (MLC), Fig. 18.2b. In Fig. 18.2c, MLCs generated elsewhere propagate through the PLC caused by the same bottleneck. An interchange in Fig. 18.2c at road 482 km produces another instance of OCT.

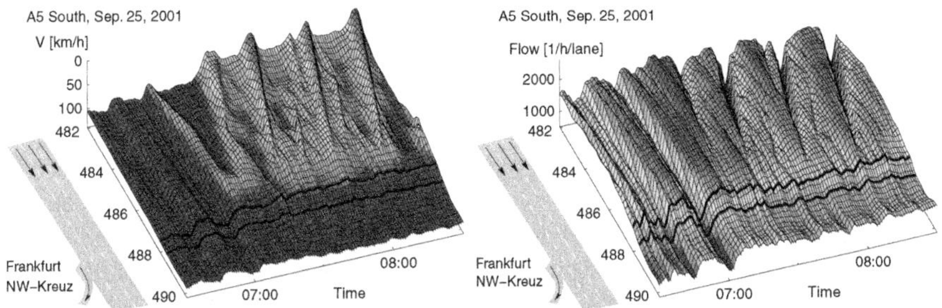

Fig. 18.3 Aggregated speed and flow data of a section of the German highway A5 showing the three factors of traffic breakdown. In most cases, the effects of the flow disturbance cannot be seen so clearly as in this "textbook case"

Figure 18.1b shows two instances of flow-conserving bottlenecks. An accident at road 43.5 km leads to a temporary closure of one lane of the three-lane highway between 17:40 h and 18:20 h. Furthermore, a hilly region ("Irschenberg") at about road 40 km produces another instance of OCT (the small junction at this location plays a comparatively minor role, here).

Disturbances caused by individual drivers. The third factor that is necessary for traffic breakdowns consists of perturbations in the traffic flow *itself*. While, theoretically, an infinitesimal perturbation suffices if traffic flow is linearly unstable, this does not apply to real traffic flow where non-congested traffic is, at most, metastable (cf. Chap. 15).[3] Disturbances in the traffic flow are realized, e.g., by inattentive drivers braking abruptly, by speeding cars, lane changes, or by *elephant races*, i.e., trucks overtaking each other at speed differences, particularly on roads with only two lanes per direction.

Due to their single-vehicle nature, such disturbances cannot be seen directly in aggregated detector data. However, they often lead to a platoon of vehicles following each other at very small time gaps. These can be identified in the data as so-called *high-flow states*: A flow peak lasting a few minutes which sometimes is associated by a comparatively small drop in the average speed. Figure 18.3 shows an example: At 7:05 h, the data show a high-flow state (right diagram) associated with a speed drop (left) to about 75 km/h. Since the initial propagation velocity of the platoon is equal to this speed, we conclude that the data show a *vehicle platoon*. When this platoon reaches the interchange "Frankfurt NW-Kreuz" at road 488 km acting as a bottleneck, the speed drop (and the platoon amplitude in terms of estimated density Q/V) increases and the platoon reverses its direction now propagating upstream at a velocity $c \approx -16$ km/h. Eventually, it becomes the first traffic wave of a triggered stop-and-go (TSG) state.

[3] This even applies to situations where a rush hour is about to increase the demand above the *static* capacity C_B of the bottleneck. So a jam is unavoidable whether traffic flow is unstable, nor not (the case described by the LWR models): Even before the static capacity is reached, a disturbance in the flow will activate the bottleneck reducing its capacity to the dynamical capacity $C_{\mathrm{B}}^{\mathrm{dyn}}$.

Summary. The first two factors, high traffic demand and bottlenecks (except those caused by accidents), are essentially deterministic effects and therefore predictable. However, since disturbances of traffic flow are stochastic in nature, we cannot predict the time of a perturbation which is powerful enough to trigger a breakdown. Moreover, in the worst case, such a perturbation can cause an accident with subsequent lane or road closure. This means that even bottlenecks may be stochastic. Consequently, we cannot make predictions about the location and time of individual traffic breakdowns. Nevertheless, we can make statements about the *probability* that a breakdown will occur on a given road section in a certain period.[4]

In contrast, once traffic has broken down, we can predict how the congested regions will evolve, including the time the jam will dissolve. This will be discussed in the Sects. 18.3 and 18.5 (cf. also Fig. 8.21 of Sect. 8.5.8), and in Problem 18.5.

The analysis of spatiotemporal local speed profiles shows that nearly all traffic breakdowns on real road networks are caused by the simultaneous action of three factors:

1. High *traffic load* (temporal aspect).
2. A *bottleneck* (spatial aspect).
3. *Local disturbances in the flow* (the trigger).

18.2 Do Phantom Traffic Jams Exist?

In the previous section, we showed that the evidence of traffic data suggests that most (if not all) traffic breakdowns on real road networks are caused by a bottleneck (together with a high traffic load and flow disturbances). This seems surprising since both traffic flow simulations and the anecdotal experience of many drivers suggests that jams without bottlenecks, so-called *phantom traffic jams*, do exist.

In simulations such as the "ring-road scenario" (see Fig. 15.8) of the authors' website[5] traffic waves form, in fact, without bottlenecks only at high traffic load (corresponding to intermediate traffic densities in closed systems such as this scenario), and by perturbations of the traffic flow caused by the different driving styles of car and truck drivers. Even when eliminating the perturbations by setting the truck percentage equal to zero, phantom traffic jams will form for appropriate densities

[4] This is similar to forecasting the weather: It is impossible to exactly predict the times and locations of individual thunderstorms/rainfalls while it is standard to predict the probability of thunderstorms/rainfalls in a certain spatiotemporal region.

[5] see: www.traffic-simulation.de

and driving parameters after a sufficiently long time.[6] Moreover, all this is theoretically explained by the stability theory of Chap. 15. Nevertheless, the observations can easily be explained by the *weakest-link principle*: If there were any bottlenecks (such as in the on-ramp scenario of the authors' simulation website), traffic flow would break down at a lower traffic demand where no phantom traffic jam would be possible. This makes true phantom traffic jams unobservable.

But how can we explain the anecdotal experience of phantom traffic jams? To this end, we consider the spatiotemporal evolution of congested traffic caused by bottlenecks such as those of the Figs. 18.1 and 18.2 *from the viewpoint of a driver*. We generate *virtual trajectories* from the data (cf. Sect. 19.6 and Fig. 19.3 below). Then, the reason of the driver's impression becomes evident: Since traffic waves propagate upstream and can become separated from "their" bottleneck by 10 km or more, the *effect* (the waves) becomes spatially and separated from the *cause* (the bottleneck). Moreover, causality seems to be violated since the driver first encounters the effect (when passing the traffic waves), and later the cause (when he or she passes the bottleneck). In case of temporary bottlenecks, the cause may even no longer exist at the time the driver encounters a traffic wave. Such a situation arises in Fig. 18.1b for road locations between 30 and 34 km and times between 8:30 and 19:00 h.

In spite of this evidence, the existence of phantom traffic jams in real traffic is discussed controversially among traffic researchers. Firstly, it is not always easy to identify a bottleneck since small inhomogeneities such as insignificant junctions or inconspicuous gradient sections may constitute a bottleneck. Secondly, such small bottlenecks typically emit triggered stop-and-go waves (TSG, see Sect. 17.2) that become macroscopically visible only one or even more kilometers further upstream providing the illusion of jams created "out of thin air" (cf. the waves triggered at an interchange at 488 km in Fig. 18.2c). With the definition of the *moving localized clusters* (MLC), *phantom traffic jams* even made it into the terminology of congested traffic patterns. However, it can be argued that each MLC is part of a TSG pattern with the source (bottleneck) outside of the investigated road section.. Finally, when interpreting accidents as events "out of thin air", there are, in fact, phantom jams. However, since the site of an accident constitutes a bottleneck, this is not at variance with our argumentation above.

18.3 Stylized Facts of Congested Traffic

Dedicated traffic-adaptive interpolation methods such as the *adaptive smoothing method* presented in Chap. 5, realize a detailed reconstruction of the spatiotemporal dynamics of the local speed. This allows us to extract the *stylized facts* of congested traffic patterns, i.e., typical empirical findings that are persistently observed on

---

[6] These are triggered either by initial conditions not perfectly representing steady-state conditions, or, ultimately, by numerical rounding errors.

various highways all over the world.[7] In the following, we summarize the relevant findings.

(1) The congestion pattern is either localized or extended. Localized patterns have constant extensions of the order of 1 km (cf. Fig. 18.2c at road 480 km)[8] while extended patterns have a time-dependent spatial extension of typically several kilometers. For example, all congestion patterns of Fig. 18.1 are extended.

(2) The downstream front is either stationary or moves upstream at a fixed velocity $\mathbf{c_{cong}}$. We denote by "downstream front" the transition zone where drivers leave the congested zone. Stationary downstream fronts are always fixed at a bottleneck while moving downstream fronts correspond to jams propagating through an essentially homogeneous road section. Moving fronts occur when (i) single waves (MLC pattern) are emitted by a bottleneck (Fig. 18.1c), (ii) a fixed downstream front of an extended state detaches itself from the bottleneck (Fig. 18.1a at 9:45 h), or (iii) the jam front starts moving when a temporary bottleneck ceases to exist, Fig. 18.1b at road 43.5 km around 18:30 h). Combining the Stylized Facts 1 and 2, we conclude that localized jams are either stationary or move upstream at velocity c_{cong} (cf. Fig. 18.2c).

(3) The upstream front of spatially extended congestion patterns has no characteristic speed. Depending on the traffic demand and the bottleneck capacity, it can propagate upstream (if the demand exceeds the capacity) or downstream (if the demand is below capacity) according to the shock-wave propagation formula (8.9). This can be seen in all extended congestion patterns of Fig. 18.1. Traffic jams dissolve if the upstream jam front meets the downstream jam front. This either happens if the upstream front propagates downstream until it reaches the stationary downstream front at the bottleneck (this is realized for two of the three traffic jams of Fig. 18.1), or if the upstream front "collides" with the moving downstream front (as in the congestion caused by the bottleneck at road 511 km in Fig. 18.1a).

(4) The propagation velocity $\mathbf{c_{cong}}$ of all internal structures is unique. All variations of speed and density inside a congestion pattern, whether due to traffic waves, variations of the bottleneck strength, or others, propagate at the same velocity as that of isolated moving jams or that of moving downstream fronts. Consequently, in graphical spatiotemporal representations of the local speed, all structures inside congested patterns are parallel to each other. This can be seen in all congested patterns of Figs. 18.1 and 18.2. This stylized fact is the empirical foundation of the adaptive smoothing method (Sect. 5.2).

(5) The frequency of traffic waves increases with the bottleneck strength. Typical periods of the internal quasi-periodic oscillations vary between about 4 min and 60 min, corresponding to wavelengths between 1 km and 15 km. In agreement

[7] The authors' website www.traffic-states.com offers a searchable image database of congested traffic patterns.

[8] Because details on scales below the distance of two detectors cannot be resolved, the extension of stationary localized jams cannot be determined exactly. For moving localized jams, Stylized Fact 4 allows to infer the extension from the time period between two waves.

with theory and simulations, the frequency of stop-and-go waves increases with the strength of the bottleneck producing them. Figure 18.1b shows an example: The accident at road 43.5 km produces traffic waves of a higher frequency than that caused by the hills of the "Irschenberg" around road 40 km representing a comparatively small bottleneck. Notice that, due to Stylized Fact 4, the wavelengths and the temporal wave frequency are coupled: Increasing frequencies correspond to decreasing wavelengths.

(6) The amplitude of traffic waves increases during propagation. As can be seen in *all* empirical extended traffic states of this textbook, traffic waves grow (sometimes to saturation) while they propagate upstream. The oscillations may already be visible at the downstream boundary (such as the jam of Fig. 18.1a produced by the bottleneck at road 511 km, or the accident-related congestion of Fig. 18.1b), or emerge further upstream (such as the extended congestions of Fig. 18.2b). This fact gives strong evidence that traffic flow inside extended congested patterns is convectively unstable (cf. Chap. 15).

(7) Light or very strong bottlenecks may cause homogeneous extended traffic patterns. The Fig. 17.6e, f show examples of homogeneous traffic caused by light and severe bottlenecks, respectively. For strong bottlenecks (typically caused by accidents), this empirical evidence is debated controversially. In particular, the oscillation periods at high congested densities (i.e., bottleneck strengths) reach the same order of magnitude as the detector aggregation interval and the smoothing time window of the interpolation method (cf. Stylized Fact 5). This makes oscillations hardly distinguishable from noise.[9]

18.4 Empirical Reality: Complex Patterns

Congested traffic flow patterns are rarely isolated from each other so well as in the textbook cases of Fig. 17.6. In most cases, they interact with each other resulting in complex traffic patterns such as these shown in Figs. 18.1 and 18.2. In the following, we qualitatively describe typical interactions.

Moving jams may propagate through other congestions essentially unaffected. The jam caused by an accident at road 43.5 km on the Autobahn A8-East (Fig. 18.1a) propagates through the congested zone previously induced by the hilly region ("Irschenberg") near road 40 km. However, it is also possible that a moving traffic wave gets "caught" by a deactivated or activated bottleneck (in the first case, the bottleneck generally becomes activated).

Moving jams may activate permanent bottlenecks. In Fig. 18.2a–c, single MLC or TSG waves interact with deactivated and activated bottlenecks (and the correspond-

[9] Moreover, speed variations between "stop and slow" may result from problems in maintaining low speeds (the accelerator and brake pedals are difficult to control in this regime), and thus are different from the collective dynamics at higher speeds.

ing congestions) further upstream. The single MLC originating from the interchange at road 488 km crosses a deactivated bottleneck at 482 km and activates it to produce a PLC pattern (Fig. 18.2c). The moving jam itself propagates essentially unaffected. In the picture of the three factors developed in Sect. 18.1, the moving jam represents the perturbation of traffic flow and acts as the final trigger to create the PLC.

Congested traffic may cause new temporary bottlenecks. This is part of the complex situation on the German Autobahn A5 depicted in Fig. 18.2b: An isolated traffic wave detaches from a PLC state at the interchange "Bad Homburg" at about 8:30 h and causes a temporary bottleneck at road 476 km (probably caused by a rear-end crash). The new bottleneck generates an extended OCT state. Later on, both the original MLC and the OCT of the temporary bottleneck intersect with an already activated bottleneck (junction "Friedberg") and its OCT pattern further upstream.

Bottleneck activation may alleviate the situation further downstream. This is also shown in Fig. 18.2b: The capacity drop due to the temporary bottleneck at road 476 km reduces the flow downstream. As a consequence, the original PLC at road 482 km dissolves.

18.5 Fundamentals of Traffic State Estimation

Real-time traffic-state estimation and short-term prognosis is the basis of dynamic routing and traffic-dependent navigation services, and of a growing number of new Intelligent Transportation Systems (ITS) such as automated jam warnings. Such applications need to estimate the present traffic situation and that of the near future at a forecasting horizon of about 30 min based on data that are available in real-time.

Since the task is macroscopic in nature and requires robust and fast models, first-order models with a triangular fundamental diagram are the model classes of choice, i.e., section-based or cell-transmission models. As an example, simple integration of the ordinary differential Eq. (8.43) for the locations of the upstream jam fronts of the section-based model allows us to calculate the actual position of the jam front from the flow data of two detectors on either side of the front. Due to the finite propagation velocities, Eq. (8.43) even allows for a forecast of the order of 15 min without further assumptions.

Since congestion patterns often are too complex to allow integration of Eq. (8.43) in an automated way, real-world applications generally use the cell-transmission model or dedicated methods based on it. In a typical application, the estimate of the actual and future traffic state is refreshed with the latest data every five minutes using a moving prediction horizon of, say, 30 min. Since each new estimation/prediction implies simulations over at least this prediction time horizon, the simulations have to run in multiple real-time speed.

In summary, the stylized facts and other regularities of the traffic flow dynamics, implicitly contained in the traffic flow models, help overcome the problems posed by the incomplete data situation:

- Most bottlenecks are infrastructure related and permanent, so the location of downstream fronts of congestions is often known a priori.
- The capacity C of homogeneous sections and the throughput of activated bottlenecks (dynamic bottleneck capacity $C_{\mathrm{B}}^{\mathrm{dyn}}$) varies little and in predictable time patterns as a function of the average efficiency of the driving style (they are higher in morning rush hours than on Sundays). This allows us to estimate the bottleneck strength ΔQ using Eq. (17.3) which, in turn, determines the predominant congested pattern (cf. Chap. 17). For example, the interchanges on the German Autobahn A9-South shown in Fig. 18.1a generally produce OCTs while the interchange "Bad Homburg" on the Autobahn A5 typically produces MLC or PLC patterns (Fig. 18.2b, c).[10]
- Temporary bottlenecks, generally caused by accidents, have their own signatures in the data which allows us to determine their location and duration, mainly by using Stylized Fact 2 (downstream fronts) and the shock-wave formula (8.9), see Problems 18.5 and 5.1.
- The propagation velocity c_{cong} of moving downstream fronts and the corresponding outflow Q_{max} per lane are constant for a given driver-vehicle composition (cf. Stylized Fact 4) and can be determined from stationary detectors without bias. Together with the mean desired speed V_0, this allows a robust online calibration of models with a triangular fundamental diagram when the latter are formulated in terms of these quantities by Eq. (8.22). Furthermore, the dynamic bottleneck capacities often are nearly constant which allows us to directly predict the future travel times (cf. Chap. 19).
- The known two values for the propagation velocity of downstream fronts (Fact 2) and the analytic relation (8.43) for the velocities of the upstream front allows a short-term prediction for the evolution and possible dissolution of jams.
- Finally, by Stylized Facts 6 and 7, we can give qualitative statements on the quality of congested flow, i.e., whether it is essentially homogeneous (near the bottleneck or for very low or high bottleneck strengths), or oscillatory.

Difficult challenges are to automatically check the new data for consistency and to fuse different data sources (cf. Sect. 5.3). There are several dedicated algorithms to tackle these problems. One could also generalize the *adaptive smoothing method* (cf. Sect. 5.3) which is only effective for offline reconstruction, to fit real-time applications. A detailed treatment of this topic is beyond the scope of this book.

Problems

18.1 Locating a temporary bottleneck

Figure 5.10 shows a situation where an accident leads to a temporary complete road closure which has been recorded by different sorts of detectors. Besides the traffic

---

[10] see: www.traffic-states.com

state (green data points correspond to free, yellow points to dense, and red points to congested traffic), it is known that the stationary detector D1 records essentially constant traffic flow during the displayed time period.

Determine the location and time at which the accident occurred, the time when the road block was lifted, and the location and time when the jam due to the accident dissolved. *Hint:* Use the stylized facts, the flow information, and the shock-wave formula (8.9).

Further Reading

- Nanthawichit, C., Nakatsuji, T., Suzuki, H.: Application of probe-vehicle data for real-time traffic-state estimation and short-term travel-time prediction on a highway. Transp. Res. Rec. J. Transp. Res. Board **1855** (2003) 49–59
- Kerner, B., Rehborn, H., Aleksic, M., Haug, A.: Recognition and tracking of spatio-temporal congested traffic patterns on freeways. Transp. Res. Part C Emerg. Technol. **12** (2004) 369–400
- Zang, Y., Papageorgiou, M.: Real-time freeway traffic state estimation based on extended kalman filter: a general approach. Transp. Res. Part B Methodol. **39** (2005) 141–167
- Mihaylova, L., Boel, R., Hegyi, A.: Freeway traffic estimation within particle filtering framework. Automatica **43** (2007) 290–300
- Treiber, M., Kesting, A., Helbing, D.: Three-phase traffic theory and two-phase models with a fundamental diagram in the light of empirical stylized facts. Transp. Res. Part B Methodol. **44**(8–9) (2010) 983–1000
- Treiber, M., Kesting, A., Wilson, R.E.: Reconstructing the traffic state by fusion of heterogenous data. Comput. Aided Civ. Infrastruct. Eng. **26** (2011) 408–419

Chapter 19
Travel Time Estimation

The trouble with life in the fast lane is that you get to the other end in an awful hurry.

John Jensen

Abstract For most persons, the *individual travel time* is the most relevant criterion when planning a route from a given origin to a given destination. Road managers and national economists are interested in the *total travel time* in a certain region over a certain time interval. Of particular interest is the total delay caused by congestion. The methods to estimate these quantities can be applied directly to stationary detector and probe-vehicle data, or to microscopic and macroscopic models. While microscopic models give the travel time directly in terms of the duration of trajectories, macroscopic models require additional evaluations, either integrating the flow at fixed positions ("virtual detectors"), or generating virtual trajectories from the local speed field $V(x,t)$. Generally, the macroscopic estimation is more robust.

19.1 Definitions of Travel Time

We define the travel time τ_{12} in the obvious way as the time a vehicle needs to pass a road section $[x_1, x_2]$. Generally, τ_{12} depends on x_1, x_2, and on time t itself. Since τ_{12} is neither a local quantity (as the flow Q) nor an instantaneous quantity (as the density ρ), the above definition is incomplete. Denoting the times when a vehicle enters and leaves the considered section by t_1 and t_2, respectively, we disambiguate the definition in two ways:

The *realized travel time* $\tau_{12}(t)$ is the time a vehicle needs to travel from x_1 to x_2 when it leaves the considered road section at time t:

$$\boxed{\tau_{12}(t_2) = t_2 - t_1.} \tag{19.1}$$

M. Treiber and A. Kesting, *Traffic Flow Dynamics*,
DOI: 10.1007/978-3-642-32460-4_19,

The *expected travel time* $\tilde{\tau}_{12}(t)$ is the time a vehicle needs to travel from x_1 to x_2 when it enters the considered road section at time t:

$$\boxed{\tilde{\tau}_{12}(t_1) = \hat{t}_2 - t_1,} \tag{19.2}$$

where $\hat{t}_2(t)$ is the estimated passing time at location x_2 if this vehicle passes the location x_1 at time t. It is evident that $\tilde{\tau}_{12}$ is the relevant quantity for routing applications. Unfortunately, it requires a short-term traffic-flow forecast to estimate t_2.

We define the *total travel time* as the cumulative time spent by all vehicles inside the spatiotemporal region $[x_1, x_2] \times [t_1, t_2]$. Denoting by $n_{12}(t)$ the number of vehicles on the section $[x_1, x_2]$ at time t, this results in the microscopic definition

$$\boxed{\tau_{\text{tot}}(x_1, x_2, t_1, t_2) = \int_{t_1}^{t_2} n_{12}(t)\,\mathrm{d}t.} \tag{19.3}$$

Macroscopically, we use the density definition $\rho = \frac{\mathrm{d}n}{\mathrm{d}x}$ to obtain

$$\boxed{\tau_{\text{tot}}(x_1, x_2, t_1, t_2) = \int_{t_1}^{t_2} \int_{x_1}^{x_2} \rho(x,t)\mathrm{d}x\,\mathrm{d}t.} \tag{19.4}$$

The total travel time can be used as an objective function to minimize congestion by traffic-flow optimization measures (cf. Chap. 21).

19.2 The Method of Trajectories

The following "microscopic" approach can be applied to microscopic simulations, probe-vehicle data, and floating-phone data (FPD). In the following, we will denote all data originating from vehicles "floating" with the traffic (probe-vehicle data and FPD) as *floating-car data* (FCD).

Microscopic simulations. In microscopic simulations, the complete information on any vehicle is available at any time, so the individual travel times can be directly determined from the trajectories:

$$\tau_{12}(t) = t_2^{\alpha} - t_1^{\alpha}, \qquad \tilde{\tau}_{12}(t) = t_2^{\beta} - t_1^{\beta}, \tag{19.5}$$

where α and β denote the last vehicles that have left and entered the investigated road section, respectively, i.e., $x^{\alpha}(t) \geq x_1 > x^{\alpha+1}$ and $x^{\beta}(t) \geq x_2 > x^{\beta+1}$.

Notice that we did not put a "hat" on t_2^β since nothing has to be estimated in the simulations.[1]

Probe-vehicle and floating-phone data. This method can also be used to estimate travel times from floating-car data. For real-time applications, however, the realized travel time $\tau_{12}(t)$ obtained in this way is generally more or less outdated if the penetration rate of probe vehicles, and thus the frequency of passing equipped vehicles, is very low. This problem aggravates when estimating the expected travel time since an estimate $\hat{t}_2^\beta$ of t_2^β implies a short-term traffic forecast. Therefore, if the penetration rate is very low (e.g., below 0.2%), it is more useful to use FCD to initialize algorithms based on stationary-detectors to obtain a continuously measured travel time (cf. Sect. 19.4 below).

19.3 The Method of Accumulated Vehicle Counts

Since stationary detectors cannot discern individual vehicles,[2] trajectory-based methods are out of the question when stationary detectors are the only data source.

Instead, one uses the complete coverage of all passing vehicles by stationary detectors to estimate the vehicle *number* (rather than its identity) by cumulating the vehicle count (integrating the flow) of at least two detectors over time. In the following, we assume a common aggregation time interval Δt for all detectors of the investigated section, and a synchronized data delivery at times $t_k = k\Delta t$. Denoting the location of detector i by x_i and the flow and vehicle number measured by detector i for the aggregation period ending at time t_k by $Q_i(t_k)$ and $n_i(t_k)$, respectively, we can calculate the cumulated vehicle number by

$$N_i(t_k) = N_i(t_0) + \int_{t_0}^{t_k} Q_i(t')\,\mathrm{d}t' \approx N_i(t_0) + \sum_{k'=1}^{k} Q_i(t_k)\Delta t, \tag{19.6}$$

or, taking the vehicle numbers directly,

$$N_i(t_k) = N_i(t_0) + \sum_{k'=1}^{k} n_i(t_{k'}). \tag{19.7}$$

The accumulated vehicle number as a function over time is called the *N-curve* of this detector. If there is no possibility of changing lanes or overtaking (as on a single-lane road with overtaking ban and without ramps), the accumulated vehicle numbers N_i have a constant offset to the vehicle index α. By appropriately setting the initial

---

[1] Unless one uses microscopic models for real-time applications which is rarely done.

[2] Except, when special features such as video-based number plate recognition are implemented. In most countries, however, this is not feasible due to privacy reasons.

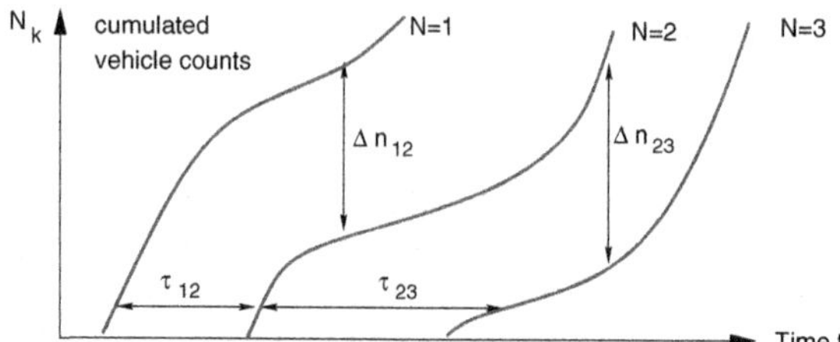

Fig. 19.1 Determining travel times between pairs of stationary detectors with the method of *N-curves*. The length of horizontal lines between the intersecting points with the *N-curves* directly gives the travel time

counting values $N_i(t_0)$, we can even identify the accumulated vehicle number with the vehicle index:

$$N_i = \alpha. \tag{19.8}$$

The fact that the accumulated numbers of all detectors i relate to the *same* vehicle can be used to determine the realized and expected travel times between any detector pair. For the detectors 1 and 2 with $x_2 > x_1$, we obtain

$$N_1(t - \tau_{12}(t)) = N_2(t), \quad N_1(t) = N_2(t + \tilde{\tau}_{12}(t)). \tag{19.9}$$

When plotting the N-curves of several detectors in a diagram, we can directly read off the travel times between the detectors by the length of horizontal lines connecting the N-curves (cf. Fig. 19.1).
In real situations there are several reasons why Eq. (19.9) is not exact:

- Due to lane-changes and overtaking maneuvers.
- Due to on-ramps and off-ramps.
- Due to detecting and counting errors.
- Since the initialization of the N-curves is not feasible.

In principle, the N-curves can be initialized at time t_0 by integrating over the density. If we assume increasing detector numbers in driving direction, $x_{i+1} > x_i$, we have

$$N_i(t_0) = N_{i+1}(t_0) + \int_{x_i}^{x_{i+1}} \rho(x, t_0)\, dx. \tag{19.10}$$

However, this is only a theoretical solution since, generally, the local density field is not known.

Moreover, in this basic form, the method of N-curves is very sensitive to systematic detection faults since the resulting counting errors accumulate over time. We show this by following example: Consider homogeneous dense traffic flow ($Q = 2{,}000$ vehicles per hour, speed $V = 100$ km/h) and two detectors that are $L = 1$ km apart from each other. Then, there are $n = \rho L = Q/VL = 20$ vehicles between the detectors. If detector 1 counts only 98 % of all vehicles and detector 1 99 %, there is a systematic bias of 20 vehicles per hour leading to a drift of the relative positions of the N-curves. Particularly, the N-curves of detector 1 and 2 intersect each other after one hour, so the method of N-curves estimates a zero travel time between the detectors (and a negative travel time afterwards). In the following section, we show how to use floating-car data to initialize the N-curves and correct the drifts at later times.

19.4 A Hybrid Method

To tackle the problem of the drifting N-curves, we propose a hybrid method that is based on the method of *N-curves* applied to the aggregated stationary detector data of the total flow over all lanes (Sect. 3.2) and additionally uses floating-car data (cf. Chap. 2) for initialization and drift compensation. In principle, the method of N-curves can be applied to single lanes or to the sum of the traffic flow over all lanes. The latter is more robust since drifts due to flow-conserving bottlenecks (e.g., a lane drop) are automatically compensated for and errors due to lane changes/passing are reduced (unless there are significant systematic speed differences between the lanes).

For usual relative counting errors of the order of 1 % and errors due to overtaking and lane changes of the same order, we need a floating car every other half hour if the road section between the detectors is sufficiently long (i.e., significantly longer than in the example at the end of the previous section). The method uses following information (cf. Fig. 19.2)

- At least two, preferably more stationary detector cross sections i providing vehicle counts $n_i(t_k) = n_i(k\Delta t)$ aggregated over time intervals Δt and summed over all lanes.
- *Floating cars (FC) j*, where the index j is ordered according to the passing times $t_{j1}$of the FCs at the first detector.[3] The method needs the times t_{ji} when FC j passes detector i.

The accumulated counts are initialized/reset by the first floating car:

$$N_i(t_{1i}) = 0. \tag{19.11}$$

In Fig. 19.2, this means $N_1(t_{11}) = N_2(t_{12}) = N_3(t_{13}) = 0$. A small complication arises because the passage times t_{1i} are not synchronized with the times t_k where the aggregation intervals of the detectors end, so the accumulated vehicle counts can only be reset at the time t_k where the next aggregation interval ends ($t_{k-1} \leq t_{1i} < t_k$). To avoid an unnecessary loss of precision due to the aggregation intervals (typically, $\Delta t = 60\,\mathrm{s}$ while the passage times t_{1i} are known to the second, at least), we reset the accumulated vehicle counts at time t_k not to zero but to a fraction of the last count assuming a steady traffic flow within each aggregation interval:

---

[3] Additional index-swapping routines are necessary for the case when two FCs overtake each other.

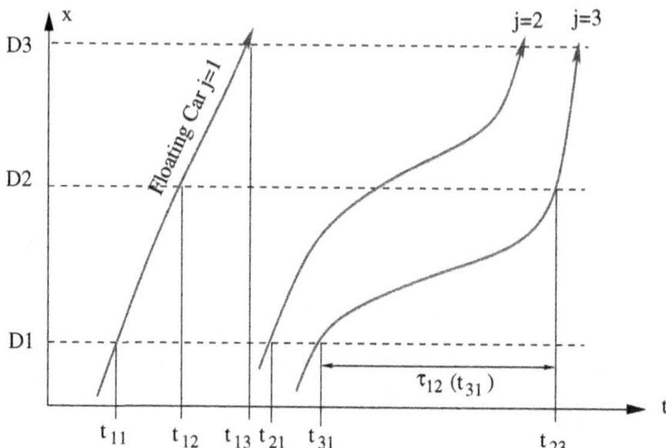

Fig. 19.2 Using several floating cars j to reset accumulated counting errors at the FC passage times t_{ji}. Notice that, in contrast to Fig. 19.1, the vertical axis denotes the location of the detector positions D_i) rather than the accumulated vehicle number (*N-curves*)

$$N_i(t_k) = n_i(t_k)\frac{t_{1k} - t_{k-1}}{\Delta t}, \quad t_{k-1} \leq t_{1i} < t_k. \tag{19.12}$$

To estimate the realized travel times between detectors i and m (where $m > i$), we can directly apply Eq. (19.9), i.e., $N_i(t - \tau_{im}(t)) = N_m(t)$. Specifically, one reads off $N_m(t_k)$ at the time $t_k \leq t$ of the last available interval and solves the condition $N_i(t') = N_m(t_k)$ for the past time t' using linear interpolation between the times $t_{k'}$, $k' \leq k$.

To tackle the drift due to measuring errors, the accumulated counts of detector i are reset whenever a floating car j crosses it (at time t_{ji}) by generalizing the initialization (19.12):

$$N_i(t_k) = n_i(t_k)\frac{t_{jk} - t_{k-1}}{\Delta t}, \quad t_{k-1} \leq t_{ji} < t_k. \tag{19.13}$$

Of course, this means, that, at a given time t_k, some detector counts are reset by more recent floating cars than others. To avoid inconsistencies when estimating the travel time with formula (19.9), each detector needs to provide not only the cumulated vehicle number after the last reset but also the accumulated vehicle numbers related to resets by previous floating cars covering, say, the last hour. Furthermore, to determine which accumulated vehicle numbers to use when evaluating Eq. (19.9), the identity of the floating cars has to be stored as well.

Notice that the correction of the N-curve travel times estimates by floating cars also provides a real-time quality control: The lower the magnitude of the jumps of the estimates before and after a floating car has passed, the lower the average total error of this method. The total error is composed mainly of detector counting errors, FC data errors, discretization errors, time-stamp errors, and errors due to lane changing and overtaking.

The N-curves method does not work directly on sections containing on-ramps, off-ramps, junctions, intersections, or other non-flow-conserving bottlenecks. If each merging or diverging link is equipped with at least one stationary detector near the merging/diverging point, the method can be generalized to include such flow-violating bottlenecks, but this will not considered here.

19.5 Virtual Stationary Detectors

This method is mainly relevant for macroscopic traffic flow models since this model class does not have output that can be directly linked to travel times. Besides the method presented here, one can also construct virtual trajectories from the speed field and determine the travel times with them (see Sect. 19.6).

In order to determine the travel times from location x_1 to $x_2 > x_1$ (which must be inside the simulated region) one "installs" two virtual detectors at these locations and generates the N-curves $N_1(t)$ and $N_2(t)$ by directly counting the passing vehicles. Since everything is known in the simulation, the initialization is straightforward and follows directly from the definition of the local density (cf. Eq. (19.10)):

$$N_1(t_0) = 0, \quad N_2(t_0) = \int\limits_{x_1}^{x_2} \rho(x, t_0)\,\mathrm{d}x.$$

Here, the integral denotes the number of vehicles between the virtual detectors at time t_0. Further corrections are not necessary unless there are merges or diverges on the road section between x_1 and x_2. In the latter case, it is better to calculate travel times by virtual trajectories instead of virtual detectors. This is discussed below.

19.6 Virtual Trajectories

If no trajectory data are available from probe vehicles or microscopic simulations but the local speed field can be estimated, one can construct *virtual trajectories* out of the speed field.

Trajectories from macroscopic simulation data. Since the complete speed field $V(x, t)$ is known in the simulated area including $[x_1, x_2]$, it is straightforward to generate *virtual trajectories* by demanding that the vehicle speed is identical to the corresponding local speed, $v_{\mathrm{FC}}(t) = V(x_{\mathrm{FC}}(t), t)$, and directly integrating the defining equation for the speed of the virtual floating cars,

$$\frac{\mathrm{d}x_{\mathrm{FC}}}{\mathrm{d}t} = V(x_{\mathrm{FC}}(t), t). \tag{19.14}$$

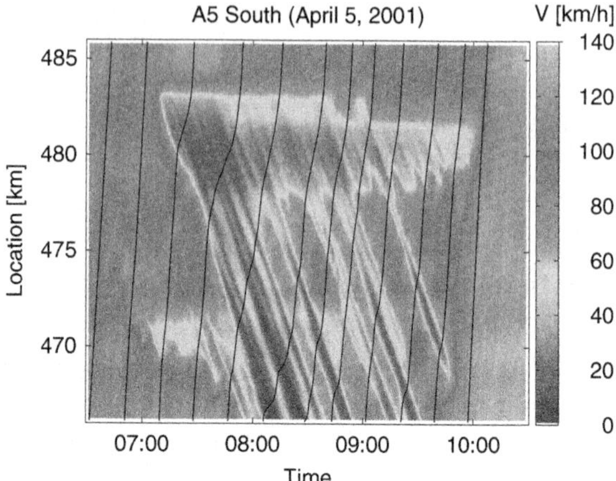

Fig. 19.3 Virtual trajectories generated from the speed field $V(x, t)$ by integrating Eq. (19.14). The speed field has been reconstructed from stationary detector data of a section on the German Autobahn A5-South during the morning hours

In order to get the expected travel time at time t, one chooses the initial condition

$$x_{\text{FC}}(t) = x_1, \tag{19.15}$$

and integrates numerically the inhomogeneous first-order ordinary differential equations until the trajectory reaches x_2 at time t_2. Then, $\tilde{\tau}_{12} = t_2 - t$. When the realized travel time $\tau_{12}(t)$ is relevant, one chooses the final condition $x_{\text{FC}}(t) = x_2$ and integrates backwards in time until the trajectory reaches x_1 at time t_1. Then, $\tau_{12} = t - t_1$.

Trajectories from stationary detector data. In principle, one can estimate travel times from speed data of real detectors via virtual trajectories in a two-step scheme (cf. Figs. 19.3 and 19.4):

- Estimate the local speed field by dedicated methods, e.g., the adaptive smoothing method (cf. Sect. 5.2).
- Generate virtual trajectories from the speed field using Eq. (19.14) and determine the passing times of this trajectory.

The Figs. 19.3 and 19.4 illustrate this method by analyzing data of a series of stationary detectors on a 20 km long section of the German Autobahn A5-South near Frankfurt/Main. As a result, we estimate that the travel time for this section increases in the morning rush hour from 10 min (corresponding to an average speed of 120 km/h) to 30 min (40 km/h). Notice that the speed field and the travel times are estimated from lane averages. When using only the detectors of the faster (slower) lanes, one would obtain lower (higher) travel-time estimates for free traffic. Due to *speed synchronization* across the lanes, there is no significant difference between the lanes for congested traffic, however.

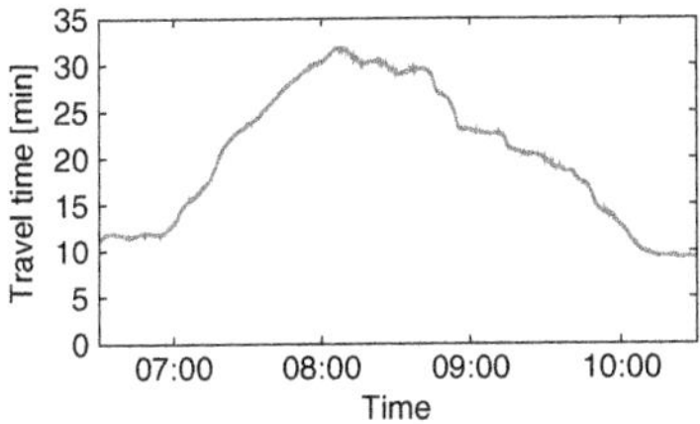

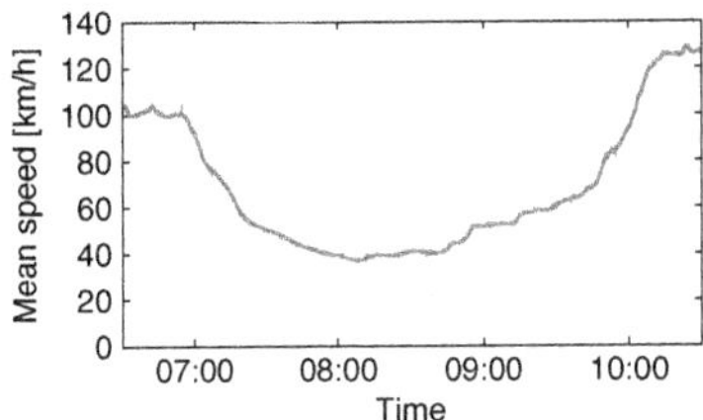

Fig. 19.4 Times series of the realized travel times estimated from the virtual trajectories of Fig. 19.3. A traffic jam in the morning rush hour leads to an increase of the travel time up to a factor of three (at about 8:00 h)

When applying this method, one has to keep in mind that aggregated stationary detector data only allow for a biased speed estimate (towards too high values, cf. Sect. 3.3). This means, travel times estimated by this method are biased towards too low values. Therefore, this method is less robust than N-curves and should only be used as a last resort when the method of N-curves cannot be applied (which is the case if the section contains on-ramps, off-ramps, or other inhomogeneities acting as traffic sources or sinks).

19.7 Instantaneous Travel Time

It is sometimes instructive to display the travel time as the simulation runs, i.e., in "real time". This is especially relevant for pedagogical simulations as that on the authors' website.[4] Rather than estimating $\hat{t}_2^{\beta}$, it is generally simpler and more robust to display the *instantaneous travel time*.

> The *instantaneous travel time* is the travel time of a hypothetical vehicle traveling through the considered section at a speed profile identical to that of the *present* local speeds.

It is more up-to-date than the realized travel time and can be calculated without predictions. It does not, however, represent realized or future travel times of any individual driver.[5] Rather, it represents a robust estimate of a typical travel time in this situation whose value is generally between τ_{12} and $\tilde{\tau}_{12}$.

Macroscopic models. In macroscopic models, the mathematical formulation follows directly from the definition:

[4] see: www.traffic-simulation.de

[5] Then, the very concept of an "instantaneous travel time" would be semantically contradictory.

$$\tau_{\text{inst}}(t) = \int_{x_1}^{x_2} \frac{1}{V(x,t)} \, \mathrm{d}x. \tag{19.16}$$

If the model is numerically integrated on a regular grid of cell size Δx and the instantaneous travel time is calculated from the beginning of cell i_1 to the beginning of cell i_2, then the integral becomes the sum

$$\tau_{\text{inst}}(t) = \sum_{i_1}^{i_2-1} \frac{\Delta x}{V(x_i,t)}. \tag{19.17}$$

To avoid divisions by zero or a significant bias towards too high travel times one introduces a finite lower limit for the vehicle speeds.

Microscopic models. Here, the fraction $\Delta x / V$ of Eq. (19.17) can be interpreted as the time headway of individual vehicles, so the instantaneous travel time is the sum of all time headways of the vehicles inside the investigated road section:

$$\tau_{\text{inst}}(t) = \sum_{\alpha=1}^{n_{12}(t)} \Delta t_\alpha = \sum_{\alpha=1}^{n_{12}(t)} \frac{\Delta x_\alpha(t)}{v_\alpha(t)}. \tag{19.18}$$

As in Eq. (19.17), one introduces a finite lower limit for the vehicle speeds v_α when using the expression to the left of the second equal sign.

Problems

19.1. Criteria for estimating travel times by N-curves
Give the conditions under which the method of N-curves is exact provided that there are no detector errors. Is it possible to approximatively initialize the accumulated counts if there are no floating cars available? If so, under which conditions?

19.2. Estimating travel times from aggregated detector data
Two stationary detectors D1 and D2 are located 4 km apart and measure following flow profile (summed over all lanes):

D1:	Time	$< 16:42$	16:42–16:50	16:50–16:58	$> 16:58$
	Q(vehicles/h)	1,800	0	3,600	1,800
D2:	Time	$<$ 16:00	16:00–16:30	16:30–17:00	$>$ 17:00
	Q (vehicles/h)	1,800	0	3,600	1,800

Furthermore, it is known that there was an accident with a temporary total road closure.

1. Estimate qualitatively the spatiotemporal evolution of the resulting congestion. Sketch the congested pattern in a spatiotemporal diagram. Also draw into the diagram the location and time of the accident, and the time where the road closure ends. *Hint:* There are at least two possible scenarios leading to the measured time series.
2. Determine the N-curves $N_1(t)$ and $N_2(t)$ of the two detectors as a function of the time difference t in seconds between the actual time and 16:00 h. Initialize the accumulated count at detector D2 by $N_2(0) = 0$. To Initialize $N_1(0)$, use the information that the two detectors measured steady state traffic flow of speed 120 km/h and flow 1,800 vehicles/h at and before 16:00 h, and that a floating car passed detector D1 at 15:58 h and D2 at 16:00 h and that this floating car reduced its speed sharply at the latter time.
3. Draw the N-curves in a diagram and determine the travel times τ_{12} and $\tilde{\tau}_{12}$ for 16:40 h, i.e., $t = 2,400$ s.
4. Give the estimated travel time $\tilde{\tau}_{12}(t)$ in a closed analytical form for the intervals $t < -120$ s and $-120\text{ s} \leq t < 2{,}520$ s. Also give $\tau_{12}(t)$ for $1{,}800\text{ s} \leq t < 3{,}000$ s. Why does $\tilde{\tau}_{12}(t)$ jump by 1,800 s at time $t = -120$ s?
5. Where and when does the accident leading to the road closure happen?

Further Reading

- Cassidy, M.J., Windover, J.R.: Methodology for assessing dynamics of freeway traffic flow. Transp. Res. Rec. J. Transp. Res. Board **1484** (1995) 73–79
- Cassidy, M.J., Bertini, R.L.: Some traffic features at freeway bottlenecks. Transp. Res. Part B Methodol. **33** (1999) 25–42
- Coifman, B.: Estimating travel times and vehicle trajectories on freeways using dual loop detectors. Transp. Res. Part A Policy Pract. **36** (2002) 351–364
- Van Lint, J.: Reliable travel time prediction for freeways. TRAIL thesis series (2004)
- Chen, M., Chien, S.: Dynamic freeway travel-time prediction with probe vehicle data: Link based versus path based. Transp. Res. Rec. J. Transp. Res. Board **1768** (2001) 157–161
- Chen, M., Chien, S.: Determining the number of probe vehicles for freeway travel time estimation by microscopic simulation. Transp. Res. Rec. J. Transp. Res. Board **1719** (2000) 61–68
- Nanthawichit, C., Nakatsuji, T., Suzuki, H.: Application of probe-vehicle data for real-time traffic-state estimation and short-term travel-time prediction on a freeway. Transp. Res. Rec. J. Transp. Res. Board **1855** (2003) 49–59

Chapter 20
Fuel Consumption and Emissions

An investment in knowledge pays the best interest.
Benjamin Franklin

Abstract Calculating fuel consumption and emissions is a typical offline analysis step that uses the data previously obtained by simulations or observations. Depending on the aggregation level and level of detail, we distinguish several global, macroscopic, and microscopic approaches. The focus of this chapter is on a microscopic physics-based model with a high level of detail. As a "modal" model, it takes speed and acceleration profiles (as obtained from microscopic simulations or real trajectory data) and the engine speed (as obtained from gear-shift schemes) as input and is parameterized by vehicle and engine attributes. The outputs are instantaneous fuel consumption and emission rates (CO_2 and others) on a single-vehicle basis. To what degree does traffic congestion increase fuel consumption? Under which conditions are roundabouts more environmentally friendly than signalized intersections? How much CO_2 emission can be saved by novel intelligent traffic systems? These are typical questions that such models—in connection with microscopic traffic flow models—can answer.

20.1 Overview

Generally, models for fuel consumption and for emissions (CO_2 but also CO, NO_x, particulate matter, and others) have the same structure, so they can be discussed together.[1] In each case, the models constitute a map from the exogenous factors (traffic demand, properties of traffic flow, vehicle composition and infrastructure) to one of two sets of endogenous variables:

[1] This becomes explicit for CO_2 emissions since there exists a 1:1 relation between fuel consumption and these emissions, see Sect. 20.1.3.

M. Treiber and A. Kesting, *Traffic Flow Dynamics*,
DOI: 10.1007/978-3-642-32460-4_20,

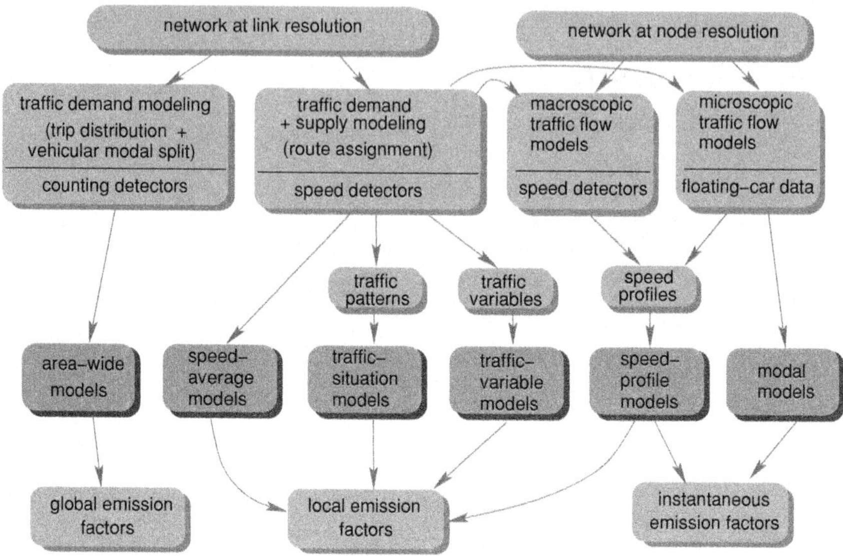

Fig. 20.1 Overview of fuel consumption and emission models. Additionally, the models are generally disaggregated with respect to vehicle classes and contain external factors such as temperature

- *Local emission factors* (often simply called "emission factors") describe fuel consumption or emissions in kg per meter (or liters per meter).
- *Instantaneous emission factors* describe fuel consumption or emissions in terms of kg per second per vehicle (or in liters per second per vehicle).[2]

Depending on the aggregation level and level of detail, there are several model categories. In the following, we shortly describe the main classes in the order of increasing level of detail and complexity (cf. Fig. 20.1).

20.1.1 Macroscopic Models

Area-wide models. In this simplest and most macroscopic approach, the only model input is the total vehicle mileage (traffic volume integrated over the total link length of the network and over time) in the investigated region in terms of *vehicle kilometers travelled* (VKT). As output, these models deliver the global fuel consumption and emissions in the investigated area. The VKT is usually disaggregated at least with respect to passenger cars and heavy-duty vehicles (trucks). Each of these categories

[2] As always, we will use the metric units kg, m, and s. In this context, this may lead to very small numerical values (particularly for the instantaneous factors), so the model results may be presented to the public in different units. However, a time-tested rule for simulator development states that one should always use the same unit system for internal calculations. Satellites have missed Mars because the navigation and controlling software used different units.

may be further disaggregated into several vehicle classes. The model input can be estimated by traffic demand models or by detector data. Since area-wide models are related to transportation planning rather than traffic flow modeling (cf. Chap. 1), we will not discuss them further.

Average-speed models. Besides the VKT disaggregated into several vehicle classes, average-speed models use as input the average speed driven on a certain link of the considered network. In addition, some of these models include external factors such as the temperature. The standard tools to obtain the speed information (generally as a function of time) are the traffic demand and route assignment models of transportation planning together with a network description including attributes such as road type and maximum speed for each link. Of course, one can also directly measure speed and traffic volume by double-loop detectors or other stationary speed-detecting devices. The model output are local emission factors, i.e., volume or mass of consumed fuel or emitted pollutant per kilometer and per vehicle, on average. To date, the majority of fuel consumption and emission investigations uses software implementing this model class, e.g., COPERT, MOBILE, MOVES or EMFAC. Since this class is more related to transportation planning than to traffic flow modeling and cannot determine the effect of jams, we will not discuss it further.

Traffic-situation models. This model class is related to the average-speed models. However, instead of a single continuous traffic-flow variable (the speed), the input consists of several distinct *driving patterns*. Most traffic-situation models define the set of driving patterns as a product set of the set of traffic-flow patterns (e.g., free, congested, stop-and-go), and the set of driving situations (e.g., highway, rural road, arterial road, residential street). The traffic-flow situation may also be defined in terms of the *level of service* (LOS) assessing the traffic flow quality on an ordinal scale from 1 (best) to 6 (worst, i.e., completely congested). As is the case for average-speed models, classical models of transportation planning or direct measurements provide the input in form of the traffic volume on a link relative to its capacity, and speed time series. Additionally, one needs a map associating the traffic volume and the speed data to a traffic flow pattern. This is straightforward if the traffic-flow patterns are defined in terms of LOS but it is more tricky if stop-and-go traffic should be distinguished from stationary situations. Software implementing this class includes HBEFA, ARTEMIS, and some versions of MOBILE.

Traffic-variable models. In contrast to the traffic-situation models relying on a finite set of qualitative (categorically scaled) traffic patterns, this model class takes as input quantitative (i.e., metrically scaled) macroscopic factors related to traffic flow such as traffic density, traffic volume relative to capacity, queue length, and, of course, speed. This information is complemented by categorically and metrically scaled infrastructure-related input such as road category, design speed, signal cycles, link length, number of lanes, and type of intersection. In contrast to the model classes above, the models of transportation planning are no longer sufficient to provide this input but they have to be complemented by microscopic or macroscopic traffic flow models. The output of this model are local emission factors, usually related to a single vehicle. Representatives of this class include Traffic Emissions and Energetics (TEE) and the queue-based Matzoros Model.

20.1.2 Microscopic Models

As a common feature, microscopic consumption and emission models need speed profiles of single vehicles at a high temporal resolution (of a few seconds or less) which can only provided by microscopic traffic flow models or by floating-car or trajectory data. As output, these models deliver local or instantaneous emission factors of single vehicles of a certain vehicle class. While traffic-situation and traffic-variable models already allow a coarse assessment to which degree congestions influence consumptions/emissions, only microscopic consumption/emission models allow us to answer questions related to individual vehicles and drivers such as the following:

- How much fuel/emissions can be saved by a fuel-efficient driving style? How does this saving potential depend on different traffic conditions? Is it possible to implement a fuel-efficient behavior into driver-assistance systems for the longitudinal driving task (adaptive cruise control)?
- What saves more fuel/emissions: Avoiding high accelerations/decelerations or driving at low engine speeds?
- Are roundabouts or signalized intersections more fuel-efficient? Does it depend on the type of roundabout, or on the origin–destination (OD) matrix characterizing traffic demand and the topology of the intersection?
- Is the savings potential and/or the optimal driving style different when switching from traditional combustion engines to modern developments such as hybrid or all-electric cars?
- What is the savings potential of recent Intelligent Transportation Systems (ITS) such as vehicle-to-vehicle or infrastructure-to-vehicle communication (e.g., a traffic light communicating its switching times to equipped cars)? How do the effects depend on the penetration rate of such ITS implementations?

Since microscopic consumption/emission models are strongly related to traffic flow models and traffic flow dynamics in general, they will be discussed in their own sections. Principally, one distinguishes two classes of microscopic models:

Speed-profile models. This model class does not use the instantaneous information provided by the simulated or measured trajectories directly. Rather, it is aggregated to several speed profile factors of a driving cycle which, in turn, determine the instantaneous consumption and emission factors.

Modal emission models. This is the most detailed model class: Instantaneous consumption and emission factors are given in terms of the instantaneous operating mode of the vehicle and the engine. This includes speed, acceleration, and the engine operating point including, at least, the *engine speed* (the revolution rate of the crankshaft) and the required power at the crankshaft.

20.1.3 *Relation Between Fuel Consumption and CO_2 Emissions*

While models for fuel consumption and emissions usually have the same form, they require different parameter sets. An exception is CO_2 emission: Due to the chemistry of combustion, there exists a strict 1:1 relationship between fuel consumption and CO_2 emissions, so a fuel consumption model is simultaneously a CO_2 emission model. Specifically, we have:

- Gasoline (98 ROZ): 2.39 kg CO_2/liter.
- Diesel fuel: 2.69 kg CO_2/liter.[3]

20.2 Speed-Profile Emission Models

The input of speed-profile models, also known as cycle-variable models, are speed profiles of single vehicles at a high temporal resolution which are obtained by floating-car data, trajectory data, or by a microscopic traffic flow simulation. At this level of detail, little infrastructure data is needed since the speed profiles implicitly contain most of this information (obvious exceptions include road gradients). As in the other models, this information has to be complemented by attributes of the considered vehicle classes and by external factors such as temperature and weather. The outputs of speed-profile models are either local or instantaneous emission factors which are related to a single vehicle. In contrast to the modal models to be discussed below, the instantaneous information of the speed profiles is not used directly. Rather, it is aggregated into several speed profile factors of a driving cycle which, in turn, determine the instantaneous consumption and emission factors. Most models of this class (e.g. MEASURE or PKE) assume a linear multivariate mapping between the speed profile factors **x** and the estimates **e** of the instantaneous emission factors:

$$\mathbf{e} = \underline{\mathbf{L}} \cdot \mathbf{x}. \tag{20.1}$$

Here, the components of the instantaneous emissions vector **e** may contain the CO_2 emission rate (which is strictly proportional to the fuel consumption rate), the rates of CO, HC, NO_x, PM, and others. The $n \times m$ matrix $\underline{\mathbf{L}}$ represents the linear relations between the m speed profile factors and the n components of the instantaneous emissions vector. The matrix $\underline{\mathbf{L}}$ is the core of models of this type.

To specify the model (i.e., identifying the relevant speed profile factors and the dimension of $\underline{\mathbf{L}}$) and calibrate it (i.e., estimating the elements of $\underline{\mathbf{L}}$), one needs

[3] The difference is mainly due to the difference in specific masses. To a good approximation, the mass of emitted CO_2 is equal to 44/12 times the mass of carbon contained in the fuel. Each carbon atom of the fuel (mass: 12 atomic units) produces one CO_2 molecule (mass: 44 atomic units) while other products containing carbon (such as CO or soot) eventually convert to CO_2 or are negligible regarding the mass balance.

Table 20.1 A selection of common speed profile factors and their effect on CO_2 emissions (– – ⇔ strongly negative, – ⇔ negative, + ⇔ positive, ++ ⇔ strongly positive)

Factor	Effect on CO_2 emissions
Constant of value 1	Reference (+++)
Fraction of time in speed class 0–25 km/h	++
Fraction of time in speed class 50–75 km/h	– –
Fraction of time in speed class 75–100 km/h	–
Fraction of time in speed class > 125 km/h	++
Standard deviation of speed	+
Average and standard deviation of acceleration	+
Average and standard deviation of deceleration	–
Frequency of acceleration–deceleration cycles	+
Fraction of time the vehicle is standing	+
Fraction of time the vehicle needs power near its maximum power	++
Fraction of road gradients greater than 5 %	+
Engine speed (crankshaft revolution rate) 1,000–2,000 rpm	– –
Engine speed (crankshaft revolution rate) > 3,500 rpm	++

instrumented vehicles measuring all relevant instantaneous emissions while driving various driving patterns. Typically, one starts by defining plausible speed profile factors as nonlinear functionals mapping the speed profile $\{v(t)\}$ to the real-valued numerical value of the factor. An example of a candidate factor is the speed variance. In the next step, one determines the n most relevant factors by a factorial analysis (cf. Table 20.1), and calibrates the coefficients of the new factors by a multivariate linear regression. Alternatively, if a microscopic traffic flow simulator and a validated modal emissions model (see Sects. 20.3 and 20.4) are available, one can specify and calibrate a speed-profile model by simulation.

Table 20.1 displays typical speed profile factors that are used by many models of this class and their influence on consumption and emission varying from strongly negative (– –) to strongly positive (++) with respect to the reference. Obviously, neutral factors defining the reference[4] do not "survive" the factorial analysis. Notice that replacing the speed-class factors by a simple factor "average speed" would lead to a mis-specified model since there is no approximatively linear relationship between the average speed and consumption/emission rates. Figure 20.6 shows that, in terms of CO_2 emissions or fuel consumption per km, vehicles emit (and consume) least at moderate speeds of 50–80 km/h, when using a continuous rather than acceleration-dominated driving style, and when driving using low engine speeds. Similar relations apply for other emission factors. The physics-based modal model of Sect. 20.4 will make this explicit.

[4] In this example, this includes the speed classes 25–50 and 100–125 km/h, and the engine speed class 2,000–3,500 rpm.

20.3 Modal Emission Models

20.3.1 General Remarks

While speed-profile models make use of the trajectory information (of floating-car data or microscopic simulations) indirectly via the profile factors, *modal emission models* use the instantaneous information directly (cf. Fig. 20.1).

At any moment, modal models calculate the vector $\mathbf{e}(t)$ of instantaneous emission factors as a function of the instantaneous "mode" of vehicle operation, essentially speed $v(t)$ and acceleration $\dot{v}(t)$. In the more refined modal models, the vehicle operation mode is complemented by a *characteristic map* describing the instantaneous operating mode of the engine. This mode is expressed by engine speed $f(t)$ (including the idling mode), power demand (or torque), and possibly other history-related factors such as engine age and temperature. At this microscopic level, the only infrastructure information that is used directly are road gradients and possibly the road surface quality.[5] Depending on the situation and model complexity, further input is necessary including local road-related variables (e.g., uphill grade ϕ), external variables (e.g., altitude, air temperature), and variables related to the engine history (e.g., engine temperature). As is the case with other microscopic models, each vehicle can be modeled individually with its own parameter set.

Models of this class are perfectly suited to be used in conjunction with time-continuous microscopic traffic flow models: The models are linked such that the endogenous variables of the traffic flow models (speed, acceleration) are exactly the main exogenous variables of the modal emission models. Consequently, most microscopic traffic flow simulators include a fuel consumption and emission module which typically runs offline on trajectory data generated by a previous simulation.

While, as the most detailed model class, modal models have the potential to give the most precise description, they also have the highest demand on data for calibration, validation, and usage. Particularly, it is notoriously difficult to measure the instantaneous emission rates on a continuous basis at a time resolution of seconds, which would be appropriate for this class. These problems are least serious (and the results most relevant) for fuel consumption rates. As a consequence, when we describe a specific modal model in detail in the next section, we will limit ourselves to fuel consumptions. Since there exists a strict 1:1 relationship between fuel consumption and CO_2 emissions (cf. Sect. 20.1.3), this also includes a model for CO_2 emissions.

Conceptionally, we can distinguish two types of modal consumption/emission models: Phenomenological models (cf. Sect. 20.3.2), and load-based models (cf. Sect. 20.3.3).

---

[5] When applying microscopic traffic flow simulations to calibrate, validate, or use modal models, detailed infrastructure and traffic information is needed indirectly as input to the microscopic traffic flow model.

20.3.2 Phenomenological Models

Models of this subclass describe the instantaneous consumption and emission rates without a recourse to the underlying physical principles. Consequently, the model parameters have no intuitive meaning. They come in the following two forms:

Statistical modal models express the consumption and emission rates as a multivariate linear function of several instantaneous ad-hoc factors that are related to the instantaneous driving mode. The model structure is similar to that of the speed-profile models, only that the factors are functions of the instantaneous speed and acceleration rather than functionals of the speed profile over a driving cycle. In the simplest case, the factors are formed by powers of the single-vehicle speed v and acceleration $\dot{v} = \frac{\mathrm{d}v}{\mathrm{d}t}$. A simple example for the instantaneous fuel consumption rate $\dot{C} = e_1$ (the first component of the instantaneous emissions vector $\mathbf{e}$) reads

$$\dot{C} = \max\left(0,\ \beta_0 + \beta_1 v + \beta_2 v^2 + \beta_3 v^3 + \beta_4 v\dot{v}\right). \tag{20.2}$$

Uphill gradients ϕ (angle in radians which is essentially 0.01 times the grade in percent) may be incorporated by adding a term $\beta_5 v\phi$. In order to avoid unphysical results, it is necessary to restrict $\dot{C}$ to non-negative values.[6] Furthermore, it is useful to treat the idling state by a separate equation $\dot{C}_{\text{idle}} = \beta_0^{\text{idle}}$. Later on, we will see that the factors of Eq. (20.2) have a physical meaning which, however, is not explicit in statistical models.

Because of their simple form and the small number of parameters, such models are relatively easy to calibrate and to use. However, they describe the effects of different driving styles only partially: While the effect of speed, acceleration, and deceleration is included, the effect of engine speed f (i.e., the selection of gears) is ignored. In principle, one could include engine speed. However, this entails a multitude of relevant new factors of the form $f^n v^m \dot{v}^k$ resulting in a model which is difficult to calibrate and prone to over-fitting. Moreover, because the parameters have no meaning, one cannot extrapolate from a database of vehicle types generated at certain test conditions to similar vehicles or to other conditions, e.g., simulating the effect of loading the vehicle with five instead of two persons, or turning on the air conditioning (A/C).

Map-based modal models drive the phenomenological approach to its extremes. Rather than on a function, such model are based on a two- or three-dimensional look-up table, or *vehicle operations map* mapping the instantaneous driving mode $(v, \dot{v}, f)$ directly to the consumption rate. In principle, this results in a parameter-free model.[7] However, generating such maps for any vehicle type and all driving

---

[6] One does not gain fuel when decelerating strongly or driving downhill.

[7] In another interpretation, the number of model parameters is equal to that of the map entries.

conditions is prohibitive. Even extensive measuring campaigns over many driving cycles will only cover a small region of the whole space of exogenous variables resulting in sparse look-up matrices.

20.3.3 Load-Based Models

In this subclass of modal consumption/emission models, the power demand or the *load* (the power demand relative to the maximum engine power) plays a central role. Firstly, the load is the most relevant influencing factor of instantaneous fuel consumption and emissions. Secondly, it is well understood in physical terms and can be analytically expressed as a function of speed, acceleration, and road gradient with intuitive parameters such as total vehicle mass, friction coefficient, and aerodynamic drag coefficient. Therefore, models of this type are also called *physics-based modal models*.

Well-known representatives of this type are the comprehensive modal emission model (CMEM), and passenger car and heavy-duty emission model (PHEM). Such models come in the following two variants.

Purely analytical models assume a constant engine efficiency (ratio of mechanical energy to chemical energy contained in the fuel) in all driving modes and essentially lead to statistical models of the form (20.2). However, the parameters β_j have an intuitive meaning, now.

Hybrid models combine the physical approach for the power demand with an engine *characteristic map* providing the engine efficiency (or emission rates) as a function of the instantaneous engine operating mode expressed by engine speed, power demand, and sometimes by additional history-related factors such as engine temperature and age. Engine characteristic maps are measured on test benches. While they are laborious to make, they require significantly less effort than the maps of the instantaneous driving modes of the map-based modal models. Furthermore, they cover the complete set of possible driving situations. A significant advantage of load-based models is their versatility: Without changes to the model or re-calibration, models of this class can describe different driving conditions, e.g., driving with one person or fully loaded, driving with heavy use of electric appliances (seat heating, A/C) and without (cf. Problem 20.6). Finally, load-based models can be easily adapted to describe various established and novel fuel-saving drive-train solutions, e.g., overrun fuel cut-off, automatic start/stop systems, and various implementations of energy recuperation during braking or downhill driving. The power-demand and energy modules of these models can even be applied to all-electric vehicles where conventional fuel consumption/emission models do not make sense. In the next section, we will describe such a model in detail.

20.4 Physics-Based Modal Consumption Model

With respect to the power demand module, the model described in the following is related to the PHEM/CMEM models. However, it is significantly simplified in the other model components.

Figure 20.2 shows an overview of the model components. The main factor determining fuel consumption is the *driving resistance* F which essentially depends (via friction, aerodynamic drag, and inertia) on the instantaneous speed v and acceleration $\dot{v}$, i.e., on the quantities resulting from the simulation of time-continuous car-following models.

Once the driving resistance is known, the required mechanical engine power to provide the oppositely directed driving force results immediately from the formula "power equals force times speed",

$$P_{\text{dyn}} = F\, v. \tag{20.3}$$

However, the overall vehicle operation requires additional power P_0 to drive the various electric appliances (radio, lights, A/C, seat heating) of modern cars, and to compensate for losses in the drivetrain. This results in the overall power demand $P = P_{\text{dyn}} + P_0$. In the next step, the fuel consumption rate is calculated from the energy density of the fuel assuming a certain efficiency factor γ. In the simple analytic version of the physics-based model γ is constant (typical values are of the order of 30 %) while it is read from the engine characteristic map in the more sophisticated model variants. Finally, for purposes of presentation, the instantaneous consumption can be extrapolated to calculate the fuel consumption per 100 km (or 100 miles) when always driving in this mode.[8]

In the following subsections, we present the model components in detail.

20.4.1 Driving Resistance

The driving resistance is the mechanical force (in Newtons) needed to maintain the instantaneous vehicle dynamics $(v, \dot{v})$ prescribed by the car-following model. We formulate it using elementary physical principles:

$$\boxed{F(v, \dot{v}) = m\dot{v} + (\mu + \phi)mg + \frac{1}{2}c_d \rho A v^2.} \tag{20.4}$$

[8] This quantity is displayed (slightly low-pass-filtered) in many modern vehicles. When the vehicle stops and the engine is idling, the display reverts to the instantaneous consumption rate.

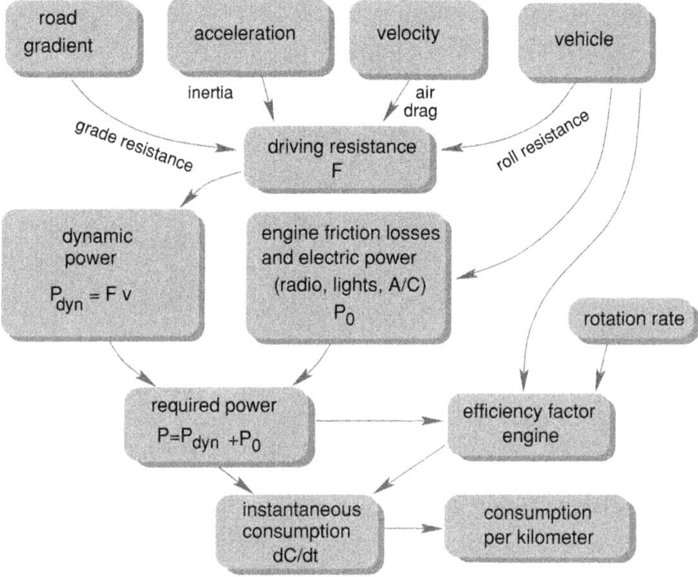

Fig. 20.2 Overview of the components of the physics-based modal consumption model

This expression contains following terms:

1. The *inertial force* $m\dot{v}$ follows from Newton's second law "force equals mass times acceleration". When decelerating ($\dot{v} < 0$), this force is negative and will be balanced with the other forces.
2. The solid-state friction force $mg\mu$ is proportional to the friction coefficient μ and the gravitational force $mg\cos\phi \approx mg$ perpendicular to the road surface (with the gravitational acceleration $g = 9.81\,\text{m/s}^2$ and the uphill gradient ϕ of the road). We assume that the dimensionless coefficient μ (which must be empirically determined and generally depends on the road surface and the tires) is speed independent. More complex models also assume an additional speed-proportional contribution.
3. The *uphill/downhill-slope force* $mg\sin\phi \approx mg\phi$ takes into account the additional gravitational forces at road gradients. Here ϕ denotes the uphill angle in radians. For our purposes, i.e., gradients $\tan\phi$ less than 20 m elevation gain or loss per 100 m, we can assume $\sin\phi \approx \tan\phi \approx \phi$, i.e., $100\,\phi$ directly denotes the gradient in percent (meter in elevation gain per 100 m of projected road length). When driving downhill, this contribution is negative and will be balanced with the other forces.
4. The *aerodynamic drag* $\frac{1}{2}c_d\rho Av^2$ is proportional to the density of air,[9] to the frontal cross section A of the vehicle (about $2\,\text{m}^2$ for cars), and to its *aerodynamic drag coefficient* c_d indicating how streamlined the vehicle is. While, in

[9] Density of air is $\rho \approx 1.3\,\text{kg/m}^3$ at ocean level and $\rho \approx 0.65\,\text{kg/m}^3$ at an altitude of 5,500 m

the 1980s, typical c_d values of cars were around 0.4–0.5 (the Volkswagen Beetle in its original design had 0.48), most modern cars (excluding experimental cars) have c_d values between 0.24 and 0.35 (SUVs have higher values). Because of its quadratic dependence on the speed, the aerodynamic drag becomes the dominant contribution of the driving resistance for high speeds (above 100–130 km/h, cf. Fig. 20.6 and Problem 20.6).

20.4.2 Engine Power

Besides the power $P_{\text{dyn}} = Fv$ to overcome the driving resistance, an additional base power or *idling power* P_0 is needed for the rest of the vehicle operations. It is composed of the power demand (i) of lights, radio, A/C and similar, (ii) to operate various actuators and electrical motors for ventilation, electric window lift, windscreen wipers etc., (iii) to compensate for electrical losses of the generator providing the electrical power, (iv) to overcome the internal friction of the engine and drivetrain.[10]

When taking into consideration that all modern vehicles have overrun fuel cut-off, i.e., no fuel is consumed if the driving resistance is negative (due to braking or downhill driving), we obtain following formula for the *overall power demand* on which the calculation of the instantaneous fuel consumption is based:

$$\boxed{P(v, \dot{v}) = \max\left[P_0 + vF(v, \dot{v}), 0\right].} \tag{20.5}$$

The maximum condition reflects a powertrain management where no mechanical energy can be "recuperated" when the overall power demand $P_0 + vF$ is negative. However, the resulting surplus of mechanical power can be used to provide the idling power P_0. In old vehicles without overrun fuel cut-off, the instantaneous consumption is at least that of the idling mode, so the overall power demand is bounded from below by P_0,

$$P(v, \dot{v}) = P_0 + \max\left[vF(v, \dot{v}), 0\right]. \tag{20.6}$$

In the ideal case of loss-free recuperation (storage) capabilities of mechanical energy during braking and downhill driving, the power balance reads

$$P(v, \dot{v}) = P_0 + vF(v, \dot{v}). \tag{20.7}$$

To reflect the more realistic case of recuperation with losses which is relevant for hybrid and all-electric vehicles, we generalize this balance equation to

$$P(v, \dot{v}) = \begin{cases} P_0 + vF(v, \dot{v}) & P_0 + vF(v, \dot{v}) \geq 0, \\ (1 - r)[P_0 + vF(v, \dot{v})] & \text{otherwise.} \end{cases} \tag{20.8}$$

---

[10] When driving rather than idling, the internal friction can be accounted for in the characteristic engine map (Sect. 20.4.4).

Here, r denotes the relative round-turn losses for one complete charging/discharging cycle. The Eqs. (20.5)–(20.8) demonstrate the flexibility of the physics-based approach to describe different forms of engine power management. It should be noted, however, that the real situation in hybrid and electric vehicles is more complex. In particular, the recuperation capabilities and losses depend on history (particularly on the charging state of the battery), on temperature, and on other factors related to the power management.

20.4.3 Consumption Rate

The translation of the required power to the fuel consumption rate is mediated by the (calorimetric) energy density of the fuel, and by the efficiency factor of the engine. Defining the *energy density*

$$w_{\text{cal}} = \frac{\Delta W_{\text{chem}}}{\Delta C} \tag{20.9}$$

as chemical energy ΔW_{chem} per volume ΔC of the fuel, and the *efficiency factor*

$$\gamma = \frac{\Delta W_{\text{mech}}}{\Delta W_{\text{chem}}} \tag{20.10}$$

as the fraction of mechanical energy ΔW_{mech} that can be converted from a certain amount ΔW_{chem} of chemical energy, we obtain the relation

$$\boxed{\Delta W_{\text{mech}} = \gamma w_{\text{cal}} \Delta C.} \tag{20.11}$$

A typical value for the energy density of gasoline is $w_{\text{cal}} = 39.6 \cdot 10^6$ J/l $= 11$ kWh/l.[11] And the efficiency factor γ is of the order of $0.25-0.35$. Taking the time derivative of Eq. (20.11) and using the identity $P = \frac{\text{d}}{\text{d}t}(\Delta W_{\text{mech}})$, we obtain a relation between the consumption rate and the overall power required according to Eq. (20.5):

$$\boxed{\dot{C} = \frac{\text{d}(\Delta C)}{\text{d}t} = \frac{P}{\gamma(P, f) w_{\text{cal}}}.} \tag{20.12}$$

In purely analytical models, the efficiency factor γ is assumed to be constant. In the more detailed hybrid models, $\gamma(P, f)$ depends on the overall power demand P and the engine speed f and is read from the engine *characteristic map* (cf. Sect. 20.4.4). The engine speed f depends directly on the driven speed v, the transmission ratio r_{t} of the selected gear (rotation rate of the engine shaft divided by the rotation rate of the tires), and the dynamical radius R_{dyn} of the tires,

---

[11] This is about fifty times larger than that of modern batteries for electric vehicles.

$$f = \frac{r_t v}{2\pi R_{dyn}}. \tag{20.13}$$

For a given gear and transmission, the efficiency factor γ depends on the speed and the required power alone. Inserting Eq. (20.13) for the engine speed, Eq. (20.5) for the power, and Eq. (20.4) for the driving force into Eq. (20.12) gives the instantaneous consumption solely as a function of endogenous variables of the microscopic flow model (speed and acceleration), the road gradient, and the characteristic map. We will now discuss characteristic maps in detail.

20.4.4 Characteristic Map for Engine Efficiency

The engine efficiency, i.e., the conversion factor from the chemical energy of the fuel to the delivered mechanical energy, depends in essential on the *operating point* characterized by engine speed f and by the *mean effective pressure* $\bar{p}$ (the pressure difference at the cylinders during one cycle). We will not consider further dependencies, e.g., on the engine and outside temperatures, here. The efficiency at different operating points is measured on engine test benches and presented in form of a two-dimensional (or multi-dimensional) table, the so-called (engine's) *characteristic map*. Characteristic maps come in different equivalent variants:

1. The exogenous variable can be defined as *specific consumption* relative to fuel mass (kg/kWh) or volume (L/kWh, cf. Fig. 20.3), as *efficiency factor* γ (in Fig. 20.4), or as the "fuel efficiency" which is defined as the inverse of the specific consumption. The efficiency factor γ and the specific consumption C_{spec} relative to volume are related by

$$\frac{1}{\gamma w_{cal}} = C_{spec}. \tag{20.14}$$

   When multiplying the volume-related specific consumption by the mass density (about 0.8 kg/l for gasoline), we obtain the mass-related specific consumption.
2. The first exogenous variable is given nearly exclusively by the engine speed f. Sometimes, this quantity is linearly transformed to the *normalized engine speed* f_{norm} such that $f_{norm} = 0$ corresponds to idling, and $f_{norm} = 1$ to the engine speed at maximum power (which generally is less than the maximum possible engine speed).
3. The second exogenous variable of the characteristic map can be the *engine power* P, its torque M, or the *mean effective pressure* $\bar{p}$, or the position of the throttle pedal.[12] Elementary relations connect the first three quantities with each other and with the engine speed f and the effective total cylinder volume V_{zyl}:

[12] The degree to which the pedal is "pushed down" is a good indicator of the mean effective pressure. Driving at "full throttle" corresponds to maximum values of $\bar{p}$, M, and P for the actual engine speed.

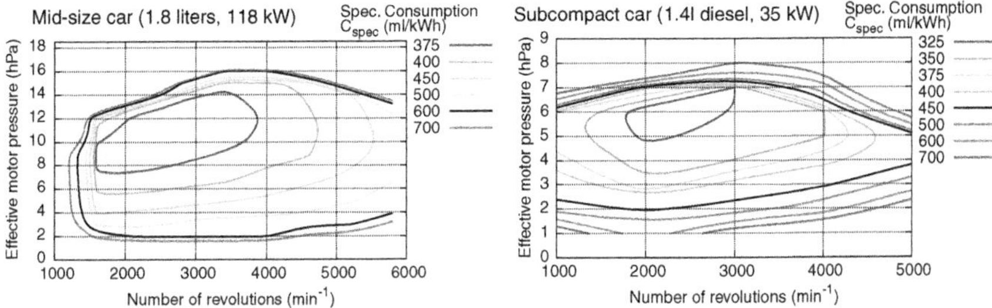

Fig. 20.3 Characteristic maps of various engines obtained from test benches. The *horizontal axis* denotes engine speed f, and the *vertical axis* the mean effective pressure $\bar{p}$. The *contour lines* indicate lines of constant specific consumption $\tilde{C}$ in units of mL fuel per kWh

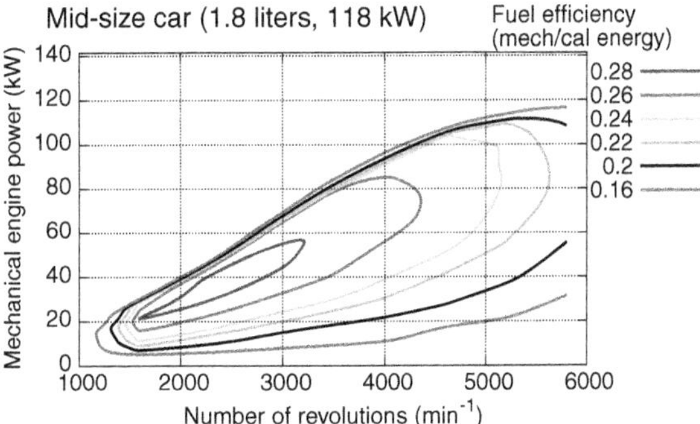

Fig. 20.4 Alternative representation of the characteristic map. The second exogenous variable is now the delivered power calculated by Eq. (20.15). The endogenous variable is now the efficiency factor γ calculated by Eq. (20.14)

$$P = 2\pi f M, \qquad \bar{p} = \frac{4\pi M}{V_{\text{zyl}}}. \tag{20.15}$$

The second relation is valid for four stroke cycle engines, only. In Fig. 20.3, we display the engine power since this is the most intuitive quantity which, moreover, is directly connected with the simulation via Eq. (20.5).

We notice that the characteristic map reflects the general fuel saving recommendations: Drive at the lowest possible engine speeds that can provide the necessary power; when increasing speed, do so at a reasonable rate (do not accelerate too slowly), and change to the highest compatible gear once cruising speed is reached. In this way, one generally operates at the left part of the characteristic map near the region of highest efficiency (cf. Problems 20.3 and 20.4).

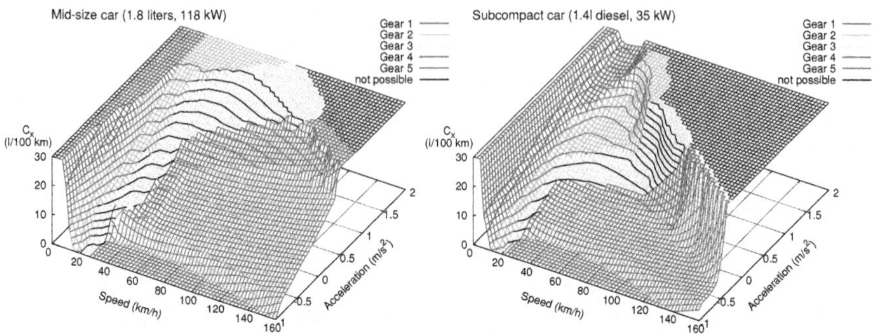

Fig. 20.5 Extrapolated consumption per 100 km according to Eq. (20.16) for the engine characteristic maps of Figs. 20.3 or 20.4, and the attributes of the corresponding cars as a function of speed and acceleration. The gears (*color-coded*) are selected to optimize fuel consumption. A *black grid* denotes forbidden regions where the power demand exceeds the engine capabilities

20.4.5 Output Quantities

With the help of the modal consumption model presented above, we can calculate intuitive quantities such as the fuel consumption or CO_2 emission per kilometer or per 100 km for arbitrary driving cycles and driving conditions (cf. Problems 20.6–20.8).

20.4.5.1 Fuel Consumption per 100 km

From the instantaneous consumption rate (20.12), we can calculate the consumption per 100 km *when driving the whole stretch in the same operating mode* by applying the chain rule of differentiation, $\frac{\partial C}{\partial t} = \frac{\partial C}{\partial x}\frac{\partial x}{\partial t} = v\frac{\partial C}{\partial x}$:

$$\boxed{C_{100} = 100\text{km}\,\frac{\mathrm{d}C}{\mathrm{d}x} = \frac{100\text{km}}{v}\frac{\mathrm{d}C}{\mathrm{d}t} = \frac{100000\text{m}}{\gamma w_{\text{cal}}}\frac{P}{v}}. \tag{20.16}$$

For a given car and engine, this quantity depends on the speed, the gear, and, via Eq. (20.5), on the acceleration.

The Figs. 20.5 and 20.6 show the vehicle *modal consumption characteristics* in terms of the consumption C_{100} per 100 km, Eq. (20.16), calculated with the vehicle parameters of Table 20.2 and the characteristic map of Fig. 20.3 using the most efficient gear for the instantaneous driving mode $(v, \dot{v})$ in question.

Figuratively, when going from the engine characteristic map to the vehicle modal characteristics, one "installs" the engine into the vehicle. Formally, the modal characteristics is equivalent to the core component of map-based modal models, the vehicle operations map (cf. Sect. 20.3.2), only that it is derived from physical principles and the engine characteristic map which is more "fundamental" and easier to determine. This means, validated physics-based models can be used to specify, calibrate, and

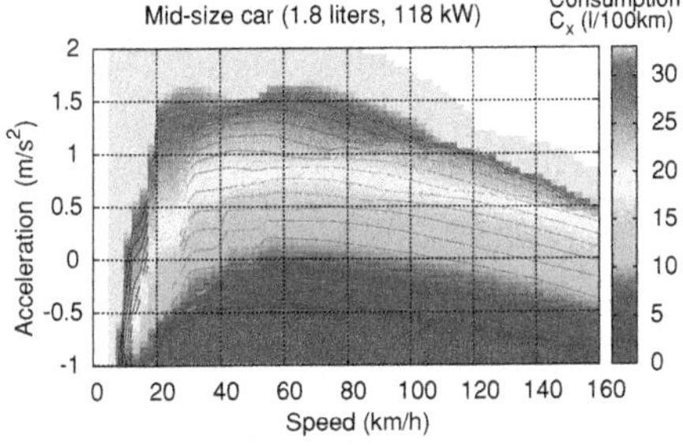

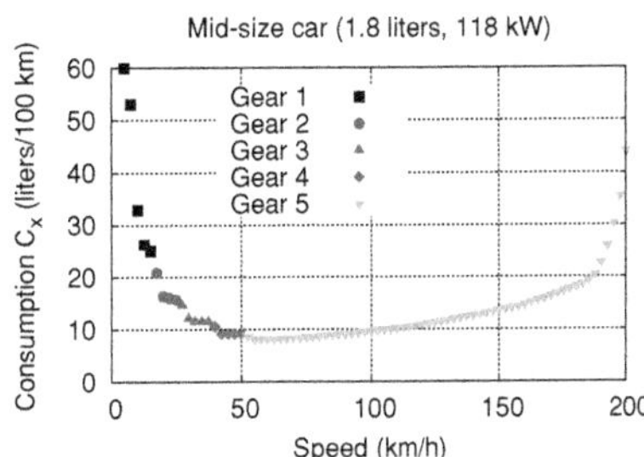

Fig. 20.6 *Left* Alternative presentation of the consumption per 100 km for the medium-class vehicle type (118 kW). *Right* Consumption when cruising at constant speed for the most efficient gear (*color-coded*). The engine characteristics used for this figure does not represent the state of the art. Modern vehicles consume significantly less

Table 20.2 Vehicular and physical parameters of the physics-based consumption model

Quantity	Symbol	Typical value for cars
Idling power	P_0	3 kW
Total mass	m	1,500 kg
Friction coefficient	μ	0.02
c_d value	c_d	0.3
Frontal cross section	A	2 m^2
Dynamic tire radius	$r_{\rm dyn}$	0.286 m
Transmission factors (1st to 5th gear)	$r_{\rm t}$	13.90, 7.80, 5.26, 3.79, 3.09
Specific consumption (if constant)	$C_{\rm spec}$	300 ml/kWh = 8.3 10^{-5} ml/J
Air density	ρ	1.3 kg/m^3
Gravitational acceleration	g	9.81 m/s^2
Fuel energy density	$w_{\rm cal}$	39.6 10^6 J/l $\approx$ 11 kWh/l

validate phenomenological map-based models, which is easier than doing this by driving-cycle experiments.

Figure 20.5 reflects following fuel-saving rule: Drive at the highest gear (i.e., at lowest engine speeds) where the engine can provide the actually needed power. i.e., only "kick down" (change to lower gears and higher engine speeds) when more power is needed (accelerating or driving uphill). From Fig. 20.6, we conclude that 40 km/h – 80 km/h is the most economic speed range regarding consumption.

20.4.5.2 Calculating the Cumulated Fuel Consumption

Most people guess that fuel consumption and CO_2 emissions are significantly increased in jams. Is this true, and if so, under which conditions? To answer this question, we combine a microscopic traffic flow model with the physics-based consumption model and simulate congested traffic (scenario I) and a reference scenario II without jams. Obviously, for a fair comparison, the same total number of vehicles must pass the considered road section in both scenarios. Therefore, to avoid the jams in scenario II, we make the driving style more efficient instead of reducing the inflow. To completely specify the scenarios we must provide the following:

- Composition of the vehicle fleet.
- Specification of demand: traffic inflow over time.
- Specification of supply (bottleneck).
- A realistic car-following model with parameterizations for all vehicle classes of the fleet for both scenarios.[13]
- Vehicle parameters and engine characteristic maps for all vehicle classes.
- A behavior rule to select the gears, e.g., always selecting the most efficient gear.

The first four items specify the traffic flow simulation, and the last two the subsequent calculation of consumption and emissions. Apart from the parameters of the car-following models, the specifications for both scenarios are identical. Making use of the definition of the instantaneous consumption rate $\dot{C}_\alpha(v_\alpha(t), \dot{v}_\alpha(t), g_\alpha(t))$ for vehicle α, we obtain for the total consumption in the simulated spatiotemporal region $[t_{\text{start}}, t_{\text{end}}] \times [x_1, x_2]$ the expression

$$\boxed{C_{\text{tot}} = \int_{t_{\text{start}}}^{t_{\text{end}}} \mathrm{d}t \sum_{\alpha=\alpha_1(t)}^{\alpha_2(t)} \dot{C}_\alpha(v_\alpha(t), \dot{v}_\alpha(t), g_\alpha(t)).} \tag{20.17}$$

The lower and upper limits $\alpha_1(t)$ and $\alpha_2(t)$ indicate the range of vehicle numbers which are inside the simulated region at a given time t. Figure 21.7 of Chap. 21 shows a calculated example. Generally, the additional percentaged increase in fuel consumption/CO_2 emissions due to jams is only a third of the percentaged increase of travel time. In extreme situations, traffic congestions can even lead to *less* consumption/emissions (cf. Problem 20.5).

Under which conditions for scenario I (congestion) and scenario II (free traffic) will a traffic breakdown lead to (i) a maximal (b) minimal (or even negative) increase of consumption and emissions?

---

[13] Such a simulation can also serve as a "reality check" for car-following models. If the accelerations become unrealistically high (as in the optimal velocity model), no consumption can be calculated since the power demand exceeds the engine capabilities (regions with black grids in Fig. 20.4).

20.4.6 Aggregation to a Macroscopic Modal Consumption Model

By aggregating over the vehicle trajectories, it is straightforward to generalize the modal consumption/emission models to a macroscopic description:

$$\dot{C}(V(x,t), \dot{V}(x,t)) = \langle C_\alpha(v_\alpha, \dot{v}_\alpha, g)\rangle. \tag{20.18}$$

The averaging $\langle\cdot\rangle$ includes a local average over the region around x ($V = \langle v_\alpha\rangle$, $\dot{V} = \langle \dot{v}_\alpha\rangle$, cf. Chap. 9), over the vehicle classes of the considered vehicle-driver composition, and over different gear-selection strategies.

By integrating over the vehicle number $n = \int \rho(x,t)\mathrm{d}x$ and over time, we obtain the total consumption/emissions produced by the traffic in a certain spatiotemporal region $[t_{\text{start}}, t_{\text{end}}] \times [x_1, x_2]$ by the double integral

$$\boxed{C_{\text{tot}} = \int\limits_{t_{\text{start}}}^{t_{\text{end}}} \mathrm{d}t \int\limits_{x_1}^{x_2} \mathrm{d}x\ \rho(x,t)\dot{C}\left[V(x,t), A(x,t)\right]} \tag{20.19}$$

where $A(x,t)$ gives the macroscopic acceleration in the comoving system (cf. Eq. 9.1).

Finally, we mention that the macroscopic speed field $V(x,t)$ can be used to determine the macroscopic factors of the traffic-variable models discussed in Sect. 20.1.1. Thus, these models can be calibrated/validated by comparing their output with expression (20.19) for the total consumption.

Problems

20.1 Coefficients of a statistical modal consumption model
A statistical modal model is characterized by following function of speed v, acceleration $\dot{v}$ and uphill gradient ϕ (cf. Eq. 20.2):

$$\dot{C} = \min\left(0,\ \beta_0 + \beta_1 v + \beta_2 v^2 + \beta_3 v^3 + \beta_4 v\dot{v} + \beta_5 v\phi\right).$$

Determine the model parameters β_0 to β_5 from the physics-based modal model with constant efficiency factor for cars described by the attributes of Table 20.2.

20.2 An acceleration model for trucks
For fully-loaded trucks, the conventional car-following models are not very useful in free-flow acceleration situations since the acceleration is restricted by the engine power. Formulate the acceleration component of a truck car-following model as a function of speed v and gradient ϕ. Plot the acceleration for a truck characterized by a total mass of 38 T, a cd value 0.8, a frontal cross section $10\,\mathrm{m}^2$, power components

$P = 310$ kW and $P_0 = 10$ kW, and other attributes as in Table 20.2. Restrict the acceleration to a maximum of $a_{max} = 1\,\text{m/s}^2$, and do not accelerate if v exceeds the desired speed $v_0 = 80$ km/h. Is the engine powerful enough to drive an uphill gradient of 2 % at 80 km/h?

20.3 Characteristic map of engine speed and power
Consider the characteristic map of Fig. 20.4. What engine power results at "full throttle" for an engine speed of 3,000 rpm? Which is the most efficient engine speed for providing a power of 60 kW?

20.4 Characteristic map of engine speed and mean effective pressure
Consider the characteristic map displayed in Fig. 20.3. (i) What power results at 2,600 rpm and full throttle? (ii) Consider a driving situation requiring a total power of 40 kW. Is it more efficient to drive at a higher gear (2,600 rpm) or at a lower gear (4,000 rpm)? At which of these engine speeds must the throttle pedal be pressed down further?

20.5 Does jam avoidance save fuel?
Why is the savings potential less when driving at high speed under non-congested conditions?

20.6 Influencing factors of fuel consumption
The following list gives some rules of how to save fuel. However, the mischievous author of this list has converted some rules to their opposite. Discuss which items are correct by comparing them with the predictions of the physics-based modal consumption model. If necessary, make a sensitivity analysis using the parameters of Table 20.2 assuming a constant efficiency factor.

1. Switching on the A/C increases the fuel consumption per kilometer, particularly in city traffic.
2. Mounting a roof rack (increases the c_d value by 0.08) has adverse effects on fuel consumption, particularly in city traffic.
3. When driving downhill, disconnecting the clutch or selecting the "neutral" mode of automatic transmission) may have adverse effects on the brakes but it saves fuel.
4. When you fill the tank only to half of its capacity, i.e., visit gas stations more frequently, you will save more than 2 % of fuel (assume a tank of 60 l).
5. Reducing the speed limit from 50 to 30 km/h in cities helps save fuel.
6. Reducing the speed limit from 130 to 110 km/h on highways helps save fuel.

20.7 Highway versus mountain pass: Which route needs more fuel?
Two route alternatives are specified as follows: (1) A level highway with driving speed of 150 km/h, (2) a mountain pass containing exclusively uphill and downhill sections of equal length with gradients of ±8 %, driving speed of 72 km/h, and overrun fuel cut-off is active.

20.8 Four-way-stops versus intersection with priority rules
Consider following two situations:

- Situation I: A city street with a series of four-way-stops.
- Situation II: Replacing the stops by a priority rule in the considered driving direction, i.e., drivers need not to decelerate or stop when going ahead.

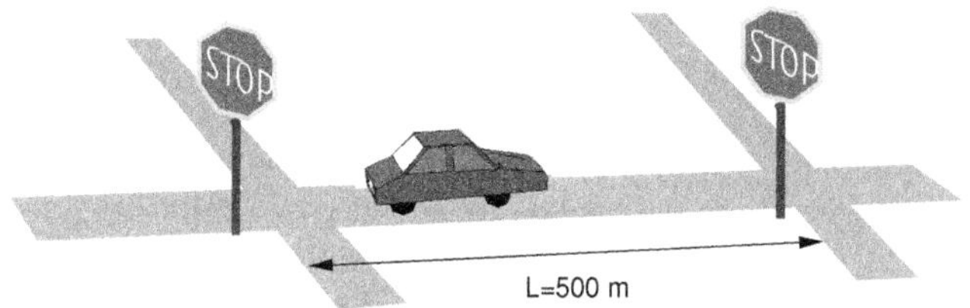

Calculate the fuel consumption on the 500 m long stretch between two intersections using the purely analytical physics-based modal model with parameters from Table 20.2 assuming (i) a free speed $v_0 = 16$ m/s, (ii) acceleration stop → free speed at 2 m/s^2 and deceleration free speed → stop at -2 m/s^2 in situation I, cruising at v_0 in situation II, (iii) no crossing traffic, i.e., immediate start after each stop in situation I.

20.9 Under which conditions do all-electric cars save CO_2 emissions?
Compare a conventional gasoline-driven vehicle characterized by a constant efficiency factor and the parameters of Table 20.2 with an all-electric car whose engine has a net efficiency factor γ_{el} (conversion of electric to mechanical energy including powertrain) of 85 %, and which can recuperate kinetic energy at round-turn losses $1 - \gamma_{rec} = 20$ % in following situations:

- Situation I: City traffic as in situation I of Problem 20.8.
- Situation II: Cruising on a highway at 130 km/h.

Calculate the total CO_2 emissions per kilometer for both vehicle types in both situations. Assume 2.39 kg CO_2 per liter gasoline in the combustion process (cf. Sect. 20.1.3), and a power-plant energy mix of 600 g CO_2 per kWh electrical energy (equalling the US average).

20.10 Fuel consumption for an OVM-generated speed profile
Consider a single vehicle entering a highway (desired speed $v_0 = 120$km/h) and accelerating according to the optimal velocity model (OVM) (cf. Sect. 10.6) at a maximum acceleration of 2 m/s^2.

1. Determine the OVM model parameter τ.
2. Calculate the required total power for the parameters of Table 20.2 as a function of the instantaneous speed v and express the result as a polynomial of the form $A_0 + A_1 v + A_2 v^2 + A_3 v^3$. Give analytical and numerical expressions for the coefficients A_0 to A_3.
3. Calculate the maximum power required during the acceleration phase. At which speed this power is needed?

20.11 Trucks at uphill gradients
In order to avoid trucks driving too slowly on uphill gradients of highways, the uphill gradients are restricted to $\phi = 4\,\%$. Should a steeper gradient of 5 % be allowed if the length L of the gradient is 500 m or less?

For modeling this situation, we assume as *worst case* a fully loaded truck whose engine just manages to maintain a speed of 80 km/h on a level road and whose maximum power does not depend on speed (a truck has many gears). The truck driver drives through the gradients at full throttle. Besides the physical parameters from Table 20.2, the relevant parameters are the following: mass 38 T, friction coefficient 0.03, cd value 0.8, and a frontal cross section of 10 m^2.

1. Determine the maximum engine power.
2. The truck enters the uphill gradient at an initial speed of 80 km/h. Give the initial decelerations for 4 and 5 % gradients.
3. Calculate the terminal speeds for the two gradients if they are sufficiently long. (You can neglect the aerodynamic drag in this calculation.)
4. We describe the decelerating process with an appropriately parameterized OVM for free traffic,
$$\frac{dv}{dt} = \frac{v_\infty - v}{\tau}.$$
Determine the parameters v_∞ and τ for the two gradients.
5. Calculate the speed and the covered distance as a function of time if the truck enters the uphill section ($x = 0$) at time $t = 0$.
6. In which of these two situations does the truck end up driving more slowly: (i) at the end of a 500 m long uphill gradient of 5 %, (ii) at the end of a 1,000 m long uphill gradient of 4 %? Assume the previously obtained (rounded) results $v_\infty = 37$ km/h, $\tau = 24$ s (uphill gradient 5 %) and $v_\infty = 42$ km/h, $\tau = 26$ s (uphill gradient 4 %). Furthermore, assume that the ends of the uphill sections are reached at 29.1 and 64.2 s in the situations (i) and (ii), respectively (these passing times can only be determined numerically).

Further Reading

- Barth, M., An, F., Norbeck, J., Ross, M.: Modal emissions modeling: A physical approach. Transportation Research Record: Journal of the Transportation Research Board **1520** (1996) 81–88

- Panis, L., Broekx, S., Liu, R.: Modelling instantaneous traffic emission and the influence of traffic speed limits. Science of the Total Environment **371** (2006) 270–285
- Cappiello, A., Chabini, I., Nam, E., Lue, A., Abou Zeid, M.: A statistical model of vehicle emissions and fuel consumption. Proceedings of the IEEE 5th International Conference on Intelligent Transportation Systems (2002) 801–809
- Ahn, K., Rakha, H., Trani, A., Aerde, M.V.: Estimating vehicle fuel consumption and emissions based on instantaneous speed and acceleration levels. Journal of Transportation Engineering **128** (2002) 182–190
- Smit R., Brown, A.L., Chan, Y.C.: Do air pollution emissions and fuel consumption models for roadways include the effects of congestion in the roadway traffic flow? Environmental Modelling & Software **23** (2008) 1262–1270
- Ericsson, E.: Independent driving pattern factors and their influence on fuel-use and exhaust emission factors. Transportation Research Part D **6** (2001) 325–345
- Treiber, M., Kesting, A., Niemann, C.: How much does traffic congestion increase fuel consumption and emissions? Applying a fuel consumption model to the NAGS trajectory data. In: TRY Annual Meeting 2008 CD-ROM, Washington, D.C., Transportation Research Board of the National Academies (2008)

Chapter 21
Model-Based Traffic Flow Optimization

When it is obvious that the goals cannot be reached, don't adjust the goals, adjust the action steps.

Confucius

Abstract By means of simulations of microscopic traffic flow models, we investigate measures to increase the efficiency and stability of traffic flow when the infrastructure and the traffic demand are fixed. Road-based measures include variable message signs (for traffic-adaptive speed limits and dynamic routing), ramp metering, and selective overtaking bans for certain vehicle classes (trucks) in certain situations (gradients). Another class of optimization measures is vehicle-based rather than road-based. At present, these measures have entered the market and are expected to have a significant market penetration (and influence) in the near future. They include semi-automated driving (adaptive cruise control), individual traffic-adaptive navigation, traffic-light assistants, and other driver-assistance systems.

21.1 Basic Principles

From the analysis of the spatiotemporal dynamics on highways and freeways in Chap. 18, we can conclude that most traffic breakdowns are caused by the simultaneous action of following three factors (cf. Sect. 18.1):

(A) High traffic load (demand in relation to road capacity).
(B) Local reduction of the road capacity (bottleneck).
(C) Local perturbations in the traffic flow itself, acting as the final trigger.

M. Treiber and A. Kesting, *Traffic Flow Dynamics*,
DOI: 10.1007/978-3-642-32460-4_21, © Springer-Verlag Berlin Heidelberg 2013

Consequently, traffic-flow optimization measures aim to remove or weaken at least one of these factors. This is tantamount to homogenizing traffic flow:

Golden Rule of Traffic Flow Optimization:
Try to homogenize traffic flow (i) with respect to time, (ii) spatially, and (iii) with respect to local speed differences.

There are several ways of implementing this principle: By *road-based* or *vehicle-based* measures focussing on different aspects of traffic homogenization, and acting on different spatial and temporal scales.

Static control of the spatial and temporal traffic demand. Measures of this class typically belong to the field of transportation planning and act mainly over longer timescales on the influencing factor (A) of the above list. Examples include constructing new roads, improving or removing existing roads, or implementing new traffic regulations.[1] Other, more politically influenced measures include vehicle tolls to enter the inner-city limits (*congestion charge*, as in London, Stockholm, or San Francisco), high-occupancy vehicle (HOV) lanes reserved for vehicles occupied by two or more persons, or initiatives to shift the modal split away from vehicular traffic, e.g., constructing new bicycle lanes or improving public transport (introducing new bus, tram, or train lines, increasing the frequency of service, or prioritizing public transport at intersections). Most measures in this category are simulated with traffic-stream models which are briefly described in Sect. 6.2.2 and which are the core of most commercial software for traffic assignment. They will not be considered here.

Dynamic control of traffic volume. In contrast to static (possibly time-dependent) control, dynamic control measures depend on the traffic situation. They include dynamic routing, i.e., rerouting by road signs or mobile devices if the principal route is congested (Sect. 21.4). Another control measure in this category is *ramp metering*: If traffic flow on the main-road is on the verge of traffic breakdown, one temporarily reduces or blocks the incoming traffic on on-ramps by traffic lights on the access lanes (cf. Sect. 21.3). Strategies in this category mainly act on the jam factor (A), but ramp metering can also be used to "level off" short-term flow peaks thereby controlling factor (C) of the above list.

Eliminating or alleviating static bottlenecks. By adding new lanes to gradient sections or redesigning interchanges and intersections, traffic engineers can reduce infrastructure-related bottlenecks thereby influencing factor (B) of the above list. Measures of this category also include more dynamic concepts such as a traffic-dependent operations management at road construction sites: Shifting the main

[1] The effects may be counter-intuitive. In an extreme case known as *Braess's paradox*, the construction of a new link may lead to longer travel times on *all* routes from a given origin to a given destination after the new user equilibrium has settled.

constructing activities to periods with low traffic volume will alleviate the construction bottleneck when it is relevant to do so.[2] Such measures will not be considered here.

Dynamic reduction of the bottleneck strength. The decisive property of a bottleneck characterizing its obstructing effect is the local reduction of the dynamic capacity. This allows to optimize traffic flow with respect to influence factor (B) by not only infrastructure-based measures but also by influencing the driving style to make it temporally and locally more effective. This includes rules for efficient behavior to be taught in driving schools as well as traffic-adaptive semi-automated driving (cf. Sect. 21.5). It also includes other driver-assistance systems from the emerging fields of *traffic telematics* and vehicle-based *Intelligent Transportation Systems* (ITS).

Homogenizing traffic flow. Measures in this category aim to reduce the traffic breakdown factor (C)—perturbations in the traffic flow itself—by suppressing local disturbances that may be caused, e.g., by abrupt lane changes, braking maneuvers, or other un-anticipated actions. The most widespread measures in this category are speed limits (Sect. 21.2), ramp metering (Sect. 21.3), and overtaking bans for trucks (Sect. 21.6) which are applied in a selective way, i.e., only, if there is high traffic volume [factor (A)] and a bottleneck is nearby [factor (B)].

In the following sections, we discuss a selection of optimization measures that are related to traffic flow modeling. Generally, suitable measures include dynamic traffic flow control rather than measures based on infrastructure or traffic demand.

21.2 Speed Limits

As a direct effect, speed limits homogenize traffic flow with respect to the speed distribution. For example, on German highways, trucks are allowed to drive at 80 km/h while there is no general speed limit for cars. This means, speed differences can reach values of 80 km/h and more, particularly between lanes (cf. Fig. 4.6). At the other extreme, imposing a global speed limit of 80 km/h leads to a very sharp speed distribution around 80 km/h since few drivers choose speeds below this value.[3]

As an indirect effect, speed limits reduce the frequency of lane changes: The majority of discretionary lane changes are no longer made since most of them are no longer associated with a significant incentive. Furthermore, the perturbations resulting from the remaining discretionary and the mandatory lane changes are weaker since fewer acceleration/braking actions are necessary to change lanes. This means,

[2] Nevertheless, this belongs to the category of static measures: Because roadworks have to be planned in advance, they can only take into account historical demand profiles, without feedback from the actual situation.

[3] In fact, Fig. 4.6 shows that the sharp "truck peak" of the speed distribution is at 89 km/h rather than 80 km/h. Obviously, no consequences threaten truck drivers up to this speed in Germany so that speed limiters are set accordingly.

speed limits help prevent or delay traffic breakdowns by reducing the traffic jam factor (C): Perturbations in the flow itself.

It is straightforward to incorporate speed limits and other vehicle-related factors limiting the speed into microscopic traffic flow models by letting the desired speed v_0^α of vehicle α depend on location and time according to

$$v_0^\alpha(x,t) = \min\left[v_0^{\text{driver}},\, v_0^{\text{speedlimit}}(x,t),\, v_0^{\text{motorization}}(x)\right]. \tag{21.1}$$

This means, the *effective* desired speed is the minimum of the driver's *actual* desired speed, speed limits (generally depending on the vehicle type, the location, people's propensity to obey speed limits, and, in case of dynamic traffic regulations, on time), and limits due to the motorization of the vehicle at uphill gradients (cf. Sect. 4.2). Notice that this interpretation of the desired speed means that the microscopic models are used as a multi-class model, i.e., every vehicle α has its own set of parameters.

When simulating speed limits and other speed-reducing factors, traffic engineers/model developers need to bear in mind the following:

- Only microscopic models are suited to reliably model the effect of speed limits. In principle, multi-class macroscopic models or mesoscopic gas-kinetic based models containing the speed distribution as part of the phase-space density are able to model speed limits. In practice, this is rather indirect and cumbersome.
- Suitable microscopic models should contain the desired speed as a model parameter.
- To avoid artificial perturbations in form of abrupt braking maneuvers caused by the speed limits themselves, they should be implemented as a continuous function of space. For example, when a general speed limit of 80 km/h is imposed for $x \ge 0$ while cars are allowed to drive at 120 km/h for $x < 0$, the speed-limit component of the desired speed function (21.1) may read

$$v_0^{\text{speedlimit}}(x,t) = \begin{cases} 120\,\text{km/h} & \text{for cars if } x < -100\,\text{m}, \\ (100 + 0.2x)\,\text{km/h} & \text{for cars if } |x| < 100\,\text{m}, \\ 80\,\text{km/h} & \text{in all other situations}, \end{cases} \tag{21.2}$$

  where x is given in units of meters. This represents the usual driver's behavior of starting braking before passing the new speed limit sign but passing the sign before the new speed is reached. If significant speed reductions are necessary (e.g., when approaching a roadworks site the speed signalization itself reflects this objective in form of so-called *speed funnels*, i.e., a series of speed limits with consecutively lower values, e.g., the sequence $120 \to 100 \to 80 \to 60$ km/h.

The reader can verify the jam-reducing effect of speed limits by interactive simulations on the authors' website.[4] When selecting the simulation scenario "lane closing", the simulation starts by default with a global speed limit of 80 km/h (Fig. 21.1, left). Since there are only insignificant speed differences between the vehicles on either

[4] see: www.traffic-simulation.de.

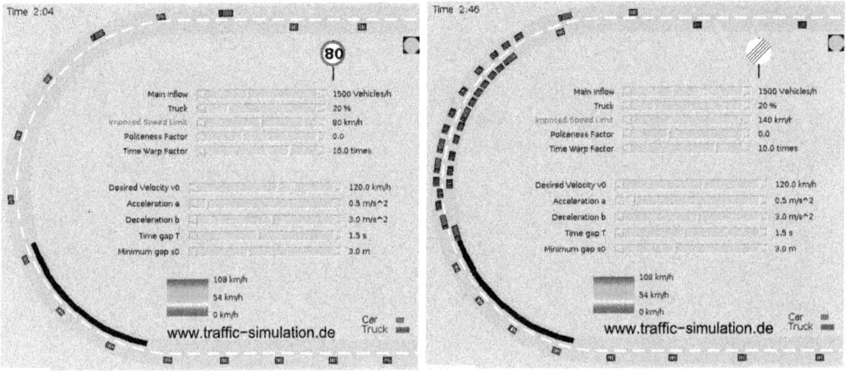

Fig. 21.1 Screenshots of microscopic simulations of speed limits. *Left* a global speed limit of 80 km/h is imposed. *Right* No speed limit for cars (desired speed 140 km/h). The traffic demand (inflow: 1,500 vehicles/h) and the lane-closing bottleneck (visualized in *black*) are the same in both simulations

lane, lane changes are easy and the drivers on the left lane manage to change lanes before arriving at the location where the left lane ends. In contrast, when interactively changing the speed limit to higher values or removing it (Fig. 21.1, right), it is harder to change lanes safely and, sooner or later, a driver gets stuck on the left lane behind the bottleneck. When this driver finally manages to change to the through lane and start at initially very low speed, he or she is likely to trigger a breakdown by this action.

We emphasize that imposing speed limits to prevent or delay breakdowns is only effective if the other two factors for a jam are present, i.e., during times of high traffic demand [rush hours, factor (A)], and near bottlenecks [factor (B)]. Therefore, speed limits for performance reasons should be imposed only temporarily and locally.[5] We can summarize the effect of speed limits as follows:

> **Jam-reducing effect of speed limits:** "Slower is sometimes faster."

21.3 Ramp Metering

Besides reducing speed differences by speed limits, traffic engineers can homogenize traffic flow on highways or other principal roads by temporarily reducing or closing on-ramp flows via access traffic lights when a flow peak arrives at the main road

[5] As a matter of fact, there are other reasons for speed limits, e.g., traffic safety or noise pollution. This will not be considered here.

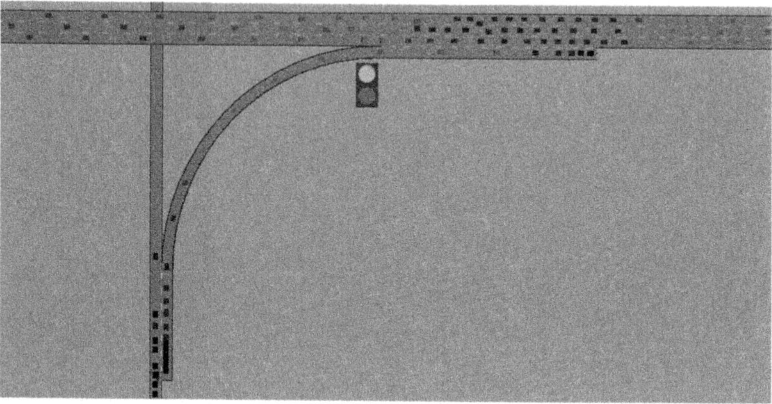

Fig. 21.2 Screenshot of the "ramp metering game" developed with the "Multi-model open-source vehicular Simulator" (MovSim). The player can control the access traffic light. Here, he or she did very badly and produced a traffic breakdown on the main-road as well as a spillover of the ramp queue onto the secondary network

(Fig. 21.2). This optimization measure, also known as *ramp metering*, is widespread in the USA while it is rarely used in most of the European countries. Although ramp metering temporarily reduces the traffic throughput, it helps increase it at later times by preventing a traffic breakdown and the associated capacity drop. We can summarize the "philosophy" behind ramp metering as follows:

Effects of ramp metering: "Less is sometimes more."

When implementing ramp metering schemes, it is crucial to prevent artificial flow peaks once the flow peak on the main-road is over and the access traffic light is switched to green. Therefore, the duration of the green phases must be sufficiently short to allow only one or a few vehicles to pass during each green phase.

Choosing an efficient switching strategy for the access traffic light is a difficult task. Firstly, the traffic situation is incompletely known in *real time*. Secondly, it takes several minutes for a certain controlling action (switching the traffic light) to have a significant effect on the traffic flow. Finally, it is easy to make things worse than the situation without control.

Microscopic traffic flow simulations such as the simulation game of Fig. 21.2 help understand the dynamics.[6] In the depicted run, the player kept the controlling traffic light red for a too long period. This leads to two adverse effects: Firstly, as soon as the player switched the light to green (displayed state), the accumulated queue of waiting ramp vehicles starts as a platoon, merges to the main-road, and initiates a

[6] see: www.movsim.org.

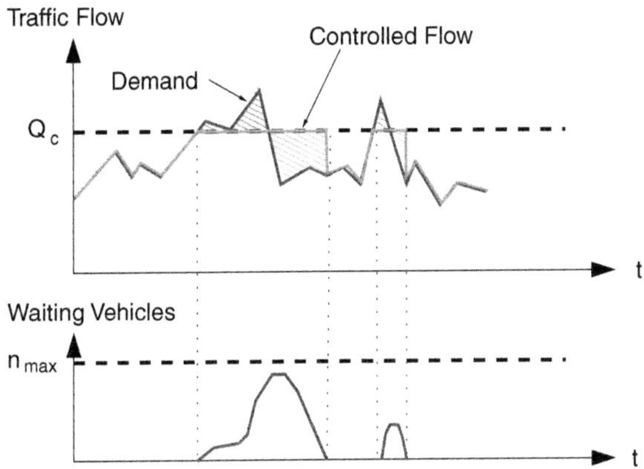

Fig. 21.3 Visualization of the capping strategy for ramp metering. *Top* Demand curve $Q_{\text{in}} + Q_{\text{rmp}}$ for the total flow that would be realized without ramp metering (*jagged line*). If the total flow exceeds the limit Q_c, it is capped to this value by the access traffic light, so a queue of n waiting vehicles forms at the on-ramp (*bottom curve*). As long as $0 < n < n_{\text{max}}$, the ramp metering is active restricting the ramp flow to $Q_{\text{rmp}} = \max(0, Q_c - Q_{\text{in}})$

breakdown on this road. Furthermore, before the queue of waiting vehicles dissolved completely, it "spilled over" onto the secondary road thereby obstructing even the drivers who did not want to enter the highway. This forced inefficient usage of road space is also called the *gridlock* phenomenon.

While a multitude of strategies has been proposed, the arguably simplest ramp-metering strategy restricts the total flow to a certain value (cf. the schematic illustration in Fig. 21.3):

Capping strategy:
Control the on-ramp flow Q_{rmp} such that the total flow $Q_{\text{in}} + Q_{\text{rmp}}$ is restricted by a certain threshold Q_{c}. If the on-ramp queue threatens to spill over to the secondary network (number of waiting vehicles $n \geq n_{\text{max}}$), discontinue ramp metering.

Figure 21.4 shows a microscopic simulation of traffic flow with the Intelligent Driver Model (Sect. 11.3) near an uphill bottleneck at road kilometer 40 without (left) and with ramp metering (right) according to the capping strategy. The on-ramp to be metered is at the upstream boundary of the simulation (road kilometer 32). The metering simulated traffic light at the on-ramp is set up to allow single passes of vehicles during one green phase, if necessary. To simulate realistic flow profiles, we have used stationary detector data of an instance of a real jam to feed the simulation. Instead of using the parameter values of Table 11.2, we have calibrated the IDM to produce a congestion that is comparable with the observations when no ramp metering is active (top left image of Fig. 21.4). The speed profile displayed in the top

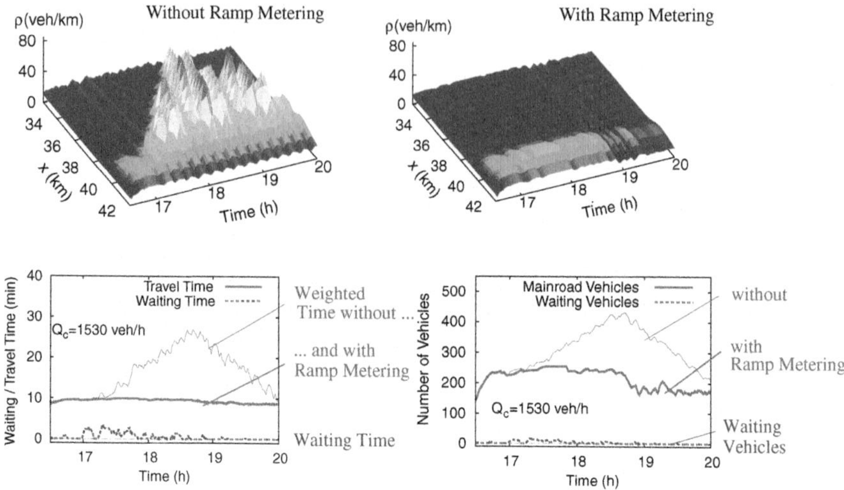

Fig. 21.4 Microscopic simulation of the ramp metering capping strategy (for details see the main text)

right image shows the "best case" that can be realized by optimizing the threshold flow Q_c for this particular case with respect to the total travel time of all vehicles on the main-road *and* on the ramp (cf. Sect. 21.7).

The time series in the lower left part of Fig. 21.4 shows the realized travel times of the main-road vehicles and, if ramp metering is active, the waiting times on the on-ramp. Obviously, the reduced travel times on the main-road more than compensate for the additional waiting times due to ramp metering. The lower right diagram of Fig. 21.4 shows the total number of vehicles on the main-road and on-ramp for a given time. In this diagram, the objective function "total traffic time" is represented by the area below the red curve plus the area below the blue curve. Here, it is explicit that the time saved by the main-road vehicles (area between the thick and thin red curves) more than compensates the total waiting time (area between the thick and thin blue curves). We emphasize that, in spite of waiting times by the metering traffic light, even most ramp vehicles benefit *individually* from the ramp metering: Except for a small time interval around 17:00 h, the waiting time is less than the time saved for covering the main-road section.

Finally, we mention that the practically applied strategies are significantly more involved than this simple capping strategy. In essence, ramp metering strategies are nonlinear functions mapping the exogenous variables (e.g., the records of nearby stationary detectors for the last fifteen minutes, the detector positions, the time of day, the past states of the traffic light) to the present state (red or green) of the metering traffic light. In order to prevent follow-up breakdowns and gridlocks in the secondary network, ramp metering is generally deactivated in oversaturated situations.

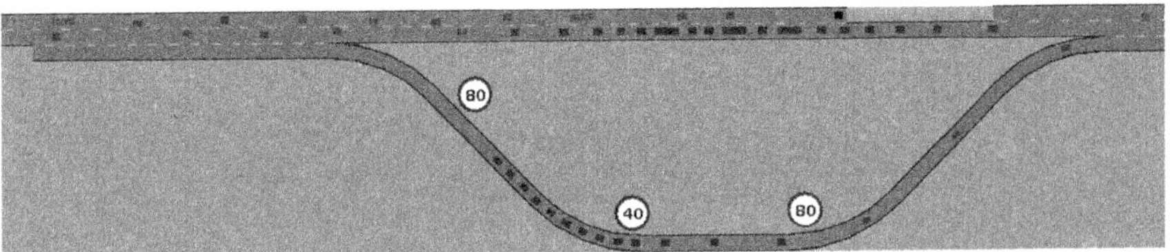

Fig. 21.5 Screenshot of the "routing game" developed with the open-source traffic simulator MovSim. The player can reroute the drivers on the *right lane* of the main-road over a deviation on the secondary network to prevent a breakdown at the lane-closing bottleneck. Here, the player adopted a sub-optimal strategy and caused jams on both routes

21.4 Dynamic Routing

One of the most obvious methods to homogenize traffic flow is distributing the traffic demand more efficiently over the network, also called *dynamic routing* and *load balancing*. Because this strategy may prevent traffic breakdowns by more efficient road usage, even a detour may lead to a shorter travel time although it would be longer in normal situations. This can be paraphrased as follows:

> **Effects of dynamic routing:** "Longer is sometimes shorter."

Traditional infrastructure-centered concepts are based on stationary detectors. The logic of such systems extracts traffic information from the detector data, estimates optimal routes for the main traffic flow directions, and transmits corresponding recommendations to the drivers via *variable message signs* (VMS). At present, alternative vehicle-based or mobile dynamic routing concepts in form of traffic-dependent navigation devices are becoming more and more relevant. Typically, these *connected* navigation devices not only receive traffic information but also send their positions in form of floating-car data in anonymized form to the traffic center of the data provider. There, the present and near-future traffic state is continuously estimated and sent back to the devices for a recalculation of the fastest routes. In future *hybrid navigation* applications, optimal routes are also (pre-) calculated directly on the server.

In contrast to forecasting the weather, dynamic routing and short-term traffic forecasts are plagued by a conceptual difficulty: The prophecy is self-destructing. The interactive "routing game" (Fig. 21.5) depicts this phenomenon in a situation where a bottleneck (lane closure) on the main road can be circumnavigated over the secondary network. The arguably simplest strategy a player can adopt is the naive and short-sighted *greedy algorithm*: Select the detour whenever it yields, at present, the shortest realized travel time. It is easy to see that this strategy, to put it mildly, is suboptimal. Firstly, the roads of the secondary network are less efficient and traffic jams are unavoidable after some time when adopting this greedy

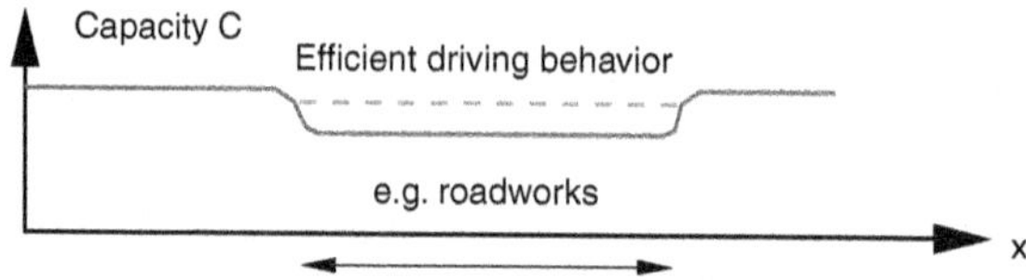

Fig. 21.6 Homogenizing the local road capacity profile and reducing the bottleneck strength by a traffic-adaptive driving style (human driving) or a space-dependent parameterization of the core acceleration controller of an ACC (semi-automated driving)

strategy. Moreover, as the reader can verify by playing this game on the web,[7] the delay of many minutes between the decision ("main-road or detour") and its consequences (traffic jams on either route) makes the greedy algorithm inherently unstable. This may even lead to a worse performance compared to ignoring the detour altogether.

Since both the penetration rate of connected dynamic navigation devices and the percentage of drivers following the recommendations are currently low, such *routing instabilities* are not yet relevant. Nevertheless, their existence suggests the need to adopt more refined routing strategies in the near future.

21.5 Efficient Driving Behavior and Adaptive Cruise Control

In which ways can changes of the (human or semi-automated) driving style influence traffic flow? To find answers we notice that the most relevant influencing factor determining the static capacity is the preferred time headway T in car-following situations (cf. Sect. 4.3). Furthermore, the stability analysis of Chap. 15 shows that an agile (responsive) driving style increases stability. Indirectly, this increases the capacity once more since an agile driving style tends to reduce the capacity drop (Eq. 17.1), i.e., it brings the relevant dynamic capacity nearer to its theoretical maximum, the static capacity.

This suggests that it should be possible to *dynamically reduce* the "capacity holes" of bottlenecks by temporarily reducing the time gap T and increasing the agility when passing a bottleneck (cf. Fig. 21.6). However, in order to ensure that this dynamic "filling up" of capacity holes does not impair safety, drivers need to be more attentive during the time they drive with reduced time gaps. Consequently, this kind of traffic flow optimization is most suited when realized by semi-automated driving via *adaptive cruise control* (ACC) rather than by the human driver.

By means of microscopic traffic flow simulations, we can investigate the consequences of different human driving styles or ACC settings near bottlenecks. Simulations of ACC-driven vehicles are expected to be particularly predictive since

[7] see: www.movsim.org.

suitable car-following models can directly serve as the core controller of ACC systems.[8]

The challenge in setting up an ACC-based driver-assistance system to enhance traffic flow efficiency consists in combining efficiency with a safe, comfortable, and understandable driving style (since, otherwise, such systems will not be accepted and used): While a comfortable and safe driving entails comparatively large time gaps and low accelerations, increasing traffic flow efficiency and stability involves reducing time gaps and increasing accelerations.

These contradictory objectives can be handled best when restricting the effective driving style to situations where this is necessary, e.g., when passing a bottleneck. If the temporary discomfort associated with the effective driving style helps prevent a traffic breakdown, then even less comfortable driving situations (stop-and-go traffic) can be avoided.

Philosophy behind driving efficiently in certain situations:
"Less driving comfort at present can lead to more comfort in the long run."

When implementing such a *traffic-adaptive driving strategy* in a semi-automated way using adaptive cruise control, the parameters of the core acceleration controller of the ACC systems are changed according to the information about the local traffic situation. Estimating the local traffic state in real-time is a major challenge. Besides autonomous range detectors to determine the immediate input of the ACC car-following model, the vehicle must communicate by different channels to obtain the necessary external information. This includes:

- Infrastructure-to-vehicle (I2V) communication, e.g., stationary detectors broadcasting their data to equipped vehicles, or traffic lights broadcasting future switching times.
- Vehicle-to-vehicle (V2V) communication, e.g., equipped vehicles broadcasting FC data of their past trajectories, or serving as relay to transmit detector or FC data generated elsewhere).
- Local infrastructure-to-infrastructure communication by means of *road-side units*.[9]
- Communications from and to traffic information centers (FCD are transmitted to the center, and dynamical maps are passed back).

All these options are examples of Intelligent Traffic Systems (ITS). This field is expected to have a significant impact on traffic flow in the near future.

Realization of a traffic-adaptive driving strategy. As an example, we will now specify a concrete strategy. This strategy distinguishes five different local traffic

[8] The ACC model (Sect. 11.3.8) has actually been implemented into the ACC systems of test cars to investigate the traffic-adaptive strategy to be described below.

[9] Vehicles pass their information via short-range communication to a road-side unit (V2I) which transmits them to a further road-side unit in the upstream direction (I2I). The latter, in turn, passes the information back to equipped vehicles (I2V).

Table 21.1 Strategy matrix for the Intelligent Driver Model (11.14), the ACC model (11.26) and related models

Situation	λ_T	λ_a	λ_b	Objective
Free traffic	1	1	1	Driving comfort (reference)
Approaching a jam	1	1	0.7	Avoiding rear-end collisions
Jammed traffic	1	1	1	Nothing can be done, so relax
Exiting a jam	0.5	2	1	Increasing the dynamic capacity
Bottleneck	0.7	1.5	1	Proactively preventing a breakdown

The matrix contains multipliers of the default parameters reflecting the reference for a given driver and vehicle

situations as seen from the vehicle: (i) free traffic, (ii) approaching congested traffic, (iii) congested traffic, (iv) leaving a congested region, and (v) passing a bottleneck.

In each state, the ACC model is parameterized in view of the major objective in the respective situation. While driving comfort has priority in free-traffic situations, a traffic-efficient driving style with not too large gaps is crucial when passing a bottleneck (whether congested or not) as depicted in Fig. 21.6. When leaving a jam, an agile driving style reduces the capacity drop thereby increases the throughput (dynamic capacity) which helps dissolve stationary and moving jams. Notice that even small relative changes of the characteristics of the traffic flow (such as an increase of the dynamic capacity by 10 %) may have significant effects since inherent nonlinearities (traffic breakdown with associated capacity drop or no breakdown) leverage these changes (cf. Fig. 21.7).

To realize the different driving styles in the ACC system, the underlying ACC model is parameterized differently for each of the five situations. In order to preserve the overall driving characteristics set by the driver,[10] we formulate strategy changes *differentially* by multiplying the default parameters for a given driver and vehicle with a state-dependent factor. When using the ACC model (11.26) [or the Intelligent Driver Model (11.14)] as the underlying car-following model, the driving style is encoded by the parameters T (increasing values reflect decreasing aggressiveness and a more conservative style), a (increasing values reflect increasing agility/responsiveness),[11] and b (increasing values reflect decreasing anticipation/risk awareness). In any case, the driver sets the speed directly.

In this way, the overall driving strategy, i.e., the adaptation of the driving behavior to each of the five traffic situations, is encoded by a 5×3 *strategy matrix* as shown in Table 21.1. The line i of this matrix contains the multipliers λ_T, λ_a, and λ_b of the parameters T, a, and b, respectively, that apply to situation i. For example, inside a bottleneck, the displayed strategy reduces the preferred time gaps in car-following mode to 70 % of the reference for free traffic to "fill" the "capacity hole". Furthermore,

[10] In ACC systems, drivers can not only set the maximum speed but also the time gap in car-following mode. Some car manufacturers also offer more "sportive" or more comfort-oriented overall settings.

[11] This must not be confused with the reaction time which is insignificant in modern ACC systems, and not contained in the ACC model.

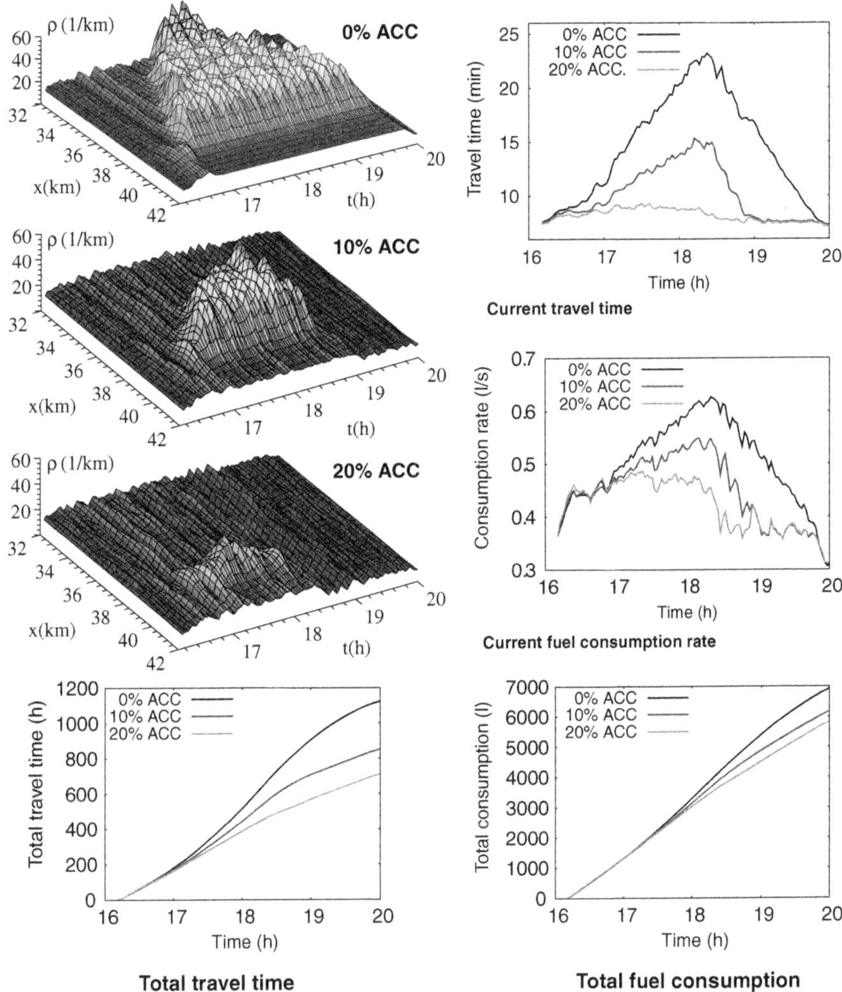

Fig. 21.7 Simulation of the jam-reducing effects of a traffic-adaptive driving style according to the strategy matrix of Table 21.1. The reference simulation (0 % equipped vehicles) has been calibrated to reflect an observed congestion on the German highway A8-East near the uphill bottleneck "Irschenberg" (cf. Fig. 18.1)

this strategy increases the agility to 150 % of the reference value to avoid traffic-flow instabilities induced by the reduced gaps (cf. Chap. 15).

Figure 21.7 shows a simulation with this strategy matrix for variable percentages of equipped vehicles (the other vehicles are simulated with all multipliers set to unity). The result show that even 10 % of vehicles driving with this strategy significantly reduce the duration and size of the congestion and, as a consequence, the travel times and the overall fuel consumptions (cf. Sect. 21.7). When running the simulation with a penetration rate of 20 % equipped vehicles, we observe only a short and insignificant congestion for the given traffic demand.

In order to assess the global utility of jam reducing measures, we compare the total difference of travel time and fuel consumption for the whole spatiotemporal window of this simulation with and without congestion, i.e., the potential which can be realized by introducing 20 % of equipped vehicles. Compared to free traffic (penetration rate 20 %), the jam resulting without equipped vehicles increases the total travel time by about 400 h (more than 50 %) and the total fuel consumption by 1.200 liters (about 15 %). Obviously, the potential in saved travel time is significantly greater than that of saved fuel.[12]

We emphasize that there is a statistical pitfall when presenting the results of such simulations which we call the *filtering dilemma*: The percentaged savings of any optimization measure in terms of travel time, fuel, or any other quantity depends on the considered spatiotemporal reference. For example when considering one year and the whole road network of a country, the percentaged savings may be minimal since, most of the time and on most roads, there is no congestion, i.e., there is nothing to optimize. Besides the potential for semi-automated driving, the simulations also point to rules for efficient driving which all non-ACC drivers can adopt, and which can be taught in driving lessons:[13]

How each of us can contribute to optimize traffic flow:
(i) Do not leave unnecessarily large gaps when passing bottlenecks.
(ii) Adopt an agile driving style when leaving jammed traffic.

21.6 Further Local Traffic Regulations

Besides speed limits, other local regulations can contribute to homogenize and optimize traffic flow. In the following, we will discuss some options for roads with two or more lanes per driving direction.

Truck overtaking bans. This regulation prevents so-called *elephant races*, i.e., trucks overtaking each other at slow speed differences. If an elephant race takes place, faster cars accumulate behind the trucks constituting a flow peak which can serve as trigger for a breakdown (cf. Fig. 18.3). This is particularly relevant in uphill or downhill sections and if the road has only two lanes per direction.

Regulations on minimum speeds. Like speed limits, such "reverse" speed limits aim to homogenize traffic flow by reducing speed differences. They are particularly

[12] As a general rule, the relative savings in travel time are about three times greater than that of fuel.

[13] Of course, safety comes always first, so the time gap should never be lower than a critical value which is necessary to avoid rear-end collisions (cf. Sect. 4.3). Furthermore, drivers should be particularly attentive in such situations.

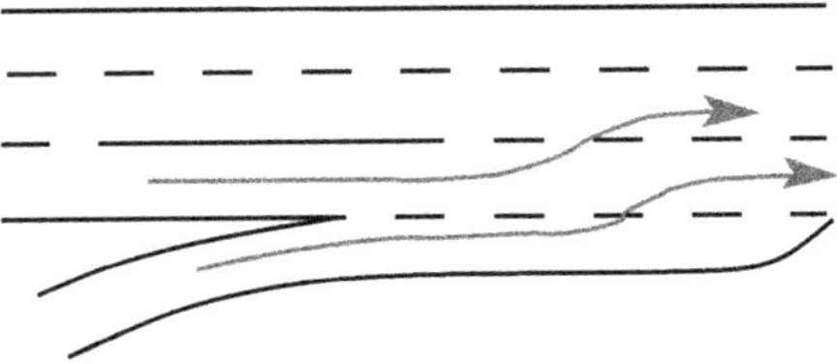

Fig. 21.8 Local ban on changing from the *right* to the *middle* (faster) lane upstream of merging lanes of major interchanges. The length of the changing ban is not to scale (it is longer in reality)

effective on the faster lanes of uphill sections and in combination with conventional speed limits and truck overtaking bans.

Stay-in-lane recommendations. This measure directly reduces perturbations caused by lane changes.

Selectively prohibiting lane changes. In certain situations, particularly upstream of on-ramps from the lane adjacent to the merging region to other lanes, this measure prevents ineffective usage of the available road space. This regulation requires sufficiently long on-ramps, so all on-ramp vehicles manage to merge in spite of the vehicles on the neighboring main-road lane. (cf. Fig. 21.8). As net effect, this measure makes sure that long acceleration lanes are used in their full length rather than merging on the first third of the acceleration lane which reduces the effective length of the merging lane thereby increasing the bottleneck strength.

Zipper merging rule. This regulation applies to regions of mandatory lane changes near lane closures if traffic is congested. In this situation, drivers on the lane to be closed must stay on this lane until reaching the location where the lane actually ends. Then, they merge zipper-like (one vehicle in every gap) to the continuous lane. This behavior contributes to a more effective usage of the road space. Furthermore, it minimizes the risk of spill-over gridlocks (see Fig. 21.2), particularly in urban networks.[14]

21.7 Objective Functions for Traffic Flow Optimization

21.7.1 Setting up the Frame

Objective functions assess quantitatively whether, and in which way, optimization measures as discussed above are successful. Since optimization measures often involve more than one road element, we consider a whole road network and formally

---

[14] As a welcome side effect for the individual driver on lanes to be closed, he or she often passes several vehicles on the neighboring lanes (in accordance with traffic regulations of most countries) when this rule is adopted.

map the cumulated arc-lengths of all links of the networks to the one-dimensional interval $[x_1, x_2]$.

Typically, it suffices to consider a single road link to assess the effects of speed limits. When investigating ramp metering (influencing the main-road and the on-ramp) or dynamic routing (influencing the main route and the deviation), two links must be considered as a minimum. More links are necessary depending on the link definitions in the simulator (e.g., when defining the upstream, merge, and downstream regions of the main-road as three separate links), or if side effects cannot be excluded, e.g., spillovers to other nodes, or gridlocks (cf. Fig. 21.2). In extreme cases, e.g., when optimizing self-organized switching strategies of traffic lights (which is not considered here), the region of influence and thus the network to be considered may consist of all the major roads in a city.

Mathematically, objective functions map the dynamical variables of traffic flow (e.g., density, local speed, trajectories) of the interesting spatiotemporal region $G = [x_1, x_2] \times [t_{\text{start}}, t_{\text{end}}]$ to a real-valued number $\mathscr{F}$ denoting, e.g., the total travel time or, more generally, a suitable *utility*.[15] Optimizing traffic flow means minimizing (or maximizing) the objective function.

21.7.2 Constraining Conditions

In order to allow equitable comparisons, we require that the initial conditions and the traffic-demand profiles of the considered road network are fixed external conditions. As initial conditions, we specify the locations and speeds of all vehicles in microscopic simulations, or, macroscopically, the initial local density and speed fields. In the simplest cases when no routing decisions are possible inside the considered network (e.g., when investigating speed limits or ramp metering without side effects), the traffic demand profile is completely defined by the inflows $Q_i^{\text{in}}(t)$ at all sources (in-flowing boundary conditions) i of the network, assuming steady-state conditions for the speed. In more complex cases, drivers make routing decisions inside the network, so information on the destination is necessary. We can specify the optimization problem in this case by disaggregating the inflows according to their destinations (sinks, exiting nodes of the network) e.g., by prescribing a (generally time dependent) *origin-destination matrix* $Q_{ij}(t)$ of drivers entering at source i and heading towards node j. To completely specify the problem for this case, we also specify the route-choice decision algorithm/heuristic of the drivers when arriving at a node.[16]

---

[15] Strictly speaking, objective functions map functions (such as the spatiotemporal local speed) onto a number, so they are, mathematically speaking, functionals rather than functions. Agreeing with the common usage, we will nevertheless speak of objective "functions".

[16] In the simplest case, the routes are statically prescribed by a succession of nodes $\{k\}$ with i being the first and j the last node. When defining route choices dynamically at each node according to the driver's subjective impression of the traffic state, things begin to get really complicated.

Furthermore, we do not allow congestion at any of the spatial and temporal boundaries of the region G. Free spatial boundaries guarantee that (i) the simulation is completely determined by the initial and upstream boundary conditions as given above,[17] (ii) traffic flow inside G is not optimized at costs of the network outside of G. A jam-free initial state guarantees consistency with the boundary conditions, and prescribing a free final state prevents optimizing traffic inside G at costs of future traffic.

21.7.3 Examples

The mathematical form and the exogenous variables of the objective function reflect the model category and the aspect of traffic flow (the "objective") to be optimized. In the following, we give some examples.

Minimizing the total travel time. In simulations of microscopic models, the total travel times is simply equal to the number of all vehicles in the network, integrated over time:

$$\mathscr{F}_\tau = \frac{1}{T_0} \int_{t_\text{start}}^{t_\text{end}} \sum_{\{l\}} n_l(t)\,\mathrm{d}t. \tag{21.3}$$

Here, $n_l(t)$ denotes the number of vehicles on link l, and the sum runs over all links of the considered network. For a single-objective optimization, the normalization constant T_0 is arbitrary and can be set to unity (then, $\mathscr{F}_\tau$ denotes the total travel time directly). For multi-purpose optimizations, T_0 serves as weighting factor to define the relative weight of travel time with respect to other criteria. Then, T_0 should have the unit of a time to make $\mathscr{F}_\tau$ dimensionless and commensurable with the objective functions characterizing other criteria.

In a macroscopic simulation, we derive the total travel time from Eq.(21.3) by applying the definition of the density "vehicles per road length" resulting in an spatial integral over the vehicle density:

$$\mathscr{F}_\tau = \frac{1}{T_0} \int_{t_\text{start}}^{t_\text{end}} \int_{x_1}^{x_2} \rho(x,t)\,\mathrm{d}x\,\mathrm{d}t. \tag{21.4}$$

Notice that, by our definition of the spatial coordinate, x runs over all links of the considered network, so there is no sum over all links l.

Maximizing the driving comfort. The main factors determining the subjective driving comfort are the acceleration $\dot{v} = \mathrm{d}v/\mathrm{d}t$ and its time derivative, the *jerk* $J = \ddot{v} = \mathrm{d}^2v/\mathrm{d}t^2$. Roughly speaking, the driving style is smooth, i.e., corresponding to a

[17] In case of traffic jams, information flow propagates upstream, so downstream boundary conditions are needed.

Fig. 21.9 Visualization of the driver's perspective in a microscopic simulation. The *coffeemeter* visualizes the discomfort due to acceleration and jerk

high driving comfort, if an uncovered cup of coffee in a cup holder inside the driving vehicle does not spill over. It is therefore highly intuitive and instructive to visualize driving comfort (or the lack thereof) by the dynamics of the surface of a cup of coffee. We may call this instrument the *coffeemeter* (cf. Fig. 21.9).[18]

In a microscopic simulation, we can formulate the "discomfort" $\mathscr{F}_{\text{comf}}$ to be minimized by

$$\mathscr{F}_{\text{comf}} = \frac{1}{T_0\, a_0^2} \int_{t_{\text{start}}}^{t_{\text{end}}} \sum_{\alpha} \left(\dot{v}_\alpha^2 + \tau_0^2 \ddot{v}_\alpha^2 \right) \, \mathrm{d}t. \tag{21.5}$$

The sum runs over all vehicles α inside the investigated region. The corresponding macroscopic formulation

$$\mathscr{F}_{\text{comf}} = \frac{1}{T_0\, a_0^2} \int_{t_{\text{start}}}^{t_{\text{end}}} \int_{x_1}^{x_2} \rho(x,t) \left[A^2(x,t) + \tau_0^2 \left(\frac{\partial A(x,t)}{\partial t} \right)^2 \right] \mathrm{d}x \, \mathrm{d}t \tag{21.6}$$

involves the local acceleration $A(x,t)$ in the comoving system, Eq. (9.1). Obviously, this macroscopic objective function only makes sense for second-order models. In any case, the square of the characteristic time τ_0 indicates how much the jerk contributes to the overall discomfort relative to acceleration. Since, in typical driving situations, longitudinal and lateral accelerations are of the order of $1\,\mathrm{m/s^2}$ and jerks are of the order of $1\,\mathrm{m/s^3}$, a characteristic time of 1 s will weigh both influencing factors evenly. The prefactor $1/a_0^2$ determines the relative weighting of the driving comfort with respect to travel time. A value $a_0 = 1\,\mathrm{m/s^2}$ means that cruising without any

[18] To simulate the coffeemeter, you do not need to apply the full three-dimensional hydrodynamics of hot coffee with mixed Dirichlet and free boundary conditions for the cup and the free coffee surface, respectively. It suffices to phenomenologically model the two orthogonal lowest-order modes (corresponding to in-phase motions of the whole surface) by a two-dimensional pendulum driven by the longitudinal and lateral accelerations. The eigenfrequencies are determined by the cup geometry, and the modes are phenomenologically damped to yield the observed decay of the oscillations inside the cup (time constant ≈ 5 s).

acceleration (situation I), and halving the travel time at the cost of permanently accelerating and decelerating at $|\dot{v}| = 1\,\text{m/s}^2$ (situation II) result in the same overall value of the objective function (disutility).

Alternatively, we can define the driving comfort by the intensity (squared-mean value of the amplitude) of the surface oscillations of the *coffeemeter*. This would lead to a similar expression as Eq. (21.6).

Minimizing total fuel consumptions and CO_2 emissions. As a matter of fact, the objective function for fuel consumption is directly proportional to the total fuel consumption in the considered spatiotemporal region. Depending on the level of detail, we can evaluate it with any fuel consumption/emissions model such as that given in Chap. 20. When calculating consumptions and emissions with the modal models of Sect. 20.3, we obtain the microscopic objective

$$\mathscr{F}_{\rm c} = \frac{1}{T_0\,\dot{C}_0} \int\limits_{t_{\rm start}}^{t_{\rm end}} \sum_\alpha \dot{C}_\alpha(v_\alpha(t), \dot{v}_\alpha(t))\,\mathrm{d}t. \tag{21.7}$$

In macroscopic simulations, we replace the sum over vehicles by the space integral over vehicle density, and the vehicle acceleration $\dot{v}_\alpha$ by the local acceleration $A(x,t)$ in the comoving system. This results in

$$\mathscr{F}_{\rm c} = \frac{1}{T_0\,\dot{C}_0} \int\limits_{t_{\rm start}}^{t_{\rm end}} \int\limits_{x_1}^{x_2} \rho(x,t)\dot{C}(V(x,t), A(x,t))\,\mathrm{d}x\,\mathrm{d}t. \tag{21.8}$$

The constant $\dot{C}_0$ indicates the relative weighting of fuel consumption with respect to travel time. For example, a value $\dot{C}_0 = 10\,\text{L/h}$ means that losing one hour of time has the same disutility as consuming ten additional liters of fuel.

Multi-objective optimization. For a multi-objective optimization, we sum the objective functions for the individual criteria:

$$\mathscr{F} = \mathscr{F}_\tau + \mathscr{F}_{\rm comf} + \mathscr{F}_{\rm c}\,. \tag{21.9}$$

The priorities and relative weightings are specified by the parameters a_0^2, τ_0^2, and $\dot{C}_0$. In principle, nothing changes when setting $T_0 = 1$. After all, $1/T_0$ appears as common factor in all components of the objective function $\mathscr{F}$. However, setting T_0 equal to the total travel time of the system before optimization allows us to compare the optimization potential of different situations with different spatiotemporal regions G. In this case, $\mathscr{F}$ is dimensionless and assumes values of the order of unity, regardless of G.

Mathematically, the task of minimizing F is similar to calibrating traffic flow models (estimating their parameters) by minimizing a function indicating the differences between observations and simulation with respect to the parameters (cf. Chap. 16).

In both cases, we have to solve the mathematical problem of multi-dimensional *non-linear optimization*. Since a single calculation of $\mathscr{F}$ generally implies a complete simulation run, the objective function is expensive to calculate and not differentiable analytically. Furthermore, due to the inherent nonlinearities of traffic flow dynamics, including *deterministic chaos*,[19] the "landscape" of the objective function is generally very jagged and contains many secondary local minima. As a consequence, simulation-based traffic flow optimization implies the use of fast and robust numerical optimization methods (cf. Sect. 16.3).

Further Reading

- Papageorgiou, M., Hadj-Salem, H., Blosseville, J.: ALINEA: A local feedback control law for on-ramp metering. Transportation Research Record **1320** (1991) 58–64
- Treiber, M., Helbing, D.: Microsimulations of freeway traffic including control measures. Automatisierungstechnik **49** (2001) 478–484
- Chatterjee, K., McDonald, M.: Effectiveness of using variable message signs to disseminate dynamic traffic information: Evidence from field trails in European cities. Transport Reviews **24** (2004) 559–585
- Hegyi, A., De Schutter, B., Hellendoorn, H.: Model predictive control for optimal coordination of ramp metering and variable speed limits. Transportation Research Part C: Emerging Technologies **13** (2005) 185–209
- Kesting, A., Treiber, M., Schönhof, M., Helbing, D.: Adaptive cruise control design for active congestion avoidance. Transportation Research Part C: Emerging Technologies **16** (2008) 668–683
- Papageorgiou, M., Kosmatopoulos, E., Papamichail, I.: Effects of variable speed limits on motorway traffic flow. Transportation Research Record: Journal of the Transportation Research Board **2047** (2008) 37–48
- Hartenstein, H., Laberteaux, K.: A tutorial survey on vehicular ad hoc networks. Communications Magazine, IEEE **46** (2008) 164–171
- Kesting, A., Treiber, M., Helbing, D.: Connectivity statistics of store-and-forward intervehicle communication. IEEE Transactions on Intelligent Transportation Systems **11** (2010) 172–181
- Treiber, M., Kesting, A.: An open-source microscopic traffic simulator. Intelligent Transportation Systems Magazine **2** (2010) 6–13

[19] Meaning exponentially sensitive dependence of the output on the initial data and model parameters.

Solutions to the Problems

Problems of Chapter 2

2.1 Floating-Car Data

GPS data provide space-time data points and (anonymized) IDs of the equipped vehicles. We can obtain their *trajectories* by connecting the data points in a space-time diagram (via a map-matching process). From the trajectories, we can infer the speed by taking the gradients. Low speeds on a highway or freeway, e.g., 30 km/h, usually indicate a *traffic jam*. Since the data provide spatiotemporal positions of the vehicles, we can deduce the location of congested zones, including their upstream and downstream boundaries, at least, if the penetration rate of equipped vehicles with activated communication is sufficiently high.[1] However, GPS measurements are only accurate to the order of 10 m and careful map-matching/error checking is necessary to exclude, for example, stopped vehicles on the shoulder or at a rest area, or vehicles on a parallel road. Moreover, due to this resolution limit, GPS data do *not* reveal lane information, nor information on *lane changes*. Since the percentage of equipped vehicles with connected devices is low, variable, and unknown, we can *not* deduce extensive quantities (traffic density and flow) from this type of data.

To wrap it up: (1) yes; (2) no; (3) no; (4) no; (5) yes; (6) yes.

2.2 Analysis of Empirical Trajectory Data

1. Flow, density, and speed: Using the spatiotemporal region $[10\,\text{s}, 30\,\text{s}] \times [20\,\text{m}, 80\,\text{m}]$ suggested for a representative free-flow situation, we obtain by trajectory counting:

[1] On major roads with high traffic flow, 0.5 % of the vehicles are typically enough; on smaller roads, we need a significantly higher percentage.

M. Treiber and A. Kesting, *Traffic Flow Dynamics*,
DOI: 10.1007/978-3-642-32460-4, © Springer-Verlag Berlin Heidelberg 2013

$$Q_{\text{free}} = \frac{11\text{ vehicles}}{20\text{ s}} = 1{,}980\text{ vehicles/h}, \quad \rho_{\text{free}} = \frac{3\text{ vehicles}}{60\text{ m}} = 60\text{ vehicles/km}.$$

The speed can be deduced either from the gradient of the trajectories or from the hydrodynamic relation Q/ρ:

$$V_{\text{free}}^{\text{gradient}} = \frac{60\text{ m}}{5\text{ s}} = 43.2\text{ km/h}, \quad V_{\text{free}}^{\text{hyd}} = \frac{Q_{\text{free}}}{\rho_{\text{free}}} = 39.6\text{ km/h}.$$

This discrepancy is tolerable in view of the reading accuracy (one may also count 12 vehicles in 20 s, yielding $Q = 2{,}160$ vehicles/h and thus $\rho = 43.2$ km/h). For congested traffic, we use, again, the suggested spatiotemporal region [50 s, 60 s] × [40 m, 100 m] and obtain analogously

$$Q_{\text{cong}} = \frac{2\text{ vehicles}}{10\text{ s}} = 720\text{ vehicles/h}, \quad \rho_{\text{cong}} = \frac{6\text{ vehicles}}{60\text{ m}} = 1{,}000\text{ vehicles/km},$$
$$V_{\text{cong}}^{\text{hyd}} = \frac{Q_{\text{cong}}}{\rho_{\text{cong}}} = 7.2\text{ km/h}.$$

2. *Propagation velocity*: The stop-and-go wave can be identified by the spatiotemporal region with nearly horizontal trajectories. First, we observe that, in the diagram, the gradient of the (essentially parallel) upstream and downstream wave boundaries are negative, i.e., the wave propagates *against* the direction of traffic. To determine the propagation velocity, we estimate from the diagram

$$c \approx -\frac{140\text{ m}}{(60-33)\text{ s}} = -5.2\text{ m/s} = -19\text{ km/h}.$$

3. *Travel time increase*: Without being obstructed by the traffic wave, the considered vehicle entering at $t = 50$ s into the investigated road section ($x = 0$) would leave the section ($x = L = 200$ m) after about 16 s. This can be deduced either by linearly extrapolating the first seconds of the trajectory, or by the quotient L/V_{free}. The *actual* vehicle leaves the investigated region at $t = 86$ s. Thus, the delay imposed by this traffic wave on the vehicle is 20 s.

4. *Lane-changing rate*: By counting all lane changes entering and leaving the considered lane in the spatiotemporal region [0 s, 80 s] × [0 m, 140 m], i.e., trajectories beginning or ending inside this region,[2] we obtain the lane-changing rate by

$$r \approx \frac{6\text{ changes}}{80\text{ s }140\text{ m}} = 0.00054\,\frac{\text{changes}}{\text{m s}} \approx 1{,}900\,\frac{\text{changes}}{\text{km h}}.$$

[2] Outside this region, a positive bias is unavoidable because real lane changes cannot be distinguished from the begin/end of recorded trajectories.

2.3 Trajectory Data of "Obstructed" Traffic Flow

1. The trajectory data shows a queue at a traffic light. The horizontal bar marks the position of the traffic light and the duration of the red light phase.
2. Flow $Q_{\text{in}} = 5$ trajectories per 20 s = 0.25 vehicles/s = 900 vehicles/h.
3. Following the trajectory which starts at $x = -80$ m, $t = -16$ s and which ends at $(x,t) = (80\,\text{m}, 0\,\text{s})$, we get the speed

$$v_{\text{in}} = \frac{160\,\text{m}}{16\,\text{s}} = 10\,\text{m/s} = 36\,\text{km/h}.$$

The density is read off the diagram as one trajectory per 40 m or is calculated using $\rho = Q/v$. Either way yields $\rho = 25$ vehicles/km.
4. Density in the congested area: 8 horizontal trajectories per 40 m $\Rightarrow \rho_{\text{jam}} = 200$ vehicles/km.
5. Outflow after the red light turns green: The best way is to count the number of lines within a 20 s interval above the blue dots marking the end of the acceleration phase, giving 10 lines per 20 s and thus $Q_{\text{out}} = 0.5$ vehicles/s = 1 800 vehicles/h. The speed is the same as in free traffic (the trajectories are parallel to those further upstream), i.e., $V = 36$ km/h. The density is obtained again by counting trajectories (two lines per 40 m) or via the hydrodynamic relation, yielding $\rho = 50$ vehicles/km.
6. Propagation velocities of the fronts can be read off the chart as the gradient of the front lines (marked by the dots) or using the continuity equation (cf. Chap. 7):

$$\text{free} \to \text{congested:} \quad v_g^{\text{up}} = \frac{\Delta Q}{\Delta \rho} = \frac{-900\,\text{vehicles/h}}{175\,\text{vehicles/km}} = -5.17\,\text{km/h},$$
$$\text{congested} \to \text{free:} \quad v_g^{\text{down}} = \frac{\Delta Q}{\Delta \rho} = \frac{1{,}800\,\text{vehicles/h}}{-150\,\text{vehicles/km}} = -12\,\text{km/h}.$$

7. Without the red light the vehicle entering at $x = -80$ m and $t = 20$ s would have reached the "end" (upper border) of the diagram ($x = 100$ m) at $t_{\text{end}} = 38$ s. De facto, it arrives at $x = 100$ m at time $t = 69$ s, thus delayed by 31 s.
8. The braking distance is $s_b = 25$ m, while the distance covered during the acceleration phase is $s_a = 50$ m. Thus

$$b = \frac{v^2}{2s_b} = 2\,\text{m/s}^2, \quad a = \frac{v^2}{2s_a} = 1\,\text{m/s}^2.$$

Alternatively, we can calculate the braking distance using the definition of the acceleration and the duration Δt of the acceleration/deceleration:

$$b = -\frac{\Delta v}{\Delta t} = -\frac{-10\,\text{m/s}}{10\,\text{s}} = 2\,\text{m/s}^2, \quad a = \frac{\Delta v}{\Delta t} = \frac{10\,\text{m/s}}{10\,\text{s}} = 1\,\text{m/s}^2.$$

Problems of Chapter 3

3.1 Data Aggregation at a Cross-Section

1. Flow and speed: With an aggregation interval $\Delta t = 30$ s and $n_1 = 6$, $n_2 = 4$ measured vehicles on lanes 1 and 2, respectively, the flow and time mean speed on the two lanes are

$$Q_1 = \frac{n_1}{\Delta t} = 0.2 \text{ vehicles/s} = 720 \text{ vehicles/h},$$

$$Q_2 = \frac{n_2}{\Delta t} = 0.133 \text{ vehicles/s} = 480 \text{ vehicles/h},$$

$$V_1 = \frac{1}{n_1} \sum_\alpha v_{1\alpha} = 25.8 \text{ m/s}, \qquad V_2 = \frac{1}{n_2} \sum_\alpha v_{2\alpha} = 34.0 \text{ m/s}.$$

2. Density: When assuming zero correlations between speeds and time headways, the covariance $\text{Cov}(v_\alpha, \Delta t_\alpha) = 0$. With Eq. (3.20), this means that calculating the true (spatial) densities by Q/V using the arithmetic (time) mean speed gives no bias:

$$\rho_1 = \frac{Q_1}{V_1} = 7.74 \text{ vehicles/km}, \quad \rho_2 = \frac{Q_2}{V_2} = 3.92 \text{ vehicles/km}.$$

3. Both lanes combined: Density and flow are *extensive quantities* increasing with the number of vehicles. Therefore, building the total quantities by simple summation over the lanes makes sense:

$$\rho_{\text{tot}} = \rho_1 + \rho_2 = 11.66 \text{ vehicles/km}, \quad Q_{\text{tot}} = Q_1 + Q_2 = 1{,}200 \text{ vehicles/h}.$$

Since speed is an *intensive* quantity (it does not increase with the vehicle number), summation over lanes makes no sense. Instead, we define the effective aggregated speed by requiring the hydrodynamic relation to be valid for total flow and total density as well:

$$V = \frac{Q_{\text{tot}}}{\rho_{\text{tot}}} = \frac{\rho_1 V_1 + \rho_2 V_2}{\rho_{\text{tot}}} = \frac{Q_{\text{tot}}}{Q_1/V_1 + Q_2/V_2} = 28.5 \text{ m/s} = 102.9 \text{ km/h}.$$

By its derivation from the hydrodynamic relation, this effective speed is the space mean speed rather than the time mean speed measured directly by the detectors. We notice that the effective speed is simultaneously the *arithmetic mean* weighted with the densities, and the *harmonic mean* weighted with the flows. However, the weighting with the densities requires that the density estimates itself are known without bias. This is the case here but not generally. Since flows can always be estimated without systematic errors from stationary detectors, the harmonic mean weighted with the flows is preferable.

4. *Fraction of trucks*: Two out of six (33 %) are in the right lane, none in the left, two out of ten (20 %) total. Notice again that the given percentages are the fraction of trucks *passing* a fixed location (*time mean*). In the same situation, we expect the fraction of trucks observed by a "snapshot" of a road section at a fixed time (*space mean*). To be higher, at least if trucks are generally slower than cars

3.2 Determining Macroscopic Quantities from Single-Vehicle Data

The distance headway $\Delta x_\alpha = 60\,\text{m}$ is constant on both lanes. All vehicles are of the same length $l = 5\,\text{m}$ and all vehicles on a given lane l (left) or r (right) have the same speed $v_\alpha^{\text{l}} = 144\,\text{km/h} = 40\,\text{m/s}$ and $v_\alpha^{\text{r}} = 72\,\text{km/h} = 20\,\text{m/s}$, respectively.

1. *Time gap / headway*: The headways $\Delta t_\alpha = \Delta x_\alpha / v_\alpha$ are

$$\Delta t_\alpha^{\text{l}} = \frac{60\,\text{m}}{40\,\text{m/s}} = 1.5\,\text{s}, \quad \Delta t_\alpha^{\text{r}} = \frac{60\,\text{m}}{20\,\text{m/s}} = 3.0\,\text{s}.$$

The time gaps T_α are equal to the headway minus the time needed to cover a distance equal to the length of the leading vehicle, $T_\alpha = \Delta t_\alpha - \frac{l_{\alpha-1}}{v_{\alpha-1}}$. Since all vehicle lengths are equal, this results in

$$T_\alpha^{\text{l}} = \frac{60\,\text{m} - 5\,\text{m}}{40\,\text{m/s}} = 1.375\,\text{s}, \quad T_\alpha^{\text{r}} = \frac{60\,\text{m} - 5\,\text{m}}{20\,\text{m/s}} = 2.75\,\text{s}.$$

2. *Macroscopic quantities*: We assume an aggregation time interval $\Delta t = 60\,\text{s}$. However, due to the stationary situation considered here, any other aggregation interval will lead to the same results. Directly from the definitions of flow, occupancy, and time-mean speed, we obtain for each lane

$$\begin{aligned}
Q^{\text{l}} &= \frac{1}{\Delta t_\alpha^{\text{l}}} = \frac{1}{1.5\,\text{s}} = 2{,}400\,\text{vehicles/h}, & Q^{\text{r}} &= \frac{1}{\Delta t_\alpha^{\text{r}}} = \frac{1}{3\,\text{s}} = 1{,}200\,\text{vehicles/h}.\\
O^{\text{l}} &= \frac{0.125}{1.5} = 0.083 = 8.3\,\%, & O^{\text{r}} &= \frac{0.25}{3.0} = 0.083 = 8.3\,\%.\\
V^{\text{l}} &= 144\,\text{km/h}, & V^{\text{r}} &= 72\,\text{km/h}.
\end{aligned}$$

Due to the homogeneous traffic situation, the arithmetic and harmonic time-mean speed are the same and directly given by the speed of the individual vehicles.

Totals and averages of both lanes: As already discussed in Problem 3.1 summing over the lanes to obtain a total quantity makes only sense for extensive quantities (Q, ρ) but not for the intensive ones (V, O).

Flow:

$$Q_{\text{tot}} = \frac{\Delta N}{\Delta t} = \frac{\Delta N^{\text{l}} + \Delta N^{\text{r}}}{\Delta t} = 3{,}600 \text{ vehicles/h}, \quad Q = \frac{Q_{\text{tot}}}{2} = 1{,}800 \text{ vehicles/h}.$$

Occupancy:

$$O = O^{\text{l}} = O^{\text{r}} = 0.083.$$

Arithmetic time mean speed:

$$V = \frac{1}{\Delta N} \sum_{\alpha} v_{\alpha} = \frac{40 \cdot 40 \text{ m/s} + 20 \cdot 20 \text{ m/s}}{60} = 120 \text{ km/h}.$$

Harmonic time mean speed:

$$V_{\text{H}} = \frac{\Delta N}{\sum 1/v_{\alpha}} = \frac{60}{\frac{40}{40 \text{ m/s}} + \frac{20}{20 \text{ m/s}}} = 108 \text{ km/h}.$$

We observe that the arithmetic mean is larger than the harmonic mean.

Which mean? In traffic flow, there are four sensible ways to average, consisting of the four combinations of (i) one of two physical ways (time mean and space mean), (ii) one of two mathematical ways (arithmetic and harmonic).

- *Time mean* means averaging at a fixed location over some time interval as done by stationary detectors.
- *Space mean* means averaging at a fixed time over some space interval (road section), e.g., when making a snapshot of the traffic flow.

For the *space mean*, we have (cf. the previous problem)

$$V = \frac{\rho_1 V_1 + \rho_2 V_2}{\rho_{\text{tot}}} = \frac{Q_{\text{tot}}}{Q_1/V_1 + Q_2/V_2}$$

while, for the time mean, we simply have

$$V = \frac{Q_1 V_1 + Q_2 V_2}{Q_{\text{tot}}}.$$

The time mean is generally larger than the space mean because, at the same partial densities, the class of faster vehicles passes the cross-section more often within the aggregation interval than the vehicles of the slower class do. The arithmetic average is generally larger than the harmonic average which can be shown for any data. Only for the trivial case of identical data, both averages agree.

Here, $\rho_1 = \rho_2$ but $Q_1 \neq Q_2$, so the simple (not weighted) arithmetic average over lanes applies for the space mean speed.

3. Speed variance between lanes: Because of identical speeds on either lane, the total variance of the speeds in the left and right lane is the same as the inter-lane variance sought after:

$$\begin{aligned}\sigma_V^2 &= \left\langle (v_\alpha - \langle v_\alpha \rangle)^2 \right\rangle \\ &= \frac{1}{60}\left(40[40 - 33.3]^2 + 20[20 - 33.3]^2\right) \\ &= 88.9\,\mathrm{m^2/s^2}.\end{aligned}$$

4. Total speed variance: We divide the speed variance

$$\sigma_V^2 = \frac{1}{N}\sum_\alpha (v_\alpha - V)^2$$

into two sums over the left and right lane, respectively:

$$\sigma_V^2 = \frac{1}{N}\left[\sum_{\alpha_1=1}^{N_1} \left(v_{\alpha_1} - V\right)^2 + \sum_{\alpha_2=1}^{N_2} \left(v_{\alpha_2} - V\right)^2\right].$$

Now we expand the two squares:

$$(v_{\alpha_1} - V)^2 = (v_{\alpha_1} - V_1 + V_1 - V)^2 = (v_{\alpha_1} - V_1)^2 + 2(v_{\alpha_1} - V_1)(V_1 - V) + (V_1 - V)^2,$$

where V_1 is the average over lane 1. We proceed analogously for $(v_{\alpha_2} - V)^2$. Inserting this into the expression for σ_V^2 and recognizing that

$$\sum_{\alpha_1}(v_{\alpha_1} - V_1)(V_1 - V) = 0, \quad \sum_{\alpha_2}(v_{\alpha_2} - V_2)(V_2 - V) = 0$$

and

$$\sigma_{V_1}^2 = \frac{1}{N_1}\sum_{\alpha_1=1}^{N_1}(v_{\alpha_1} - V_1)^2$$

and similarly for $\sigma_{V_2}^2$, we obtain

$$\sigma_V^2 = \frac{N_1}{N}\left[\sigma_{V_1}^2 + (V_1 - V)^2\right] + \frac{N_2}{N}\left[\sigma_{V_2}^2 + (V_2 - V)^2\right].$$

With $p_1 = N_1/N$ and $p_2 = N_2/N = 1 - p_1$, we get the formula of the problem statement. If $p_1 = p_2 = 1/2$ we have $V = (V_1 + V_2)/2$ and thus

$$\sigma_V^2 = \frac{1}{2}\left[\sigma_{V_1}^2 + \sigma_{V_2}^2\right] + \frac{(V_1 - V_2)^2}{4}.$$

Notice that this mathematical relation can be applied to both space mean and time mean averages.

Problems of Chapter 4

4.1 Analytical Fundamental Diagram

We have to distinguish between free and congested traffic.

Free traffic:

$$V^{\text{free}}(\rho) = V_0 = \text{const.}$$

Flow by using the hydrodynamic relation:

$$Q^{\text{free}}(\rho) = \rho V^{\text{free}}(\rho) = \rho V_0.$$

Congested traffic: The speed-dependent equilibrium gap between vehicles, $s(v) = s_0 + vT$, leads to the gap-dependent equilibrium speed $V^{\text{cong}}(s)$:

$$V^{\text{cong}}(s) = \frac{s - s_0}{T}.$$

Using the definition of the density ρ, we replace the gap s:

$$\begin{aligned}\rho &= \frac{\text{number of vehicles}}{\text{road length}} \\ &= \frac{\text{one vehicle}}{\text{one distance headway (front-to-front distance)}} \\ &= \frac{1}{\text{vehicle length + gap (bumper-to-bumper distance)}} = \frac{1}{l + s}.\end{aligned}$$

Thus $s(\rho) = \frac{1}{\rho} - l$ and therefore

$$V^{\text{cong}}(\rho) = \frac{s - s_0}{T} = \frac{1}{T}\left[\frac{1}{\rho} - (l + s_0)\right].$$

The flow-density relation is obtained again by the hydrodynamic relation:

$$Q^{\text{cong}}(\rho) = \rho V^{\text{cong}}(\rho) = \frac{1}{T}\left[1 - \rho(l + s_0)\right].$$

The sum $l_{\text{eff}} = l + s_0$ of vehicle length and minimum gap can be interpreted as an *effective* vehicle length (typically 7 m in city traffic, somewhat more on highways). Accordingly, the maximum density is

$$\rho_{\text{max}} = \frac{1}{l + s_0} = \frac{1}{l_{\text{eff}}}.$$

To obtain the critical density ρ_C separating free and congested traffic, we determine the point where the free and congested branches of the fundamental diagram intersect:

$$Q^{\text{cong}}(\rho) = Q^{\text{free}}(\rho) \Rightarrow \rho V_0 T = 1 - \rho(l + s_0) \Rightarrow \rho_C = \frac{1}{V_0 T + l_{\text{eff}}}.$$

This is the "tip" of the triangular fundamental diagram, and the corresponding flow is the capacity C (the maximum possible flow):

$$C = Q^{\text{cong}}(\rho_C) = Q^{\text{free}}(\rho_C) = \frac{1}{T}\left(\frac{1}{1 + \frac{l_{\text{eff}}}{V_0 T}}\right). \tag{2.1}$$

The capacity C is of the order of (yet always less than) the inverse time gap T. The lower the free speed V_0, the more pronounced the discrepancy between the "ideal" capacity $1/T$ and the actual value.

Given the numeric values stated in the problem, we obtain the following values for ρ_{max}, ρ_C, and C:

$$\rho_{\text{max}} = 143 \text{ vehicles/km}, \quad \rho_C = 16.6 \text{ vehicles/km},$$
$$C = 0.552 \text{ vehicles/s} = 1{,}990 \text{ vehicles/h}.$$

4.2 Flow-Density Diagram of Empirical Data

The free velocity can be read off the speed-density diagram (at low densities):

$$V_{\text{A8}}^{\text{free}} = 125 \text{ km/h}, \quad V_{\text{A9}}^{\text{free}} = 110 \text{ km/h}.$$

(Dutch police is very rigorous in enforcing speed limits and uses automated systems to do so. This explains why few people drive faster than 110 km/h.)

The flow-density diagram immediately shows the maximum density (the density where the flow data drop to zero at the right-hand side) and the capacity (maximum flow):

$$\rho_{\text{A8}}^{\text{max}} = 80 \text{ vehicles/km}, \quad \rho_{\text{A9}}^{\text{max}} = 110 \text{ vehicles/km},$$
$$C_{\text{A8}} = 1{,}700 \text{ vehicles/h}, \quad C_{\text{A9}} = 2{,}400 \text{ vehicles/h}.$$

The headways can be calculated by solving the capacity equation (2.1) for T,

$$T = \frac{1 - \frac{l_{\text{eff}} C}{V_0}}{C} = \frac{1 - \frac{C}{V_0 \rho_{\text{max}}}}{C},$$

and thus

$$T_{\text{A8}} = 1.91\,\text{s}, \quad T_{\text{A9}} = 1.27\,\text{s}.$$

One cautionary note is in order: If the data of the scatter plots are derived from arithmetic time mean speeds via the relation $\rho = Q/V$ (as it is the case here), the density of congested traffic flow, and in the consequence T, will be underestimated. To a lesser extent, this also applies to harmonic averages (cf. Fig. 4.10). In the Sects. 8.5.3 and 16.3, we will learn about more robust estimation methods based on propagation velocities.

Problems of Chapter 5

5.1 Reconstruction of the Traffic Situation Around an Accident

Part 1: In the space-time diagram below, thin dashed green lines mark confirmed free traffic while all other information is visualized using thicker lines and different colors. The respective information is denoted in the key. The signal "zero flow" means "I do not know; either empty road or stopped traffic".

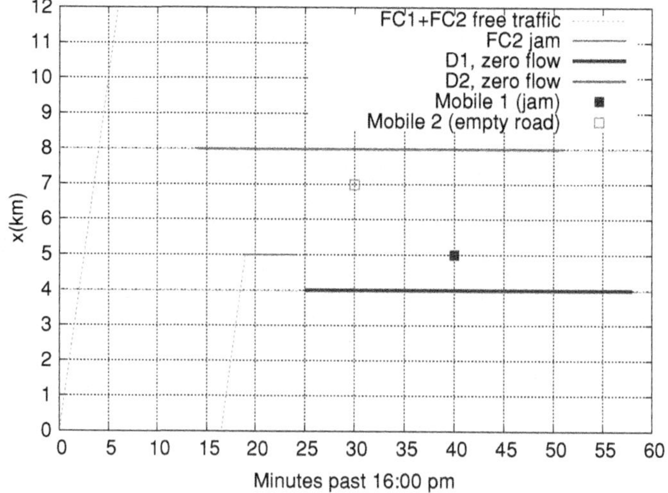

Part 2: The information of the first floating car (FC1) tells us the speed in free traffic, $V_{\text{free}} = 10\,\text{km}/5\text{min} = 120\,\text{km/h}$. From the second floating car (FC2) we know that an upstream jam front passes $x = 5$ km at 4:19 pm.

The stationary detectors D1 at $x = 4$ km and D2 at 8 km both report zero flows in a certain time interval but this does not tell apart whether the road is maximally congested or empty. However, we additionally know by the two mobile phone calls that the road is fully congested at 5 km while it is empty at 7 km. The congestion at 5 km is also consistent with the trajectory of the second floating car. Since downstream jam fronts (transition jam $\rightarrow$ free traffic) are either stationary or propagate upstream at velocity $c \approx -15$ km/h but never downstream (apart from the special case of a moving bottleneck), we know that the missing vehicle counts of D1 are the consequence of standing traffic while that of D2 reflect an empty road (at least when ignoring the possibility that there might be another obstruction more downstream causing a second jam).

With this information, we can estimate the motion of the upstream jam front. Assuming a constant propagation velocity c^{up}, we determine this velocity from the spatiotemporal points where detector D1 and the second floating car encounter congestion, respectively:

$$c^{\text{up}} = \frac{-1\,\text{km}}{6\,\text{min}} = -10\,\text{km/h}.$$

The motion of this front is another strong evidence that D2 does not measure a transition from free to fully congested traffic but from free traffic to no traffic at all at $x = 8$ km and $t = 4{:}14$ pm: Otherwise, the propagation velocity c^{up} would be $-4\,\text{km}/5\,\text{min} = -48\,\text{km/h}$ in this region which is not possible even if we do not require c^{up} to be constant: The largest possible negative velocity c^{up}, realized under conditions of maximum inflow against a full road block, is only insignificantly larger in magnitude than $|c^{\text{down}}| \approx 15$ km/h.

Now we have enough information to determine location and time of the initial road block (accident). Intersecting the line

$$x^{\text{up}}(t) = 4\,\text{km} + c^{\text{up}}t - 25\,\text{min} = 4\,\text{km} - \frac{t - 25\,\text{min}}{6}\,\text{km/min}$$

characterizing the upstream front with the trajectory $x_{\text{last}}(t)$ of the last vehicle that made it past the accident location,

$$x_{\text{last}}(t) = 8\,\text{km} + v_0(t - 14\,\text{min}) = 8\,\text{km} + (t - 14\,\text{min})\,2\,\text{km/min}$$

yields location and time of the road block:

$$x_{\text{crash}} = 6\,\text{km}, \quad t_{\text{crash}} = 4{:}13\,\text{pm}.$$

Part 3: After the accident site is cleared, the initially stationary downstream jam front (fixed at the accident site) starts moving at the characteristic velocity $c^{\text{down}} =$

$-15\,\text{km/h} = -1\,\text{km}/4\,\text{min}$. Since the detector at $x = 4\,\text{km}$ (D1) detects non-zero traffic flow from 4:58 pm onwards, the front is described by

$$x^{\text{down}}(t) = 4\,\text{km} + c^{\text{down}}(t - 58\,\text{min}).$$

Obviously, the accident location ($x_{\text{crash}} = 6\,\text{km}$) is cleared exactly at the time where the moving downstream jam front crosses the accident site (cf. the figure), i.e., at $t_{\text{clear}} = 4{:}50\,\text{pm}$.

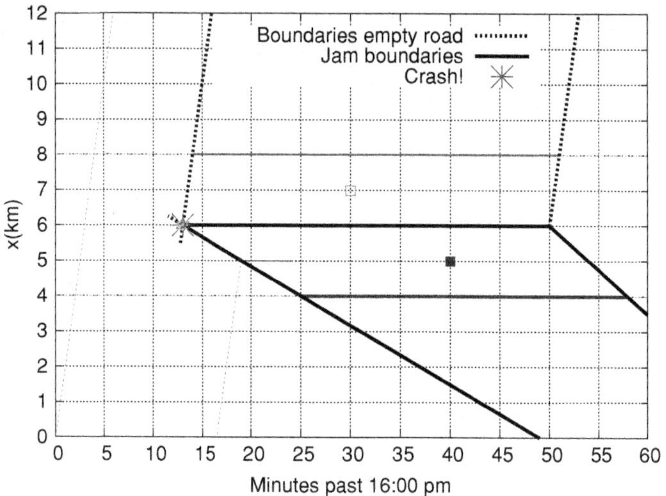

5.2 Dealing with Inconsistent Information

Using equal weights, $V = \frac{1}{2}(V_1 + V_2)$, the error variance is

$$\sigma_V^2 = \frac{1}{4}\left(\sigma_1^2 + \sigma_2^2\right) = \frac{1}{4}\left(\sigma_1^2 + 4\sigma_1^2\right) = \frac{5}{4}\sigma_1^2,$$

assuming negligible systematic errors and independent random errors. Consequently, the error increases by a factor of $\sqrt{5/4}$ due to the inclusion of the noisy floating-car data. Using optimal weights,

$$V_{\text{opt}} = \frac{1}{5}(4V_1 + V_2),$$

yields the error variance

$$(\sigma_V^2)_{\text{opt}} = \frac{1}{25}\left(16\sigma_1^2 + \sigma_2^2\right) = \frac{1}{25}\left(16\sigma_1^2 + 4\sigma_1^2\right) = \frac{4}{5}\sigma_1^2.$$

This means, adding floating-car data with a small weight to the stationary detector data *reduces* the uncertainty by a factor of down to $\sqrt{4/5}$, in spite of the fourfold variance of the floating-car data compared to the stationary detector data.

Problems of Chapter 6

6.1 Speed Limit on the German Autobahn?

The *safety aspect* of speed limits cannot be modeled or simulated by traffic-flow models simply because these models are calibrated to normal situations (including traffic jams). However, accidents are typically the consequence of a series of unfortunate circumstances and extraordinary driving behavior which is not included in the models. In contrast, the effect on *fuel consumption* can be modeled and simulated reliably by combining microscopic or macroscopic traffic flow models with the corresponding models for fuel consumption or emissions, cf. Chap. 20. To assess the *economic effect* of speed limits, including social welfare or changed traffic patterns, one needs models for traffic demand and route choice, i.e., models of the domain of transportation planning. Traffic flow models are suited, however, to investigate certain environmental and societal aspects on a smaller scale. For example, traffic flow models in connection with consumption/emission models describe the direct effect of speed limits on emissions. Furthermore, since speed limits change the propensity for traffic breakdowns and traffic flow/emission models can describe this influence as well as the changed emissions in the jammed state, these models also describe the indirect effect via traffic breakdowns.

Problems of Chapter 7

7.1 Flow-Density-Speed Relations

We require

$$Q_{\text{tot}} = \sum_i Q_i = \sum_i \rho_i V_i = \rho_{\text{tot}} V$$

which we can fulfil by suitably defining the effective average V of the local speed across all lanes. Solving this condition for V directly gives

$$V = \sum_i \frac{\rho_i}{\rho_{\text{tot}}} V_i = \sum_i w_i V_i,$$

i.e., the definition (7.6) of the main text.

7.2 Conservation of Vehicles
In a closed ring road, the vehicle number $n(t)$ is the integral of the total vehicle density $\rho_{\rm tot} = I(x)\rho$ over the complete circumference of the ring. Applying Eq. (7.15) for $\nu_{\rm rmp} = 0$ and the hydrodynamic relation $\rho V = Q$, we obtain for the rate of change of the vehicle number

$$\frac{\mathrm{d}n}{\mathrm{d}t} = -\oint \left(I(x)\frac{\partial Q(x,t)}{\partial x} + Q(x,t)\frac{\mathrm{d}I}{\mathrm{d}x} \right) \mathrm{d}x = -\oint \frac{\partial (I(x)Q(x,t))}{\partial x} \mathrm{d}x = IQ|_{x_{\rm a}}^{x_{\rm e}}.$$

If the road is closed, we have $x_{\rm e} = x_{\rm a}$, so $\frac{\mathrm{d}n}{\mathrm{d}t} = 0$ and the total vehicle number n does not change over time.

7.3 Continuity Equation I
(i) The continuity equation for $x < 0$ or $x > L = 300\,\mathrm{m}$, i.e., outside of the merging region, has no source terms and reads

$$\frac{\partial \rho}{\partial t} + \frac{\partial Q}{\partial x} = 0.$$

In the merging region $0 \le x \le L$, we have an additional source term $\nu_{\rm rmp}(x,t)$ that generally depends on space and time:

$$\frac{\partial \rho}{\partial t} + \frac{\partial Q}{\partial x} = \nu_{\rm rmp}(x,t).$$

Now we assume that $\nu_{\rm rmp}$ is constant with respect to x over the merging region L. In view of the definition of ν, this means that the differential merging rate on a small segment of the merging lane, divided by the number of main-road lanes, is constant. We obtain the connection between the ramp flow $Q_{\rm rmp}$ and $\nu_{\rm rmp}$ by integrating ν over the merging region and requiring that the result is equal to the ramp flow (if there are several ramp lanes, the total ramp flow) divided by the number I of main-road lanes. Thus,

$$\frac{Q_{\rm rmp}}{I} = \int_0^L \nu_{\rm rmp}\, dx = \nu_{\rm rmp} \int_0^L dx = \nu_{\rm rmp} L,$$

so

$$\nu_{\rm rmp} = \frac{Q_{\rm rmp}}{IL} = \frac{1}{2}\,\frac{600\ \text{vehicles/h}}{300\,\text{m}} = 1\ \text{vehicle/m/hour}.$$

(ii) It is easy to generalize the source term $\nu_{\rm rmp}(x)$ to inhomogeneous differential merging rates. In the most general case, we prescribe a distribution function of the merging points[3] over the length of the merging lane by its probability density $f(x)$:

[3] Later on, when we explicitly model lane changes by microscopic models, we will assume instantaneous lane changes, so the merging point is well-defined. For real continuous changes, one can

$$\nu_{\text{rmp}}(x,t) = \frac{Q_{\text{rmp}}(t)}{I} f(x).$$

For case (i) (constant differential changing rates),

$$f_{\text{uniform}}(x) = \begin{cases} 1/L & \text{if } 0 \le x \le L, \\ 0 & \text{otherwise,} \end{cases}$$

i.e., the merging points are uniformly distributed over the interval $[0, L]$. To model drivers who, in their majority, merge in the first half of the length of the merging lane, we prescribe a distribution $f(x)$ which takes on higher values at the beginning than near the end of the lane, e.g., the triangular distribution

$$f_{\text{early}}(x) = \begin{cases} \frac{2(L-x)}{L^2} & \text{if } 0 \le x \le L, \\ 0 & \text{otherwise.} \end{cases}$$

If we want to describe a behavior where drivers change to the main road near the end (which applies for some situations of congested traffic, we mirror $f_{\text{early}}(x)$ at $x = L/2$ to arrive at

$$f_{\text{late}}(x) = \begin{cases} \frac{2x}{L^2} & \text{if } 0 \le x \le L, \\ 0 & \text{otherwise.} \end{cases}$$

Remark A temporal dependency is modeled directly by a time-dependent ramp flow $Q_{\text{rmp}}(t)$.

7.4 Continuity Equation II

A stationary traffic flow is characterized by zero partial time derivatives, particularly,

$$\frac{\partial \rho(x,t)}{\partial t} = 0, \qquad \frac{\partial Q(x,t)}{\partial t} = 0.$$

This simplifies the continuity equation (7.15) for the effective (lane-averaged) flow and density for the most general case including ramps and variable lane numbers to

$$\frac{\mathrm{d}Q}{\mathrm{d}x} = -\frac{Q(x)}{I(x)} \frac{\mathrm{d}I}{\mathrm{d}x} + \nu_{\text{rmp}}(x). \tag{2.2}$$

By the condition of stationarity, the partial differential equation (7.15) for $\rho(x,t)$ and $Q(x,t)$ with the independent variables x and t changes to an ordinary differential

define x to be the first location where a vehicle crosses the road marks separating the on-ramp from the adjacent main-road lane.

equation (ODE) for Q as a function of x. Stationarity also implies that the traffic inflow at the upstream boundary is constant, $Q(x=0,t)=Q_0$.

(i) We can solve the ODE (2.2) for $\nu_{\text{rmp}}=0$ by the standard method of *separating the variables*:

$$\frac{\mathrm{d}Q}{Q}=-\frac{\mathrm{d}I}{I}\frac{\mathrm{d}x}{\mathrm{d}x}=-\frac{\mathrm{d}I}{I}.$$

Indefinite integration of both sides with respect to the corresponding variable yields $\ln Q=-\ln I+\tilde{C}$ with the integration constant $\tilde{C}$. Applying the exponentiation on both sides results in

$$Q(x)=\frac{C}{I(x)}$$

where $C=\exp(\tilde{C})$. The new integration constant is fixed by the spatial initial conditions $C=I(x=0)Q(x=0)=I_0Q_0$ where I_0 is the number of lanes at $x=0$. This also determines the spatial dependency of the flow:

$$Q(x)=\frac{I_0Q_0}{I(x)}. \tag{2.3}$$

Notice that this is consistent with the stationarity condition $Q_{\text{tot}}=I(x)Q(x)=I_0Q_0=\text{const}$.
(ii) To describe an on-ramp or off-ramp merging to or diverging from a main-road with I lanes with a constant differential rate, we set

$$\nu_{\text{rmp}}=\frac{Q_{\text{rmp}}}{IL}=\text{const},$$

where $Q_{\text{rmp}}<0$ for off-ramps. Applying the condition of stationarity to the continuity equation (7.15) assuming a constant number I of lanes results in the ODE

$$\frac{\mathrm{d}Q}{\mathrm{d}x}=\begin{cases}\nu_{\text{rmp}} & \text{parallel to merging/diverging lanes,}\\ 0 & \text{otherwise,}\end{cases}$$

with prescribed and constant $Q(x=0)=Q_0=Q_{\text{tot}}/I$ at the upstream boundary. In our problem with an off-ramp upstream of an on-ramp (which is the normal configuration at an interchange), we have

$$\frac{\mathrm{d}Q}{\mathrm{d}x}=\begin{cases}-Q_{\text{off}}/(IL_{\text{off}}) & \text{if } 300\,\text{m}\le x<500\,\text{m},\\ Q_{\text{on}}/(IL_{\text{on}}) & \text{if } 700\,\text{m}\le x<100\,\text{m},\\ 0 & \text{otherwise.}\end{cases}$$

where $L_{\text{off}}=200\,\text{m}$, and $L_{\text{on}}=300\,\text{m}$. We calculate the solution to this ODE by simple integration:

$$Q(x) = \begin{cases} Q_0 & \text{if } x < 300\,\text{m}, \\ Q_0 - Q_{\text{off}}(x - 300\,\text{m})/(I L_{\text{off}}) & \text{if } 300\,\text{m} \le x < 500\,\text{m}, \\ Q_0 - Q_{\text{off}}/I & \text{if } 500\,\text{m} \le x < 700\,\text{m}, \\ Q_0 - Q_{\text{off}}/I + Q_{\text{on}}(x - 700\,\text{m})/(I L_{\text{on}}) & \text{if } 700\,\text{m} \le x < 1{,}000\,\text{m}, \\ Q_0 + (Q_{\text{on}} - Q_{\text{off}})/I & \text{if } x \ge 1{,}000\,\text{m}. \end{cases}$$

7.5 Continuity Equation III

The highway initially has $I_0 = 3$ lanes, and a lane drop to 2 lanes over the effective length L:

$$I(x) = \begin{cases} 3 & x < 0, \\ \left(3 - \frac{x}{L}\right) & 0 \le x \le L, \\ 2 & x > L. \end{cases}$$

Since the traffic demand (inflow) is constant, $Q_{\text{in}} = Q_{\text{tot}}(0) = 3{,}600$ vehicles/h, and there is no other explicit time dependence in the system, the traffic flow equilibrates to the stationary situation characterized by $\frac{\partial}{\partial t} = 0$:

$$\frac{\mathrm{d}Q}{\mathrm{d}x} = -\frac{Q(x)}{I(x)}\frac{\mathrm{d}I}{\mathrm{d}x}.$$

1. The solution for the section with a variable lane number reads [cf. Eq. (2.3)]:

$$Q(x) = \frac{I_0 Q_0}{I(x)} = \frac{Q_{\text{tot}}}{I(x)}.$$

Upstream and downstream of the lane drop, we have $I(x) = \text{const.}$, i.e.,

$$Q(x) = \frac{Q_{\text{tot}}}{I}.$$

In summary, this results in

$$Q(x) = \begin{cases} Q_{\text{tot}}/3 & x < 0, \\ \frac{Q_{\text{tot}}}{3 - \frac{x}{L}}, & 0 \le x \le L \\ Q_{\text{tot}}/2 & x > L. \end{cases}$$

Furthermore, the hydrodynamic relation $Q = \rho V$ with $V = 108$ km/h gives the density

$$\rho(x) = \frac{Q}{V} = \begin{cases} 11.11\,/\text{km} & x < 0, \\ \frac{33.33}{3 - \frac{x}{L}}\,\frac{1}{\text{km}} & 0 \le x \le L, \\ 16.67\,/\text{km} & x > L. \end{cases}$$

2. We insert the relation $I(x) = 3 - x/L$ and $dI/dx = -1/L$ for the lane drop and $Q(x) = Q_{\text{tot}}/(3 - x/L)$ for the flow into the right-hand side of Eq. (2.2):

$$\frac{\mathrm{d}Q}{\mathrm{d}x} = -\frac{Q_{\text{tot}}}{I^2(x)}\frac{\partial I}{\partial x} = \frac{Q_{\text{tot}}}{L\left(3 - \frac{x}{L}\right)^2}.$$

The right-hand side can be identified with the searched-for effective ramp term:

$$\nu_{\text{rmp}}^{\text{eff}} = \frac{Q_{\text{tot}}}{L\left(3 - \frac{x}{L}\right)^2}.$$

We determine the effective ramp flow corresponding to $\nu_{\text{rmp}}^{\text{eff}}(x)$ from the point of view of the two remaining through lanes:

$$\begin{aligned} Q_{\text{rmp}}^{\text{eff}} &= 2\int_0^L \nu_{\text{rmp}}^{\text{eff}}\,\mathrm{d}x = \frac{2Q_{\text{tot}}}{L}\int_0^L \frac{\mathrm{d}x}{\left(3 - \frac{x}{L}\right)^2} = \frac{2Q_{\text{tot}}}{L}\left(\frac{1}{3 - \frac{x}{L}}\right)\Bigg|_0^L \\ &= \frac{Q_{\text{tot}}}{3} = 1{,}200 \text{ vehicles/h/lane.} \end{aligned}$$

Here, we used the indefinite integral

$$\int \frac{\mathrm{d}x}{\left(3 - \frac{x}{L}\right)^2} = \frac{L}{3 - \frac{x}{L}}.$$

7.6 Continuity Equation for Coupled Maps

Assuming the steady-state condition, $\rho_k(t + \Delta t) = \rho_k(t)$ for all road cells k, we obtain from Eq. (7.16)

$$\begin{aligned} 0 &= Q_k^{\text{up}}\left(1 + \frac{I_{\text{up}} - I_{\text{down}}}{I_{\text{down}}}\right) - Q_k^{\text{down}} + \frac{Q_{\text{k,rmp}}}{I_{\text{down}}} \\ &= Q_k^{\text{up}}\frac{I_{\text{up}}}{I_{\text{down}}} - Q_k^{\text{down}} + \frac{Q_{\text{k,rmp}}}{I_{\text{down}}} \end{aligned}$$

and, after multiplying with I_{down},

$$0 = Q_k^{\text{up}} I_{\text{up}} - Q_k^{\text{down}} I_{\text{down}} + Q_{\text{k,rmp}}$$

which is the flow balance given in the problem statement. This balance means that, in steady-state conditions, the ramp flow is equal to the total outflow $Q_k^{\text{down}} I_{\text{down}}$ from a cell minus the total inflow $Q_k^{\text{up}} I_{\text{up}}$ which is consistent with vehicle conservation.

7.7 Parabolic Fundamental Diagram

For the fundamental diagram

$$Q(\rho) = \rho V(\rho) = \rho V_0 \left(1 - \frac{\rho}{\rho_{\max}}\right),$$

the maximum flow (capacity per lane) $Q_{\max}$ is at a density ρ_C. We determine ρ_C, as usual, by setting the gradient $Q'(\rho)$ equal to zero:

$$Q'(\rho_C) = V_0 - 2\frac{V_0 \rho_C}{\rho_{\max}} = 0 \quad \Rightarrow \quad \rho_C = \frac{1}{2}\rho_{\max}.$$

Hence

$$Q_{\max} = Q(\rho_C) = \frac{\rho_{\max} V_0}{4}.$$

Problems of Chapter 8

8.1 Propagation Velocity of a Shock Wave Free → Congested

The triangular fundamental diagram is semi-concave, i.e., the second derivative $Q_e''(\rho)$ is non-positive, and the first derivative $Q_e'(\rho)$ is monotonously decreasing. This means, any straight line connecting two points on the fundamental diagram always lies below or, at most, on the fundamental diagram $Q_e(\rho)$. Consequently, the slope $c_{12} = (Q_2 - Q_1)/(\rho_2 - \rho_1)$ of this line cannot be greater than $Q'(0) = V_0$ and not less than $Q'(\rho_{\max}) = c$ proving the statement. Since this argumentation only relies on the semi-concavity of the fundamental diagram, it can also be applied to the parabolic fundamental diagram of Problem 7.7 leading to $c_{12} \in [-V_0, V_0]$.

8.2 Driver Interactions in Free Traffic

There are not any in this model. If there were interactions, the followers would react to the leaders, so the information of the shock wave would propagate at a lower velocity than the vehicle speed, contrary to the fact.

8.3 Dissolving Queues at a Traffic Light

When the traffic light turns green, the traffic flow passes the traffic light in the maximum-flow state. For the triangular fundamental diagram, the speed at the maximum-flow state is equal to the desired speed and the transition from the waiting queue (density $\rho_{\max}$) to the maximum-flow state propagates backwards at a velocity $c = -l_{\text{eff}}/T$ corresponding to the congested slope of the fundamental diagram. In the microscopic picture, every follower starts a time interval T later than its leader and instantaneously accelerates to V_0 (Fig. 8.12). This suggests to interpret T as the reaction time of each driver, so $|c|$ is simply the distance between two queued vehicles divided by the reaction time.

We emphasize, however, that the LWR model does not contain any reaction time. Moreover, the above microscopic interpretation no longer holds for LWR models with other fundamental diagrams. Therefore, another interpretation is more to the point. As above, the driver instantaneously starts from zero to V_0 which follows directly from the sharp macroscopic shock fronts. However, the drivers only start their "rocket-like" acceleration when there is enough time headway at V_0. Thus, $|c|$ is the distance between two queued vehicles divided by the desired time gap T in car-following mode. Similar considerations apply for concave fundamental diagrams (such as the parabola-shaped of Problem 7.7). This allows following general conclusion:

The fact that not all drivers start simultaneously at traffic lights is not caused by reaction times but by the higher space requirement of moving with respect to standing vehicles: It simply takes some time for the already started vehicles to make this space.

8.4 Total Waiting Time During One Red Phase of a Traffic Light

The total waiting time in the queue is equal to the number $n(t)$ of vehicles waiting at a given time, integrated over the duration of the queue: Defining $t = 0$ as the begin of the red phase and $x = 0$ as the position of the stopping line, this means

$$\tau_{\text{tot}} = \int_0^{\tau_r+\tau_{\text{diss}}} n(t)\mathrm{d}t = \int_0^{\tau_r+\tau_{\text{diss}}} \int_{x_u(t)}^{x_o(t)} \rho_{\max}\, \mathrm{d}x\, \mathrm{d}t = \rho_{\max} A,$$

i.e., the total waiting time is equal to the jam density times the area of the queue in space-time (cf. the following diagram).

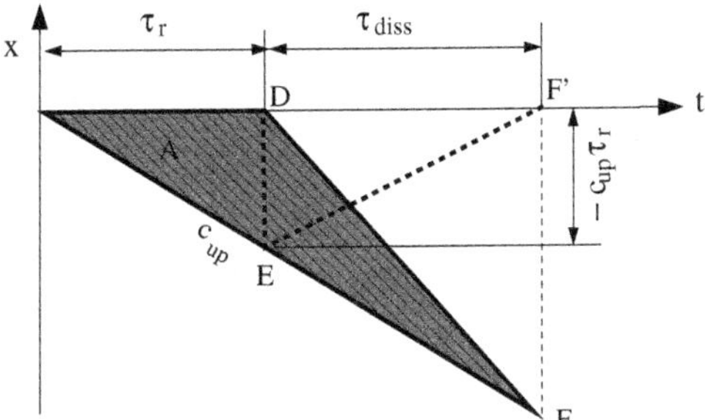

The area of the congested area is equal to the sum of the area of the two right-angled triangles with the legs $(\tau_r, -c_{up}\tau_r)$ and $(\tau_{diss}, -c_{up}\tau_r)$, respectively:

$$\tau_{tot} = \frac{1}{2}\rho_{max}\left(-c_{up}\tau_r^2 - c_{up}\tau_r\tau_{diss}\right).$$

To obtain the second right-angled triangle DEF', we have shifted the point F of the original triangle DEF to F' which does not change the enclosed area. Furthermore, we have the geometrical relation (cf. the figure above)

$$c_{up}\tau_r = (c_{cong} - c_{up})\tau_{diss},$$

i.e., $\tau_{diss} = c_{up}\tau_r/(c_{cong} - c_{up})$. Inserting this into the expression for τ_{tot} finally gives

$$\tau_{tot} = \frac{1}{2}\rho_{max}\tau_r^2 \frac{c_{up}c_{cong}}{c_{up} - c_{cong}}$$

with

$$c_{up} = \frac{Q_{in}}{Q_{in}/V_0 - \rho_{max}}, \quad c_{cong} = -\frac{1}{\rho_{max}T}.$$

The total waiting time increases with the *square* of the red time.

8.5 Jam Propagation on a Highway I: Accident

Subproblem 1: With the values given in the problem statement, the capacity per lane reads

$$Q_{max} = \frac{V_0}{V_0 T + l_{eff}} = 2{,}016 \text{ vehicles/h}.$$

The total capacity of the road in the considered driving direction without accident is just twice that value:

$$C = 2Q_{max} = 4{,}032 \text{ vehicles/h}.$$

This exceeds the traffic demand 3,024 vehicles/h at the inflow ($x = 0$), so no jam forms before the accident, and only road section 1 exists. Since there are neither changes in the demand nor road-related changes, traffic flow is stationary and the flow per lane is constant:

$$Q_1 = \frac{Q_{\text{in}}}{2} = 1{,}512 \text{ vehicles/h}, \quad V_1 = V_0 = 28 \text{ m/s}, \quad \rho_1 = \frac{Q_1}{V_0} = 15 \text{ vehicles/km}.$$

This also gives the travel time to traverse the $L = 10$ km long section:

$$t_{\text{trav}} = \frac{L}{V_0} = 357 \text{ s}.$$

Subproblem 2: At the location of the accident, only one lane is open, so the bottleneck capacity

$$C_{\text{bottl}} = Q_{\text{max}} = 2{,}016 \text{ vehicles/h}$$

does not meet the demand any more, and traffic breaks down at this location. This means, there are now three regions with different flow characteristics:

- Region 1, free traffic upstream of the congestion: Here, the situation is as in Subproblem 1.
- Region 2, congested traffic at and upstream of the bottleneck.
- Region 3, free traffic downstream of the bottleneck.

From the propagation and information velocities of perturbations in free and congested traffic flow, and from the fact that the flow but not the speed derives from a conserved quantity, we can deduce following general rules:

Free traffic flow is controlled by the flow at the upstream boundary, congested traffic flow and the traffic flow downstream of "activated" bottlenecks is controlled by the bottleneck capacity.

For the congested region 2 upstream of the accident (both lanes are available), this means

$$Q_2 = \frac{C_{\text{bottl}}}{2} = 1{,}008 \text{ vehicles/h}.$$

To determine the traffic density, we invert the flow-density relation of the congested branch of the fundamental diagram,[4]

$$\rho_2 = \rho_{\text{cong}}(Q_2) = \frac{1 - Q_2 T}{l_{\text{eff}}} = 72.5 \text{ vehicles/km}.$$

[4] *Beware*: The fundamental diagram and derived quantities (as $\rho_{\text{cong}}(Q)$) are always defined for the lane-averaged effective density and flow.

Subproblem 3: To calculate the propagation velocity of the shock (discontinuous transition free → congested traffic), we apply the shock-wave formula:

$$c^{\text{up}} = c_{12} = \frac{Q_2 - Q_1}{\rho_2 - \rho_1} = -8.77\,\text{km/h}.$$

Subproblem 4: After lifting the lane closure, the capacity is, again, given by $C = 2Q_{\text{max}} = 4{,}032$ vehicles/h, everywhere. In the LWR models, the outflow from congestions is equal to the local capacity, so the outflow region 3 is characterized by $Q_3 = C/2 = Q_{\text{max}}$, $V_3 = V_0$, and $\rho_3 = Q_3/V_0 = 20$ vehicles/h. Furthermore, the transition from regions 2 to 3 (downstream jam front) starts to move upstream at a propagation velocity again calculated by the shock-wave formula:

$$c^{\text{down}} = c = c_{23} = \frac{Q_3 - Q_2}{\rho_3 - \rho_2} = -19.2\,\text{km/h}.$$

The jam dissolves if the upstream and downstream jam fronts meet. Defining t as the time past 15:00 h, x as in the figure of the problem statement, and denoting the duration of the bottleneck by $\tau_{\text{bottl}} = 30$ min, we obtain following equations of motion for the fronts,

$$x_{\text{up}}(t) = L + c^{\text{up}}\,t,$$
$$x_{\text{down}}(t) = L + c(t - \tau_{\text{bottl}}).$$

Setting these positions equal results in the time for complete jam dissolution:

$$t_{\text{dissolve}} = \tau_{\text{bottl}}\,\frac{c}{c - c^{\text{up}}} = 3{,}312\,\text{s}.$$

The position of the last vehicle to be obstructed at obstruction time is equal to the location of the two jam fronts when they dissolve:

$$t_{\text{dissolve}} = L + c^{\text{up}}\,t_{\text{dissolve}} = 1{,}936\,\text{m}.$$

Subproblem 5: In the spatiotemporal diagram, the congestion is restricted by three boundaries:

- Stationary downstream front at the bottleneck position $L = 10$ km for the times $t \in [0, \tau_{\text{bottl}}]$,
- Moving downstream front for $t \in [\tau_{\text{bottl}}, t_{\text{dissolve}}]$ whose position moves according to $x_{\text{down}}(t) = L + c(t - \tau_{\text{bottl}})$,
- Moving upstream front for $t \in [0, t_{\text{dissolve}}]$ whose position moves according to $x_{\text{up}}(t) = L + c^{\text{up}}\,t$

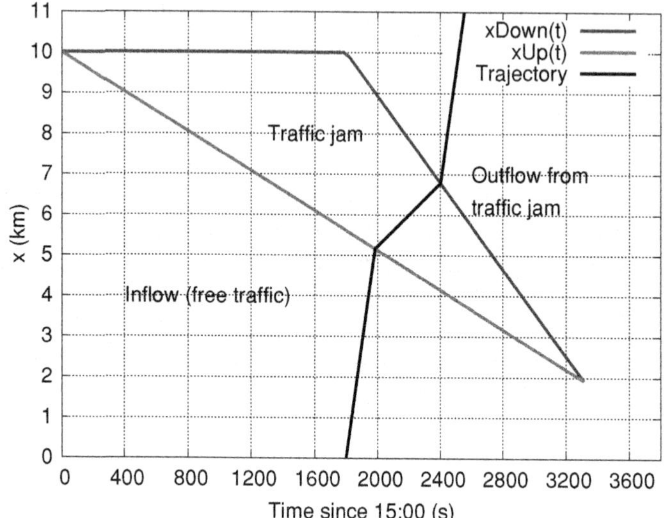

Subproblem 6: We follow the vehicle trajectory starting at time $t = t_0 = 1{,}800$ s at the upstream boundary $x = 0$ by piecewise integrating it through the three regions (cf. the diagram):

1. *Traversing the inflow region*: The vehicle moves at constant speed V_0 resulting in the trajectory $x(t) = V_0(t - t_0)$.
2. *Traversing the jam*: To calculate the time t_{up} of entering the jam, we intersect the free-flow trajectory with the equations of motion $x_{up}(t) = L + c^{up}\,t$ for the upstream front:

$$t_{up} = \frac{L + V_0 t_0}{V_0 - c^{up}} = 1{,}984\,\text{s}.$$

   The corresponding location $x_{up} = V_0(t_{up} - t_0) = 5{,}168$ m. Hence, the trajectory reads

$$x(t) = x_{up} + v_{cong}(t - t_{up}), \quad v_{cong} = \frac{Q_2}{\rho_2} = 3.86\,\text{m/s}.$$

3. *Trajectory after leaving the jam*: Since, at time t_{up}, the bottleneck no longer exists, we calculate the exiting time by intersecting the trajectory calculated above with the equations of motion of the moving downstream front. This results in

$$t_{down} = \frac{L - x_{up} - ct_0 + v_{cong} t_{up}}{v_{cong} - c} = 2{,}403\,\text{s},$$
$$x_{down} = x_{up} + c_{cong}(t_{down} - t_{up}) = 6{,}783\,\text{m}.$$

   After leaving the jam, the vehicle moves according to trajectory $x(t) = x_{down} + V_0(t - t_{down})$, so the vehicle crosses the location $x = L = 10$ km at time

$$t_{\text{end}} = t_{\text{down}} + \frac{L - x_{\text{down}}}{V_0} = 2{,}518\,\text{s}.$$

In summary, we obtain for the total travel time to traverse the $L = 10\,\text{km}$ long section

$$\tau = t_{\text{end}} - t_0 = 718.1\,\text{s}.$$

8.6 Jam Propagation on a Highway II: Uphill Grade and Lane Drop

Subproblem 1: As in the previous problem, we calculate the capacities with the capacity formula of the triangular fundamental diagram:

$$Q_{\text{max}} = \frac{V_0}{V_0 T + l_{\text{eff}}} = 2{,}000\,\text{vehicles/h},$$
$$Q_{\text{max}}^{\text{III}} = \frac{V_{03}}{V_{03} T_3 + l_{\text{eff}}} = 1{,}440\,\text{vehicles/h}.$$

Subproblem 2: For the *total* quantities, lane drops, gradients, and other flow-conserving bottlenecks are irrelevant, and the continuity equation reads

$$\frac{\partial \rho_{\text{tot}}}{\partial t} + \frac{\partial Q_{\text{tot}}}{\partial x} = \frac{\partial Q_{\text{tot}}}{\partial x} = 0.$$

Since the inflow is constant, $Q_{\text{in}} = 2{,}000$ vehicles/h, and less the minimum capacity $C^{\text{III}} = 2Q_{\text{max}}^{\text{III}} = 2{,}880$ vehicles/h, this amounts to stationary free traffic flow in all four regions I–IV with $Q_{\text{tot}} = Q_{\text{in}} = \text{const}$. From this information, we calculate the effective flow of all regions by dividing by the respective number of lanes, and the density by the free part of the fundamental diagram:

	V (km/h)	Q_{tot} (vehicles/h)	Q (vehicles/h/lane)	ρ_{tot} (vehicles/km)	ρ (vehicles/km/lane)
Region I	120	2,000	667	16.7	5.55
Region II	120	2,000	1,000	16.7	8.33
Region III	60	2,000	1,000	33.3	16.7
Region IV	120	2,000	1,000	16.7	8.33

Subproblem 3: Traffic breaks down if the local traffic flow is greater than the local capacity. Thus, the jam forms at a location and at a time where and when this condition is violated, for the first time. Since the capacities in the four regions are given by 6,000, 4,000, 2,880, and 4,000 vehicles/h, respectively, the interface between regions II and III at $x = 3$ km is the first location where the local capacity can no longer meet the new demand $Q_{\text{in}} = 3{,}600$ vehicles/h. Traffic breaks down if the information of

the increased demand reaches $x = 3\,\text{km}$. This information propagates through the regions I and II at $c_{\text{free}} = V_0 = 120\,\text{km/h}$, or at 2 km per minute, so

$$x_{\text{breakd}} = 3\,\text{km}, \quad t_{\text{breakd}} = 16{:}01{:}30\,\text{h}.$$

Subproblem 4: To determine density, flow, and speed of congested traffic in the regions I and II, we, again, adhere to the rule that free traffic flow is controlled by the upstream boundary while the total flow of congested regions and of regions downstream of "activated" bottlenecks are equal to the bottleneck capacity at some earlier times determined by the information propagation velocities c_{free} and c_{cong}, respectively. Furthermore, densities inside congestions are calculated with the congested branch of the fundamental diagram while the free branch is used in all other cases. Denoting with regions Ib and IIb the congested sections of regions I and II, respectively, and with regions Ia and IIa the corresponding free-flow sections, this leads to following table for the traffic-flow variables:

	V (km/h)	Q_{tot} (vehicles/h)	Q (vehicles/h/lane)	ρ_{tot} (vehicles/km)	ρ (vehicles/km/lane)
Ia ($I = 3$)	16	3,600	1,200	30	10
Ib ($I = 3$)	120	2,880	960	180	60
IIa ($I = 2$)	36	3,600	1,800	30	15
IIb ($I = 2$)	120	2,880	1,440	80	40
III ($I = 2$)	60	2,880	1,440	48	24
IV ($I = 2$)	120	2,880	1,440	24	12

Notice that the local vehicle speed inside congested two-lane regions is more than *twice* that of three-lane regions.[5]

Subproblem 5: To calculate the propagation velocities of the upstream jam front in the regions I and II, we use, again, the shock-wave formula together with the table of the previous subproblem:

$$v_g = \frac{\Delta Q}{\Delta \rho} = \begin{cases} -240/(60-10)\,\text{km/h} = -4.8\,\text{km/h} & \text{Interface Ia-Ib, situation (i)}, \\ -360/(40-15)\,\text{km/h} = -14.4\,\text{km/h} & \text{Interface IIa-IIb, situation (ii)}. \end{cases}$$

8.7 Diffusion-Transport Equation

Inserting the initial conditions into (8.57) results in

[5] When being stuck inside jams without knowing the cause, this allows to draw conclusions about the type of bottleneck, e.g., whether it is a three-to-two, or three-to-one lane drop.

$$\rho(x,t) = \rho_0 \int_0^L \mathrm{d}x' \frac{1}{\sqrt{4\pi Dt}} \exp\left[\frac{-(x-x'-\tilde{c}t)^2}{4Dt}\right].$$

Since the integrand is formally identical to the density function $f_N(x')$ of a (μ, σ^2) Gaussian distribution (with space and time dependent expectation $\mu(t) = x - \tilde{c}t$ and variance $\sigma^2(t) = 2Dt$), we can the above integral write as[6]

$$\rho(x,t) = \rho_0 \int_0^L \mathrm{d}x' f_N^{(\mu,\sigma^2)}(x').$$

Since the integrand is the density of a Gaussian distribution function, the integral itself can be expressed in terms of the (cumulated) Gaussian or normal distribution

$$F_N(x) = \int_{-\infty}^{x} f_N(x')\mathrm{d}x',$$

resulting in

$$\rho(x,t) = \rho_0 \left[F_N^{(\mu,\sigma^2)}(L) - F_N^{(\mu,\sigma^2)}(0)\right].$$

Since this is no elementary function, we express the result in terms of the tabulated standard normal distribution $\boldsymbol{\Phi}(x) = F_N^{(0,1)}(x)$ by using the relation $F(x) = \boldsymbol{\Phi}((x-\mu)/\sigma)$ taught in statistics courses. With $\mu = x - \tilde{c}t$ and $\sigma^2(t) = 2Dt$, this results in

$$\begin{aligned}
\rho(x,t) &= \rho_0 \left[\boldsymbol{\Phi}\left(\frac{L-\mu}{\sigma}\right) - \boldsymbol{\Phi}\left(\frac{-\mu}{\sigma}\right)\right] \\
&= \rho_0 \left[\boldsymbol{\Phi}\left(\frac{L-x+\tilde{c}t}{\sqrt{2Dt}}\right) - \boldsymbol{\Phi}\left(\frac{-x+\tilde{c}t}{\sqrt{2Dt}}\right)\right] \\
&= \rho_0 \left[\boldsymbol{\Phi}\left(\frac{x-\tilde{c}t}{\sqrt{2Dt}}\right) - \boldsymbol{\Phi}\left(\frac{x-\tilde{c}t-L}{\sqrt{2Dt}}\right)\right],
\end{aligned}$$

In the last line, we have used the symmetry relation $\boldsymbol{\Phi}(x) = 1 - \boldsymbol{\Phi}(-x)$. In the limiting case of zero diffusion, the two standard normal distribution functions degenerate to jump functions with jumps at the positions $\tilde{c}t$ and $L + \tilde{c}t$. This is consistent with the analytic solution of the section-based model, $\rho(x,t) = \rho_0(x - \tilde{c}t)$ where $\rho_0(x)$ denotes the initial density given in the problem statement. For finite diffusion constants, the initially sharp density profiles smear out over time (cf. Fig. 8.26).

---

[6] When doing the integral, watch out that the variable to be integrated is x' rather than x.

Problems of Chapter 9

9.1 Ramp Term of the Acceleration Equation

Macroscopically, the total derivative $\frac{dV}{dt}$ of the local speed denotes the rate of change of the average speed of all $n = \rho\,\Delta x$ vehicles in a (small) road element of length Δx comoving with the local speed V,

$$\frac{dV}{dt} = \frac{d\langle v_\alpha \rangle}{dt} = \frac{d}{dt}\left(\frac{1}{n}\sum_{\alpha=1}^{n} v_\alpha\right). \tag{2.4}$$

Without acceleration of single vehicles $\left(\frac{dv_\alpha}{dt} = 0\right)$, the rate of change is solely caused by vehicles entering or leaving this road element at a speed $V_{\text{rmp}} \neq V$ (cf. Fig. 9.5). Assuming that the position of the merging vehicles is uniformly distributed over the length L_{rmp} of the merging region, the rate of change of the vehicle number is given by

$$\frac{dn}{dt} = q = Q_{\text{rmp}} \frac{\Delta x}{L_{\text{rmp}}} \tag{2.5}$$

whenever the moving road element is parallel to the merging section of an on-ramp. When evaluating the time derivative (2.4), we notice that both the prefactor $\frac{1}{n}$ and the sum itself depend explicitly on time. Specifically,

$$\frac{d}{dt}\left(\frac{1}{n}\right) = -\frac{q}{n^2}, \quad \frac{d}{dt}\left(\sum_\alpha v_\alpha\right) = q\,V_{\text{rmp}}.$$

The second equation follows from the problem statement that all vehicles enter the road at speed V_{rmp} and no vehicles (including the ramp vehicles) accelerate ($v_\alpha = \text{const}$). Using these relations and $\sum_\alpha v_\alpha = nV$, we can write the rate of change of the local speed as

$$\begin{aligned}
A_{\text{rmp}} &= \frac{d\langle v_\alpha \rangle}{dt} = \frac{d}{dt}\left(\frac{1}{n(t)}\sum_{\alpha=1}^{n(t)} v_\alpha\right) \\
&= -\frac{qnV}{n^2} + \frac{q\,V_{\text{rmp}}}{n} \\
&= \frac{q(V_{\text{rmp}} - V)}{n} \\
&= \frac{Q_{\text{rmp}}\Delta x(V_{\text{rmp}} - V)}{n L_{\text{rmp}}} \\
&= \frac{Q_{\text{rmp}}(V_{\text{rmp}} - V)}{I\rho L_{\text{rmp}}}.
\end{aligned}$$

In the last step, we have used $n = I\rho\,\Delta x$.

9.2 Kinematic Dispersion

Subproblem 1: The lane-averaged local speed is given by

$$V = \frac{1}{\rho_1 + \rho_2}\left(\rho_1 V_1 + \rho_2 V_2\right).$$

First, we calculate the initial speed variance across the lanes ($k = 1$ and 2 for the left and right lanes, respectively):

$$\begin{aligned}\sigma_V^2(x,0) &= \left\langle (V_k(x,0) - V)^2 \right\rangle \\ &= \frac{\rho_1(V_1 - V)^2 + \rho_2(V_2 - V)^2}{\rho_1 + \rho_2},\end{aligned}$$

or, for the special case $\rho_1 = \rho_2$,

$$\sigma_V^2(x,0) = \frac{(V_1 - V_2)^2}{4} = 100\,(\text{m/s})^2 = \text{const.}$$

Notice that these expressions give the true spatial (instantaneous) variance. In contrast, when determining the time mean variance at a given location from data of a stationary detector station, we would obtain for lane 1 the weighting factor $Q_1/(Q_1 + Q_2) = 1/3$ instead of the correct value $\rho_1/(\rho_1 + \rho_2) = 1/2$ resulting in a biased estimate for the true variance (cf. Chap. 4).

Subproblem 2: The kinematic part $P_{\text{kin}} = \rho\sigma_V^2$ of the pressure term leads to a following contribution of the local macroscopic acceleration,

$$A_{\text{kin}} = -\frac{1}{\rho}\frac{\partial P}{\partial x} = -\frac{1}{\rho}\frac{\mathrm{d}}{\mathrm{d}x}\left(\rho(x,t)\sigma_V^2\right) = \begin{cases} \frac{0.01 s^{-2}}{\rho} & 0 \le x \le 100\,\text{m}, \\ 0 & \text{otherwise.} \end{cases}$$

(The factor $0.01\,\text{s}^{-2}$ result from the gradient $\frac{\partial\rho}{\partial x} = 10^{-4}\,\text{m}^{-2}$ multiplied by the variance $100\,(\text{m/s})^2$.) Consequently, a finite speed variance implies that a negative density gradient leads to a positive contribution of the macroscopic acceleration. This will be discussed at an intuitive level in the next subproblem.

Subproblem 3: If there is a finite variance σ_V^2 *and* a negative density gradient (a transition from dense to less dense traffic), then the vehicles driving faster than the local speed V go from the region of denser traffic to the less dense region while the slower vehicles are transported backwards (in the comoving system!) to the denser region. Due to the density gradient, the net inflow of faster vehicles is positive and that of slower vehicles negative. This is illustrated in the following figure, the upper graphics of which depicting the situation in the stationary system, and the lower one in a system comoving with V. As a result, the averaged speed V is increasing although *not a single vehicle* accelerates while the total number of vehicles in the element, i.e., the density, is essentially constant.

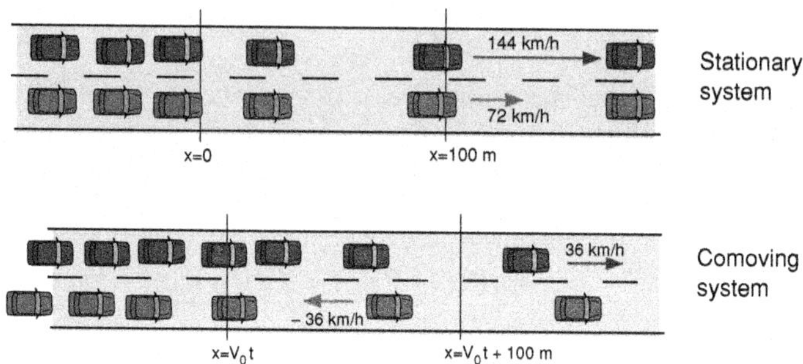

Subproblem 4: Assuming that higher actual speeds are positively correlated with higher desired speeds, the mechanism described in Subproblem 3 leads to a *segregation* of the desired speeds such that the fast tail of the desired speed distribution tends to be found further downstream than the slow tail. This is most conspicuous in multi-lane queues of city traffic waiting behind a red traffic light when the light turns green: If there is one lane with speeding drivers, these drivers will reach first a given position downstream of the stopping line of the traffic light. At this moment, the traffic composition at this point consists exclusively of speeding drivers. Macroscopically, this can only be modeled by multi-class macroscopic models where the desired speed $V_0(x,t)$ becomes another dynamical field with its own dynamical equation. Because of their complexity, such *Paveri-Fontana models* are rarely used.

9.3 Modeling Anticipation by Traffic Pressure

Subproblem 1: Since, by definition, the traffic density is equal to the number of vehicles per distance, one vehicle distance, i.e., the distance headway d, can be expressed by the density:

$$d = \frac{1}{\rho((x+x_\mathrm{a})/2,t)} \approx \frac{1}{\rho(x,t)}.$$

The first expression to the right of the equal sign is accurate to second order in $x_\mathrm{a} - x = d$. The second expression $1/\rho(x,t)$ is accurate to first order which is sufficient in the following.

Subproblem 2: We expand the nonlocal part $V_e(\rho(x_\mathrm{a},t)) = V_e(\rho(x+d,t))$ of the adaptation term to first order around x:

$$\begin{aligned} V_e(\rho(x+d,t)) &= V_e(\rho(x,t)) + \frac{\mathrm{d}V_e(\rho(x,t))}{\mathrm{d}x} d + \mathcal{O}(d^2) \\ &= V_e(\rho(x,t)) + \frac{1}{\rho}\frac{\mathrm{d}V_e(\rho(x,t))}{\mathrm{d}x} + \mathcal{O}(d^2). \end{aligned}$$

Inserting this into the speed adaptation term results in

$$\begin{aligned} \left(\frac{\mathrm{d}V}{\mathrm{d}t}\right)_{\text{relax+antic}} &\approx \frac{V_e(\rho(x,t)) - V(x,t)}{\tau} + \frac{1}{\rho\tau}\frac{\mathrm{d}V_e(\rho(x,t))}{\mathrm{d}x} \\ &\overset{!}{=} \frac{V_e(\rho(x,t)) - V(x,t)}{\tau} - \frac{1}{\rho}\frac{\mathrm{d}P(x,t)}{\mathrm{d}x}, \end{aligned}$$

where, in the last step, we set the result equal to the general expression for the acceleration caused by P. The comparison yields

$$P(x,t) = -\frac{V_e(\rho(x,t))}{\tau}.$$

Subproblem 3: According to the problem statement, the density profile obeys (with $\rho_0 = 20$ vehicles/km $= 0.02$ vehicle/m, $c = 100$ vehicles/km^2 $= 10^{-4}$ vehicle/m^2)

$$\rho(x,t) = \begin{cases} \rho_0 & x < 0, \\ \rho_0 + cx & 0 < x \le 200\,\text{m}, \\ 2\rho_0 & x > 200\,\text{m}. \end{cases} \tag{2.6}$$

(i) Acceleration by anticipation when using the original relaxation term:

$$\left(\frac{\mathrm{d}V}{\mathrm{d}t}\right)_{\text{relax}} = \frac{V_e(\rho(x + 1/\rho(x,t), t)) - V_e(\rho(x,t))}{\tau},$$

where $V_e(\rho) = V_0(1 - \rho/\rho_{\max})$ is given in the problem statement (such a relation is rather unrealistic; it serves to show the principle in the easiest possible way). Inserting Eq. (2.6), we obtain

$$\left(\frac{\mathrm{d}V}{\mathrm{d}t}\right)_{\text{relax}} = \begin{cases} 0 & x \le -1/\rho_0 \text{ or } x > 200\,\text{m}, \\ \frac{-V_0 c}{\tau\rho_{\max}}\left(\frac{1}{\rho_0} + x\right) & -\frac{1}{\rho_0} < x \le 0, \\ \frac{-V_0 c}{\tau\rho_{\max}\rho} & 0 < x \le 200\,\text{m} - \frac{1}{2\rho_0}, \\ \frac{-V_0}{\tau\rho_{\max}}(2\rho_0 - \rho) & 200\,\text{m} - \frac{1}{2\rho_0} < x \le 200\,\text{m}, \end{cases}$$

where the criterion separating the last two cases is only approximatively valid.
(ii) Expressing the acceleration contribution by the pressure term, we obtain

$$\left(\frac{dV}{dt}\right)_{\text{pressure}} = \begin{cases} 0 & x \le 0 \text{ or } x > 200\,\text{m}, \\ \frac{-V_0 c}{\tau \rho_{\max} \rho} & 0 < x \le 200\,\text{m} - \frac{1}{2\rho_0}. \end{cases}$$

Except for the transition regions at the beginning and end of the density gradient, this agrees with the acceleration derived from the original relaxation term. However, in contrast to the pressure term, the nonlocal anticipation term provides "true" anticipation everywhere, including the region $-1/\rho_0 \le x < 0$ where the local approximation by the pressure term does not "see" anything. In summary, the nonlocal route to modeling anticipation is more robust.

9.4 Steady-State Speed of the GKT Model

In the steady state on homogeneous roads, all spatial and temporal derivatives vanish, so the GKT acceleration equation (9.24) reduces to $V = V_e^*$. Furthermore, the homogeneity associated with the steady state implies $V_a = V$ and $\rho_a = \rho$, and the Boltzmann factor is given by $B(0) = 1$. Using these conditions and the definition (9.27) for V_e^*, we can write the condition $V = V_e^*$ as

$$\frac{V}{V_0} = 1 - \frac{\alpha(\rho)}{\alpha(\rho_{\max})} \left(\frac{\rho_a V T}{1 - \rho_a/\rho_{\max}}\right)^2.$$

This is a quadratic equation in V. Its positive root reads

$$V = V_e(\rho) = \frac{\tilde{V}^2}{2V_0}\left(-1 + \sqrt{1 + \frac{4V_0^2}{\tilde{V}^2}}\right)$$

with the abbreviation

$$\tilde{V} = \sqrt{\frac{\alpha(\rho_{\max})}{\alpha(\rho)} \frac{(1 - \rho/\rho_{\max})}{\rho T}}.$$

For densities near the maximum density we have $\tilde{V} \ll V_0$ and $\alpha(\rho) \approx \alpha(\rho_{\max})$. With the micro-macro relation $s = 1/\rho - 1/\rho_{\max}$, we can write the steady-state speed in this limit as

$$V_e(\rho) \approx \tilde{V} \approx \frac{(1 - \rho/\rho_{\max})}{\rho T} = \frac{s}{T}.$$

Notice that this implies that T has the meaning of a (bumper-to-bumper) time gap in heavily congested traffic.

9.5 Flow-Conserving form of Second-Order Macroscopic Models

We start by setting $V = Q/\rho$ in the continuity equation:

$$\frac{\partial \rho}{\partial t} + \frac{\partial Q}{\partial x} = -\frac{Q}{I}\frac{\mathrm{d}I}{\mathrm{d}x} + \nu_{\mathrm{rmp}}.$$

Multiplying the acceleration equation (9.11) by ρ and inserting $V = Q/\rho$ gives the intermediate result

$$\rho\frac{\partial V}{\partial t} + Q\frac{\partial V}{\partial x} = \frac{\rho V_e^* - Q}{\tau} - \frac{\partial P}{\partial x} + \frac{\partial}{\partial x}\left(\eta\frac{\partial (Q/\rho)}{\partial x}\right) + \rho A_{\mathrm{rmp}}.$$

Now we substitute the time derivative of the local speed by a time derivative of the flow. The left-hand side of the last equation then reads

$$\begin{aligned}\rho\frac{\partial V}{\partial t} + Q\frac{\partial V}{\partial x} &= \frac{\partial Q}{\partial t} - V\frac{\partial \rho}{\partial t} + Q\frac{\partial V}{\partial x}\\ &= \frac{\partial Q}{\partial t} + V\frac{\partial Q}{\partial x} + V\frac{Q}{I}\frac{\mathrm{d}I}{\mathrm{d}x} - V\nu_{\mathrm{rmp}} + Q\frac{\partial V}{\partial x}\\ &= \frac{\partial Q}{\partial t} + \frac{\partial (QV)}{\partial x} + V\frac{Q}{I}\frac{\mathrm{d}I}{\mathrm{d}x} - V\nu_{\mathrm{rmp}}.\end{aligned}$$

Substituting again $V = Q/\rho$ and grouping the spatial derivatives together, we obtain

$$\frac{\partial Q}{\partial t} + \frac{\partial}{\partial x}\left[\frac{Q^2}{\rho} + P - \eta\frac{\partial}{\partial x}\left(\frac{Q}{\rho}\right)\right] = \frac{\rho V_e^* - Q}{\tau} + \frac{Q^2}{\rho I}\frac{\mathrm{d}I}{\mathrm{d}x} - \frac{Q\nu_{\mathrm{rmp}}}{\rho} + \rho A_{\mathrm{rmp}}.$$

9.6 Numerics of the GKT Model

Neglecting the pressure term (its maximum relative influence is of the order of $\sqrt{\alpha} = 10\,\%$), the first CFL condition (9.39) for the convective numerical instability reads

$$\Delta t < \frac{\Delta x}{V_0} = 1.5\,\mathrm{s}. \tag{2.7}$$

Since the GKT model does not contain diffusion terms, the second CFL condition is not relevant. However, the relaxation instability must be tested: The characteristic equation $\det(\underline{\mathbf{L}} - \lambda\underline{\mathbf{1}}) = 0$ for the eigenvalues of the matrix $\underline{\mathbf{L}}$ of the linear equation (9.36) reads

$$-\lambda(L_{22} - \lambda) = -\lambda\left[\frac{1}{\tau}\left(-1 + \rho\frac{\partial \tilde{V}_e(\rho, Q)}{\partial Q}\right) - \lambda\right] = 0$$

resulting in the eigenvalues

$$\lambda_1 = 0, \quad \lambda_2 = -\frac{1}{\tau}\left(1 - \rho \frac{\partial \tilde{V}_e(\rho, Q)}{\partial Q}\right),$$

where

$$\tilde{V}_e(\rho, Q) = V_e^*(\rho, Q, \rho, Q) = V_0 \left[1 - \frac{\alpha(\rho)}{\alpha_{\max}} \left(\frac{Q_e T}{1 - \rho/\rho_{\max}}\right)^2\right].$$

With this result, the condition $\Delta t < |\lambda_2^{-1}|$ to avoid relaxation instability becomes

$$\Delta t < \frac{\tau}{1 + \frac{2\alpha(\rho) V_0 \rho Q_e}{\alpha_{\max}} \left(\frac{T}{1-\rho/\rho_{\max}}\right)^2},$$

i.e., Eq. (9.44) of the main text. In the limit of high densities $\rho \approx \rho_{\max}$ we make use of the approximate relation (cf. Problem 9.4)

$$V_e(\rho) \approx \frac{1}{T}\left(\frac{1}{\rho} - \frac{1}{\rho_{\max}}\right)$$

to arrive at

$$\Delta t \left(1 + 2\frac{V_0}{V_e}\right) < \tau$$

which is Eq. (9.45) of the main text. Inserting $\rho_{\max,\text{sim}} = 0.1\,\text{m}^{-1}$ from the problem statement and $V_e(\rho_{\max,\text{sim}}) = 4.14$ m/s (watch out for the units! If in doubt, always use the SI units m, kg, and s), we finally obtain

$$\Delta t < \frac{1}{|\lambda_2|} = 1.32\,\text{s}. \tag{2.8}$$

The definitive limitation of the time step is given by the more restrictive one of the conditions (2.7) and (2.8), so $\Delta t < 1.32$ s.

The expression (9.48) for the numerical diffusion of both equations at $V = 20$ m/s and $\Delta t = 1$ s (i.e., the conditions for linear numerical stability are satisfied) evaluates to

$$D_{\text{num}} = V\frac{\Delta x}{2}\left(1 - V\frac{\Delta t}{\Delta x}\right) = 300\,\text{m/s}^2.$$

This is only about $1/30$ of the (real) diffusion introduced to the Kerner-Konhäuser model by the term proportional to D_v (assuming standard parameterization).

Problems of Chapter 10

10.1 Dynamics of a Single Vehicle Approaching a Red Traffic Light

Subproblem 1 (parameters). The free acceleration is the same as that of the OVM. Hence, v_0 is the desired speed and τ the adaptation time. If the model decelerates, it does so with the deceleration b. Since, at this deceleration, the kinematic braking distance to a complete stop is given by $\Delta x_{\text{brake}} = v^2/(2b)$, the vehicle stops at a distance s_0 to the (stopping line of) the red traffic light. This explains the meaning of the last parameter. Notice that, in this model, vehicles would follow any leading vehicle driving at a constant speed $v_l < v_0$ at the same gap s_0, i.e., the model does not include a safe gap. Nor does it contain a reaction time. The model is accident-free with respect to stationary obstacles, but not when slower vehicles are involved.

Subproblem 2 (acceleration). Here, the first condition of the model applies, so we have to solve the ordinary differential equation (ODE) for the speed

$$\frac{dv}{dt} = \frac{v_0 - v}{\tau} \quad \text{with } v(0) = 0.$$

The exponential ansatz $e^{\lambda t}$ for the homogeneous part $\frac{dv}{dt} = -v/\tau$ gives the solvability condition $\lambda = 1/\tau$. Furthermore, the general solution for the full inhomogeneous ODE reads

$$v(t) = Ae^{-t/\tau} + B.$$

The asymptotic $v(\infty) = B = v_0$ yields the inhomogeneous part B. Determining the integration constant A by the initial condition $v(0) = A + B = A + v_0 = 0$ gives $A = -v_0$, so the speed profile reads

$$v(t) = v_0\left(1 - e^{-\frac{t}{\tau}}\right).$$

Once $v(t)$ is known, we determine the trajectory $x(t)$ by integrating over time. With $x(0) = 0$, we obtain

$$\begin{aligned} x(t) &= \int_0^t v(t')\,dt' = v_0 \int_0^t \left(1 - e^{-\frac{t'}{\tau}}\right) dt' \\ &= v_0\left[t' + \tau e^{-\frac{t'}{\tau}}\right]_{t'=0}^{t'=t} = v_0 t + v_0\tau\left(e^{-\frac{t}{\tau}} - 1\right). \end{aligned}$$

By identifying parts of this expression with $v(t)$, this simplifies to

$$x(t) = v_0 t - v(t)\tau.$$

Finally, to obtain the acceleration profile, we either differentiate $v(t)$, or insert $v(t)$ into the right-hand side of the ODE. In either case, the result is

$$\dot{v} = \frac{v_0 - v}{\tau} = \frac{v_0}{\tau}\,\mathrm{e}^{-1/\tau}.$$

Subproblem 3 (braking phase). The red traffic light represents a standing virtual vehicle of zero length at the stopping line, so $\Delta v = v$. This phase starts at a distance

$$s_c = s_0 + \frac{v^2}{2b} = 50.2\,\mathrm{m}$$

to the stopping line, and the vehicle stops at a distance s_0 to this line.

Subproblem 4 (trajectory). For the accelerating phase, the trajectory has already been calculated. The deceleration phase begins at the location

$$x_c = L - s_c = L - s_0 - \frac{v^2}{2b} \approx 450\,\mathrm{m}.$$

To approximatively determine the time t_c at which the deceleration phase begins, we set $v(t_c) = v_0$ to obtain

$$x_c(t_c) = v_0 t_c - v(t_c)\tau \approx v_0(t_c - \tau) \quad \Rightarrow \quad t_c = \frac{x_c}{v_0} + \tau = 32.4\,\mathrm{s} + 5.0\,\mathrm{s} = 37.4\,\mathrm{s}.$$

With the braking time v_0/b, this also gives the stopping time

$$t_{\text{stop}} = t_c + \frac{v_0}{b} = 44.3\,\mathrm{s}.$$

In summary, the speed profile $v(t)$ can be expressed by (cf. the graphics below)

$$v(t) = \begin{cases} v_0\left(1 - \mathrm{e}^{-t/\tau}\right) & 0 \le t < t_c, \\ v_0 - b(t - t_c) & t_c \le t \le t_{\text{stop}}, \\ 0 & \text{otherwise.} \end{cases}$$

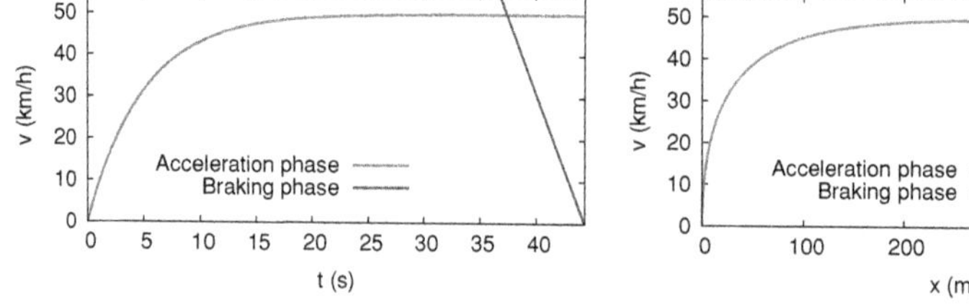

10.2 OVM Acceleration on an Empty Road

(i) The maximum acceleration $a_{\max} = v_0/\tau$ is reached right at the beginning, $t = 0$. (ii) Prescribing $a_{\max} = 2\,\text{m/s}^2$ and a desired speed $v_0 = 120\,\text{km/h}$ determines the speed relaxation time by

$$\tau = \frac{v_0}{a_{\max}} = 16.7\,\text{s}.$$

(iii) We require that, at a time t_{100} to be determined, the speed should reach the value $v_{100} = 100\,\text{km/h}$:

$$v(t_{100}) = v_{100} = v_0 \left(1 - \mathrm{e}^{-\frac{t_{100}}{\tau}}\right).$$

Solving this condition for t_{100} gives

$$\frac{v_{100}}{v_0} = 1 - \mathrm{e}^{-\frac{t_{100}}{\tau}} \quad \Rightarrow \quad t_{100} = -\tau \ln \left(1 - \frac{100}{120}\right) \approx 29.9\,\text{s}.$$

10.3 Optimal Velocity Model on a Ring Road

The problem describes a situation with evenly spaced identical vehicles on a ring road which, initially, are at rest. This means, traffic flow is not stationary (since the initial gaps are greater than the minimum gap) but homogeneous: Since the road is homogeneous, and the vehicle fleet consists of identical vehicles, the homogeneity imposed by the initial conditions is not destroyed over time. For microscopic models, homogeneity implies that the dynamics depend neither on x nor on the vehicle index α. So, dropping α, the OVM reads

$$\frac{\mathrm{d}v}{\mathrm{d}t} = \frac{v_{\text{opt}}(s(0)) - v}{\tau}.$$

The solution to this ODE is analogously to Problem 10.2, only v_0 is replaced by the steady-state speed $v_e = v_{\text{opt}}(s(0))$.

10.4 Full Velocity Difference Model

General plausibility arguments require the steady-state speed $v_{\text{opt}}(s)$ to approach the desired speed v_0 when the gap s tends to infinity. However, for an arbitrarily large distance to the red traffic light modeled by a standing virtual vehicle ($\Delta v = v$), the FVDM vehicle accelerates according to

$$\dot{v} = \frac{v_0 - v}{\tau} - \lambda v = \frac{v_0}{\tau} - \left(\frac{1}{\tau} + \lambda\right) v.$$

From this it follows that the acceleration $\dot{v}$ becomes zero for a terminal speed

$$v^* = \frac{v_0}{1 + \lambda\tau}.$$

This is the maximum speed an initially standing Full Velocity Difference Model (FVDM) vehicle can reach in this situation. It is significantly lower than v_0. For the parameter values of the problem statement, $v^* = 13.5$ km/h which agrees with Fig. 10.6.

10.5 A Simple Model for Emergency Braking Maneuvers

Subproblem 1 (identifying the parameters). T_r = denotes the reaction time, and b_{max} is the maximum deceleration in emergency cases.

Subproblem 2 (braking and stopping distance). Assuming a fixed reaction timer T_r and a constant deceleration b_{max} in the braking phase, elementary kinematic relations yield following expressions for the braking and stopping distances $s_B(v)$ and $s_{\text{stop}}(v) = vT_r + s_B(v)$, respectively:

$$s_B(v) = \frac{v^2}{2b_{\text{max}}}, \quad s_{\text{stop}}(v) = vT_r + s_B(v)$$

with the numerical values

$$\begin{aligned} v = 50\,\text{km/h}: &\quad s_B(v) = 12.1\,\text{m}, \quad s_{\text{stop}}(v) = 25.9\,\text{m}, \\ v = 70\,\text{km/h}: &\quad s_B(v) = 23.6\,\text{m}, \quad s_{\text{stop}}(v) = 43.1\,\text{m}. \end{aligned}$$

Subproblem 3 (emergency braking). At first, we determine the initial distance such that a driver driving at $v_1 = 50$ km/h just manages to stop before hitting the child:

$$s(0) = s_{\text{stop}}(v_1) = 25.95\,\text{m}.$$

Now we consider a speed $v_2 = 70$ km/h but the same initial distance $s(0) = 25.95$ m as calculated above. At the end of the reaction time, the child is just

$$s(T_r) = s(0) - v_2 T_r = 6.50\,\text{m}$$

away from the front bumper. Now, the driver would need the additional braking distance $s_B(v_2) = 23.6$ m for a complete stop. However, only 6.50 m are available resulting in a difference $\Delta s = 17.13$ m. With this information, the speed at collision can be calculated by solving $\Delta s = (\Delta s)_B(v) = v^2/(2b_{\text{max}})$ for v, i.e.,

$$v_{\text{coll}} = \sqrt{2b_{\max}\Delta s} = 16.56\,\text{m/s} = 59.6\,\text{km/h}.$$

Remark This problem stems from a multiple-choice question of the theoretical exam for a German driver's licence. The official answer is 60 km/h.

Problems of Chapter 11

11.1 Conditions for the Microscopic Fundamental Diagram

The plausibility condition (11.5) is valid for any speed v_l of the leading vehicle. This also includes standing vehicles where Eq. (11.5) becomes $a_{\text{mic}}(s, 0, 0) = 0$ for $s \le s_0$. This corresponds to the steady-state condition $v_e(s) = 0$ for $s \le s_0$.

Conditions (11.1) and (11.2) are valid for any speed v_l of the leader as well, including the steady-state situation $v_l = v$ or $\Delta v = 0$. For the alternative acceleration function $\tilde{a}(s, v, \Delta v)$, this means

$$\frac{\partial \tilde{a}(s, v, 0)}{\partial s} \ge 0, \quad \frac{\partial \tilde{a}(s, v, 0)}{\partial v} < 0.$$

Along the one-dimensional manifold of steady-state solutions $\{v_e(s)\}$ for $s \in [0, \infty[$, we have $\tilde{a}(s, v_e(s), 0) = 0$, so the differential change $\mathrm{d}\tilde{a}$ along the equilibrium curve $v_e(s)$ must vanish as well:

$$\mathrm{d}\tilde{a} = \frac{\partial \tilde{a}(s, v_e(s), 0)}{\partial s}\mathrm{d}s + \frac{\partial \tilde{a}(s, v_e(s), 0)}{\partial v} v_e'(s)\mathrm{d}s = 0,$$

hence

$$v_e'(s) = \frac{-\partial \tilde{a}(s, v, 0)/\partial s}{\partial \tilde{a}(s, v, 0)/\partial v} \ge 0.$$

If the leading vehicle is outside the interaction range, we have $v_e'(s) = 0$ [second condition of Eq. (11.2)]. Finally, the condition $\lim\limits_{s\to\infty} v_e(s) = v_0$ follows directly from the second part of condition (11.1).

11.2 Rules of Thumb for the Safe Gap and Braking Distance

Subproblem 1. One mile corresponds to 1.609 km. However, the US rule does not give explicit values for a vehicle length. Here, we assume 15 ft = 4.572 m. In any case, the gap s increases linearly with the speed v, so the time gap $T = s/v$ is independent of speed. Implementing this rule, we obtain

$$T = \frac{s}{v} = \frac{15\text{ft}}{10\,\text{mph}} = \frac{4.572\,\text{m}}{16.09\,\text{km/h}} = \frac{4.572\,\text{m}}{4.469\,\text{m/s}} = 1.0\,\text{s}.$$

Notice that, in the final result, we rounded off generously. After all, this is a rule of thumb and more significant digits would feign a non-existent precision.[7] Notice that this rule is consistent with typically observed gaps (cf. Fig. 4.8).

Subproblem 2. Here, the speedometer reading is in units of km/h, and the space gap is in units of meters. Again, the quotient, i.e., the time gap T is constant and given by (watch out for the units)

$$T = \frac{s}{v} = \frac{\frac{1}{2}\text{m}\left(\frac{v}{\text{km/h}}\right)}{v} = \frac{\frac{1}{2}\text{m}}{\text{km/h}} = \frac{0.5\,\text{h}}{1{,}000} = \frac{1{,}800\,\text{s}}{1{,}000} = 1.8\,\text{s}.$$

Subproblem 3. The kinematic *braking distance* is $s(v) = v^2/(2b)$, so the cited rule of thumb implies that the braking deceleration does not depend on speed. By solving the kinematic braking distance for b and inserting the rule, we obtain (again, watch out for the units)

$$b = \frac{v^2}{2s} = \frac{v^2}{0.02\,\text{m}}\left(\frac{\text{km}}{\text{h}\,v}\right)^2 = \frac{50}{3.6^2}\,\text{m/s}^2 = 3.86\,\text{m/s}^2.$$

For reference, comfortable decelerations are below $2\,\text{m/s}^2$ while emergency braking decelerations on dry roads with good grip conditions can be up to $10\,\text{m/s}^2$, about $6\,\text{m/s}^2$ for wet conditions, and less than $2\,\text{m/s}^2$ for icy conditions. This means, the above rule could lead to accidents for icy conditions but is okay, otherwise.

11.3 Reaction to Vehicles Merging into the Lane

Reaction for the IDM. For $v = v_0/2$, the IDM steady-state space gap reads

$$s_e(v) = \frac{s_0 + vT}{\sqrt{1 - \left(\frac{v}{v_0}\right)^\delta}} = \frac{s_0 + \frac{v_0 T}{2}}{\sqrt{1 - \left(\frac{1}{2}\right)^\delta}}.$$

The prevailing contribution comes from the prescribed time headway (for $s_0 = 2\,\text{m}$ and $\delta = 4$, the other contributions only make up about 10 %). This problem assumes that the merging vehicle reduces the gap to the considered follower to half the steady-

[7] There is also a more conservative variant of this rule where one should leave one car length every five mph corresponding to the "two-second rule" $T = 2.0\,\text{s}$.

state gap, $s = s_e/2 = v_0 T/4$, while the speed difference remain zero. The new IDM acceleration of the follower (with $a = 1$ m/s^2 and $\delta = 4$) is therefore

$$\begin{aligned}
\dot{v}_{\text{IDM}} &= a\left[1-\left(\frac{v}{v_0}\right)^{\delta}-\left(\frac{s_0+vT}{s}\right)^2\right] \\
&\overset{(v=v_0/2,\, s=s_e/2)}{=} a\left[1-\left(\frac{1}{2}\right)^{\delta}-\left(\frac{s_0+v_0T/2}{s_e/2}\right)^2\right] \\
&\overset{s_e(v)=s_e(v_0/2)}{=} -3a\left[1-\left(\frac{1}{2}\right)^{\delta}\right]=-\frac{45}{16}\,\text{m/s}^2=-2.81\,\text{m/s}^2.
\end{aligned}$$

Reaction for the simplified Gipps' model. For this model, the steady-state gap in the car-following regime reads $s_e(v) = v\Delta t$. Again, at the time of merging, the merging vehicle has the same speed $v_0/2$ as the follower, and the gap is half the steady-state gap, $s = (v\Delta t)/2 = v_0\Delta t/4$. The new speed of the follower is restricted by the safe speed v_{safe}:

$$v(t+\Delta t) = v_{\text{safe}} = -b\Delta t + \sqrt{b^2(\Delta t)^2 + \left(\frac{v_0}{2}\right)^2 + \frac{bv_0\Delta t}{2}} = 19.07\,\text{m/s}.$$

This results in an effective acceleration

$$\left(\frac{\mathrm{d}v}{\mathrm{d}t}\right)_{\text{Gipps}} = \frac{v(t+\Delta t)-v(t)}{\Delta t} \approx -0.93 m/s^2.$$

We conclude that the Gipps' model describes a more relaxed driver reaction compared to the IDM. Notice that both the IDM and Gipps' model would generate significantly higher decelerations for the case of slower leading vehicles (dangerous situation).

11.4 The IDM Braking Strategy

A braking strategy is self-regulating if, during the braking process, the *kinematically necessary deceleration* $b_{\text{kin}} = v^2/(2s)$ approaches the comfortable deceleration b. In order to show this, we calculate the rate of change of the kinematic deceleration (applying the quotient and chain rules of differentiation when necessary) and set $\dot{s} = -v$ and $\dot{v} = -b_{\text{kin}}^2/b = -v^4/(4bs^2)$, afterwards. This eventually gives Eq. (11.19) of the main text:

$$\begin{aligned}
\frac{\mathrm{d}b_{\text{kin}}}{\mathrm{d}t} &= \frac{\mathrm{d}}{\mathrm{d}t}\left(\frac{v^2}{2s}\right) = \frac{4vs\dot{v}-2v^2\dot{s}}{4s^2} \\
&= \frac{v^3}{2s^2}\left(1-\frac{v^2}{2sb}\right) = \frac{v\,b_{\text{kin}}}{s\,b}\left(b-b_{\text{kin}}\right),
\end{aligned}$$

11.5 Analysis of a Microscopic Model

Subproblem 1 (parameters). For interaction-free accelerations, $v_{\text{safe}} > v_0$, so v_{safe} is not relevant. Hence v_0 denotes the desired speed, and a the absolute value of the acceleration and deceleration for the cases $v < v_0$ and $v > v_0$, respectively. The steady-state conditions $s = \text{const.}$ and $v = v_l = v_e = \text{const.}$ give

$$v_e = \min(v_0, v_{\text{safe}}).$$

Without interaction, $v_{\text{safe}} > v_0$, so $v_e = v_0$. With interactions, the safe speed becomes relevant and the above condition yields

$$v_e = v_{\text{safe}} = -aT + \sqrt{a^2T^2 + v_e^2 + 2a(s - s_0)}$$

which can be simplified to

$$s = s_0 + v_e T.$$

Thus, s_0 is the minimum gap for $v = 0$, and T the desired time gap. The model produces a deceleration $-a$ not only if $v > v_0$ (driving too fast in free traffic) but also if $v > v_{\text{safe}}$ (driving too fast in congested situations). Furthermore, the model is symmetrical with respect to accelerations and decelerations. Obviously, it is not accident free.

Subproblem 2 (steady-state speed). We have already derived the steady-state condition

$$v_e(s) = \min\left(v_0, \frac{s - s_0}{T}\right).$$

Macroscopically, this corresponds to the triangular fundamental diagram

$$Q_e(\rho) = \min\left(v_0\rho, \frac{1 - \rho l_{\text{eff}}}{T}\right)$$

where $l_{\text{eff}} = 1/\rho_{\text{max}} = l + s_0$. The capacity per lane is given by $Q_{\text{max}} = (T + l_{\text{eff}}/v_0)^{-1} = 1{,}800$ vehicles/h at a density $\rho_C = 1/(l_{\text{eff}} + v_0 T) = 25$ /km. For further properties of the triangular fundamental diagram, see Sect. 8.5.

Subproblem 3. The acceleration and braking distances to accelerate from 0 to 20 m/s or to brake from 20 m/s to 0, respectively, are the same:

$$s_a = s_b = \frac{v_0^2}{2a} = 200\,\text{m}.$$

At a minimum gap of 3 m and the location $x_{\text{stop}} = 603$ m of the stopping line of the traffic light, the acceleration takes place from $x = 0$ to $x_1 = 200$ m, and the deceleration from $x_2 = 400$ m to $x_3 = 600$ m. The duration of the acceleration and deceleration phases is $v_0/a = 20s$ while the time to cruise the remaining stretch of 200 m at v_0 amounts to 10 s. This completes the information to mathematically describe the trajectory:

$$x(t) = \begin{cases} \frac{1}{2}at^2 & t \le t_1 = 20\,\text{s}, \\ x_1 + v_0(t - t_1) & t_1 < t \le t_2 = 30\,\text{s}, \\ x_2 + v_0(t - t_2) - \frac{1}{2}a(t - t_2)^2 & t_2 < t \le t_3 = 50\,\text{s}, \end{cases}$$

where $t_1 = 20$ s, $t_2 = 30$ s and $t_3 = 50$ s.

11.6 Heterogeneous Traffic

The simultaneous effects of heterogeneous traffic and several lanes with lane-changing and overtaking possibilities results in a curved free part of the fundamental diagram even for models that would display a triangular fundamental diagram for identical vehicles and drivers (as the Improved Intelligent Driver Model, IIDM). This can be seen as follows: For heterogeneous traffic, each vehicle-driver class has a different fundamental diagram. Particularly, the density ρ_C at capacity is different for each class, so a simple weighted average of the individual fundamental diagrams would result in a curved free part and a rounded peak. However, without lane-changing and overtaking possibilities, all vehicles would queue up behind the vehicles of the slowest class resulting in a straight free part of the fundamental diagram with the gradient representing the lowest free speed.[8] So, both heterogeneity and overtaking possibilities are necessary to produce a curved free part of the fundamental diagram.

[8] Even when obstructed, drivers can choose their preferred gap (in contrast to the desired speed), so the congested branch of the fundamental diagram is curved even without overtaking possibilities.

11.7 City Traffic in the Improved IDM

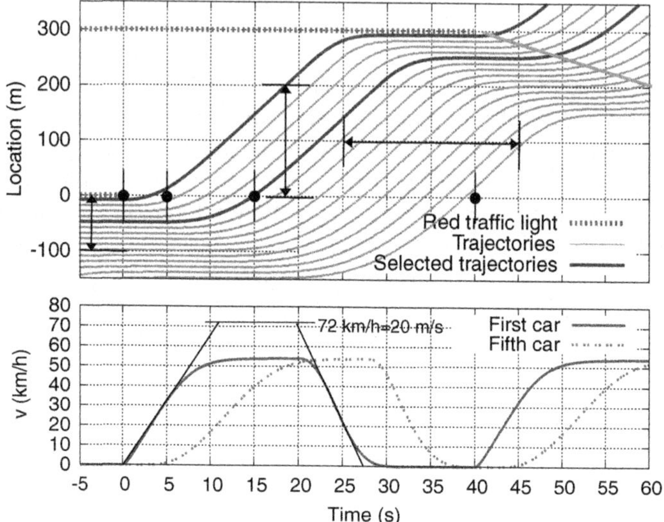

1. For realistic circumstances, the maximum possible flow is given by the *dynamic* capacity, i.e., the outflow from moving downstream congestion fronts. In our case, the "congestion" is formed by the queue of standing vehicles behind a traffic light. Counting the trajectories (horizontal double-arrow in the upper diagram) yields

$$C = Q_{\text{max}} \approx \frac{9\text{ vehicles}}{20\text{ s}} = 1{,}620\text{ vehicles/h}.$$

2. Counting the trajectories passing $x = 0$ for times less than 5, 15, and 40 s (black bullets in the upper diagram) gives

$$n(5) = 1,\ n(15) = 5,\ n(40) = 15,$$

respectively. We determine β by the average time headway after the first vehicles have passed,

$$\beta = \frac{1}{C} = \frac{40\text{ s} - 15\text{ s}}{15 - 5} = 2.5\text{ s/vehicles}.$$

We observe, that β denotes the inverse of the capacity. The obtained value agrees with the result of the first subproblem within the "measuring uncertainty" of one vehicle.[9] This also gives the *additional time* until the first vehicle passes: $\tau_0 = 15\text{ s} - 5\beta = 2.5\text{ s}$. (Notice that this is *not* a reaction time since the IIDM does not have one.)

---

[9] One could have calculated β as well using the pairs $\{n(15), n(5)\}$ or $\{n(40), n(5)\}$ with similar results.

3. The propagation velocity of the position of the starting vehicles in the queue is read off from the upper diagram:

$$c_{\text{cong}} = -\frac{100\,\text{m}}{20\,\text{s}} = -5\,\text{m/s} = -18\,\text{km/h}.$$

4. We estimate the desired speed by the maximum speed of the speed profile (lower diagram): $v_0 = 15\,\text{m/s} = 54\,\text{km/h}$. The effective length l_{eff} is equal to the distance between the standing vehicles in the upper diagram: $\rho_{\max} = 1/l_{\text{eff}} = 10\,\text{vehicles}/100\,\text{m} = 100\,\text{vehicles/km}$, i.e., $l_{\text{eff}} = 10\,\text{m}$. Since the steady state of this model corresponds to a triangular fundamental diagram, the time gap parameter T is determined by the propagation speed and the maximum density: $T = -l_{\text{eff}}/c = 2\,\text{s}$. Finally, the maximum acceleration a and the comfortable deceleration b can be read off the lower diagram by estimating the maximum and minimum gradient of the speed profile:

$$a = \frac{20\,\text{m/s}}{10\,\text{s}} = 2\,\text{m/s}^2, \quad b = \frac{20\,\text{m/s}}{7\,\text{s}} = 2.9\,\text{m/s}^2.$$

Problems of Chapter 12

12.1 Statistical Properties of the Wiener Process

To determine the expectation $\langle w(t)w(t')\rangle$ from the given formal solution $w(t)$ to the stochastic differential equation of the Wiener process, we insert the formal solution into $\langle w(t)w(t')\rangle$ carefully distinguishing the arguments t and t' from the formal integration variables t_1 and t_2. This gives the double integral

$$\langle w(t)w(t')\rangle = \frac{2}{\tilde{\tau}} \int\limits_{t_1=-\infty}^{t} \int\limits_{t_2=-\infty}^{t'} \mathrm{e}^{-(t-t_1+t'-t_2)/\tilde{\tau}} \langle \xi(t_1)\xi(t_2)\rangle \, \mathrm{d}t_1 \mathrm{d}t_2.$$

Notice that the operations of integration and averaging (expectation value) are exchangeable. We now consider the case $t > t'$. Setting $\langle \xi(t_1)\xi(t_2)\rangle = \delta(t_1 - t_2)$ and using the definition $\int f(t)\delta(t)\mathrm{d}t = f(0)$ of the Dirac δ-distribution to eliminate the integral over t_1[10] yields

$$\langle w(t)w(t')\rangle = \frac{2}{\tilde{\tau}} \int\limits_{t_2=-\infty}^{t'} \mathrm{e}^{(2t_2-t-t')/\tilde{\tau}} \, \mathrm{d}t_2$$

[10] Since $t \geq t'$ and the above integration property of the δ-distribution only applies if the integration interval includes zero (i.e., $t_1 = t_2$), we cannot use this property to eliminate the integral over t_2.

which can be analytically solved resulting in

$$\langle w(t)w(t')\rangle = \mathrm{e}^{-(t-t')/\tau}.$$

If $t < t'$, the derivation proceeds analogously resulting in $\langle w(t)w(t')\rangle = \mathrm{e}^{-(t'-t)/\tau}$. Consolidating these two cases, we arrive at

$$\langle w(t)w(t')\rangle = \mathrm{e}^{-|t-t'|/\tau}.$$

As important special case, we obtain the variance $\langle w^2(t)\rangle = 1$. Finally, when we apply the averaging operation $\langle\cdot\rangle$ to the formal solution $w(t)$ itself using the condition $\langle\xi(t)\rangle = 0$, we obtain $\langle w\rangle = 0$, i.e., the second condition (12.7). This concludes the derivation of the statistical properties of the Wiener process.

12.2 Consequences of Estimation Errors

Overestimating the gap by 10 %, i.e., by the factor 1.1 results in a *smaller* steady-state gap. With the values of the problem statement, we obtain

$$1.1 s_e = s_0 + vT \quad \Rightarrow \quad s_e = \frac{s_0 + vT}{1.1} = 28.3\,\mathrm{m}$$

instead of the "true" steady-state $s_e = 31.1\,\mathrm{m}$. When there is a constant additive acceleration component $\Delta a = 0.4\,\mathrm{m/s^2}$, the steady-state condition reads

$$\dot{v} = \Delta a + \frac{\frac{s_e - s_0}{T} - v}{\tau} \stackrel{!}{=} 0,$$

or $s_e = s_0 + vT - \tau a_z = 30.9\,\mathrm{m}$. Notice that the surprisingly small amount of change can be tracked back to the "rigidity" of the OVM reaction caused by the small relaxation time τ.

12.3 Multi-Anticipation for the IDM

Applying the general equation (12.17) for multi-anticipative effects to the IDM gives

$$c \sum_{j=1}^{n_\mathrm{a}} a \left(\frac{s_0 + vT}{js}\right)^2 = a \left(\frac{s_0 + vT}{s}\right)^2,$$

hence $c = 1/(\sum_{j=1}^{n_\mathrm{a}} \frac{1}{j^2})$, i.e., Eq. (12.18). Instead of introducing c, it is obviously possible for the IDM to *renormalize* the parameters s_0 and T by multiplying them with a common factor. For this purpose, we write the left-hand side of above equation as

$$\sum_{j=1}^{n_\mathrm{a}} a \left(\frac{\sqrt{c}s_0 + v\sqrt{c}T}{js}\right)^2,$$

so the factor $\sqrt{c}$ of the problem statement is evident. The factor $\sqrt{c}$ assumes values between 1 (no multi-anticipation) and $\sqrt{6}/\pi \approx 0.78$ (multi-anticipation to infinitely many leaders). This means, the numerical values of s_0 and T are reduced by no more than 22 %.

Problems of Chapter 13

13.1 Dynamic Properties of the Nagel-Schreckenberg Model

To obtain physical units, we multiply the dimensionless desired speed of the NSM with $\Delta x/\Delta t$. Thus, $v_0 = 2$ (city traffic) corresponds to 54 km/h, and $v_0 = 5$ (highways) to 135 km/h. Likewise, multiplying the dimensionless accelerations with $\Delta x/(\Delta t)^2 = 7.5\,\text{m/s}^2$ yields the physical accelerations. In the deterministic NSM, $a = 1$, so $a_{\text{phys}} = 7.5\,\text{m/s}^2$ resulting in an acceleration time of

$$\tau_{0\to 100} = \frac{v_{\text{phys}}}{a_{\text{phys}}} = 3.7\,\text{s}.$$

In the stochastic model, the acceleration a is realized only with a probability $(1-p)$. So, the average acceleration time increases by a factor $(1-p)^{-1}$ to 6.2 s.

13.2 Approaching a Red Traffic Light

The driver approaches with the desired speed v_0 until the distance to the traffic light falls below the "interaction distance" $g = v_0$. Then, there are two possibilities for the deceleration process: (i) stopping in one step (if, after crossing the interaction point, the gap in the next time step is already zero), (ii) stopping in two steps $v_0 \to v_1 \to 0$ (if the gap after crossing the interaction point is $v_1 > 0$). If $v_0 = 2$, the realized decelerations are (i) $-15\,\text{m/s}^2$ or (ii) $-7.5\,\text{m/s}^2$.

13.3 Fundamental Diagram of the Deterministic NSM

Without stochastic components, the steady-state speed as a function of the gap g is well-defined:

$$v_e(g) = \max(v_0, g)$$

meaning that the macroscopic fundamental diagram (in physical units) has the well-known triangular shape given by[11]

$$Q_e(\rho) = \min\left[V_0^{\text{phys}}\rho, \frac{1}{T}\left(1 - \frac{\rho}{\rho_{\max}}\right)\right].$$

The values of its three parameters are

[11] We drop the superscripts "phys" denoting physical quantities where no confusion is possible, i.e., for ρ and Q. We will retain the superscripts for speeds and velocities.

$$V_0^{\text{phys}} = v_0 \Delta x / \Delta t = \begin{cases} 54\,\text{km/h} & \text{cities} \\ 135\,\text{km/h} & \text{highways,} \end{cases}$$

$$T = \Delta t = 1\,\text{s}, \quad \rho_{\max} = \frac{1}{\Delta x} = 133\,\text{vehicles/km}.$$

In the stochastic model ($p > 0$), the average flow $\langle Q(\rho)\rangle$ as a function of the local density is below that of the deterministic case. The fundamental diagram is no longer triangular, and also the gradients at zero and maximum density are different. In the following two problems, we will derive

$$V_0^{\text{phys}} = Q_e'(0) = (v_0 - p)\frac{\Delta x}{\Delta t}, \quad c_{\text{cong}}^{\text{phys}}(\rho_{\max}) = Q_e'(\rho_{\max}) = -(1-p)\frac{\Delta x}{\Delta t}.$$

13.4 Macroscopic Desired Speed

Without interaction and after a sufficient time, the vehicle speed is either v_0 (no dawdling in the last time step), or $v_0 - 1$ (dawdling). If $v = v_0$, then the speed will be reduced in the next time step to $v_0 - 1$ with probability p. If $v = v_0 - 1$, the speed in the next time step will reach v_0 with probability $1 - p$. The situation is stationary in the stochastic sense if expectation values do not change over time, $\langle v(t+1)\rangle = \langle v(t)\rangle$. Here, this means that the probabilities for the speeds v_0 and $v_0 - 1$ do not change over time, i.e., the unconditional "probability fluxes" from v_0 to $v_0 - 1$ and from $v_0 - 1$ to v_0 balance to zero.[12] Setting up the balance for the speed state $v = v_0$ and denoting by θ the probability for this state, the probability flux $v_0 \to v_0 - 1$ away from v_0 is $-\theta p$ (probability θ times conditional probability p; negative sign because the flux is outflowing). The "inflowing" probability flux $v_0 - 1 \to v_0$ is $(1-\theta)(1-p)$ (probability $1-\theta$ times conditional probability $1-p$). So, stationarity implies

$$\frac{\mathrm{d}}{\mathrm{d}t}\left(\text{Prob}(v = v_0)\right) = -\theta p + (1-\theta)(1-p) \stackrel{!}{=} 0 \;\Rightarrow\; \theta = 1 - p,$$

or

$$\langle v\rangle = \theta v_0 + (1-\theta)(v_0 - 1) = v_0 - p.$$

In physical units, this means $V^{\text{phys}} = \langle v\rangle \Delta x/\Delta t$, i.e., the result displayed above in the solution to Problem 13.4.

13.5 Propagation Velocity of Downstream Jam Fronts

Assume a queue of standing vehicles where only the first vehicle has space to accelerate. This first vehicle will accelerate with probability $1 - p$ (with probability

[12] Mathematically, this balance of probability fluxes is called a *Master equation*.

p, dawdling occurs). Only if this vehicle accelerates, the next vehicle in the queue has the possibility to accelerate in the next time step, which it does, again, with probability $1-p$. This means, the "starting wave" propagates at an average velocity $c_{\text{cong}} = -(1-p)\Delta x/\Delta t$. For $p = 0.4$, this yields the reasonable value $c = -16.2$ km/h.

Problems of Chapter 14

14.1 Why the Grass is Always Greener on the Other Side?

We assume two lanes with staggered regions of highly congested traffic (ρ_1, V_1) and less congested traffic $(\rho_2 < \rho_1, V_2 > V_1)$ of the same length: Whenever there is highly congested traffic on lane 1, congestion is less on lane 2, and vice versa (cf. the figure in the problem statement). Since traffic in both regions is (more or less) congested and the fundamental diagram is triangular by assumption, the transitions from region 1 and 2 and from 2 to 1 remain sharp and propagate according to the shock-wave formula (8.9) at a constant velocity

$$c = \frac{Q_2 - Q_1}{\rho_2 - \rho_1} = -\frac{l}{T} = -5\,\text{m/s}.$$

The fraction of time in which drivers are stuck in the highly congested regions is obviously equal to the fraction of time spent in regions of type 1. Denoting by τ_i the time intervals τ_i to pass one region $i = 1$ or 2, we express this fraction by

$$p_{\text{slower}} = p_1 = \frac{\tau_1}{\tau_1 + \tau_2}.$$

When evaluating τ_i, it is crucial to realize that the regions propagate in the opposite direction to the vehicles, so the relative velocity $V_i + |c|$ is relevant. Assuming equal lengths L for both regions, the passage times are $\tau_i = L/(V_i + |c|)$, so

$$p_1 = \frac{\frac{L}{V_1+|c|}}{\frac{L}{V_1+|c|} + \frac{L}{V_2+|c|}} = \frac{V_2 + |c|}{V_2 + V_1 + 2|c|}.$$

For example, if $V_1 = 0$ and $V_2 = 10$ m/s, the fraction is

$$p_1 = \frac{10+5}{10+10} = \frac{3}{4},$$

i.e., drivers are stuck in the slower lane 75 % of the time—regardless which lane they choose or of whether they change lanes or not.

Alternatively, one picks out a vehicle at random. Since the less and highly congested regions have the same length, the fraction of vehicles in the highly congested region, i.e., the probability of picking one from this region, is given by

$$p_1 = \frac{\rho_1}{\rho_1 + \rho_2} = \frac{200}{200 + 200/3} = \frac{3}{4}.$$

14.2 Stop or Cruise?

We distinguish two cases: (i) Drivers can pass the traffic light at unchanged speed in the yellow phase, i.e.,

$$s < s_1 = v\tau_y.$$

(ii) When cruising, drivers would pass the traffic light in the red phase, so stopping is mandatory. In this case, drivers need a reaction time T_r to perceive the signal, make a decision, and stepping on the braking pedal. Afterwards, we assume that they brake at a constant deceleration b so as to stop just at the stopping line. This results in the *stopping distance*

$$s = vT_r + \frac{v^2}{2b}. \tag{2.9}$$

Obviously, the *worst case* for the initial distance s to the stopping line at switching time green-yellow is the threshold $s = s_1$ between (i) and (ii), i.e., cruising is just no more legal. Inserting $s = s_1$ into Eq. (2.9) and solving for b gives

$$b = \frac{v}{2(\tau_y - T_r)} = 3.47\,\text{m/s}^2.$$

This is a significant, though not critical, deceleration. It is slightly below the deceleration $3.86\,\text{m/s}^2$ implied by the braking distance rule "speedometer reading in km/h squared divided by 100" (Problem 11.2) but above typical comfortable decelerations of the order of $2\,\text{m/s}^2$. We conclude that the legal minimum duration of yellow phases is consistent with the driver and vehicle capabilities.

14.3 Entering a Highway with Roadwork

In this situation, we can apply both the safety criterion (14.4) of the general lane-changing model or the safety criterion of the decision model (14.27) for entering a priority road. With the notations of Fig. 14.6, we obtain for a defensive driver ($b_{\text{safe}} = 0$)

$$s_{\text{f}} > s_{\text{safe}}(v_{\text{f}}, v) = s_{\text{opt}}(v_{\text{f}}) = v_{\text{f}}T = 20\,\text{m}.$$

Here, we have dropped all hats, consistent with the convention adopted in Sect. 14.5. In the "worst case", the driver decides to merge if the vehicle on the highway is just $s_{\text{safe}} = 20\,\text{m}$ away. Now we calculate the minimum deceleration b_{min} the driver on the main-road has to adopt to avoid a crash. Applying the kinematic braking distance $s = \Delta v^2/(2b)$ to the critical distance $s_{\text{safe}} = 20\,\text{m}$, the initial speed difference $\Delta v = v_{\text{f}} = 20\,\text{m/s}$, and the relative deceleration $b = b_{\text{min}} + a$ (assuming that the merging vehicle accelerates at $a = 2\,\text{m/s}^2$), and solving for b_{min} results in

$$b_{\text{min}} = \frac{v_{\text{f}}^2}{2s} - a = 8\,\text{m/s}^2.$$

We observe that, in spite of the very conservative assumption $b_{\text{safe}} = 0$ in the decision model, the *actually* necessary deceleration of the main-road vehicle corresponds to an emergency braking maneuver. This discrepancy can be traced back to the OVM whose braking strategy is inconsistent with kinematic constraints and does not contain the speed difference although this is a crucial exogenous factor (in fact, the OVM simulation will lead to crashes in this situation). As shown in the next problem, drivers modeled by the Gipps' model or the IDM family will make a consistent decision in this situation.

14.4 An IDM Vehicle Entering a Priority Road
In this situation (Fig. 14.6 for $v_\alpha = 0$), the IDM safety criterion (14.27) reads

$$s_{\text{f}} > s_{\text{safe}}^{\text{IDM}}(v_{\text{f}}, 0) = \frac{s_0 + v_{\text{f}}T + \frac{v_{\text{f}}^2}{2\sqrt{ab}}}{\sqrt{\frac{a_{\text{free}}(v_{\text{f}})}{a} + \frac{b_{\text{safe}}}{a}}},$$

or, with $s_0 = 0$, $v_{\text{f}} = v_0$, and $a = b = b_{\text{safe}}$ (then, the square root is equal to 1)

$$s_{\text{f}} > s_{\text{safe}}^{\text{IDM}} = v_{\text{f}}T + \frac{v_{\text{f}}^2}{2b_{\text{safe}}}.$$

This means, the minimum gap to allow merging corresponds to the *stopping distance* of the follower on the main-road (braking distance $v^2/(2b)$ plus distance vT_r driven during the reaction time) when the desired time headway is set equal to the reaction time, $T = T_r$. Consequently, the IDM safety criterion for merging is consistent with the driver capabilities and kinematic constraints. For $T = 1\,\text{s}$, $b_{\text{safe}} = 2\,\text{m/s}^2$, and $v_0 = 50\,\text{km/h}$, we obtain as safe distance for merging $s_{\text{safe}}^{\text{IDM}} = 31\,\text{m}$.

Problems of Chapter 15

15.1 Characterizing the Type of Instability
The displayed traffic flow is locally stable since, after a sufficient time, each driver reverts to the steady state (he or she stops, at most, once). At the same time, the dynamics is string unstable since the amplitude of the oscillations increase from vehicle to vehicle. The string instability is convective (of the upstream type) since

there is only a single traffic wave: After a sufficiently long time, traffic flow reverts to the steady state at any fixed location.

15.2 Propagation Velocity of Traffic Waves in Microscopic Models

At first, we transform the microscopic propagation velocity given in the comoving (Lagrangian) coordinate system to a stationary coordinate system:

$$\tilde{c} = v_e + \tilde{c}_{\text{rel}} = v_e - (s_e + l)v_e'(s_e) = v_e - \frac{v_e'(s_e)}{\rho_e}.$$

Here, we used the relation $s_e + l = 1/\rho_e$. Now we express the microscopic gradient $v_e'(s)$ by the corresponding macroscopic quantity $V_e'(\rho)$. Using the identity $v_e(s) = V_e(\rho(s))$ and the micro-macro relation $\rho = 1/(s+l)$, we obtain

$$v_e'(s) = \frac{\mathrm{d}v_e}{\mathrm{d}s} = \frac{\mathrm{d}V_e}{\mathrm{d}\rho}\frac{\mathrm{d}\rho}{\mathrm{d}s} = -\frac{V_e'(\rho)}{(s+l)^2} = -\rho^2 V_e'(\rho). \tag{2.10}$$

Inserting this into the expression for $\tilde{c}$ gives the final result

$$\tilde{c} = v_e - \frac{v_e'(s_e)}{\rho_e} = V_e + \rho_e V_e'(\rho_e) = \frac{\mathrm{d}}{\mathrm{d}\rho_e}(\rho_e V_e) = Q_e'(\rho_e).$$

15.3 Instability Limits for the Full Velocity Difference Model

Subproblem 1: The local stability criterion is satisfied if

$$\tilde{a}_v + \tilde{a}_{\Delta v} = -\frac{1}{\tau} - \gamma \le 0 \quad \Rightarrow \quad \gamma \ge -\frac{1}{\tau} = -0.2\,\mathrm{s}^{-1}.$$

This is true even for slightly negative values of the sensitivity γ to speed differences (although this implies accelerations in response to positive approaching rates which is no reasonable behavior). As such, it reflects the result that all reasonable (and even some unreasonable) models without explicit delays (reaction times) are locally stable.

Subproblem 2: For $v \ge v_0$, there are no interactions and, therefore, no instabilities. If $v < v_0$, we have $v_e'(s) = 1/T$, $\tilde{a}_v = -1/\tau$, and $\tilde{a}_{\Delta v} = -\gamma$. Inserting these relations into condition (15.25) for an oscillation-free local car-following characteristics yields

$$\frac{1}{T} \le \frac{1}{4\tau}(1 + \gamma\tau)^2.$$

Solving this quadratic inequality for γ results in

$$\gamma \ge -\frac{1}{\tau} \pm \frac{2}{\sqrt{T\tau}} = 0.69\,\mathrm{s}^{-1}.$$

Here, we used the general plausibility condition $\gamma \geq 0$ to select the positive sign of the square root when calculating the numerical value.

Subproblem 3: To determine the limits of string instability, we use criterion (15.69). Solving the resulting inequality for γ yields

$$\gamma > \frac{1}{T} - \frac{1}{2\tau} = 0.9\,\mathrm{s}^{-1}.$$

We observe that car-following schemes may be string unstable even if they do not produce *any* kind of oscillations (damped or otherwise) when following a single leader. Here, this applies to the parameter range $0.69\,\mathrm{s}^{-1} < \gamma \leq 0.9\,\mathrm{s}^{-1}$. This is highly relevant when investigating the effects of adaptive cruise control systems on traffic flow.

15.4 Stability Properties of the Optimal Velocity Model Compared to Payne's Model

The OVM criterion for string stability reads $v_e'(s) \leq 1/(2\tau)$, and the corresponding criterion for flow stability in Payne's model $-V_e'(\rho) \leq 1/(2\rho^2\tau)$. Using the micro-macro relation $v_e'(s) = 1/\rho_e^2 V_e'(\rho)$ already needed for Problem 15.2, we show the equivalence by direct substitution:

$$\frac{1}{2\tau} \geq v_e'(s_e) = -\rho_e^2 V_e'(\rho_e(s_e)) \quad \Rightarrow \quad -V_e'(\rho) \leq \frac{1}{2\rho^2\tau} \quad \text{q.e.d.}$$

15.5 Flow Instability in Payne's Model and in the Kerner-Konhäuser Model

Subproblem 1: We have solved the general flow stability problem for Payne's model already in Problem 15.4. For the triangular fundamental diagram as specified in the problem formulation, the gradient of the speed-density relation reads

$$V_e'(\rho) = \begin{cases} 0 & \rho \leq \rho_C, \\ -\frac{1}{\rho^2 T} & \rho > \rho_C, \end{cases}$$

with the density at capacity $\rho_C = 1/(v_0 T + l_{\text{eff}}) = 20$ vehicles/km. For free traffic ($\rho < \rho_C$) there are no interactions ($V_e'(\rho) = 0$) and therefore unconditional stability. Congested traffic flow ($\rho \geq \rho_C$) is stable if

$$\tau < \frac{T}{2}.$$

This means, Payne's model describes stable congested traffic for unrealistically small adaptation times τ, only.

Subproblem 2: For the Kerner-Konhäuser model, the flow stability criterion reads $(\rho V_e'(\rho))^2 < \theta_0$. Inserting the steady-state relation $V_e(\rho) = \max[V_0, 1/T(1/\rho - 1/\rho_{\max})]$ gives, again, unconditional stability for free traffic flow ($\rho < \rho_C = 20$/km) which is consistent with the requirements of the problem formulation. For congested traffic ($\rho > \rho_C$), we have

$$\frac{1}{\rho^2 T^2} < \theta_0.$$

From this condition, we determine θ by demanding that congested traffic flow should be unstable for densities below $\rho_3 = 50$ /km, and stable above. With $T = 1.1$ s and $\rho_3 = 50$ /km, we finally obtain (cf. the figure below)

$$\theta_0 = \frac{1}{\rho_3^2 T^2} = 331 \frac{\mathrm{m}^2}{\mathrm{s}^2}.$$

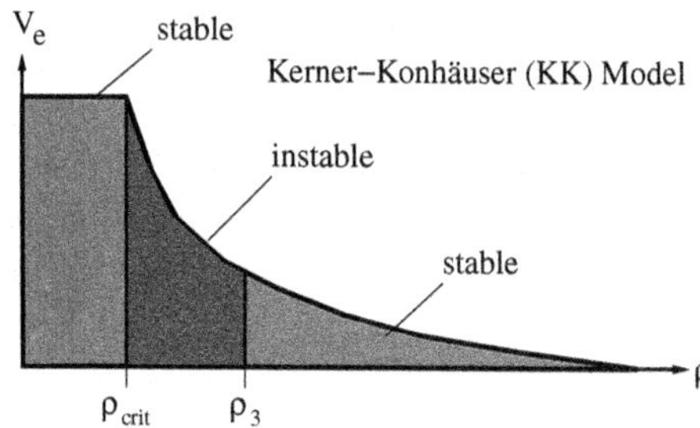

15.6 Flow Instability of the GKT Model

For high densities near the maximum density, we can approximate the GKT steady-state flow by $Q_e \approx 1/T(1 - \rho/\rho_{\max})$, or $V_e'(\rho) = -1/(T\rho^2) \approx -1/(T\rho_{\max})^2$. Without anticipation ($\gamma = 0$) and assuming a constant speed variance prefactor $\alpha_{\max} = \alpha(\rho) \approx \alpha(\rho_{\max})$ which is equivalent to $P_e' \approx \sigma_V^2 \approx \alpha_{\max} V_e^2$, the GKT stability criterion (15.83) becomes

$$(\rho V_e')^2 - P_e' = \frac{1}{T^2 \rho^2} - \alpha_{\max} V_e^2 \le 0.$$

Since, for $\rho \to \rho_{\max}$, the expression $(T\rho)^{-2}$ tends to the squared propagation velocity c^2 of moving downstream jam fronts while the speed variance $\alpha_{\max} V_e^2$ tends to zero, the stability criterion cannot be satisfied: Without anticipation, the GKT model is unconditionally unstable for sufficiently high densities!

For a finite anticipation range $s_a = \gamma V_e T$, however, the third term of the stability condition (15.83) can stabilize traffic flow. Sufficiently close to the maximum density, we can approximate the full GKT flow stability criterion to an analytically tractable condition. If $\rho \approx \rho_{\max}$, we have, up to linear order in V_e

$$\rho V_e' \approx -\frac{1}{T\rho}, \quad P_e' \approx \alpha_{\max} V_e^2 \approx 0, \quad s_a(V_0 - V_e) \approx \gamma V_e V_0 T.$$

Furthermore, with $\rho_{\max}/(\rho_{\max} - \rho) \approx (\rho V_e T)^{-1}$, we can approximate the bracket of the last term of Eq. (15.83) by

$$\left[\frac{\rho_{\max}}{\rho_{\max} - \rho} - \frac{\rho V_e'}{\sigma_V \sqrt{\pi}}\right] \approx \frac{1}{\rho V_e T}\left(1 + \frac{1}{\sqrt{\alpha_{\max}}}\right).$$

Inserting all this into the GKT stability condition (15.83), we find that the GKT model is string stable for densities near the maximum density if the anticipation factor γ fulfils

$$\gamma > \frac{\tau}{2T^2 \rho_{\max} V_0 \left[1 + (\alpha_{\max}\pi)^{-1/2}\right]}$$

which is condition (15.84).

15.7 IDM Stability Class Diagram for other Parameter Values

In the following, we will denote the scaled dimensionless quantities with a tilde. According to the problem formulation, the scaled time and space coordinates as well as derived variables (speed, acceleration) are related to the unscaled quantities as

$$t = \sqrt{\frac{s_0}{b}}\tilde{t}, \quad x = s_0 \tilde{x}, \quad v = \sqrt{b s_0}\tilde{v}, \quad \frac{\mathrm{d}v}{\mathrm{d}t} = b\frac{\mathrm{d}\tilde{v}}{\mathrm{d}\tilde{t}}.$$

Inserting this transformation into the IDM equations results in

$$\frac{\mathrm{d}\tilde{v}}{\mathrm{d}\tilde{t}} = \frac{a}{b}\left[1 - \left(\frac{\sqrt{b s_0}\tilde{v}}{v_0}\right)^4 - \left(\frac{\tilde{s}^*}{\tilde{s}}\right)^2\right],$$

$$\tilde{s}^* = \frac{s^*}{s_0} = 1 + \sqrt{\frac{b}{s_0}}T\tilde{v} - \sqrt{\frac{b}{a}}\frac{\tilde{v}\Delta\tilde{v}}{2}.$$

As a consequence, the prefactors of the different new terms are dimensionless as well. Moreover, they come in only three combinations of the original IDM parameters which we can identify as the new model parameters:

$$\tilde{v}_0 = \frac{v_0}{\sqrt{b s_0}}, \quad \tilde{a} = \frac{a}{b}, \quad \tilde{T} = T\sqrt{\frac{b}{s_0}}.$$

Thus, the scaled IDM equations read

$$\frac{\mathrm{d}\tilde{v}}{\mathrm{d}\tilde{t}} = \tilde{a}\left[1 - \left(\frac{\tilde{v}}{\tilde{v}_0}\right)^4 - \left(\frac{\tilde{s}^*}{\tilde{s}}\right)^2\right], \quad \tilde{s}^* = 1 + \tilde{v}\tilde{T} - \frac{\tilde{v}\Delta\tilde{v}}{2\sqrt{\tilde{a}}}.$$

This allows a powerful conclusion[13]: Changing the five physical IDM parameters such that $\tilde{v}_0$, $\tilde{a}$, and $\tilde{T}$ remain unchanged does not change the scaled IDM equations, nor the local dynamics. This allows to reduce the five-dimensional IDM parameter space spanned by V_0, T, a, b, and s_0 to the three-dimensional space $(\tilde{V}_0, \tilde{T}, \tilde{a})$ spanned by the dimensionless parameters. However, the stability class depends not only on the local dynamics but also on the vehicle length influencing the macroscopic fundamental diagram $Q_e(\rho)$ and the sign of propagation velocities. Therefore, to ensure the same stability class, a forth dimensionless parameter

$$\tilde{l} = l/s_0$$

must be kept constant.

When applying these insights to the concrete problem of where to read off the stability class in the a-T-class diagram when other IDM parameters are changed, we observe that this is only possible if speed is changed proportionally to changes of $\sqrt{bs_0}$ and the vehicle length changes proportionally to s_0. Only then, the two scaled parameters $\tilde{v}_0$ and $\tilde{l}$ containing neither a nor T remain unchanged. This is fulfilled here since s_0 does not change anyway and the new values $v_0^* = 139\,\text{km/h}$ and $b^* = 2\,\text{m/s}^2$ of the desired speed and time headway, respectively, satisfy $\tilde{v}_0 = v_0(bs_0)^{-1/2} = v_0^*(b^*s_0)^{-1/2} = \text{const}$. In order to make sure that $\tilde{a}$ and $\tilde{T}$ remain unchanged as well, we read off the old diagram at the coordinate $(T^*, a^*) = (\tau T, \alpha a)$ rather than at (T, a). We fix the scaling factors τ and α to fulfill the conditions

$$\frac{a}{b} = \frac{a^*}{b^*} = \frac{\alpha a}{b^*}, \quad T\sqrt{\frac{b}{s_0}} = T^*\sqrt{\frac{b^*}{s_0}} = \tau T\sqrt{\frac{b^*}{s_0}}$$

resulting in

$$\tau = \sqrt{\frac{b}{b^*}} = 0.87, \quad \alpha = \frac{b^*}{b} = 1.33.$$

This means, for the new values of v_0 and b, one reads of the class diagram at 0.87 times the original T coordinate and 1.33 times the original a coordinate.

15.8 Fundamental Diagram with Hysteresis

Subproblem 1: Maximum free-traffic flow:

$$Q_{\text{max}}^{\text{free}} = v_0 \rho_{\text{max}}^{\text{free}} = 2{,}400\ \text{vehicles/h}.$$

[13] In hydrodynamics, such scale relations are the basis to measure the hydrodynamics of big objects (ships, planes etc.) by observing a scaled-down (physical) model of the object in a wind or water channel rather than observing/measuring the real thing.

Subproblem 2: The congested part of the fundamental diagram corresponds to the congested part of the triangular fundamental diagram for the parameters $Q_{\text{cong}}(\rho) = 1/T(1 - \rho l_{\text{eff}})$ where $l_{\text{eff}} = l + s_0 = 6.67\,\text{m}$. This congested branch intersects the free branch at the same point $(\rho_C, Q_{\max}^{\text{dyn}})$ that would correspond to the maximum of the triangular diagram without hysteresis:

$$\rho_{\min}^{\text{cong}} = \rho_C = \frac{1}{v_0 T + l_{\text{eff}}} = 16.67\ \text{vehicles/km}.$$

Subproblems 3 and 4: The jam outflow is characterized by the dynamic capacity $Q_{\max}^{\text{dyn}} = v_0 \rho_C = 2{,}000$ vehicles/h. This describes a *capacity drop* of

$$\Delta Q = Q_{\max}^{\text{free}} - Q_{\max}^{\text{dyn}} = 400\ \text{vehicles/h} \quad (\text{or } 16.7\,\%).$$

The density of the outflow is given by ρ_C. This means, hysteretic effects can take place in the density range $\rho \in [\rho_C, \rho_{\max}^{\text{free}}]$, or, numerically, $\rho \in [16.67$ vehicles/h, 20 vehicles/h$]$.

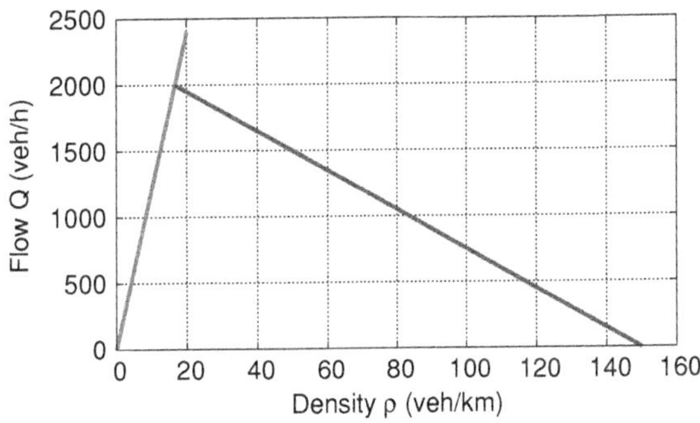

Problems of Chapter 16

16.1 Influence of Serial Correlations on Measures of Parsimony

Given are twenty data points (s_i, v_i) of which the first 10 and the second 10 are identical. This means, Model (i) can fit, at least, ten data points exactly, while Model (ii) fits the data to 100 %, after calibration. In contrast, if the data were not serially correlated, Models (i) and (ii) will fit at least one and two data points, respectively. This result does not depend on the specific model, since the same would apply to, say, the models (i): $\hat{v}(s) = \ln(\beta_0/s)$ and (ii): $\hat{v}(s) = \ln(\beta_0/s + \beta_1)$. This means, even completely nonsensical models fit one additional data point per parameter if the data points are independent, but they fit ten additional points if they are serially correlated as in this example. Now assume a nonsensical addition to a given model introducing one new parameter. For iid errors in the data, a parsimony test such as the likelihood-ratio test would yield a negative result for the augmented model since each parameter can always fit one additional data point without increasing its predictive value which such tests take care of. However, for correlated data as above, this parameter explains ten additional points. For robustness tests assuming iid errors, this corresponds to nine nontrivial fits which such tests may erroneously interpret as worth the additional parameter.

Problems of Chapter 17

17.1 Phase Diagram for Stability Class 3

For class 3, there are no traffic-flow instabilities and no hysteresis. Therefore, one just distinguishes between free traffic and homogeneous congested traffic:

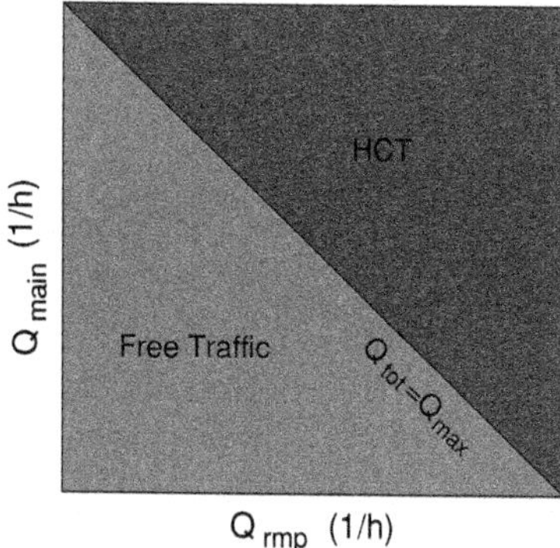

Furthermore, since no instability also implies no hysteresis effects, there is only one phase diagram valid for both small and large initial perturbations.

17.2 Boundary-Induced Phase Diagram

The kind of extended congested pattern (TSG, OCT, HCT) is directly defined by the supply restriction Q_{out}. Because of the oscillatory nature of the congested states for comparatively high values of Q_{out} (OCT and TSG), significant perturbations arrive at the upstream boundary activating the "inflow-bottleneck" if $Q_{\text{in}} > C_{\text{dyn}}$ (remember that we have $I = 1$ lane). If, additionally, $Q_{\text{out}} \geq C_{\text{dyn}}$, the potential outflow is higher than the inflow restricted by the activated inflow-bottleneck. This means there is free traffic in the bulk of the investigated road section corresponding to the maximum-flow state. If, however, $Q_{\text{out}} < C_{\text{dyn}}$, congested traffic will arise everywhere and the inflow-bottleneck (whether activated or not) is no longer relevant. Therefore, the maximum-flow state requires both $Q_{\text{in}} > C_{\text{dyn}}$ and $Q_{\text{out}} \geq C_{\text{dyn}}$. This results in following boundary-induced phase diagram:

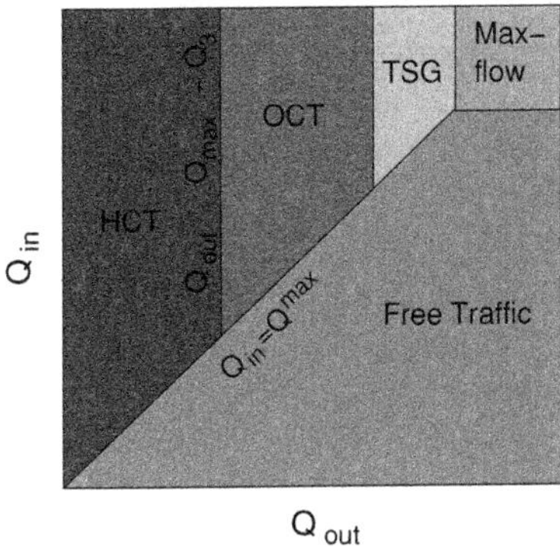

The situation with an activated inflow bottleneck corresponds to the stationary front of a bottleneck in inhomogeneous systems. However, since no upstream region is simulated here, the stationary front appears as a "standing wave". In simulations, activated inflow bottlenecks are a serious problem since they introduce bottlenecks not corresponding to anything in reality. Dedicated and very complex upstream boundary conditions (not discussed here) are necessary to avoid them.

Problems of Chapter 18

18.1 Locating a Temporary Bottleneck

From the data of floating car 3, we know that this car leaves a jam, i.e., crosses its downstream boundary, at the spatiotemporal point A depicted in the diagram below. From the data of detector 2 (point B), we know that this front is moving. We can exclude that the transition from congested to free traffic recorded by detector 2 at point B corrresponds to a downstream moving upstream front because (i) detector D1 records essentially constant traffic flow, (ii) the data of the detector D2 and the floating car 3 imply an upstream propagating upstream jam front, i.e., a growing jam. Hence, the upstream front is propagating backwards as long as it exists.

From Stylized Fact 2 we know that downstream fronts are either stationary or move at a constant velocity c_{cong}. Hence, the set of possible spatiotemporal points indicating when and where the road closure is lifted, lies on a line connecting the points A and B at a position $x > x_{\text{A}}$. The end of the road block not only sets the downstream jam front into motion but also leads to a transition empty road $\rightarrow$ maximum-flow state. The shockwave formula (8.9) implies that this front propagates with the desired free-flow speed v_0. Microscopically, this transition is given by the first car passing the accident site. This car is recorded as first trajectory of the trajectory data from the bridge at point C. Assuming, for simplicity, an instantaneous acceleration to the speed v_0, another set of possible spatiotemporal points for the removal of the road block is given by a line parallel to the first trajectory and touching it at point C (dashed line in the diagram). Intersecting the lines $\overline{AB}$ and the line parallel to the trajectories and going through C gives us the location and time of the lifting of the road block by the intersecting set of the two lines (point D), and also the location of the accident.

To estimate the time when the accident occurred, we determine the intersection F of the line $x = x_D$ of the temporary bottleneck, and the line representing the extrapolation of the last trajectory (point E) to locations further upstream (dashed line). Finally, because of the constant inflow recorded by D1, we know from the shockwave formula that the upstream jam front propagates essentially at a constant velocity, i.e., it is given by the line intersecting F and G. The jam dissolves when the upstream and downstream fronts meet at point H.

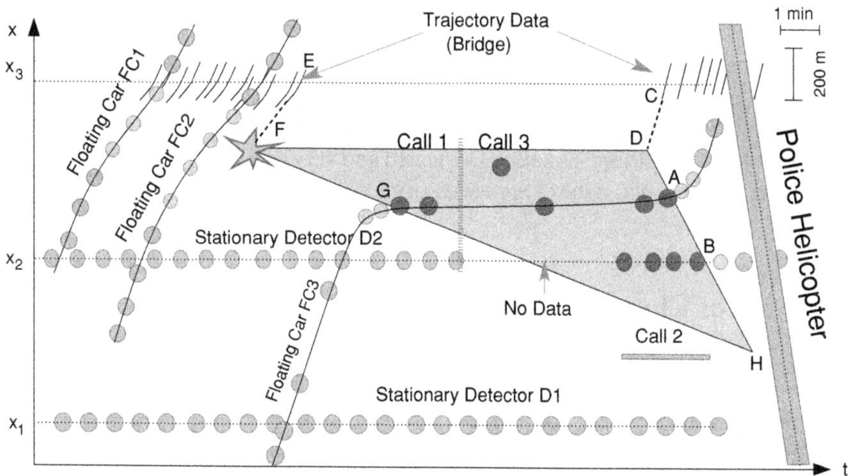

Problems of Chapter 19

19.1 Criteria for Estimating Travel Times by N-Curves

This method works exactly on roads with a single lane per driving direction and without ramps or other non-flow-conserving bottlenecks. Without floating cars, initialization and corrections of the cumulated vehicle numbers are possible on a heuristic basis, only. If the past speed data indicate free-flow speed and there are no fast-growing differences $N_i - N_j$ between the N-curves of the detectors i and j, one assumes that there is free traffic in between, and initializes/corrects the N-curves by Eq. (19.10) with the density estimated by the average flow divided by the average speed. During the evolution of a jam (indicated by fast growing differences $N_i - N_{i+1}$), there are no correction possibilities without floating cars.

19.2 Estimating Travel Times from Aggregated Detector Data

Subproblem 1: The following figure displays two possibilities leading to the observed zero traffic flow at detector D2 between 16:00 and 16:30 h: (1) The accident happens upstream of D2 causing a temporarily empty road ($\rho = 0$, $Q = 0$) near the detector location D2. (2) The accident happens downstream of D2 causing temporarily blocked traffic ($\rho = \rho_{max}$, $Q = 0$) near the detector location D2.

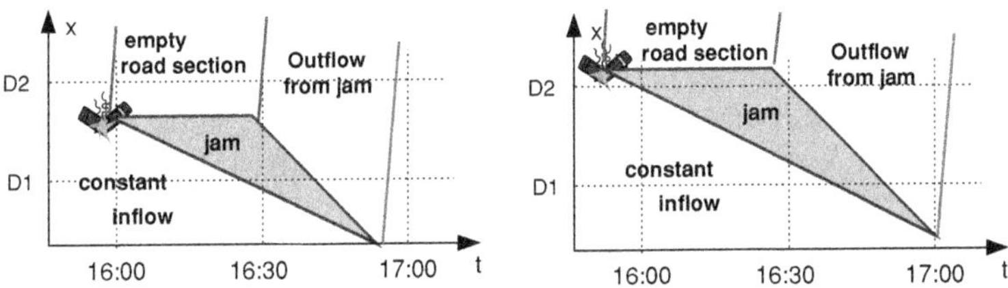

Subproblem 2: Assuming a free-flow speed of 120 km/h, it takes $\tau_{12} = 2$ min $= 120$ s to pass the 4 km long section between the detectors D1 and D2. In this time interval, $\Delta n = 60$ vehicles have passed D1. Setting the cumulated vehicle count $N_1(0) = 0$ for the time 16:00 h (corresponding to $t = 0$), we obtain $N_2(0) = 60$. With this initialization, we calculate the cumulated vehicle count as a function of time, i.e., the N-curves, $N_1(t)$ and $N_2(t)$, by piecewise integration of the flows given in the problem statement:

$$N_1(t) = \begin{cases} 60 + 0.5\,t & t < 2520, \\ 1320 & 2520 \le t < 3000, \\ 1320 + (t - 3000) = t - 1680 & 3000 \le t < 3480, \\ 1800 + 0.5\,(t - 3480) = 60 + 0.5\,t & t \ge 3480, \end{cases}$$

and

$$N_2(t) = \begin{cases} 0.5\,t & t < 0, \\ 0 & 0 \le t < 1800, \\ t - 1800 & 1800 \le t < 3600, \\ 0.5\,t & t \ge 3600. \end{cases}$$

Subproblem 3: Sketch of the N-curves:

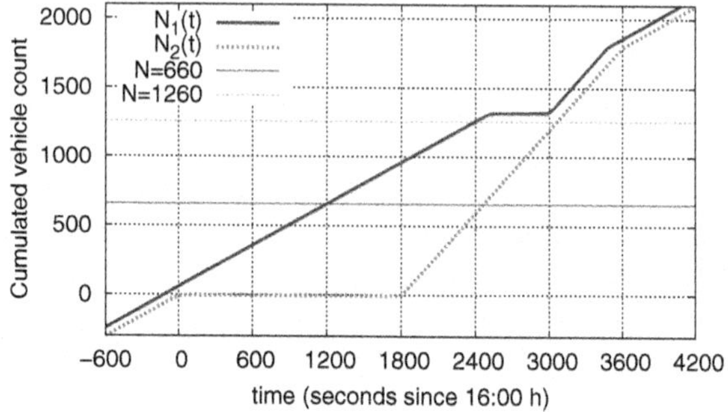

The realized travel time $\tau_{12}(t)$ at time $t = 2{,}400$ s can be read from the diagram by the length of the horizontal line at height $N = N_2(2400) = 600$ s intersecting the curves $N_1(t)$ and $N_2(t)$: $\tau_{12}(t = 2{,}400) \approx 1{,}300$ s (exactly: 1,260, see below). The expected travel time $\tilde{\tau}_{12}(t)$ at time $t = 2{,}400$ s is equal to the length of the horizontal line at height $N_1(2400) = 1{,}250$ intersecting the two N-curves: $\tilde{\tau}_{12}(t) \approx 600$ s (exactly: 660).

Subproblem 4: The diagram of the *N-curves* shows that, when estimating $\tilde{\tau}_{12}(t)$ within the time interval $-120\,\text{s} \leq t < 2{,}520$ s, the horizontal line intersecting the N-curves has a height N between $N_1(-120) = 0$ and $N_1(2{,}520) = 1{,}320$. We determine its length between the intersections with the N-curves using the results of subproblem 2:

$$N_1(t) = N_2(t + \tilde{\tau}_{12})$$

$$60 + \frac{t}{2} = (t + \tilde{\tau}_{12}) - 1{,}800 \quad \Rightarrow \quad \tilde{\tau}_{12} = 1{,}860 - \frac{t}{2}.$$

For $t < -120$ s, we have $\tilde{\tau}_{12} = 120$ s, i.e., equal to the free-flow travel time. At $t = -120$ s, there is a jump from 120 to 1,920, i.e., by 1,800 s or 30 min. This corresponds to the waiting time difference between the last vehicle that can pass before the road closure becomes active (possibly the car causing the accident), and the car after it having to wait the full duration of the road block.

For $\tau_{12}(t)$ within the time interval $1{,}800\,\text{s} \leq t < 3120$ s, we obtain analogously

$$N_1(t - \tau_{12}) = N_2(t)$$

$$60 + \frac{1}{2}(t - \tau_{12}) = t - 1{,}800 \quad \Rightarrow \quad \tau_{12} = 3{,}720\,\text{s} - t.$$

Subproblem5: Since, according to the problem statement, the floating car slows down sharply when passing D2 at 16:00 h, the accident happened downstream of D2 somewhat before 16:00 h. This corresponds to situation (2) discussed in subproblem 1 above.

Problems of Chapter 20

20.1 Coefficients of a Statistical Modal Consumption Model

Assuming a constant specific consumption $C_{\text{spec}} = 1/(\gamma w_{\text{cal}})$ (purely analytic physics-based model) and inserting Eqs. (20.14), (20.5), and (20.4) into Eq. (20.12) gives following function for the instantaneous model consumption:

$$\begin{aligned}\dot{C} = C_{\text{spec}} P &= C_{\text{spec}} \max\,[0,\, P_0 + Fv] \\ &= C_{\text{spec}} \max\left[0,\, P_0 + mv\dot{v} + m(\mu + \phi)gv + \frac{1}{2}c_d \rho A v^3\right].\end{aligned}$$

Apart from the maximum condition, this is a parameter-linear function whose parameters β_j can be easily estimated by conventional multivariate regression. Comparing this function with the statistical model specified in the problem statement and using Table 20.2 gives following relations and values for the model parameters:

$$\begin{aligned}\beta_0 &= C_{\text{spec}} P_0 &&= 25.0 \cdot 10^{-3}\,\text{l/s}, \\ \beta_1 &= C_{\text{spec}} m g \mu &&= 24.5 \cdot 10^{-6}\,\text{l/m}, \\ \beta_2 & &&= 0, \\ \beta_3 &= \tfrac{1}{2} C_{\text{spec}} c_d \rho A &&= 32.5 \cdot 10^{-9}\,\text{l}\,\text{s}^2/\text{m}^3, \\ \beta_4 &= C_{\text{spec}} m &&= 125 \cdot 10^{-6}\,\text{l}\,\text{s}^2/\text{m}^2, \\ \beta_5 &= C_{\text{spec}} m g &&= 1.23 \cdot 10^{-3}\,\text{l/s}.\end{aligned}$$

The following figure gives a plot of this function:

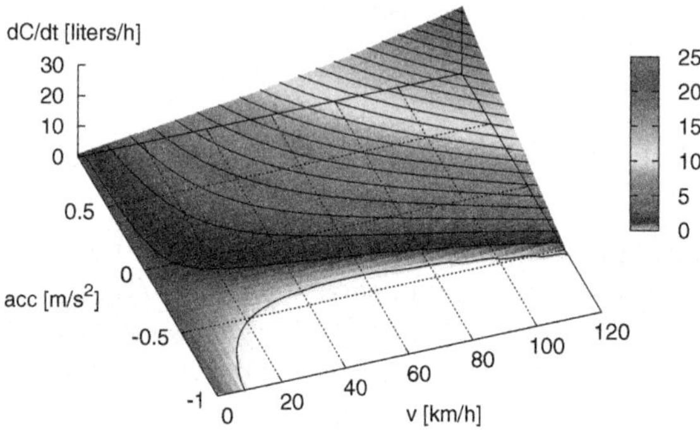

20.2 An Acceleration Model for Trucks

To solve this problem, we only need the power module of the physics-based modal consumption/emissions model. Assuming a constant engine power P and solving Eq. (20.5) for the acceleration gives

$$\dot{v}_P(v, \boldsymbol{\phi}) = \frac{P - P_0}{mv} - g(\mu + \boldsymbol{\phi}) - \frac{1}{2} c_d \rho A v^2.$$

To include the restraints "maximum acceleration $a_{\max}$" and "no positive acceleration at speed $v > v_0$", we obtain the final form of the free-flow truck acceleration model:

$$\dot{v}(v, \boldsymbol{\phi}) = \begin{cases} \min\left(a_{\max}, \dot{v}_P(v, \boldsymbol{\phi})\right) & v \le v_0 \\ \min\left(0, \dot{v}_P(v, \boldsymbol{\phi})\right) & v > v_0. \end{cases}$$

Following plot shows that the engine power is not sufficient to drive the truck at 80 km/h along a 2 % uphill gradient:

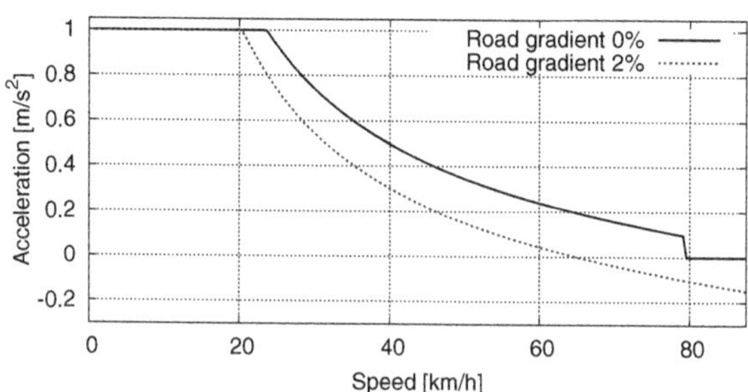

20.3 Characteristic Map of Engine Speed and Power

"Full throttle" corresponds to the top part of the allowed operating region for a given engine speed, i.e., to the top contourline. At 3,000 rpm it corresponds to 70 kW. For a power demand of 60 kW, an engine speed $f = 3{,}300$ rpm (or the speed nearest to this value allowed by the transmission) results in most efficient fuel usage.

20.4 Characteristic Map of Engine Speed and Mean Effective Pressure
(i) 60 kW, (ii) An engine speed of $2{,}600\,\text{min}^{-1}$ results in a specific consumption below 375 ml/kWh while, at $4{,}000\,\text{min}^{-1}$, the specific consumption is above 400 ml/kWh. The first option is more efficient.

20.5 Does Jam Avoidance Save Fuel? At high vehicle speeds, the aerodynamic drag becomes dominant and the consumption per kilometer increases nearly quadratically. Therefore, the savings potential decreases and can even become negative (when comparing homogeneously flowing congested traffic with high-speed free traffic).

20.6 Influencing Factors of Fuel Consumption Combining Eqs. (20.16), (20.14), (20.5), and (20.4) for the purely analytical physics-based model (constant specific consumption), we obtain following relation for the consumption per travel distance:

$$C_x = \frac{\mathrm{d}C}{\mathrm{d}x} = C_{\text{spec}} \max\left[0, \left(\frac{P_0}{v} + m\dot{v} + (\mu + \phi)mg + \frac{1}{2}c_d \rho A v^2\right)\right]. \quad (2.11)$$

1. Air condition: Correct. The additional power ΔP_0 results in an additional consumption $\Delta C_x = C_{\text{spec}} \Delta P_0 / v$ which increases for decreasing speed. (Numerical values for $\Delta P_0 = 4$ kW: 3 l/100 km at 40 km/h and 1.5 l/100 km at 80 km/h.)

2. Roof rack: False. It is true that the increased c_d value increases the consumption per distance by $\Delta C_x = C_{\text{spec}} \Delta c_d \rho A v^2 / 2$. However, this increase grows *quadratically* with the vehicle speed, i.e., it is lowest for city traffic. (Numerical values for $\Delta c_d = 0.08$: 0.43 l/100 km at 80 km/h and 1.71 l/100 km at 160 km/h)

3. Disconnecting the clutch when driving downhill: False. If the clutch is disconnected, the driving shaft is decoupled from the generator and the overrun fuel cut-off cannot operate. In this case, the instantaneous consumption rate is given by the idling consumption rate $\dot{C}_0 = C_{\text{spec}} P_0$ leading to $C_{x0} = C_{\text{spec}} P_0 / v$ for the consumption per distance. With the clutch connected, the fuel consumption C_x is less than C_{x0} if $F < 0$, and the overrun fuel cut-off is fully operative, i.e., $C_x = 0$, if $F < -P_0/v$. (Numerical values at 50 km/h: $C_{x0} = 1.8\,\text{l}/100\,\text{km}$; downhill gradient where the driving resistance F is equal to zero: $-2.5\,\%$; downhill gradient where the overrun fuel cut-off is fully operative: $-4.0\,\%$.)

4. Only use half the capacity of the tank: False. At a tank capacity of 60 l, the average fuel volume is 30 l for the cycle full-empty-full etc, and 15 l for the cycle half filled-empty-half-filled etc. This corresponds, on average, to a savings of the total mass by $\Delta m < 15$ kg (since the specific mass of fuels is less than 1 kg/l). The resulting effect on the consumption per distance, $\Delta C_x = -C_{\text{spec}} \Delta m g \mu < 0.025\,\text{l}/100\,\text{km}$, is independent of the speed v and negligible (but the risk to run out of fuel increases).

5. *Reduce speed from 50 to 30 km/h*: False. At speeds below the optimal value of about 50–60 km/h (cf. the figure below), the consumption (2.11) per distance increases with decreasing speed. Specifically, $C_x = 5.7\,\text{l}/100\,\text{km}$ at 30 km/h and 4.9 l/100 km at 50 km/h.

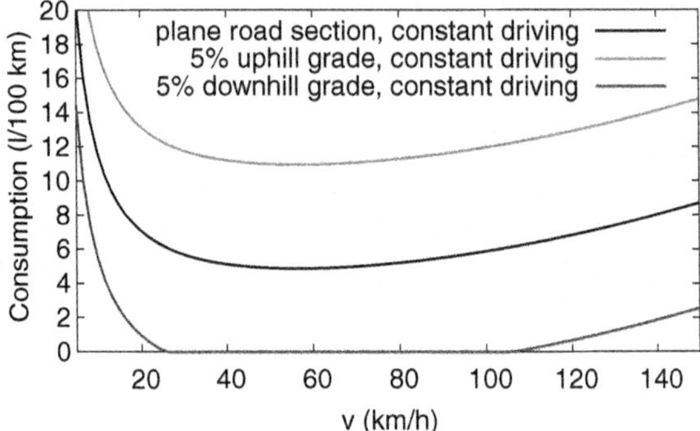

6. *Reduce speed from 150 to 130 km/h*: Correct. $C_x = 7.4\,\text{l}/100\,\text{km}$ at 130 km/h and 8.7 l/100 km at 130 km/h (cf. the figure above).

20.7 Highway Versus Mountain Pass: Which Route needs More Fuel? When choosing alternative 1, i.e., driving the level highway at 150 km/h, one needs 8.7 l/100 km (cf. Solution to Problem 20.6).
When choosing alternative 2, i.e., driving the mountain pass at 72 km/h, one needs fuel only for the 50 % of the route going uphill while the downhill gradient of 8 % is more than enough to fully activate the overrun fuel cut-off (cf. the figure at the solution to Problem 20.6). With Eq. (2.11), we obtain for the uphill sections ($\phi = 0.08$) a consumption $C_x = 14.8\,\text{l}/100\,\text{km}$. For the complete mountain pass (uphill and downhill), the consumption halves to $C_x = 7.4\,\text{l}/100\,\text{km}$ which is less than the consumption on the highway! (The balance tips over to the other side for gradients of more than 10 % or when driving more slowly on the highway.)

20.8 Four-way-stops Versus Intersection with Priority Rules
The analysis of situation II (constant speed $v_0 = 16\,\text{m/s}$) is easy: With Eq. (2.11), we obtain for the 500 m long stretch between two intersections

$$C_{\text{II}} = LC_x = 24.2\,\text{ml}.$$

For situation I, we separate the driving cycle between two intersections into three driving modes: (i) accelerating from zero to v_0, (ii) cruising at v_0, and (iii) decelerating to a full stop at the next intersection.

(i) Acceleration phase. With $\dot{v} = a = 2\,\text{m/s}^2$, this phase lasts a time interval of $t_a = 8\,\text{s}$ during which a distance of $L_a = v_0^2/2a = 64\,\text{m}$ is covered. Because both C_x and $\dot{C}$ are variable during the acceleration phase, explicit integration is necessary. We choose integration over time. With Eq. (20.12) and $C_{\text{spec}} = 1/(\gamma w_{\text{cal}})$, the integrand $\dot{C}$ reads

$$\dot{C}(t) = \frac{\mathrm{d}C}{\mathrm{d}t} = C_{\text{spec}}\left(P_0 + m\dot{v}v(t) + (\mu + \phi)mgv(t) + \frac{1}{2}c_d\rho A v^3(t)\right). \quad (2.12)$$

With $v(t) = at$, the integration can be evaluated analytically:

$$\begin{aligned} C_{\text{acc}} &= \int_0^{t_a} \dot{C}(t)\mathrm{d}t \\ &= C_{\text{spec}} \int_0^{t_a} \left(P_0 + ma^2 t + \mu m g a t + \frac{1}{2}c_d \rho A a^3 t^3\right) \mathrm{d}t \\ &= C_{\text{spec}}\left(P_0\, t_a + \frac{1}{2} m a (a + \mu g) t_a^2 + \frac{1}{8} c_d \rho A a^3 t_a^4\right). \end{aligned}$$

Using $t_a = v_0/a$ and $L_a = \frac{1}{2}at_a^2$, we simplify this expression to

$$C_{\text{acc}} = C_{\text{spec}} W_{\text{acc}} = C_{\text{spec}}\left(P_0\, t_a + \frac{1}{2} m v_0^2 + m\mu g L_a + \frac{1}{4} c_d \rho A v_0^2 L_a\right) = 19.8\,\text{ml}.$$

The terms in the parenthesis of the last equation have the following meanings: $P_0 t_a$ is the energy necessary to operate all the secondary appliances during the acceleration phase, $\frac{1}{2}mv_0^2$ is the kinetic energy at the end of this phase, $m\mu g L_a$ is the energy lost (or, more precisely, transformed to heat) by the solid-state friction, and $\frac{1}{4}c_d\rho A v_0^2 L_a$ is the energy lost by the aerodynamic drag.

(ii) Cruising phase. Since both the acceleration and deceleration phases cover a road section of $L_a = 64$ m, a distance $L_c = L - 2L_a = 372$ m remains for the cruising phase. Correspondingly,

$$C_{\text{cruise}} = C_{\text{spec}} W_{\text{cruise}} = L_c C_x(v_0, \dot{v} = 0) = 18.0\,\text{ml}.$$

(iii) Deceleration phase. Due to overrun fuel cutoff, no fuel is consumed in this phase, so $C_{\text{brake}} = 0$.[14]

Result for situation I: The total consumption between two intersections for the traffic rules of situation I equals

$$C_{\text{I}} = C_{\text{acc}} + C_{\text{cruise}} + C_{\text{brake}} = 37.9\,\text{ml}$$

which has to be compared with $C_{\text{II}} = 24.2$ ml.

[14] Strictly speaking, this is not true for the very last part of the deceleration phase when the speed (ignoring aerodynamic drag) drops below $v_c = P_0/[m(|\dot{v}| - \mu g)] \approx 7$ km/h. This is neglected here.

In summary, the fuel saving potential of changing the traffic rules from that of situation I to that of situation II is about 35 % which is massive.

20.9 Under Which Conditions Do All-Electric Cars Save CO_2 emissions?

The gasoline vehicle is treated as in Problem 20.8 with an increased cruising speed for situation II: Per kilometer, i.e., doubling the values between two intersections, we obtain following fuel consumptions for situations I and II:

$$C_x^{\mathrm{I}} = 0.076\,\mathrm{l/km}, \quad C_x^{\mathrm{II}} = 0.074\,\mathrm{l/km},$$

or, with 2.39 kg CO_2/l, the CO_2 emissions

$$E_x^{\mathrm{I,\,gas}} = 181\,\mathrm{g/km}, \quad E_x^{\mathrm{II,\,gas}} = 176\,\mathrm{g/km}.$$

For the all-electric car in situation I, we need per kilometer (i.e., for two start-stop cycles between intersections) a total electrical energy of

$$W_{\mathrm{el}}^{\mathrm{I}} = 2\left[\frac{1}{\gamma_{\mathrm{el}}}\left(W_{\mathrm{acc}} + W_{\mathrm{cruise}} + \gamma_{\mathrm{rec}} W_{\mathrm{brake}}\right)\right] = 794 \cdot 10^3\,\mathrm{Ws}$$

where the required mechanical work for the acceleration and cruising phases of the driving cycle between two intersections is calculated as in Problem 20.8, and the mechanical work for the braking phase is that of the acceleration phase minus the double kinetic energy at cruising speed, $W_{\mathrm{brake}} = W_{\mathrm{acc}} - m v_0^2$:

$$W_{\mathrm{acc}} = 238 \cdot 10^3\,\mathrm{Ws}, \quad W_{\mathrm{cruise}} = 216 \cdot 10^3\,\mathrm{Ws}, \quad W_{\mathrm{brake}} = -146 \cdot 10^3\,\mathrm{Ws}.$$

The required total electrical energy per kilometer in situation II (cruising at 130 km/h) is calculated from the mechanical energy "power times time",

$$W_{\mathrm{el}}^{\mathrm{II}} = \frac{1}{\gamma_{\mathrm{el}}}\frac{P}{v}\,1\,\mathrm{km} = 1\,042 \cdot 10^3\,\mathrm{Ws}.$$

With an energy mix of 600 g CO_2 per kWh electrical energy, the global CO_2 emissions due to operating the electric car in the two situations amounts to

$$E_x^{\mathrm{I,\,el}} = 132\,\mathrm{g/km}, \quad E_x^{\mathrm{II,\,el}} = 174\,\mathrm{g/km}.$$

Comparing this with the emissions of the gasoline car, we conclude that the all-electrical car emits significantly less CO_2 in city traffic but the global CO_2 emissions differ insignificantly when driving on highways. Of course, the results depend on many assumptions. For example, the battery adds about 150 kg to the electric vehicle (which is neglected here), the energy mix in Europe is more favorable, etc. However, these assumptions are all transparent in the physics-based consumption/emission models.

20.10 Fuel Consumption for an OVM-Generated Speed Profile

Subproblem 1: The OVM free acceleration $\dot{v} = (v_0 - v)/\tau$ is maximal at $v = 0$. Prescribing $\dot{v}_{\text{max}} = v0/\tau = 2\,\text{m/s}^2$ gives the relaxation time $\tau = v_0/a = 16.67\,\text{s}$.[15]

Subproblem 2: We calculate the instantaneous power at a given speed v from Eq. (20.5) with Eq. (20.4):

$$P(v, \dot{v}) = \frac{\dot{C}(v, \dot{v})}{C_{\text{spec}}} = P_0 + m\dot{v}v + (\mu + \phi)mgv + \frac{1}{2}c_d \rho A v^3.$$

Inserting the OVM free acceleration $\dot{v} = (v_0 - v)/\tau$, we obtain $P_{\text{OVM}}(v) = A_0 + A_1 v + A_2 v^2 + A_3 v^3$ where

$$\begin{aligned} A_0 &= P_0 = 3\,\text{kW}, & A_1 &= m\left(g\mu + \frac{v_0}{\tau}\right) = 3{,}294\,\text{W s/m}, \\ A_2 &= -\frac{m}{\tau} = -90\,\text{W (s/m)}^2, & A_3 &= \frac{1}{2}c_d \rho A = 0.39\,\text{W (s/m)}^3. \end{aligned}$$

Subproblem 3: As usual, we calculate extremal values by setting the derivative with respect to the interesting variable (here, the speed v) equal to zero:

$$\frac{\mathrm{d}P_a}{\mathrm{d}v} = A_1 + 2A_2 v + 3A_3 v^2 \stackrel{!}{=} 0.$$

This quadratic equation has two solutions and corresponding extremal power requirements P_{OVM}:

$$\begin{aligned} v_1 &= 132.6\,\text{m/s} = 477.4\,\text{km/h}, & P_{\text{OVM}}(v_1^*) &= -233\,\text{kW}, \\ v_2 &= 21.2\,\text{m/s} = 76.4\,\text{km/h}, & P_{\text{OVM}}(v_2^*) &= 36.1\,\text{kW}. \end{aligned}$$

Obviously, the second solution is the correct one since the power (and the OVM acceleration) is negative for the first one.[16] The maximum power during the acceleration phase is reached at 76 km/h. Its value is 36.1 kW.

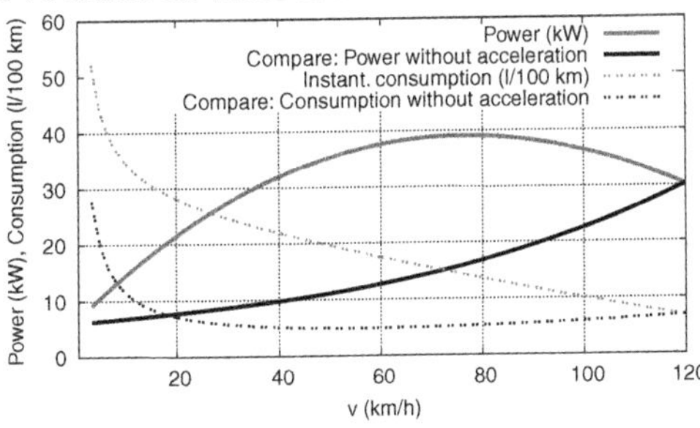

20.11 Trucks at Uphill Gradients

---

[15] At this value, the OVM is extremely unstable and cannot be used for simulating interacting or congested traffic.

[16] This solution represents the minimum power requirement.

1. Engine power: With Eqs. (20.5) and (20.4), we obtain for the necessary power to maintain a speed $v = v_{\text{limit}} = 80\,\text{km/h}$ at level roads

$$P_{\text{dyn}} = P - P_0 = vF(v) = \mu m g v + \frac{1}{2} c_d \rho A v^3 = 249\,\text{kW} + 57\,\text{kW} = 306\,\text{kW}.$$

2. Initial deceleration: The two new forces entering the balance are the uphill-slope force and the inertial force. Since, initially, all other forces remain unchanged, the two new forces must cancel each other, i.e.,

$$\phi g + \dot{v} = 0 \Rightarrow \dot{v} = -\phi g = \begin{cases} -0.49\,\text{m/s}^2 & \text{at } 5\,\%\text{ gradient,} \\ -0.39\,\text{m/s}^2 & \text{at } 4\,\%\text{ gradient.} \end{cases}$$

3. Terminal speed: Equation (20.5) also delivers the terminal speed at a gradient ϕ by setting $\dot{v} = 0$ and solving for v. Neglecting the aerodynamic drag, we obtain

$$P_{\text{dyn}} = (\mu + \phi) g m v \;\Rightarrow\; v_\infty = \frac{P_{\text{dyn}}}{(\mu + \phi) g m} = \begin{cases} 10.2\,\text{m/s} & \text{at } 5\,\%\text{ gradient,} \\ 11.7\,\text{m/s} & \text{at } 4\,\%\text{ gradient.} \end{cases}$$

4. Estimating the OVM parameters in the uphill section: Since the OVM speed approaches asymptotically the desired speed v_∞, we can set the OVM "desired speed" in the uphill sections equal to the terminal speed v_∞. The initial accelerations calculated in subproblem 2 at the desired speed v_0 of the *level* section serve to estimate τ via $\dot{v} = (v_\infty - v_0)/\tau$, i.e., $\tau = (v_\infty - v_0)/\dot{v}$, resulting in

$$\tau = \begin{cases} 24.4\,\text{s} & \text{at } 5\%\text{ gradient,} \\ 26.8\,\text{s} & \text{at } 4\,\%\text{ gradient.} \end{cases}$$

5. Speed and distance over time: The solution to the inhomogeneous ordinary differential equation $\frac{dv}{dt} = (v_\infty - v)/\tau$ for the initial condition $v(0) = v_0$ reads (cf. the following figure)

$$\begin{aligned} v(t) &= v_\infty + (v_0 - v_\infty)\, e^{-t/\tau}, \\ x(t) &= v_\infty (t - \tau) + v_0 \tau - (v_0 - v_\infty)\, \tau e^{-t/\tau}. \end{aligned}$$

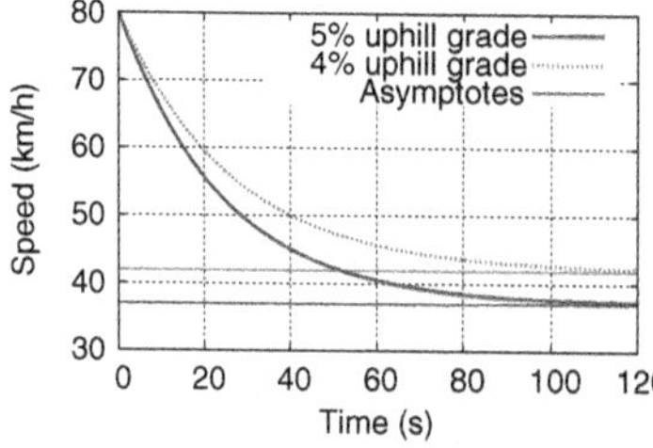

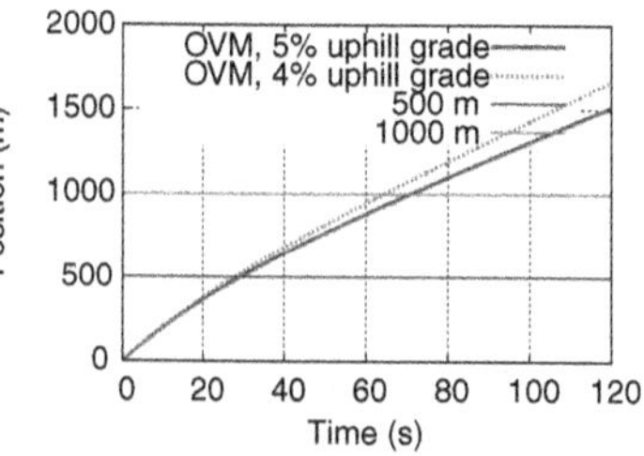

For the uphill section 1 of length $L_1 = 500\,\text{m}$ and gradient $\phi_1 = 5\,\%$, we obtain for the time $t = t_1 = 29.1\,\text{s}$ given in the problem statement:

$$v(t_1) = 13.8\,\text{m/s} = 49.8\,\text{km/h}.$$

(Test: $x(t_1) = 500.5\,\text{m}$.) Analogously, we obtain for the uphill section 2 of length $L_2 = 1{,}000\,\text{m}$ and gradient $\phi_2 = 4\%$ at time $t = t_2 = 64.2\,\text{s}$:

$$v(t_2) = 12.6\,\text{m/s} = 45.2\,\text{km/h}.$$

(Test: $x(t_2) = 1000.2\,\text{m}$.)

Discussion: Although the uphill section 1 is steeper, the speed of the trucks at its end is higher than at the end of the less steep but longer uphill section 2. Therefore, it makes sense to allow higher gradients on shorter uphill sections.

Index

M. Treiber and A. Kesting, *Traffic Flow Dynamics*,
DOI: 10.1007/978-3-642-32460-4, © Springer-Verlag Berlin Heidelberg 2013

CPSIA information can be obtained
at www.ICGtesting.com
Printed in the USA
BVOW07*0216221116
468576BV00003B/4/P